Neurological Disorders and Imaging Physics, Volume 3

Application to autism spectrum disorders and Alzheimer's

Neurological Disorders and Imaging Physics, Volume 3

Application to autism spectrum disorders and Alzheimer's

Ayman El-Baz
Louisville University, Louisville, Kentucky, USA

Jasjit S Suri
AtheroPoint, CA, USA

IOP Publishing, Bristol, UK

ISBN   978-0-7503-1793-1 (ebook)
ISBN   978-0-7503-1764-1 (print)
ISBN   978-1-64327-648-9 (mobi)

DOI   10.1088/978-0-7503-1793-1

Version: 20191101

IOP ebooks

British Library Cataloguing-in-Publication Data: A catalogue record for this book is available from the British Library.

Published by IOP Publishing, wholly owned by The Institute of Physics, London

IOP Publishing, Temple Circus, Temple Way, Bristol, BS1 6HG, UK

US Office: IOP Publishing, Inc., 190 North Independence Mall West, Suite 601, Philadelphia, PA 19106, USA

With love and affection to my mother and father, whose loving spirit sustains me still.
Ayman El-Baz
To my late loving parents, immediate family, and children.
Jasjit S Suri

Contents

Preface

This volume of the book covers two significant neurological disorders: autism spectrum disorder (ASD) and Alzheimer's disease (AD). Autism spectrum disorder is a neurological disorder that is characterized by difficulties with social skills, repetitive behaviors, speech and nonverbal communication, as well as unique strengths and differences. According to the Centers for Disease Control and Prevention (CDC), the estimate of ASD prevalence in the United States is 1 in 68 children (i.e. 1 in 42 boys and 1 in 189 girls). Additionally, an estimated 50 000 teens with autism become adults—and lose school-based autism services—each year. Alzheimer's disease is a chronic neuro-degenerative disorder marked by cognitive and behavioral impairments. Statistically, 42% of AD sufferers are over 85 years of age with the percentage decreasing to only 6% for those who are 70–74 years old. Although the probability is small, younger individuals may also be affected.

As it is one of the most crucial neuro-developmental disorders, several state-of-the-art machine learning techniques for the early diagnosis of ASD are presented in this book. Such techniques utilize various imaging modalities, such as diffusion tensor imaging (DTI), structural magnetic resonance imaging (sMRI), and functional magnetic resonance imaging (fMRI). Also, nuclear neurology, genetics, and parental behavior will be covered in this volume. In addition, various studies are discussed in this book to demonstrate the formation, causes, and medical treatment of AD. This involves state-of-the-art machine learning techniques, neuropathology, neuroimaging, retinal imaging, and genetics. AD is viewed from different perspectives, in addition to demonstrating the obstacles that still face researchers, in particular relating to accurate understanding and early diagnosis of the disease.

In summary, the main aim of this volume is to help advance scientific research within the broad fields pertaining to the detection of both ASD and AD. The book focuses on major trends and challenges in this area, and it presents work aimed at identifying new achievements and their use in biomedical analysis.

Ayman El-Baz
Jasjit S Suri

Acknowledgments

The completion of this book could not have been possible without the participation and assistance of so many people whose names cannot all be mentioned. Their contributions are sincerely appreciated and gratefully acknowledged. However, the editors would like to express their deep appreciation and indebtedness in particular to Dr Ali H Mahmoud and Fatma El-Zahraa A El-Gamal for their endless support.

Ayman El-Baz
Jasjit S Suri

Editor biographies

Ayman El-Baz

Ayman El-Baz is a Professor, University Scholar, and Chair of the Bioengineering Department at the University of Louisville, KY. Dr El-Baz earned his BSc and MSc degrees in electrical engineering in 1997 and 2001, respectively. He earned his PhD in electrical engineering from the University of Louisville in 2006. In 2009 Dr El-Baz was named a Coulter Fellow for his contributions to the field of biomedical translational research. Dr El-Baz has 17 years of hands-on experience in the fields of bio-imaging modeling and non-invasive computer-assisted diagnosis systems. He has authored or coauthored more than 500 technical articles (133 journals, 25 books, 57 book chapters, 212 refereed-conference papers, 143 abstracts, and 27 US patents and disclosures).

Jasjit S Suri

Jasjit S Suri is an innovator, scientist, visionary, industrialist, and an internationally known world leader in biomedical engineering. Dr Suri has spent over 25 years in the field of biomedical engineering/devices and management. He received his PhD from the University of Washington, Seattle and his Business Management Sciences degree from Weatherhead, Case Western Reserve University, Cleveland, OH. Dr Suri received the President's Gold medal in 1980 and was made a Fellow of the American Institute of Medical and Biological Engineering for his outstanding contributions. In 2018 he was awarded the Marquis Life Time Achievement Award for his outstanding contributions and dedication to medical imaging and its management.

List of contributors

Saman Sarraf
IEEE, USA

Gilberto Sousa Alves
Federal University of Ceará, Brazil

Leonardo Caixeta
Federal University of Goiás, Brazil

Lucas Schilling
The Pontifical Catholic University of Rio Grande do Sul, Brazil

MD Umur Kayabasi
Uskudar University, Istanbul

Javier Santabárbara
Universidad de Zaragoza, Spain

Patricia Gracia-García
Universidad de Zaragoza, Spain

Anais Sevil-Pérez
Universidad de Zaragoza, Spain

Beatriz Villagrasa
Universidad de Zaragoza, Spain

Raúl López-Antón
Universidad de Zaragoza, Spain

Ola Eid
National Research Centre, Egypt

Maha Eid
National Research Centre, Egypt

Osman Farooq
State University of New York at Buffalo, USA

Robert Miletich
State University of New York at Buffalo, USA

Michelle Hartley-McAndrew
State University of New York at Buffalo, USA

Ana-Maria Bratu
National Institute for Laser, Plasma and Radiation Physics, Romania

Cristina Popa (Achim)
National Institute for Laser, Plasma and Radiation Physics, Romania

Mioara Petrus
National Institute for Laser, Plasma and Radiation Physics, Romania

Dan C Dumitras
National Institute for Laser, Plasma and Radiation Physics, Romania

Diana Delgado
University of Memphis, TN, USA

Kimberly Frame
Savannah State University, TN, USA

Laura Casey
University of Memphis, TN, USA

C S Sandeep
University of Kerala, India

A Sukesh Kumar
University of Kerala, India

Fatma El-Zahraa El-Gamal
University of Louisville, KY, USA

Mohammed Elmogy
University of Louisville, KY, USA

Hassan Hajjdiab
Abu Dhabi University, UAE

Ashraf Khalil
Abu Dhabi University, UAE

Mohammed Ghazal
Abu Dhabi University, UAE

Ali Mahmoud
University of Louisville, KY, USA

Hassan Soliman
Mansoura University, Egypt

Ahmed Atwan
Mansoura University, Egypt

Gregory Barnes
University of Louisville, KY, USA

Ayman El-Baz
University of Louisville, KY, USA

Madam Yasmeen Farouk
Ain Shams University, Egypt

Sherine Rady
Ain Shams University, Egypt

Yaser ElNakieb
University of Louisville, KY, USA

Ahmed Shalaby
University of Louisville, KY, USA

Fatma Taher
Zayed University, UAE

Ahmed Soliman
University of Louisville, KY, USA

Robert Keynton
University of Louisville, KY, USA

Ali Mahmoud
University of Louisville, KY, USA

Omar Dekhil
University of Louisville, KY, USA

Reem Haweel
University of Louisville, KY, USA

Olfa Ben-Ahmed
University of Poitiers, France

Christine Fernandez-Maloigne
University of Poitiers, France

Adrien Julian
University of Poitiers, France

Marc Paccalin
University of Poitiers, France

Said Ghniemy
University of Ain Shams, Egypt

IOP Publishing

Neurological Disorders and Imaging Physics, Volume 3
Application to autism spectrum disorders and Alzheimer's
Ayman El-Baz and Jasjit S Suri

Chapter 1

Machine learning applications to recognize autism and Alzheimer's disease

Saman Sarraf

Brain disorders such as autism spectrum disorder (ASD) and Alzheimer's disease (AD) have been of increasing interest to neuroscience researchers over the past few decades. As novel findings and technologies become available in the field of computational neuroscience, hope increases for the early prediction of such brain disorders in order to reduce patient impact, including symptoms interfering with daily tasks. Researchers have shown that machine learning techniques enable them to distinguish, with high accuracy, patients with these brain disorders from normal control subjects using automatic approaches. To extract patterns from neuro-imaging data, various statistical methods and machine learning algorithms have been explored for the diagnosis of AD among older adults in both clinical and research applications. However, distinguishing between AD and healthy brain data has been challenging in older adults (age > 75) due to the highly similar patterns of brain atrophy and image intensities. Recently, cutting-edge deep learning technologies have rapidly expanded in numerous fields, including medical image analysis. In this review, we will describe ASD and AD in their various aspects while providing information on the cutting-edge application of machine learning pipelines and techniques to classify and recognize these brain disorders. Deep learning applications for AD have been more promising than those for ASD.

1.1 Introduction

Autism spectrum disorder (ASD) is one of a cluster of cerebral diseases involving impairments in communication/social interactive skills, mood, attention, cognitive and adaptive skills, and cognitive functions. That is, a set of neurodevelopmental impairments causing difficulty in connecting with other people. ASD is characterized by repetitive, cyclic, and obstructive behaviors with symptoms stemming from a convoluted genotype–phenotype relationship wherein pre-existing neurodevelopmental

doi:10.1088/978-0-7503-1793-1ch1

liabilities interact with the child's environment. In responsive modification, the child typically develops compensatory tactics and defense mechanisms. Studies of children at high genetic ASD risk, defined by an older diagnosed sibling, are discovering developmental corridors to phenotype manifestation [1]. ASD is severe in terms of incidence, morbidity, and societal impact, and while the precise organic origins of the brain disorder remain a mystery, the principle discoveries suggest that both genes and environment influence the development of autistic behavior. Environmental factors are thought to interact with the infant's genes and cause anomalous deformities in cerebral and neuronal development and operative connectivity [1].

Children on the autism spectrum display concurrent sense-processing complications and are clinically treated using self-modifying mediation. Contemporary therapy utilizes sensory interventions utilizing various hypothetical paradigms that have differing goals, deploying a multiplicity of sensory modalities consisting of remarkably disparate procedures. Earlier evaluations studied the effects of sensory interventions without recognizing such empirical contradictions [2]. ASD diagnoses are typically delayed, resulting in unrealized treatment opportunities during the formative period of development. Our investigation extrapolates previous assessments of age-related factors in ASD diagnosis, offering clinical research recommendations, programs, and early detection methods [3]. Mutations affect typical neurodevelopment *in utero* through to adolescence via gene composites involved in exuberant synaptogenesis and axon motility. Recent advancements in neuroimaging investigation offer crucial knowledge on pathological brain deformities in ASD *in vivo* patients. The amygdala is the limbic system's central element, involving the affective loop of the cortico-striato-thalamo-cortical circuit in cognition. The nucleus accumbens is the second most important structure related to the social-reward response in ASD, hence ASD's popularity in neuropathological and neuroimaging studies [4]. A higher rate of ASD diagnosis is consistently found in males over females, despite which egregiously little research has sought the causes of this inconsistency, the understanding of which could prevent or treat ASD in both sexes [5].

Studies of heredity have unearthed hundreds of gene deviations in autism with radically differing risk effects habitually associated with similar conditions. However, numerous variations coalesce into mutual biological pathways, indicating characteristically pervasive autism traits including aetiological heterogeneity, variable penetrance, and genetic pleiotropy [5].

1.2 Brain disorders

1.2.1 Autism spectrum disorder (ASD)

Researchers believe that typical ASD symptoms in children should be considered as ASD, while adult symptoms should not. Few behavioral indicators are diagnosed in the child's first year, most emerging in the second [1]. According to the *Diagnostic and Statistical Manual of Mental Disorders, Fifth Edition* (DSM-5), ASD sufferers react inappropriately to conversational cues and engage in abnormal routines and inappropriate obsessions and can struggle with relationships [6]. ASD patients present the gamut of cognitive aptitudes, from acute intellectual retardation to remarkable intelligence. The DSM-5 does not include postponement of lingual acquisition as a core ASD symptom as not all ASD sufferers display this trait. Among disorders on the autism spectrum, the eponymous affliction is the most acute, differing from other neurodevelopmental disorders such as Asperger's syndrome (AS) and pervasive developmental disorder not otherwise specified (PDD-NOS) by the said delay in language expansion and the severity of cerebral/ behavioral deformities. About 20 per 10 000 children suffer from ASD. Research utilizing functional magnetic resonance imaging (fMRI) on the brains of ASD sufferers suggests a substantial decrease in long-distance connectivity. Microstructurally, the disturbance of brain development is caused by the atypical adaptation of cell division, apoptosis, and elevated inflammation of neurons. Recent studies have observed both hypo- and hyper-connectivity issues in autistic children's brains depending on age-related factors, for example a child of three months at high risk of developing autism displays elevated connectivity in contrast to low-risk children, a variant which dwindles between the ages of six and nine months. Findings imply that autistic brains suffer from morphological deformities including premature overgrowth of brain organs, including the frontal cortex, amygdala, and cerebellum; at six months the cephalic circumference of ASD infants accretes rapidly when compared to typical infants, but declines at adolescence, resulting in a typical adult cerebral mass and volume. Mounting evidence suggests the significance of mirror neurons—brain cells activated when an individual executes and observes a motor action. Mirror neurons affect other individuals' recognition of motor acts and the regulation of social, emotional, and cognitive tasks. The mirror neuron system allows individuals to recognize others' motor actions, engendering social cognitive facilities such as empathy, sympathy, compassion, and regret, while enabling coordination of the motor cortex and higher visual processing brain areas implicated in speech, memory, and motion planning. Evidence of deficiencies in the mirror neuron system in ASD children derives from a range of imaging techniques, including fMRI, electroencephalography (EEG), and electromyography (EMG). Studies have proved that mirror neuron activity is affected in ASD children, obstructing their understanding and recognition of others' motor activity.

The genetic origin of this neurodevelopmental disorder has proven to be complex via whole-exome sequencing (WES) and cytogenetics. Likewise, studies of twins and families suggest that ASD's heritability is more than 80%. The principal ASD-associated syndromes are fragile X syndrome (FXS) and tuberous sclerosis (TS), both of which have similar pathophysiological processes to those of ASD, including deviant mRNA translation and elevated synthesis of protein. FXS is an X-linked genetic disease caused by an inconsistent increase of the FMR1 gene's multiple CGG repeat and is characterized by abnormal facial features as well as variously severe cognitive deficiencies. TS is an autosomal-dominant disorder caused by mutations in either the TSC1 or TSC2 genes, presenting as epilepsy, learning challenges, and social interactive issues. More than 40% of patients with TS also suffer from ASD, hence the elevated incidence of epileptic seizures in both ASD and TS sufferers.

Whole-exome sequencing (WES), chromosomal microarray, and selective-candidate gene-analysis are the most common methods for identifying ASD-predisposition genes. WES recognizes new or rare genetic flaws in various heterogeneous disorders such as ASD. A recent study of 928 patients showed that ASD is linked to intensely disruptive *de novo* mutations in brain-expressed genes [7].

A fundamental component of ASD-related behavioral/functional performance is faulty sensory processing. In 1974 Ornitz reasoned that defective sensory modulation results in the stereo-typic or repetitive activities of ASD children in their attempt to heighten arousal (sensory-seeking) or to self-calm. Clinicians ascribe repetitive behaviors including twirling, rocking, and spinning to sensory-processing problems, finding that children with ASD and other stereotypical behaviors suffered far greater sensory-processing problems ($d = 2.0$) than controls. Rigidity, e.g. the refusal to switch to a new activity or behavior with a preference for regimen/sameness, might also be triggered by hyper-/hyporeactivity.

ASD-related sensory-processing issues may also affect children's diurnal functional performance, including eating, sleeping, and bath-/bedtime behaviors. Selective eating in children is often accompanied by gustatory and/or olfactory over-sensitivities leading to specific food antipathies. Hyperreactivity and taste-aversion often cause anxiety/rigidity with regard to ingestion, evolving into anxious/disruptive eating behaviors. Sensory-processing issues may also upset the patient's sleep cycle, as sensory modulation issues in ASD children are linked to unstable sleeping patterns, specifically related to entering REM sleep, with 50%–80% of children with ASD afflicted. Additional studies are required to determine how sensory processing and hypo-/hyperreactivity influence self-modulation and stimulation [2].

Autism is distinguished by: (i) a qualitative deficiency in societal communication via non-verbal behaviors including eye contact, hand gestures, and physical deportment, leading to a failure to bond and a lack of spontaneity, interest-sharing, or social/emotional reciprocity; (ii) qualitative impairments in social communication manifested by retarded lingual development without non-verbal compensation, difficulty initiating and maintaining dialog, repetitive and stereotypical language, and a lack of creative invention, imagination, and imitative play; (iii) a limited interest repertoire, object or topic obsession, a slavery to dysfunctional ritual/regimen/routine, stereotyped motor mannerisms, and a fixation on object parts or aspects rather than the gestalt. Sensory aberrations are common, including hypo-/hypersensitivity and preoccupation with certain sensations. The lack of imaginative play implies problems with idea generation, which is essential for human bonding.

Numerous factors have caused a decline in the referral age and diagnosis of autism: (i) heightened detection by healthcare professionals of early autism symptoms, leading to earlier recommendation to pediatric and child-development experts, and (ii) increased public and media interest involving the publication of memoirs and biographical journalism, including the depiction of pediatric ASD behavior, leading to parental help outreach.

Screening methods applied to both referred and general populations (the Checklist for Autism in Toddlers (CHAT)) have identified autism as early as

18 months. However, in the sole general-population study to date, while the CHAT screening presented a high positive predictive significance, its sensitivity was insignificant and it cannot be recommended for general-population screening at a single point in time.

The diagnosis of autism at the age of two years is less accurate/stable than that of related ASD, but the preliminary response should not be impulsive as working diagnoses are in most cases refined over time in conference with parents. Medical evaluations must identify difficulties in early non-verbal interactions characteristic of children with ASD from an age of two years onwards. The particular syndrome presenting in a two-year-old with ASD may vary widely from a more exemplar four- or five-year-old. In particular, explicit repetitive and/or stereotyped behaviors may be present in a minority of cases, although when concordant with social/interactive deformity strongly indicate ASD [8].

The cerebral architecture in ASD patients, in both the frontal and temporal lobes, is notably disrupted, in particular the amygdala which is influential in cognition as proved by numerous neuropathological and neuroimaging studies. The medial temporal lobe anterior to the hippocampal construction is essential for declarative memory (conscious recollection of facts and events) and determines anti-social/social behavior in ASD patients. The amygdala is the core of the limbic system and the affective loop of the cortico-striato-thalamo-cortical circuit, determining eye contact and facial recognition/motion. Amygdala injury manifests in changes in fear processing, memory modulation with emotive content, and eye contact with the human visage. The effects in individuals with amygdala lesions mimic ASD phenomena as the amygdala processes somatosensory, visual, auditory, visceral, and synesthesia inputs, channeling through the chief efferent conduits, the stria terminalis and ventral amygdalofugal pathway. The amygdala comprises 13 nuclei which, histochemically, are divided into three subgroups: the basolateral (BL), centromedial (CM), and superficial groups. The BL group functions as a node-connecting sensory stimulus for higher social cognition, linking the CM and superficial groups with reciprocity to the orbitofrontal cortex, anterior cingulate cortex (ACC), and medial prefrontal cortex (mPFC). The BL group is neurologically responsive to other people's facial and bodily actions, but this sensitivity is absent in the remaining two groups of amygdala. The CM group is comprised of the central, medial, and cortical nuclei and periamygdaloid complex, innervating the majority of the brain stem's visceral, autonomic, and effector regions while providing a vital output to the hypothalamus, thalamus, and reticular and ventral tegmental regions. The superficial group includes the nucleus of the lateral olfactory tract.

Neurochemical investigations of the amygdala have revealed both an elevated opiate and benzodiazepine/GABAa-receptor density including cholinergic, dopaminergic, noradrenergic, and serotonergic cell bodies and pathways. Because a population of aggressive patients with temporal epilepsy experienced decreased aggression following the bilateral stereotactic ablation of basal and corticomedial amygdaloid nuclei, amygdala function in emotional processing, specifically rage, has been studied with confirmation of amygdala deficiency in ASD patients. Postmortem investigations have revealed amygdala disease in ASD patients in comparison to age- and sex-matched controls. Neuronal smallness and heightened cell density in the central, cortical, and medial nuclei of the amygdala were found in patients with ASD [4].

The often cited ratio of 4:1 for male/female prevalence is an extrapolated average from international studies. The male predominance is not exclusive to ASD, however, as studies dependably document greater prevalence of attention deficit/hyperactivity and similar developmental disorders in males than females. The ratio's variance is in fact due to variance in identification methods, as the numbers vary radically from 2:1 to 7:1. Interaction with IQ also determines variability, with a lower instance of sex-bias in cohorts with a lower mean IQ than in 'high-functioning' groups with a higher IQ, an interaction exacerbated by a lower mean IQ in ASD subjects compared to males, further inflating gender prejudice.

Lastly, studies tracking the younger siblings of ASD patients suggest investigative prejudice causes an overemphasis on gender-bias, particularly in the high-functioning control. Self-advocates have declared timely and precise diagnoses an 'essential need'. Despite diagnostic variance, the gender disparity in ASD predominance remains at 2:1–3:1, indicating the need for the investigation of sexual dimorphism in the symptomatology of ASD.

The 'female protective effect' (FPE) suggests that ASD females are at less risk of certain ASD symptoms than affected males. The FPE has in other disorders, such as clubfoot, been ascribed to a clear gender prejudice, and genetic research of ASD cohorts has discovered a greater liability for *de novo* copy number variation (CNV) and *de novo* loss of function-point aberration in ASD females than in male counterparts. In addition, inherited small CNVs are transmitted more often from unafflicted mothers than counterpart fathers [5] (table 1.1).

The clinical benefits of early detection imply the importance of parental intervention for autistic children. Heightened parental intervention dates back three decades, encouraging learning in ASD children. Parents trained as 'co-therapists' with consistent interventional handling and modification of the child's ASD behaviors enhance formative pediatric social interactions, including heightened skills and confidence, in addition to lessening parental/filial stress. Group parental skill coaching has been proven to promote communal support. The metrics of parental intervention in pediatric development must include: pediatric developmental improvement, parental–filial communication, parental understanding, outlook and anxiety, familial cohesion, and cost–benefit analysis.

The majority of assessments that involve timely parental intercession have not been systematic, thus reducing their scope and validity as they include uncontrolled studies that rely on single case studies without clinical precision, applicable universality, or methodological compliance, often omitting statistics that did not benefit their prognosis. Smith employed a restrictive foundation for result comparison by favoring children's cerebral performance, while knowing that the majority of autistic children are mentally retarded [2].

The New York State Department of Health performed the most responsible clinical study of timely autism intervention, allegedly to develop scientific procedural

Table 1.1. Phenotype summary, M: Male, F: Female. ADOS score: † means the information was not available [9].

Site	ASD		TC		FD	
	Age avg. (SD)	ADOS (SD)	Count	Age avg. (SD)	Count	
CALTECH	27.4 (10.3)	13.1 (4.7)	M 15, F 4	28 (10.9)	M 14, F 4	0.07
CMU	26.4 (5.8)	13.1 (3.1)	M 11, F 3	26.8 (5.7)	M 10, F 3	0.29
KKI	10.0 (1.4)	12.5 (3.6)	M 16, F 4	10.0 (1.2)	M 20, F 8	0.17
LEUVEN	17.8 (5.0)	† (†)	M 26, F 3	18.2 (5.1)	M 29, F 5	0.09
MAX MUN	26.1 (14.9)	9.5 (3.6)	M 21, F 3	24.6 (8.8)	M 27, F 1	0.13
NYU	14.7 (7.1)	11.4 (4.1)	M 65, F 10	15.7 (6.2)	M 74, F 26	0.07
OHSU	11.4 (2.2)	9.2 (3.3)	M 12, F 0	10.1 (1.1)	M 14, F 0	0.1
OLIN	16.5 (3.4)	14.1 (4.1)	M 16, F 3	16.7 (3.6)	M 13, F 2	0.18
PITT	19.0 (7.3)	12.4 (3.3)	M 25, F 4	18.9 (6.6)	M 23, F 4	0.15
SBL	35 (10.4)	9.2 (1.7)	M 15, F 0	33.7 (6.6)	M 15, F 0	0.16
SDSU	14.7 (1.8)	11.2 (4.3)	M 13, F 1	14.2 (1.9)	M 16, F 6	0.09
STANFORD	10.0 (1.6)	11.7 (3.3)	M 15, F 4	10.0 (1.6)	M 16, F 4	0.11
TRINITY	16.8 (3.2)	10.8 (2.9)	M 22, F 0	17.1 (3.8)	M 25, F 0	0.11
UCLA	13.0 (2.5)	10.9 (3.6)	M 48, F 6	13.0 (1.9)	M 38, F 6	0.19
UM	13.2 (2.4)	† (†)	M 57, F 9	14.8 (3.6)	M 56, F	0.14

by-laws, but flagrantly failed to investigate parental intervention. Other self-motivated clinicians made ostensibly methodical studies of randomized controlled trials of parent-arbitered timely intervention by revealing only child-related results, knowing that parental intervention would hurt their medical practice by removing their own importance [10].

In an attempt to understand the genetic systems engendering ASD risk, researchers have studied various afflicted families, including those with con-sanguinity and single- (simplex family) and multi-affected family members with ASD, often spanning numerous generations. Employing WES in ASD-affected families, particular abnormalities were discovered in AMT, MECP2, NLGN4X, PAH, PEX7, POMGNT1, SYNE1, and VPS13B24. MECP2, NLGN4X, and SYNE1 have been traditionally associated with ASD. A recent investigation employing CMA and WES by the Autism Genome Project (AGP) on 2147 ASD patients found that 4.6% ($n = 99$) carried a *de novo* rare CNV29. Studies by the Simons Simplex Collection proved that the incidence of *de novo* rare CNVs increases to over 10% when limited to simplex syndrome. Likewise, a study of 1532 families with multiple affected patients from the Autism Genetic Resource Exchange (AGRE) proved that both rare *de novo* and hereditary CNVs further ASD progression. While the incidence of *de novo* CNVs discovered in the AGRE study as expected proved lower than in the simplex cases, a higher burden of large, rare CNVs, including hereditary deviations, in ASD patients contrasted with their unafflicted siblings. Compellingly, in over two thirds of families in which a high-risk ASD-related CNV was identified, this CNV did not manifest in all afflicted siblings, underscoring the intra-familial inherited heterogeneity of ASD [11] (table 1.2).

Table 1.2. Common systems to describe the levels of evidence and the criteria used to analyze studies in psychology and occupational therapy [2].

Randomized controlled trial (RCT) criteria [12, 13]	Types of studies [12, 13]	PEDro scale (Physiotherapy Evidence Database), scores range from 0 to 10	Levels of evidence (Center for Evidence-Based Medicine)
Should include: • Comparison groups with random assignment. • Blinded assessments. • Clear inclusion and exclusion criteria. • Standardized assessment. • Adequate sample size for statistical power. • Intervention manual. • Fidelity measure. • Clearly described statistical methods. • Follow-up measures.	• Type 1: the most rigorous, randomized, prospective clinical trial that meets all criteria. • Type 2: a clinical trial in which at least one aspect of the type 1 study is missing. • Type 3: a clinical trial that is methodologically limited, for example, a pilot study or open trial. • Type 4: a review of published data, for example, meta-analyses. • Type 5: reviews that do not include secondary data analyses. • Type 6: case studies, essays, and opinion papers.	• Random allocation used. • Allocation concealed. • Groups comparable at baseline. • Blinding of participants. • Blinding of all study therapists. • Blinding of all assessors who measured at least one key outcome. • Outcome measures obtained from more than 85% of the initial sample. • Intent-to-treat analyses used. • Between-group statistical comparisons reported. • Pre/post-measures and measures of variability reported (or effect sizes reported).	• Level I: a systematic review (of RCTs) or RCT conducted. • Level II: a systematic review of cohort studies; a low quality RCT; an individual cohort study; outcomes research. • Level III: a systematic review of case-control studies; an individual case-control study. • Level IV: case series; poor quality case-control studies. • Level V: expert opinion without explicit critical appraisal.

1.2.2 Alzheimer's disease (AD)

Alzheimer's disease is an irreversible, progressive neurological brain disorder and multifaceted disease that slowly destroys brain cells, causing memory and thinking skill losses, and ultimately loss of the ability to carry out even the simplest tasks. The cognitive decline caused by this disorder ultimately leads to dementia. The disease begins with mild deterioration and grows progressively worse as a neurodegenerative type of dementia. Diagnosing Alzheimer's disease requires very careful medical assessment, including patient history, a mini mental state examination (MMSE), and physical and neurobiological examinations [14, 15]. In addition to these evaluations, structural magnetic resonance imaging (sMRI) and resting-state functional magnetic resonance imaging (rs-fMRI) offer non-invasive methods of studying the structure of the brain, functional brain activity, and changes in the brain. During scanning using both sMRI (anatomical) and rs-fMRI techniques, patients remain prone on the MRI table and do not perform any tasks. This allows data acquisition to occur without any effects from a particular task on functional activity in the brain [16–18]. Alzheimer's disease causes shrinkage of the hippocampus and cerebral cortex and enlargement of ventricles in the brain. The level of these effects is dependent on the stage of disease progression. In the advanced stage of AD, severe shrinkage of the hippocampus and cerebral cortex, as well as significantly enlarged ventricles, can easily be recognized in MR images. This damage affects the brain regions and networks related to thinking, remembering (especially short-term memory), planning, and judgment. Since brain cells in the damaged regions have degenerated, MR image (or signal) intensities are low in both MRI and rs-fMRI techniques [19–21]. However, some of the signs found in the AD imaging data are also identified in normal aging imaging data. Identifying the visual distinction between AD data and images of older subjects with normal aging effects requires extensive knowledge and experience, which must then be combined with additional clinical results in order to accurately classify the data (i.e. MMSE) [14]. The development of an assistive tool or algorithm to classify MR-based imaging data, such as structural MRI and rs-fMRI data, and, more importantly, to distinguish brain disorder data from healthy subjects, has always been of interest to clinicians [14]. A robust machine learning algorithm such as deep learning, which is able to classify Alzheimer's disease, will assist scientists and clinicians in diagnosing this brain disorder and will also aid in the accurate and timely diagnosis of Alzheimer's patients [22].

1.2.3 Mild cognitive impairment

Cognitive impairment and related conditions that lie between normal ageing and very early dementia have been studied in the field of neuroscience for several years. Certain researchers have recently proposed that mild cognitive impairment should be considered as an early state of cognitive impairment but still abnormal. Since then, mild cognitive impairment has been studied extensively in clinical and research projects. Many epidemiological research projects have shown an accelerated rate of progression to dementia and Alzheimer's disease (AD) in mild cognitive impairment subjects. However, a broadly constant definition of mild cognitive impairment that is acceptable for researchers has not been obtained yet due to the controversy over the concept of mild cognitive impairment. Scientists have proposed that the diagnosis of mild cognitive impairment should be performed in the same fashion as Alzheimer's disease. Machine learning algorithms (discussed later in this work) have been developed to assist the clinician in identifying subjects and sub-classifying them into the various types of mild cognitive impairment [23–25].

1.3 Deep learning

Hierarchical or structured deep learning is a modern branch of machine learning that was inspired by the human brain. This technique has been developed based on complicated algorithms that model high-level features and extract these abstractions from the data by using an architecture similar to neural networks, which are actually much more complicated. Neuroscientists have discovered that the 'neocortex', which is part of the cerebral cortex concerned with sight and hearing in mammals, processes sensory signals by propagating them through a complex hierarchy over time. This served as the primary motivation for the development of deep machine learning, which focuses on computational models for information representation which exhibit characteristics similar to those of the neocortex [26–28].

Convolutional neural networks (CNNs) that are inspired by the human visual system are similar to classic neural networks. This architecture has been specifically designed based on the explicit assumption that raw data are comprised of two-dimensional images that enable certain properties to be encoded while also reducing the amount of hyper parameters. The topology of CNNs utilizes spatial relationships to reduce the number of parameters that must be learned, thus improving on general feed-forward backpropagation training [29, 30]. Equation (1.1) demonstrates how the gradient component for a given weight is calculated in the backpropagation step, where E is error function, y is the neuron $N_{i,j}$, x is the input, l represents layer numbers, w is filter weight with a and b indices, N is the number of neurons in a given layer, and m is the filter size:

$$\frac{\partial E}{\partial w_{ab}} = \sum_{i=0}^{N-m}\sum_{j=0}^{N-m} \frac{\partial E}{\partial x_{ij}^l}\frac{\partial x_{ij}^l}{\partial w_{ab}} = \sum_{i=0}^{N-m}\sum_{j=0}^{N-m} \frac{\partial E}{\partial x_{ij}^l} y_{(i+a)(j+b)}^{l-1}. \tag{1.1}$$

In convolutional layers, the gradient component of a given weight is calculated by applying the chain rule. Partial derivatives of the error for the cost function with respect to the weight are calculated and used to update the weight.

Equation (1.2) describes the backpropagation error for the previous layer using the chain rule. This equation is similar to the convolution definition, where $x_{(i+a)(j+b)}$ is replaced by $x_{(i-a)(j-b)}$. It demonstrates the backpropagation results in convolution while the weights are rotated. The rotation of the weights derives from a delta error in the convolutional neural network:

$$\frac{\partial E}{\partial y_{ij}^{l-1}} = \sum_{a=0}^{m-1}\sum_{b=0}^{m-1} \frac{\partial E}{\partial x_{(i-a)\,(j-b)}^l}\frac{\partial x_{(i-a)\,(j-b)}^l}{\partial y_{ij}^{l-1}} = \sum_{a=0}^{m-1}\sum_{b=0}^{m-1} \frac{\partial E}{\partial x_{(i-a)\,(j-b)}^l} w_{ab}. \tag{1.2}$$

The error of backpropagation for the previous layer is calculated using the chain rule. This equation is similar to the definition of convolution, but the weights are rotated.

In CNNs, small portions of the image (called local receptive fields) are treated as inputs to the lowest layer of the hierarchical structure. One of the most important features of CNNs is that their complex architecture provides a level of invariance to shift, scale, and rotation, as the local receptive field allows the neurons or processing units access to elementary features, such as oriented edges or corners. This network is primarily comprised of neurons with learnable weights and biases, forming the convolutional layer. It also includes other network structures, such as a pooling layer, a normalization layer, and a fully connected layer. As briefly mentioned above, the convolutional layer (or conv layer) computes the output of neurons that are connected to local regions in the input, each computing a dot product between its weight and the region it is connected to in the input volume. The pooling layer, also known as the pool layer, performs a downsampling operation along the spatial dimensions. The normalization layer, also known as the rectified linear units (ReLU) layer, applies an element-wise activation function, such as $\max(0, x)$ thresholding at zero. This layer does not change the size of the image volume [26, 27, 31]. The fully connected (FC) layer computes the class scores, resulting in the volume of the number of classes. As with ordinary neural networks, and as the name implies, each neuron in this layer is connected to all of the numbers in the previous volume [27, 31]. The convolutional layer plays an important role in CNN architecture and is the core building block in this network. The conv layer's parameters consist of a set of learnable filters. Each filter is spatially small but extends through the full depth of the input volume. During the forward pass, each filter is convolved across the width and height of the input volume, producing a 2D activation map of that filter. During this convolving, the network learns of filters that activate when they see some specific type of feature at some spatial position in the input. Next, these activation maps are stacked for all filters along the depth dimension, which forms the full output volume. Every entry in the output volume can thus also be interpreted as an output from a neuron that only examines a small region in the input and shares parameters with neurons in the same activation map [27, 32]. A pooling layer is usually inserted between successive conv layers in CNN architecture. Its function is to reduce (down sample) the spatial size of the representation in order to minimize network hyper parameters, and hence also to control overfitting. The pooling layer operates independently on every depth slice of the input and resizes it spatially using the max operation [26, 27, 31–33]. In convolutional neural network architecture, the conv layer can accept any image (volume) of size $W_1 \times H_1 \times D_1$ that also requires four hyper parameters, which are: K, the number of filters; F, their spatial extent; S, the size of stride; and P, the amount of zero padding. The conv layer outputs the new image, whose dimensions are $W_2 \times H_2 \times D_2$, calculated as

$$W_2 = (W_1 - F)/S + 1$$
$$H_2 = (H_1 - F)/S + 1 \qquad\qquad (1.3)$$
$$D_2 = D_1.$$

An understanding of how the conv layer produces new output images is important to realize the effect of filters and other operators, such as stride (S), on input images.

LeNet-5 was first designed by Y LeCun *et al* [26]. This architecture successfully classified digits and was applied to hand-written check numbers. The application of this fundamental but deep network architecture expanded into more complicated problems by adjusting the network's hyper parameters. The LeNet-5 architecture, which extracts low- to mid-level features, includes two conv layers, two pooling layers, and two fully connected layers, as shown in figure 1.1.

More complex CNN architectures were developed to recognize numerous objects derived from high volume data, including AlexNet (ImageNet) [34], ZF Net [35], GoogleNet [31], VGGNet [36], and ResNet [37]. GoogleNet, which was developed by Szegedy *et al* [31], is a successful network that is broadly used for object recognition and classification. This architecture is comprised of a deep, 22-layer network based on a modern design module called Inception. One of the fundamental approaches to improving the accuracy of CNN architecture is to increase the size of layers. However, this straightforward solution causes two major issues. First, a large number of hyper parameters requires more training data and may also result in overfitting, particularly in the case of limited training data. On the other hand, uniform increases in network size dramatically increase interactions with computational resources, which affect the timing performance and the cost of providing infrastructure. One of the optimized solutions for both problems would be the development of a sparsely connected architecture rather than a fully connected network. Strict mathematical proofs demonstrate that the well-known Hebbian principle of neurons firing and wiring together created the Inception architecture of GoogleNet. The Inception module of GoogleNet, as shown in figure 1.2, was developed by discovering the optimal local sparse structure to construct convolutional blocks. The Inception architecture allows for a significant increase in the number of units at each layer, while computational complexity remains under control at later stages, which is achieved through global dimensionality reduction prior to costly convolutions with larger patch sizes.

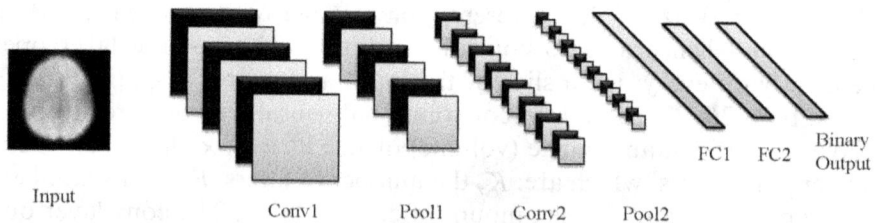

Figure 1.1. LeNet-5 includes two conv, two pool, and two FC layers. The original version of this network classified ten digits. In this work, the architecture was optimized for binary outputs, which were Alzheimer's disease (AD) and normal control (NC), respectively.

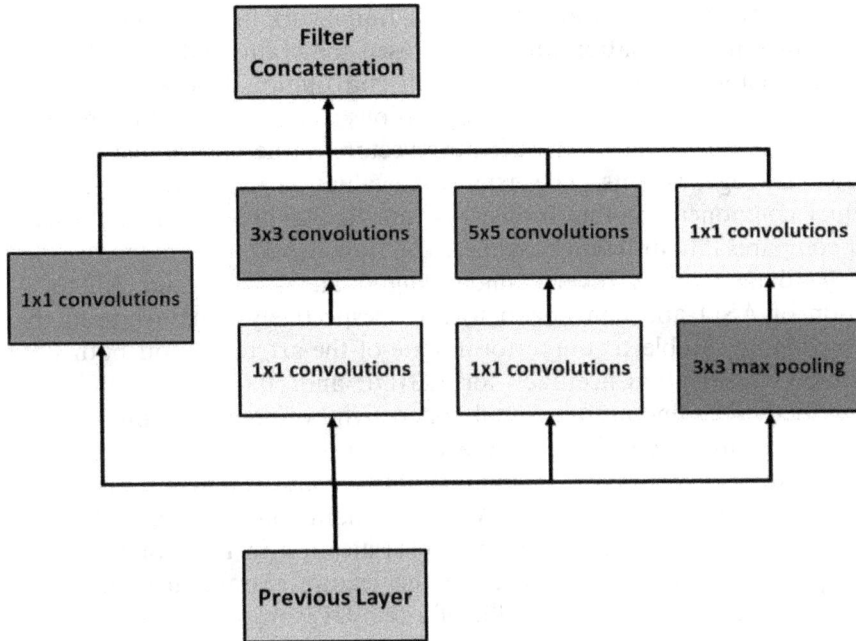

Figure 1.2. The Inception module with dimensionality reduction in GoogleNet architecture.

1.3.1 ASD and deep learning

An innovative deep learning approach was developed by [38] that directly learns the regulatory code from large-scale chromatin and transcription factor profiling data to enabling *de novo* prediction of the regulatory effects of sequence alterations with single-nucleotide sensitivity. We demonstrate its value by focusing on the causal contribution of noncoding variants to clinical outcomes in ASD. Even relatively sophisticated traditional analyses fail to find such signals because many noncoding mutations are not functional and there are enormous statistical challenges in such studies. The proposed architecture for stereotypical motor movement detection in the static feature space utilized a three-layer CNN. The first CNN layer receives the one-second-long time series of several IMU sensors at time t and transfers it to the first level reduced feature map. The second CNN layer uses the first level reduced feature map as its input and transfers it to the second-level reduced feature map. The third CNN layer which is a reduced feature map of this layer is reshaped to the learned feature vector using the flattening operation. The learned feature vector is fed to a fully connected layer followed by a SoftMax layer to classify the samples to SMM and no-SMM classes [39, 40]. Other researchers developed a pipeline by using T1w MR images and the Destrieux atlas including predefined regions in order to extract individual networks, followed by structural connectivity feature extraction. The stacked auto-encoder was used to perform the feature extraction in the previous step. In the next step, the top features are selected and a deep neural network architecture is utilized to perform the classification task [41]. Parisot *et al* [42] presented a thorough evaluation of a generic framework that leverages both imaging and non-imaging information and can be used for brain analysis in large populations. This methodology explored a graph convolutional network (GCN) based approach that involved representing populations as a sparse graph, where its nodes are associated with imaging-based feature vectors, while phenotypic information is integrated as edge weights. The extensive evaluation explores the effect of each individual component of this framework on disease prediction performance and further compares it to different baselines. The framework performance was tested on two large datasets with diverse underlying data, ABIDE and ADNI, for the prediction of ASD and conversion to AD, respectively. They showed that their novel pipeline was able to outperform state-of-the-art results on both databases, with 70.4% classification accuracy for ABIDE and 80.0% for ADNI. A feature selection method of finding functional connectivity of regions of interest (ROIs) as predefined regions followed by using a denoising auto-encoder was adopted. The extracted features were used to train and validate various machine learning models such as support vector machines (SVMs), random forest (RF), and deep neural networks. However, the performance of classification in none of the models was promising. This might be explained by the feature extraction section not being completely successful [9]. An ensemble of a two-layer neural net that shares hidden nodes across tissues with a maximum of 30 hidden variables and also with sigmoidal non-linearities and SoftMax output was designed by Xiong *et al* to predict RNA splicing levels from DNA sequences [43]. This proposed CAD system started with brain segmentation into CWM and Cx. Shape analysis was then performed on the reconstructed meshes, from which eight features were extracted that were then fused to classify autistic and control brains. Next, a deep fusion classification network (DFCN) was utilized in order to obtain a global diagnosis from the collected shape features, eight per lobe per hemisphere (64 per scan). This could be done using the raw features in a vertex-wise manner, but this is inefficient since the number of nodes for each mesh is 48K. The average accuracies for CWM and Cx segmentation in infant subjects using the proposed approach were 94.7% and 93.8%, respectively [44]. Dvornek *et al* [45] presented several methodologies in their work for incorporating phenotypic data with rs-fMRI into a single deep learning framework for classifying ASD. The trained model was tested using a cross-validation framework on the large, heterogeneous first cohort from the Autism Brain Imaging Data Exchange. In their best model, the accuracy rate reached 70.1%. In another work, Li *et al* proposed a new whole-brain fMRI-analysis scheme to identify ASD and explored biological markers in ASD classification. The idea was to utilize both spatial and temporal information in fMRI along with the potential benefits of using a sliding window over time to measure temporal statistics and using 3D convolutional neural networks (CNNs) to capture spatial features. The sliding window created two-channel images, which were used as inputs to the 3D CNN. From the outputs of the 3D CNN convolutional layers, ASD-related fMRI spatial features were directly deciphered. Compared with traditional machine learning classification models, the proposed 2CC3D method increased mean F-scores over 8.5% [46]. Andrews *et al* [47] reviewed the traditional and certain novel techniques to classify

ASD data. Guo *et al* [48] developed a deep neural network based pipeline where the DNN with a novel feature selection method (DNN-FS) was designed for the high dimensional whole-brain resting-state FC pattern classification of ASD patients versus typical development (TD) controls. The feature selection method was able to help the DNN generate low dimensional high-quality representations of the whole-brain FC patterns by selecting features with high discriminating power from multiple trained sparse auto-encoders. They also showed that the best classification accuracy of 86.36% was obtained from the DNN-FS approach with their hidden layers and 150 hidden nodes. A novel multi-kernel support vector machine classification framework was designed by Jin *et al* [49] using the connectivity features gathered from WM connectivity networks, which were generated via multiscale regions of interest (ROIs) and multiple diffusion statistics such as fractional anisotropy, mean diffusivity, and average fiber length. The framework achieved an accuracy of 76% and an area of 0.80 under the receiver operating characteristic curve (AUC), in comparison to the accuracy of 70% and the AUC of 70% provided by the best single-parameter single-scale network.

1.3.2 Alzheimer's and deep learning

Changes in brain structure and function caused by Alzheimer's disease have proved of great interest to numerous scientists and research groups. In diagnostic imaging in particular, classification and predictive modeling of the stages of AD have been broadly investigated. Suk *et al* [50–52] developed a deep learning based method to classify AD magnetic current imaging (MCI) and MCI–converter structural MRI and PET data, achieving accuracy rates of 95.9%, 85.0%, and 75.8% for the mentioned classes, respectively. In their approach, Suk *et al* developed an auto-encoder network to extract low- to mid-level features from images. Next, classification was performed using multi-task and multi-kernel support vector machine (SVM) learning methods. This pipeline was improved by using more complicated SVM kernels and multimodal MRI/PET data. However, the best accuracy rate for Suk *et al* remained unchanged [53]. Payan *et al* [54] of Imperial College London designed a predictive algorithm to distinguish AD MCI from normal healthy control subject imaging. In this study, an auto-encoder with 3D convolutional neural network architecture was utilized. Payan *et al* obtained an accuracy rate of 95.39% in distinguishing AD from NC subjects. The research group also tested a 2D CNN architecture, and the reported accuracy rate was nearly identical in terms of value. Additionally, a multimodal neuroimaging feature extraction pipeline for multiclass AD diagnosis was developed by Liu *et al* [55]. This deep learning framework was developed using a zero-masking strategy to preserve all possible information encoded in imaging data. High-level features were extracted using stacked auto-encoder (SAE) networks, and classification was performed using SVM against multimodal and multiclass MR/PET data. The highest accuracy rate achieved in that study was 86.86%. Aversen *et al* [56], Liu *et al* [57], Siqi *et al* [58], Brosch *et al* [59], Rampasek *et al* [60], De Brebisson *et al* [61], and Ijjina *et al* [62] also demonstrated the application of deep learning in the automatic classification of Alzheimer's disease from structural MRI, where AD, MCI, and NC data were classified (table 1.3).

Table 1.3. The table below summarizes the related works and percentage accuracies reported in each reference. Using MRI and PET data has been of interest to researchers and different techniques were utilized to classify AD as distinct from normal control (NC) or mild cognitive impairment (MCI). As can be seen, rs-fMRI and a full deep learning based pipeline were used as a novel approach. Also, the end-to-end pipeline improved the accuracy rate for structural MRI data. The abbreviations in the table are as follows: CAE—convolutional auto-encoder, SAE—sparse auto-encoder, CNN—convolutional neural networks, MSLF—matrix-similarity based loss function, MTFS—multi-task feature selection, DL—deep learning and MFE—multiview feature extraction. *—slice level classification, **—subject level classification followed by the decision making algorithm.

Reference	Modality	Method	AD/MCI/NC	AD + MCI/NC	AD/NC	AD/MCI	NC/MCI
Suk et al [50]	PET + MRI + CSF	SAE + SVM	N/A	N/A	95.9	N/A	85
Suk et al [53]	PET + MRI	SAE + SVM	N/A	N/A	95.4	N/A	85.7
Zhu et al [63]	PET + MRI + CSF	MSLF + SVM	N/A	N/A	95.9	N/A	82
Zu et al [64]	PET + MRI	MTFS + SVM	N/A	N/A	96	N/A	80.3
Liu et al [55]	PET + MRI	SAE + SVM	53.8	N/A	91.4	N/A	82.1
Liu et al [65]	MRI	MFE + SVM	N/A	N/A	93.8	N/A	89.1
Li et al [66]	PET + MRI + CSF	PCA + SVM	N/A	N/A	91.4	70.1	77.4
Payan et al [54]	MRI	2D-SAE	89.4	N/A	95.4	86.8	92.1
Hossein-Asl [67]	MRI	3D-CAE	89.1	90.3	97.6	N/A	N/A
Sarraf et al [68]	rs-fMRI	DL-CNN	N/A	N/A	99.9	N/A	N/A
Sarraf et al [69]	MRI	DL-CNN	N/A	N/A	98.84	N/A	N/A
Sarraf et al [68]	rs-fMRI	DL-CNN	N/A	N/A	97.77	N/A	N/A
Sarraf et al [69]	MRI	DL-CNN	N/A	N/A	100	N/A	N/A

DeepAD, developed by Sarraf *et al* [69], is a deep learning based approach in which aggressive preprocessing steps against functional and structural MRI data were utilized and enabled the CNN model to learn the patterns from the data more efficiently. Classification of AD images and normal, healthy images required several steps, from preprocessing to recognition, which resulted in the development of an end-to-end pipeline. Three major modules formed this recognition pipeline: (a) preprocessing; (b) data conversion; and (c) classification, respectively. Two different approaches were used in the preprocessing module, as preprocessing of 4D rs-fMRI and 3D structural MRI data requires different methodologies, which will be explained later in this chapter. After the preprocessing steps, the data were converted from medical imaging to a lossless portable network graphics (PNG) format to input into the deep learning based classifier. Finally, the CNN-based architecture receiving images in its input layer was trained and tested (validated) using 75% and 25% of the dataset, respectively. In practice, two different pipelines were developed, each of which was different in terms of preprocessing but similar in terms of data conversion and classification steps, as demonstrated in figure 1.3.

A common strategy employed is to visualize the weights of filters to interpret the conv layer results. These are usually most interpretable on the first conv layer, which directly examines the raw pixel data, but it is also possible to find the filter weights deeper in the network. In a well-trained network, smooth filters without noisy patterns are usually discovered. A smooth pattern without noise is an indicator that the training process is sufficiently long, and likely no overfitting occurred. In addition, visualization of the activation of the network's features is a helpful technique to explore training progress. In deeper layers, the features become more sparse and localized, and visualization helps to explore any potential dead filters (all zero features for many inputs). Filters and features of the first layer for a given fMRI and MRI trained LeNet model were visualized using an Alzheimer's brain and a

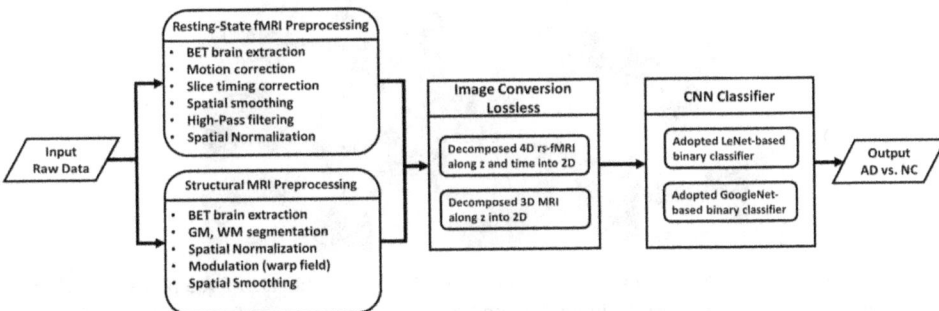

Figure 1.3. End-to-end recognition based on deep learning CNN classification methods is comprised of three major components: the preprocessing, image conversion, and classification modules. In the preprocessing step, two different submodules are developed for rs-fMRI and structural data. Next, the lossless image conversion module creates PNG images from medical imaging data using the algorithm described in the following section of this paper. The final step is to recognize AD from NC samples using CNN models, which is performed by training and testing models using 75% and 25% of the samples, respectively.

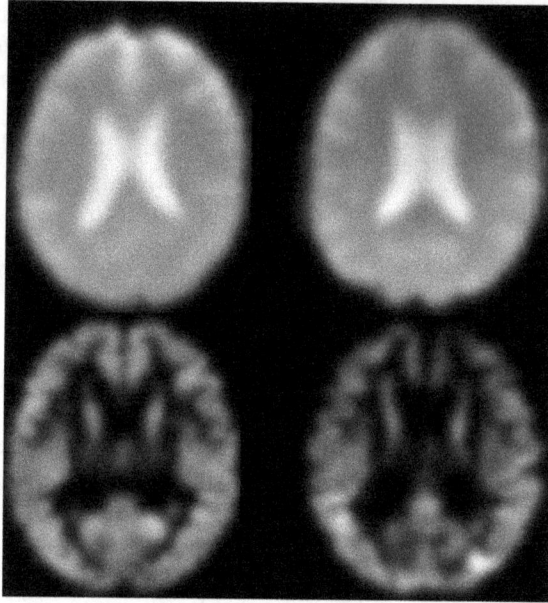

Figure 1.4. A middle cross-section of fMRI data (22, 27, 22) with thinness of 4 mm, representing a normal healthy brain (top-left) and an Alzheimer's brain (top-right). A middle cross-section of structural MRI (45, 55, 45) with a thickness of 2 mm, representing a normal brain (bottom-left) and an Alzheimer's subject (bottom-right) are also demonstrated. In both fMRI and MRI modalities, different brain patterns and signal intensities are identified.

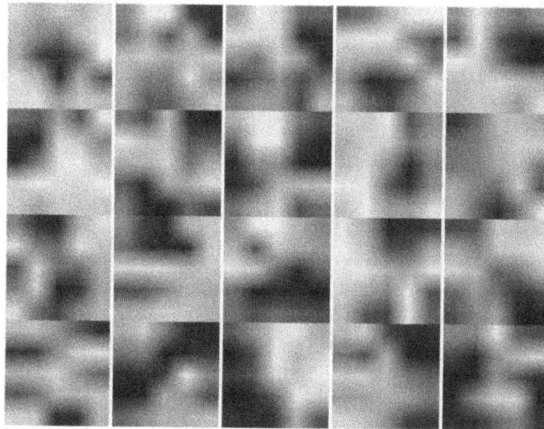

Figure 1.5. In the first layer of LeNet in a given trained fMRI model, 20 filters of 5×5 pixels were visualized. The weights shown were applied to the input data and produced activation, or features, of a given sample.

normal control brain (figure 1.4). Figures 1.5 and 1.6 demonstrate 20 filters of 5×5 pixels for fMRI and MRI models, respectively. Additionally, 20 features of 24×24 pixels in figures 1.7 and 1.8 reveal various regions of the brain that were activated in AD and NC samples.

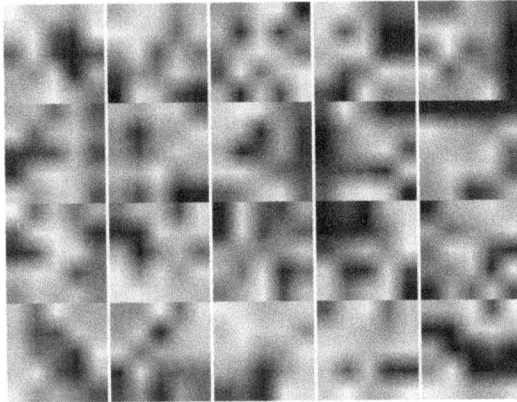

Figure 1.6. In a trained LeNet model, 20 filters with a kernel of 5 × 5 were visualized for the first layer. The filters shown were generated from a model in which MRI data smoothed by sigma = 3 mm were used for training.

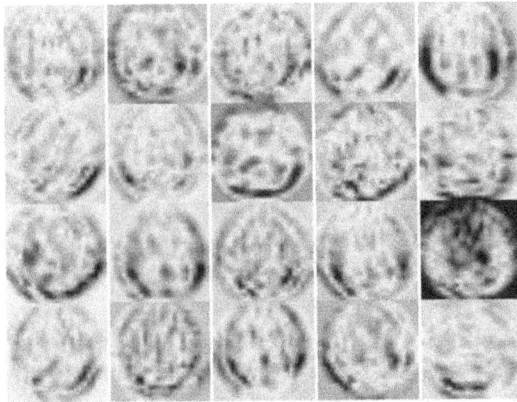

Figure 1.7. 20 activations (features) of the first layer of LeNet trained using MRI data were displayed for a given AD MRI sample (45, 55, 45). A smooth pattern without noise reveals that the model was successfully trained.

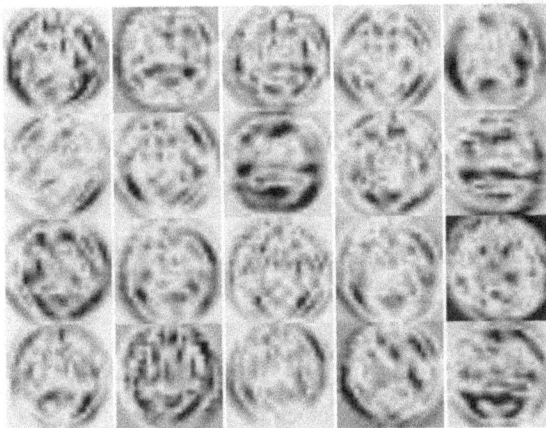

Figure 1.8. Features of the first layer of the same MRI trained model were displayed for a normal control (NC) brain slice (45, 55, 45). A basic visual comparison reveals significant differences between AD and NC samples.

1.4 Conclusion

In this chapter the crucial brain disorders ASD and AD were discussed from various points of view, such as clinical symptoms and diagnosis through imaging techniques. As discussed, computational neuroscience and machine learning techniques have been recently utilized to solve certain prediction and classification tasks in the field of brain imaging. Although the developed techniques and pipelines based on machine learning are not able to predict the disorders with complete accuracy, the improvement in accuracy rates for those methods are promising and significant. However, such systems should not replace scientists and specialists in the field but rather enable them to make more reliable decisions in recognizing these brain disorders in the early stages.

References

[1] Jones E J *et al* 2014 Developmental pathways to autism: a review of prospective studies of infants at risk *Neurosci. Biobehav. Rev.* **39** 1–33

[2] Case-Smith J, Weaver L L and Fristad M A 2015 A systematic review of sensory processing interventions for children with autism spectrum disorders *Autism* **19** 133–48

[3] Daniels A M and Mandell D S 2014 Explaining differences in age at autism spectrum disorder diagnosis: a critical review *Autism* **18** 583–97

[4] Park H R *et al* 2016 A short review on the current understanding of autism spectrum disorders *Exp. Neurobiol.* **25** 1–13

[5] Halladay A K *et al* 2015 Sex and gender differences in autism spectrum disorder: summarizing evidence gaps and identifying emerging areas of priority *Mol. Autism* **6** 36

[6] American Psychiatric Association 2013 *Diagnostic and Statistical Manual of Mental Disorders (DSM-5)* 5th edn (Arlington, VA: American Psychiatric Publishing)

[7] Fakhoury M 2015 Autistic spectrum disorders: a review of clinical features, theories and diagnosis *Int. J. Dev. Neurosci.* **43** 70–7

[8] Charman T and Baird G 2002 Practitioner review: diagnosis of autism spectrum disorder in 2-and 3-year-old children *J. Child Psychol. Psychiatry* **43** 289–305

[9] Heinsfeld A S *et al* 2018 Identification of autism spectrum disorder using deep learning and the ABIDE dataset *NeuroImage* **17** 16–23

[10] McConachie H and Diggle T 2007 Parent implemented early intervention for young children with autism spectrum disorder: a systematic review *J. Eval. Clin. Pract.* **13** 120–9

[11] Vorstman J A *et al* 2017 Autism genetics: opportunities and challenges for clinical translation *Nat. Rev. Genet.* **18** 362

[12] Chambless D L and Hollon S D 1998 Defining empirically supported therapies *J. Consult. Clin. Psychol.* **66** p 7

[13] Nathan P E 2007 Efficacy, effectiveness, and the clinical utility of psychotherapy research *The Art and Science of Psychotherapy* pp 69–83

[14] Vemuri P, Jones D T and Jack C R 2012 Resting state functional MRI in Alzheimer's disease *Alzheimer;s Res. Ther.* **4** 2

[15] He Y *et al* 2007 Regional coherence changes in the early stages of Alzheimer's disease: a combined structural and resting-state functional MRI study *Neuroimage* **35** 488–500

[16] Sarraf S and Sun J 2016 Advances in functional brain imaging: a comprehensive survey for engineers and physical scientists *Int. J. Adv. Res.* **4** 640–60

[17] Grady C *et al* 2016 Age differences in the functional interactions among the default, frontoparietal control, and dorsal attention networks *Neurobiol. Aging* **41** 159–72

[18] Saverino C *et al* 2016 The associative memory deficit in aging is related to reduced selectivity of brain activity during encoding *J. Cogn. Neurosci.* **28** 1331–44

[19] Warsi M A 2012 The fractal nature and functional connectivity of brain function as measured by BOLD MRI in Alzheimer's disease *PhD Dissertation*

[20] Grady C L *et al* 2003 Evidence from functional neuroimaging of a compensatory prefrontal network in Alzheimer's disease *J. Neurosci.* **23** 986–93

[21] Grady C L *et al* 2001 Altered brain functional connectivity and impaired short-term memory in Alzheimer's disease *Brain* **124** 739–56

[22] Raventós A and Zaidi M 2015 Automating neurological disease diagnosis using structural MR brain scan features

[23] Petersen R C 2004 Mild cognitive impairment as a diagnostic entity *J. Intern. Med.* **256** 183–94

[24] Mitchell A J and Shiri-Feshki M 2009 Rate of progression of mild cognitive impairment to dementia–meta-analysis of 41 robust inception cohort studies *Acta Psychiatr. Scand.* **119** 252–65

[25] Fitzpatrick-Lewis D *et al* 2015 Treatment for mild cognitive impairment: a systematic review and meta-analysis *CMAJ Open* **3** E419

[26] LeCun Y *et al* 1998 Gradient-based learning applied to document recognition *Proc. IEEE* **86** 2278–324

[27] Jia Y *et al* 2014 CAFFE:convolutional architecture for fast feature embedding *Proc. of the 22nd ACM International Conference on Multimedia* (New York: ACM)

[28] Ngiam J *et al* 2011 Multimodal deep learning *Proc. of the 28th International Conference on Machine Learning (ICML-11)*

[29] Erhan D *et al* 2010 Why does unsupervised pre-training help deep learning? *J. Mach. Learn. Res.* **11** 625–60

[30] Schmidhuber J 2015 Deep learning in neural networks: an overview *Neural Netw.* **61** 85–117

[31] Szegedy C *et al* 2015 Going deeper with convolutions *Proc. of the IEEE Conference on Computer Vision and Pattern Recognition*

[32] Wang L *et al* 2015 Object-scene convolutional neural networks for event recognition in images *Proc. of the IEEE Conference on Computer Vision and Pattern Recognition Workshops*

[33] Arel I, Rose D C and Karnowski T P 2010 Deep machine learning-a new frontier in artificial intelligence research *IEEE Comput. Intell. Mag.* **5** 13–8

[34] Krizhevsky A, Sutskever I and Hinton G E 2012 Imagenet classification with deep convolutional neural networks *NIPS'12 Proc. of the 25th Int. Conf. on Neural Information Processing Systems* vol **1** pp 1097–105

[35] Lowe D G 2004 Distinctive image features from scale-invariant keypoints *Int. J. Comput. Vis.* **60** 91–110

[36] Simonyan K and Zisserman A 2014 Very deep convolutional networks for large-scale image recognition, arXiv: 1409.1556

[37] He K *et al* 2016 Deep residual learning for image recognition *Proc. of the IEEE Conference on Computer Vision and Pattern Recognition.*

[38] Zhou J, Theesfeld C and Troyanskaya O 2018 Decoding the role of noncoding genome in neurological disease with deep learning *Biol. Psychiatry* **83** S82

[39] Rad N M *et al* 2018 Deep learning for automatic stereotypical motor movement detection using wearable sensors in autism spectrum disorders *Signal Process.* **144** 180–91

[40] Rad N M *et al* 2017 Deep learning for automatic stereotypical motor movement detection using wearable sensors in autism spectrum disorders, arXiv: 1709.05956

[41] Kong Y *et al* 2019 Classification of autism spectrum disorder by combining brain connectivity and deep neural network classifier *Neurocomputing* **324** 63–8

[42] Parisot S *et al* 2018 Disease prediction using graph convolutional networks: application to autism spectrum disorder and Alzheimer's disease *Med. Image Anal.* **48** 117–30

[43] Xiong H Y *et al* 2015 The human splicing code reveals new insights into the genetic determinants of disease *Science* **347** 1254806

[44] Ismail M *et al* 2017 A new deep-learning approach for early detection of shape variations in autism using structural MRI *2017 IEEE Int. Conf. on Image Processing (ICIP)* (Piscataway, NJ: IEEE)

[45] Dvornek N C, Ventola P and Duncan J S 2018 Combining phenotypic and resting-state fMRI data for autism classification with recurrent neural networks *2018 IEEE 15th Int. Symp. on Biomedical Imaging (ISBI 2018)* (Piscataway, NJ: IEEE)

[46] Li X *et al* 2018 2-channel convolutional 3D deep neural network (2CC3D) for fMRI analysis: ASD classification and feature learning *2018 IEEE 15th Int. Symp. on Biomedical Imaging (ISBI 2018)* (Piscataway, NJ: IEEE)

[47] Andrews D S *et al* 2018 Using pattern classification to identify brain imaging markers in autism spectrum disorder *Biomarkers in Psychiatry* (Berlin: Springer)

[48] Guo X *et al* 2017 Diagnosing autism spectrum disorder from brain resting-state functional connectivity patterns using a deep neural network with a novel feature selection method *Front. Neurosci.* **11** 460

[49] Jin Y *et al* 2015 Identification of infants at high-risk for autism spectrum disorder using multiparameter multiscale white matter connectivity networks *Hum. Brain Mapp.* **36** 4880–96

[50] Suk H -I and Shen D 2013 Deep learning-based feature representation for AD/MCI classification *Int. Conf. on Medical Image Computing and Computer-Assisted Intervention* (Berlin: Springer)

[51] Suk H I *et al* 2015 Latent feature representation with stacked auto-encoder for AD/MCI diagnosis *Brain Struct. Funct.* **220** 841–59

[52] Suk H I, Shen D and Initiative A S D N 2015 Deep learning in diagnosis of brain disorders *Recent Progress in Brain and Cognitive Engineering* (Berlin: Springer) pp 203–13

[53] Suk H I *et al* 2014 Hierarchical feature representation and multimodal fusion with deep learning for AD/MCI diagnosis *NeuroImage* **101** 569–82

[54] Payan A and Montana G 2015 Predicting Alzheimer's disease: a neuroimaging study with 3D convolutional neural networks, arXiv: 1502.02506

[55] Liu S *et al* 2015 Multimodal neuroimaging feature learning for multiclass diagnosis of Alzheimer's disease *IEEE Trans. Biomed. Eng.* **62** 1132–40

[56] Arvesen E 2015 Automatic classification of Alzheimer's disease from structural MRI *MSc Thesis*

[57] Liu F and Shen C 2014 Learning deep convolutional features for MRI based Alzheimer's disease classification, arXiv: 1404.3366

[58] Liu S *et al* 2014 High-level feature based PET image retrieval with deep learning architecture *J. Nucl. Med.* **55** 2028

[59] Brosch T, Tam R and Initiative A S D N 2013 Manifold learning of brain MRIs by deep learning *Int. Conf. on Medical Image Computing and Computer-Assisted Intervention* (Berlin: Springer)

[60] Rampasek L and Goldenberg A 2016 Tensorflow: biology's gateway to deep learning? *Cell Syst.* **2** 12–4

[61] de Brebisson A and Montana G 2015 Deep neural networks for anatomical brain segmentation *Proc. of the IEEE Conf. on Computer Vision and Pattern Recognition Workshops*

[62] Ijjina E P and Mohan C K 2016 Hybrid deep neural network model for human action recognition *Appl. Soft Comput.* **46** 936–52

[63] Zhu X, Suk H -I and Shen D 2014 A novel matrix-similarity based loss function for joint regression and classification in AD diagnosis *NeuroImage* **100** 91–105

[64] Zu C *et al* 2016 Label-aligned multi-task feature learning for multimodal classification of Alzheimer's disease and mild cognitive impairment *Brain Imaging Behav.* **10** 1148–59

[65] Liu M *et al* 2016 Inherent structure-based multiview learning with multitemplate feature representation for Alzheimer's disease diagnosis *IEEE Trans. Biomed. Eng.* **63** 1473–82

[66] Li F *et al* 2015 A robust deep model for improved classification of AD/MCI patients *IEEE J. Biomed. Health Inform.* **19** 1610–6

[67] Hosseini-Asl E, Keynto R and El-Baz A 2016 Alzheimer's disease diagnostics by adaptation of 3D convolutional network, arXiv: 1607.00455

[68] Sarraf S and Tofighi G 2016 Deep learning-based pipeline to recognize Alzheimer's disease using fMRI data *2016 Future Technologies Conf. (FTC)*

[69] Sarraf S *et al* 2017 *DeepAD: Alzheimer's Disease Classification via Deep Convolutional Neural Networks using MRI and fMRI* (bioRxiv) https://doi.org/10.1101/070441

IOP Publishing

Neurological Disorders and Imaging Physics, Volume 3
Application to autism spectrum disorders and Alzheimer's
Ayman El-Baz and Jasjit S Suri

Chapter 2

Neuropathology and neuroimaging of Alzheimer's disease

Gilberto Sousa Alves, Leonardo Caixeta and Lucas Schilling

Abbreviations

AD	Alzheimer's disease
CC	corpus callosum
DA	axial diffusivity
FA	fractional anisotropy
GM	gray matter
MD	mean diffusivity
MCI	mild cognitive impairment
RD	radial diffusivity
DTI	diffusion tensor imaging
WM	white matter
TBSS	tract-based spatial statistics

Since the first pathological depiction by Alois Alzheimer, a great deal of effort has been made to understand the neuropathological underpinnings of Alzheimer's disease (AD). The so-called neuroimaging biomarkers may reflect pathophysiological characteristics of the disease, such as axonal damage, neurofibrillary tangle pathology, and amyloid deposition, and encompass magnetic resonance imaging (MRI) and positron emission tomography (PET). In the current chapter, the main underpinnings of AD, including cortical and white matter (WM) alterations, are analyzed in the light of more recent advances achieved through neuroimaging evidence. Recent PET and MRI studies have attempted to look beyond the classical amyloid hypothesis, trying to relate the patterns of AD pathological progression with specific network disconnection syndromes and underlying genetic characteristics. In addition, neuroimaging biomarkers shed light on some putative events involved in the neuroprogression of AD, enabling the discussion of a few models, including the retrogenesis, Wallerian degeneration, and tau deposition hypotheses.

doi:10.1088/978-0-7503-1793-1ch2

Imaging biomarkers have made it possible to move the concept of AD from a clinical–pathological entity toward a laboratorial–biological one. Following these achievements, structural MRI has enabled the understanding of WM disconnection patterns and how they interact with gray matter (GM) atrophy. The adoption of multimodal neuroimaging thus offers an exciting opportunity to clarify the importance of the current models of retrogenesis, Wallerian degeneration, and tau deposition, establishing their clinical importance in the field of age-related cognitive disorders.

2.1 Alzheimer's disease: history, concept, clinical picture, and neurobiology

2.1.1 Brief history and concept

Alzheimer's disease (AD) is not currently the same condition as when it was first described by Alois Alzheimer in 1907 and introduced into nosology by Kraepelin, who claimed that AD was a *sui generis* disease. The concept of AD has shifted from a rare, early-onset, atypical dementia, characterized mainly by psychiatric symptoms in Alois Alzheimer's time, toward the most common senile dementia mainly characterized by episodic amnesia in the modern era. The adoption of the 'cognitive paradigm' at the beginning of the twentieth century has moved the core of the concept of dementia from behavior to cognitive function, in particular memory impairment [1, 2].

In the twenty-first century the concept of AD has expanded, advancing to encompass also preclinical or prodromal states (preclinical AD and/or prodromal AD) [3]. At this time, the diagnosis of AD is no longer based on clinical symptomatology and instead can be extrapolated from *in vivo* neurobiological findings, such as the presence of biological markers using amyloid positron emission tomography (PET) or by the search for tau proteins in cerebrospinal fluid (CSF). These new neurobiological data have moved the concept of AD from a clinical–pathological entity, where its diagnosis required the presence of dementia with distinct pathologic features, toward a laboratorial–biological one that recognizes AD as a pathological spectrum that ranges from normal cognition (with abnormal laboratorial findings) to the dementia stage [4, 5].

2.1.2 Clinical presentation

Memory impairment is the earliest and most pervasive feature in classical AD. The early involvement of the entorhinal cortex and CA1 of the hippocampus in AD supports the observed decline in episodic memory even at the initial stages of the disease [6]. In addition to episodic memory, topographical and temporal orientation, attention, praxis, language, executive function, visuoperceptive, and visuoconstructive deficits appears *pari passu* with the pathological spread in cortical areas. In other words, the clinical presentation of AD correlates with the distribution and pattern of atrophic changes in the brain [7]. Therefore, the intensity, severity, and nature of cognitive changes and behavior in AD depend greatly on the clinical–pathological stage of the disease.

Alzheimer's disease is a heterogeneous disease, with many possible different clinical presentations or phenotypes. In addition to the classical phenotype with short-term amnesia, AD can present several less common variants (many of them without amnesia), according to the initial neuroanatomical site involved in the degenerative process [8]. It can assume different clinical–pathological associations resulting in specific phenomenologies, as follows [9]:

1. Short-term episodic amnesia related pathologically to hippocampal atrophy (the main classical phenomenology).
2. Language disorder featuring a primary progressive aphasia (PPA) presentation related to left frontoparietal atrophy (the logopenic variant).
3. Visual disorder related to posterior cortical atrophy (the posterior cortical atrophy variant).
4. Disexecutive presentation related to prefrontal atrophy (the frontal variant).
5. Corticobasal syndrome related to asymmetric motor and extrapyramidal symptoms as well as exuberant apraxia (the corticobasal variant).

In summary, AD pathology is not uncommonly described in cases with atypical cortical syndromes, suggesting that the diagnosis of AD needs to be considered even in patients who present with focal dementia without significant memory loss, particularly in cases with visual, language, frontal/disexecutive, and apraxic/motor presentations [10]. The use of specific biomarkers is therefore useful in order to improve *in vivo* diagnosis of the pathological substrate in such cases.

2.1.3 Neurobiology and physiopathology of Alzheimer's disease

The majority of evidence regarding the physiopathology of most degenerative dementias, including AD, points to anomalous proteins with altered physiochemical and neurotoxic properties. These abnormal proteins have a conformational rearrangement that endows them with a tendency to aggregate and became deposited within tissues or cellular compartments, leading to early neuronal death. This is why these dementias are called 'conformational diseases' [11].

The neuropathological features of Alzheimer's disease are the presence of neuritic (senile) plaques and neurofibrillary tangles (NFT), along with neuronal loss, dystrophic neurites, and gliosis in histological postmortem examinations [12]. Neuritic (or senile) plaques are extracellular lesions, where their main component is the amyloid $\beta42$ protein (A$\beta42$). Neurofibrillary tangles are intracellular lesions and are mostly composed of hyperphosphorylated tau protein. Despite the controversial results in different research, the progression of the clinical syndrome of AD dementia follows the pattern of progression of these lesions in the brain. Patterns of gray matter loss are correlated mainly with the neurofibrillary tangle (NFT) pathology seen in AD (figure 2.1).

The amyloid precursor protein (APP) is a transmembrane protein, and one of the most abundant proteins in the central nervous system, and is also expressed in

transentorhinal	limbic	isocortical
I - II	III - IV	V - VI

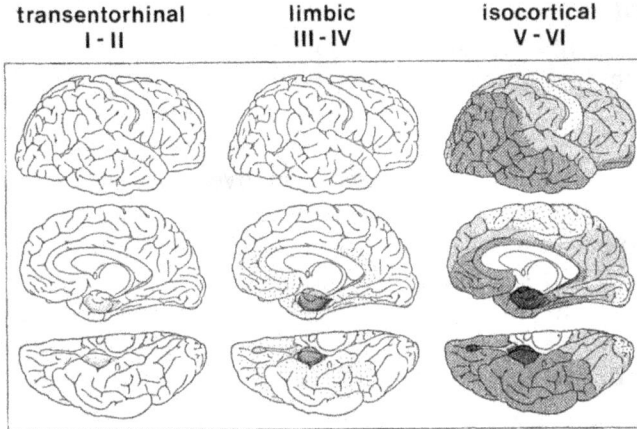

Figure 2.1. Neuropathological staging of Alzheimer-related changes. Six stages (I–VI) can be distinguished. Stages I and II show alterations that are practically confined to a single layer of the transentorhinal region (transentorhinal I–II). The main characteristic of stages III–IV is the severe impairment of the pre-alfa entorhinal and transentorhinal layers (limbic III–IV). Stages V–VI are marked by isocortical destruction (isocortical V–VI). The increase in density of shading indicates the increase in severity of neuropathological changes [7].

peripheral tissues, epithelium, and blood cells [13]. APP is metabolized by two distinct pathways: the α-secretase pathway (or non-amyloidogenic pathway) and the β-secretase pathway (or amyloidogenic pathway). In the α-secretase pathway, APP is cleaved by the α-secretase enzyme, releasing a soluble N-terminal fragment (sAPPα) and a C-terminal fragment (C83). The latter is cleaved by γ-secretase activity, releasing a C-terminal fragment of 3KDa (C3). Cleavage of APP by α-secretase occurs in the region containing the Aβ peptide opposing the formation of this peptide [14]. Alternatively, APP can be cleaved by the enzyme β-secretase by releasing an N-terminal fragment (sAPPβ) and a C-terminus (C99). The latter is cleaved by the enzyme γ-secretase yielding the amyloid β-peptide (Aβ).

There are several types of β-amyloid peptides, and those with 40 and 42 amino acids (Aβ40 and Aβ42) are the most abundant in the brain [15]. The types of Aβ are released as monomers which progressively aggregate into dimers, oligomers, and protofibrils, and finally fibril deposition occurs and the amyloid plaques are formed. Despite their similarities, Aβ42 is more prone to aggregation and fibrillation, and therefore is fundamental in the pathogenesis of AD [16].

Aβ oligomers are considered the most toxic form of β-amyloid peptide [17]. They interact with neurons and glial cells, leading to the activation of proinflammatory cascades, mitochondrial dysfunction, increased oxidative stress [18], insufficiency of intracellular signaling pathways and synaptic plasticity, increased phosphorylation of tau, increased GSK3β activity, deregulation of calcium metabolism, induction of neuronal apoptosis, and cell death [15].

Under physiological conditions, APP is metabolized preferentially by the non-amyloidogenic pathway and there is a balance between Aβ peptide production and the clearance of the brain [19]. Currently, two proteins are closely involved in clearance of Aβ peptides in the brain: apolipoprotein E (apoE) and insulin-degrading enzyme (IDE). However, pathological conditions cause predisposition to the APP metabolism of the amyloidogenic pathway and/or decrease the Aβ brain clearance, therefore facilitating Aβ accumulation in the nervous tissue [20].

Aβ accumulation triggers several deleterious events for neurons, for example mitochondrial dysfunction and increased oxidative abnormality and inflammatory response, decreased neurotrophic support, decreased neuroplasticity and neurogenesis, tau protein hyperphosphorylation, apoptosis, and calcium metabolism dysfunction. These events occur in a positive feedback manner, amplifying Aβ neurotoxicity and, ultimately, neuronal death [18]. Previously, deposits of large aggregates were thought to disrupt normal neuronal function, leading to neuronal dysfunction and death. More recently, however, it has been recognized that soluble oligomers of the proteins are the most toxic species, causing synaptic dysfunction, which is the major pathological correlate to disease progression [21]. In this line of thought, the formation of larger, insoluble aggregates may in fact represent a neuroprotective response, and may be a strategy to isolate these toxic proteins from the rest of the nervous tissue.

Another neuropathological feature of Alzheimer's disease is the presence of intraneuronal lesions called neurofibrillary tangles (NFT) [22]. The main component of the NFT is the matched helical filaments (PHF) formed by the hyperphosphoamrylated tau protein. Abnormal phosphorylation of tau protein at serine/threonine residues near the binding region of tau favors tubulin disaggregation and formation of PHF which gives rise to NFT [23]. Because of the importance of tau in maintaining neuronal stability and homeostasis, its hyperphosphorylation leads to a cascade of neuronal events that eventually causes neuronal death.

Despite the strong evidence supporting the major role of any Aβ peptides or hyperphosphorylated tau proteins in the pathogenesis of Alzheimer's disease, neither hypothesis fully accounts for the full range of pathological processes associated with this disease. Thus, some alternative and complementary hypotheses have been proposed to explain the pathophysiology of Alzheimer's disease, some of them addressed in the following sections. Most of these hypotheses involve the activity of protein and enzymes in cascades involved in the regulation of Aβ/APP and tau metabolism.

The enzyme GSK3β is a key enzyme in the regulation of cellular metabolism, therefore its activity level is fundamental in regulating the phosphorylation state of the tau protein in the neurons, that is, when it is hyperactive, it can cause hyperphosphorylation of tau protein, which is one of the AD markers. Deregulation of GSK3β activity is involved in several other pathological events associated with AD, for example, increased Aβ peptide production, increased apoptosis, reduced neurogenesis, and synaptic plasticity [24].

Excessive GSK3β activity is the initial event in the pathophysiology of Alzheimer's disease, triggering a cascade of sequential events, culminating in increased Aβ peptide production and tau hyperphosphorylation. Although the mechanisms involved in the GSK3β hypothesis encompass, in a broader sense, both the β-amyloid and the tau metabolism dysfunction hypotheses, they lack consistent empirical evidence to go beyond the current hypotheses on the patho-physiology of AD [5].

As previously mentioned, Alzheimer's disease is, in essence, a disease probably caused by abnormal protein metabolism (overproduction) and deposition (impaired clearance) in the brain cortex. This 'proteinopathy' occurs both in the extracellular space and inside the neuron. Extracellular beta-amyloid deposition occurs in the form of neuritic plaques and intraneuronal tau protein deposition in the form of intraneuronal NFTs. This pathological phenomenon probably begins 20 years before the appearance of clinical features. It seems that the 'start ignition' is the overproduction and impaired clearance of beta-amyloid and, in sequence, tau hyperphosphorylation and neuronal toxicity.

Neurofibrillary tangle burden and neuronal loss show a strong association with global cognitive impairment [25]. Neurofibrillary tangles are not pathognomonic of AD and can be identified in other dementia subtypes. It is rarely observed in non-primate mammals and even nonhuman primates barely develop a pathology comparable to that observed in patients with AD. Neurofibrillary degeneration is not randomly distributed throughout the brain in AD. It primarily affects areas of the brain that have become more and more predominant during the evolutionary process of encephalization. There is a clear and well-established association between neuropathological progression in specific cortical areas and the symptomatological course in AD. During disease progression, it affects cortical areas in a stereotypic sequence that recapitulates inversely the ontogenetic development of the brain. The specific distribution of cortical pathology in AD seems, moreover, to be determined by the modular organization of the cerebral cortex, which is basically a structural reflection of its ontogeny [26].

2.1.4 Clinical and physiopathological feature assessment

Alzheimer's disease, as discussed above, is characterized by a set of impairment symptoms of cognitive abilities, in particular memory. The diagnosis of this disease is based on clinical criteria, which include a detailed clinical history, physical examination, neuropsychological evaluation, and complementary laboratory and imaging tests. Despite the clinical findings characteristic of the disease, it is important to note that since the initial description of this condition for more than a century, the confirmation of the disease required the evaluation of the histopatho-logical aspects (senile plaques and neurofibrillary tangles) in the postmortem analysis of the cerebral cortex. Albeit the very clear identification of the clinical and physiopathological features of the disease, the absence of indicators capable of detecting the development of this pathological process placed great limitations on the scientific advances in the field until a few years ago. The development of accurate and reliable biomarkers for Alzheimer's disease has made a fundamental contribu-tion to clinical and scientific advances in the field, allowing the assessment and documentation *in vivo* of several pathological processes that previously could be only studied in postmortem analysis, leading to a new biological framework definition of the disease.

2.2 Biomarkers

2.2.1 CSF biomarkers

The concept of biomarkers encompasses a characteristic that is objectively measured and evaluated as an indicator of normal biology, pathological processes, or pharmacological responses to a therapeutic intervention. The AD biomarkers can be obtained by imaging or in cerebrospinal fluid (CSF). The first well-established and validated biomarkers for diagnosing AD were based on CSF analysis, and include measures of total tau (t-tau), phosphorylated tau-181 (p-tau), and amyloid-$\beta1$–42 (Aβ 42). These biomarkers reflect pathophysiological characteristics of the disease, such as axonal damage, neurofibrillary tangle pathology, and amyloid deposition, respectively. In AD, the CSF signature is reduced Aβ-42 (<500 pg ml^{-1}) and increased values of t-tau (>450 pg ml^{-1} for 51–70 year old subjects and >600 pg ml^{-1} for older than 71 years old subjects) and p-tau (>60 pg ml^{-1}) [27]. The combination of these three biomarkers has a sensitivity of $>95\%$ and a specificity of $>85\%$ for AD diagnosis [28–30]. The major limitation of CSF biomarkers occurs due to the invasive procedure required for collection, necessitating adequate infrastructure to perform the required lumbar puncture and laboratory analysis. Another limitation of CSF biomarker information is the lack of topographical information of the pathological processes, as only the fluid quantification of the neuropathological proteins is expressed.

2.2.2 Neuroimaging biomarkers

2.2.2.1 MRI biomarkers

Neuroimaging biomarkers are capable of identifying structural and metabolic changes in AD, allowing diagnosis and assessing progression of the disease. Structural biomarkers include the results of magnetic resonance imaging (MRI), focusing on brain atrophy as evidence of neurodegeneration related to symptom severity [31, 32]. Typical features of the disease, in particular hippocampal and entorhinal cortex atrophy, can be identified by MRI, which can contribute to AD diagnosis and can also predict a higher rate of conversion to dementia in patients with mild cognitive impairment [33, 34].

2.2.2.2 PET biomarkers

Positron emission tomography (PET) allows for molecular quantification of several neuropathological processes through injection of specific radiotracers, such as glucose hypometabolism, and amyloid and tau accumulation. The development of Pittsburgh compound-B (PiB), an ^{11}C-labeled ligand that binds to fibrillar amyloid, allowed for the first time the imaging *in vivo* of amyloid deposition in the brain [35]. In AD patients, the typical PiB uptake shows amyloid deposition in the frontal, medial, and lateral posterior parietal cortices, the precuneus, the occipital and lateral temporal cortices, as well as in the striatum (figure 2.1) [36, 37]. Despite PiB being considered a standardized PET amyloid radiotracer, the very short half-life of the carbon molecule has limited its clinical and scientific use, leading to the development of ^{18}F-labeled amyloid ligands. Due to their longer half-life, the radiotracers Florbetapir, Flutemetamol, and Florbetapen have allowed amyloid imaging application on a large scale [38]. The interpretation of the amyloid imaging and the definition of cut-off point values is still widely discussed, however, a standardized uptake value ratio (SUVR) of 1,42 has been proposed as an accurate single cut-off point for amyloid positivity [39].

The recent development of radiotracers with high affinity for tau fibrils has contributed to refined and detailed interpretation of the non-amyloid physiopathological changes in AD and also in other neurodegenerative diseases classified as tauopathies, such as cases of frontotemporal dementia, progressive supranuclear palsy, and primary progressive aphasia [40]. The pattern of ^{18}F]T807 (Flortaucipir) accumulation shows concordance with Braak staging, with evidence of tau neurofibrillary tangles across the frontal, temporal, and parietal cortices, and also the entorhinal region and hippocampus [41]. The accuracy and applicability of these radioligands are still debated, but their application could contribute not only to identifying tau hyperphosphorylation and accumulation, but also its correlation with disease severity—previous studies showed elevated SUVR in AD compared to MCI and healthy controls—and possibly to monitor responses to future tau-targeting therapies [42]. In this line, PET 18-fluordeoxyglucose has a major role on the assessment of neurodegeneration and metabolic changes observed in the progression of disease. The glucose metabolisms assessed by the radiotracer have been interpreted as indicators of synaptic density and neuronal activity [43, 44]. The typical metabolic signature of cerebral hypometabolism in AD includes changes in the posterior cingulate, precuneus, parietal, frontal, and medial temporal cortices, with relative preservation of the primary sensory motor and visual cortices, basal ganglia, thalamus, and cerebellum (figures 2.2 and 2.3) [45, 46].

2.2.2.3 Biological framework: AT(N) classification system

The recognition that AD includes several neuropathological events—made possibly by the advent of biomarkers—supports the concept of the disease progression as a continuum that encompasses different clinical and pathological stages, including preclinical and prodromic phases [9, 47, 48]. In this direction, AD biomarkers have made a strong contribution and have led to the new definition of AD as a biological framework grouping β amyloid deposition, pathological tau, and neurodegeneration in the AT(N) classification system by the Alzheimer's Association (NIA-AA) research framework (table 2.1) [49]. The objective of the NIA-AA research framework is the diagnosis of AD in living persons using biomarkers. In the AT(N) classification system the PET amyloid and CSF $A\beta_{42}$ and $A\beta_{42}/A\beta_{40}$ ratio are used as amyloid biomarkers, PET tau and CSF p-tau as tau biomarkers, and ^{18}F] fluorodeoxyglucose positron emission tomography (^{18}F]FDG PET), CSF t-tau, and anatomic MRI as neurodegeneration biomarkers.

2.2.2.4 Clinical application of AD biomarkers

Clinically, the use of AD biomarkers is widely discussed and should only be requested by specialists, as it can include serious ethical issues. The recommendation

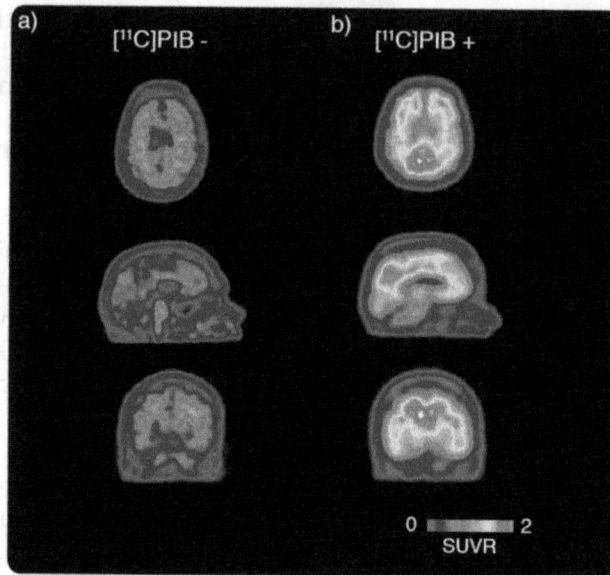

Figure 2.2. Representative [^{11}C]PIB PET images showing white matter uptake of [^{11}C]PIB in a patient with corticobasal syndrome ((a) age 74, MMSE 23), and extensive cortical uptake in a patient with AD ((b) age 70, MMSE 28) [37].

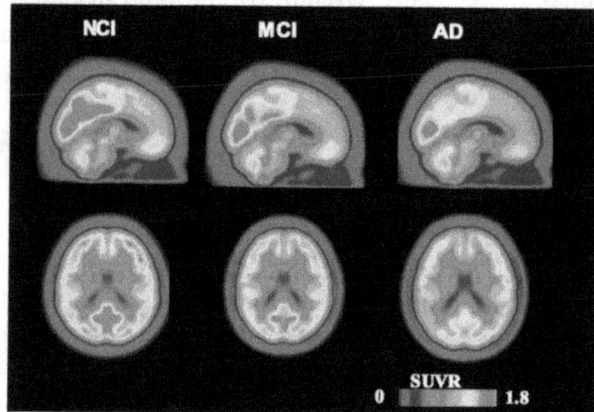

Figure 2.3. Representative PET [^{18}F]FDG images obtained from a total of 103 individuals with normal cognition (NCI, $N = 17$), mild cognitive impairment (MCI, $N = 52$), and dementia due to AD (AD, $N = 27$) obtained from a total of 103 structural MRI and [^{18}F]FDG scans from the Alzheimer's Disease Neuroimage Initiative (ADNI) database. The hot color scale represents the magnitude of [^{18}F]FDG standardized uptake value ratio (SUVR), proportional to the glucose uptake. Note the lower [^{18}F]FDG SUVRs in MCI and AD compared to controls. High SUVRs are particularly reduced in the posterior cingulate, precuneus, and the prefrontal (medial and dorsolateral) cortices [37].

Table 2.1. AT(N) biomarker classification system.

AT(N) system	Imaging and CSF biomarkers
Amyloid	**Amyloid PET** $[^{11}\text{C}]$PIB $[^{18}\text{F}]$Flutemetamol (Vyzamil®) $[^{18}\text{F}]$Florbetapir (Amivid®) $[^{18}\text{F}]$Florbetapen (Neuraceq®) **CSF** $A\beta_{42}$ $A\beta_{42}/A\beta_{40}$ ratio
Tau	**Tau PET** $[^{18}\text{F}]$T807/$[^{18}\text{F}]$AV-1451(Flortaucipir) $[^{18}\text{F}]$FDDNP $[^{18}\text{F}]$THK5351 $[^{18}\text{F}]$THK5117 $[^{11}\text{C}]$PBB3 **CSF** Phosphorylated tau (p-tau)
Neurodegeneration	**PET** $[^{18}\text{F}]$FDG **CSF** Total tau (t-tau) **MRI** Anatomic MRI

is to consider AD biomarkers as a supplement to clinical evaluation, particularly in uncertain, atypical, and/or early-onset dementia cases, in order to identify or exclude AD as the etiology of the clinical picture [50, 51]. Due to these recommendations and specifications, the clinical application of those biomarkers should only be considered in the tertiary care level and requested by specialists.

2.3 Understanding AD progression through structural imaging

2.3.1 Mapping AD neuropathology through neuronal circuits: what is the point?

The neuropathology of AD can also be understood in terms of disconnection syndromes and functional distributed networks underlying cognitive abilities [52, 53].

The investigation of AD neuropathology has developed substantially with the advent of neuroimaging techniques in the last few decades. The widespread use of magnetic resonance imaging (MRI) and new post-processing techniques of volumetric measures, functional MRI, diffusion tensor imaging, and tracer based nuclear medicine methods such as single positron emission tomography (SPECT) and [^{18}F] FDG allowing measures in metabolism, have been of great importance in understanding the neuro-biology of AD [54, 55]. Evidence from AD studies with subjects at the pre-dementia stage have shown that reductions in the medial temporal lobe and hippocampus may prove to be the best neuroimaging phenotypes to predict conversion to dementia [56]. While early approaches were basically focused on volumetric, morphometric, and region of interest (ROI) investigations, newer methods, including diffusor tensor imaging (DTI) analysis, offer higher accuracy for white matter (WM) registration between subjects. DTI is sensitized to the random motion of water molecules as they interact within tissues, thus reflecting characteristics of their immediate structural surroundings. DTI enables the definition of major WM tracts, through the anatomical quantification of WM microstructure [52], thus providing a comprehensive investigation of brain circuitry integrity. DTI has also provided important insights on the mechanisms of AD pathology, particularly in terms of gradient of progression.

Overall, age has been determined to be one of the most important risk factors for DTI changes [57, 58] and cognitive alterations associated with DTI changes might be related to early age alterations during neurodevelopmental stages. An increasing number of DTI studies indicate that age effects may follow an anterior to posterior gradient in WM changes [59–63].

2.3.2 Cortical myelination throughout life

2.3.2.1 Gradient of myelination

Compared to lower mammals, whose neocortical areas are reduced and consist basically of primary fields, human evolution was accompanied by a dramatic expansion of association areas, enabling the processing of an increasing amount of data. Interestingly, neocortical evolution was accompanied by increasing internal differentiation and functional sophistication of the primary fields, such as the primary visual field [64]. Ultimately, it is possible to find in the human species the greatest differential gradient between highly refined primary fields and the relatively simple high order sensory association areas and prefrontal regions. Such matura-tional differences among regional areas reflect the biological principles that sustain the formation of myelin and lipofuscin deposits throughout the human cortex [65]. Myelin may be a living, metabolically active part of the neuronal axon, with a membrane running through it, which is an extension of the cell (axonal membrane). Mitogenic activation is involved in cell plasticity and there is consistent evidence showing that mitogenic pathways in neurons are erroneously activated early during AD [65]. Distinct mechanisms may be associated with such mitogenic pathways [65], including hypoxia and β-amyloid deposition [66, 67], deficiency of vitamin B12 levels or folate, increased serum homocysteine levels, and increased serum methylmalonic acid levels [68, 69], although their interaction in the process of myelin degeneration awaits further elucidation (for a thorough review see Arendt [65]).

Evidence based on human neuropathological studies suggests that the brain regions most metabolically active in AD might also be the most capable of responding to mitogenic stimulus and, consequently, those with the highest vulnerability to degenerate [70]. One useful term for characterizing the pattern of neuronal vulnerability for the retrogenic process is 'arboreal entropy' [71]. According to this model, the greater the neuroprotection, the less vulnerable is the myelin and axon. Conversely, neuronal fibers may be attacked from their inside by neurofibrillary and neurotubular changes secondary to hyperphosphorylation, which ultimately may lead to axonal injury and myelin loss [70].

Atherosclerosis and cerebrovascular disease are other risk factors associated with AD, which have been primarily associated with myelin disruption. Therefore, the entire retrogenesis process implicated in AD neuropathology may comprise myelin, as well as the neuronal reactivation of mitogenic factors. The process of myelination is now known to continue well into the latter portion of life [72, 73]. Possibly, myelin plays a role not only in the conduction of electrical impulses in the neuron, but also in protection and maintenance of the oligodendroglia, myelin, and axonal relationship [72, 73]. Accordingly, early-myelination neurons may become increasingly more thickly myelinated over the years. Consequently, the most recently affected and, as a result, most thinly myelinated brain regions may be the most vulnerable to injury.

The susceptibility of neuronal fibers to the interactions of myelin breakdown, axonal damage and swelling, and other microstructural events may be more deeply appreciated through DTI studies. Indeed, evidence from DTI has been considered a surrogate marker of neuronal loss and synaptic disruption and, in addition to cerebrospinal fluid and PET techniques, may be incorporated in the multimodal staging of dementia. Moreover, DTI may help in mapping the progression of circuit disruption along AD evolution, enabling the establishment of patterns of subclinical features associated with disrupted neuronal pathways. However, current neuroimaging methods for dementia still need to achieve with greater accuracy and reliability the complex interpretation of multimodal techniques, from well controlled subjects to daily clinical routine.

2.3.3 Models of the neuropathological progression of AD

2.3.3.1 Neuropathological basis of WM abnormalities

Microstructural WM abnormalities in AD have often been interpreted as myelin breakdown and axonal damage [74]. Pathological models for AD—particularly the NIA-AA model for the preclinical asymptomatic phase of AD—provided an unquestionable advance, although a lot of unsolved issues remain regarding the neuropathological basis of AD. The field sill wrestles with issues regarding the role of amyloid as a sufficient marker for AD, the threshold of pathology needed for its diagnosis, and clinical factors mediating the rate of progression of the disease [75]. The typical patterns of regional spread lead to distinct neuropsychological syndromes, including the AD-type and frontotemporal dementia (FTD), each of these being influenced by characteristic patterns of neuronal damage, network disruption, and genetic factors [76]. Misfold disease proteins seem to aggregate within small, particularly vulnerable populations of neurons, most probably in specific areas, representing the brain signature of clinical deficits [76].

The term clinical–anatomical convergence has been used to describe the correspondence of clinical syndromes and the underpinning pathological entities [76]. One important issue is whether convergence occurs at the regional, clinical, or even neuronal level [76]. For example, different tauopathies could converge at the clinical network level simply because they reached disease specific nodes within the same circuitry network [76]. In the case of the retrogenesis hypothesis, there is staggered multifocal onset, with no direct connectivity between areas, such as the case of the frontal callosal fibers and the posterior cingulate, which nevertheless could explain the spatial patterning of neurodegenerative disease [77].

2.3.3.2 Cortical gradients of neurodegeneration

Evidence has shown that some neuronal cells (or even circuits) are more prone to develop AD-related pathology and seem to exhibit a particular inclination to neurodegeneration, while others offer more resistance to it [64]. High order processing regions seem to more affected [64, 78]. Most of the vulnerable neuronal fibers are those phylogenetically late-appearing elements that achieve functional maturity late in life and retain a high degree of structural plasticity, and show signs of immaturity (for instance, slow dendritic growth, lower regulation of neuronal differentiation) that persists throughout the adult years [64, 78]. This is the case of human neocortical projection neurons of the prefrontal and sensory association areas [64, 72]. Conversely, neurons more susceptible to developing hyperphosphorylated tau are usually projection neurons, with a localized dendritic arbor and long axon; neurons with short sized axons, with the exception of large cholinergic local circuit neurons of the striatum and chandelier cells in the basal temporal neocortex, are spared from AD pathology. Axons are also protected by the myelin sheath, which provides a mechanical barrier against bacteria or viruses by isolating the axon from the extraneuronal space. In addition, oligodendroglia cells offer additional protection and support, while energy balance and contact with the vascular system is possible due to astrocytes [79, 80].

Multiple molecular mechanisms may be involved in projection neurons with immaturely myelinated axons; conversely, neurons not involved in basic brain survival functions, but in highly demanding cognitive activities, can be more susceptible to both higher turnover and energy consumption, leading to greater exposure to chronic oxidative stress [81, 82].

2.3.4 Hypothetical models of neurodegeneration

Two different pathological models have been suggested to account for the microstructural changes of WM: retrogenesis and Wallerian degeneration.

2.3.4.1 Retrogenesis theory

Retrogenesis assumes primary white matter atrophy through myelin breakdown and axonal damage [70, 73, 74, 83]. It has been suggested that fibers more susceptible to neurodegeneration due to the retrogenesis process are those with small-diameter cortico-cortical axons [13, 57, 84], namely from the temporal lobe and neocortical areas. Previous studies reported a correlation between gray matter (GM) temporal atrophy and the reduced volume of corpus callosum (CC) posterior segments [74], while in others cortical atrophy failed to show an association with anterior CC fibers [74]. Structural changes, including the breakdown and dissolution of both the axonal cytoskeleton and myelin, and ultimately the elimination of myelin and other debris by Schwann cells and macrophages, are pathological events involved in secondary degeneration, which is in turn induced by amyloid deposition [85–87]. In fact, it has also been demonstrated that the genu of the CC is a region where fibers myelinate later in neurodevelopment [88]; these regions contain the highest density of small-diameter fibers, whereas fibers of the splenium of the CC myelinate earlier in life [88].

Recent DTI studies have addressed the progression of WM disruption in the CC based on the retrogenesis hypothesis. The CC comprises late-myelinating fibers in the genu [88, 89] and early-myelinating fibers in the splenium. The posterior CC sub-region receives axons directly from the temporo-parietal lobe, which are the same brain regions primarily affected by AD pathology [88]. Conversely, late-myelinating fibers connect the frontal lobes to the limbic system [90]. One study [91] reported lower fractional anisotropy (FA) and higher radial diffusivity (RD) in the body of the CC. Mounting evidence supporting the retrogenesis gradient in early AD has also been found for the genu of the CC. Other investigations also showed a similar FA profile in the CC between AD and MCI participants who later converted to clinical AD [92]. The CC also exhibited large clusters of voxels with significant differences between MCI converters and non-converters, in particular a decreased FA and an accompanying increase in RD [92]. Taken together, it seems plausible to suppose that in the CC both mechanisms (i.e. myelin breakdown and Wallerian degeneration) may be associated with WM disconnection. Furthermore, these mechanisms may be involved in region-specific illness effects.

Hence, the pattern of WM disruption in amnestic MCI takes place initially in limbic and commissural tracts and, later in clinically established dementia, may progress to the two remaining tracts—projection and association fibers. Similar results for these fibers have been previously reported [91, 93, 94]. These preliminary findings also suggest that cortical atrophy and progression of WM disruption from amnestic MCI to AD may follow a cortical thinning pattern, spreading over time from the temporal and limbic cortices to frontal and occipital cortices [91].

2.3.4.2 Wallerian degeneration hypothesis

There is consistent evidence suggesting that WM alterations could reflect Wallerian degeneration as a secondary product of cortical pathology [95]. The pathological basis for investigating Wallerian degeneration has been largely demonstrated by experimental animal models, such as those with the sciatic nerve of the frog [87]. In fact, amyloid deposition around neuronal cells or neurofibrillary tangles in the cell

CC-ant

	FA	DR	DA	MD
Vernooij 2008				
Acosta-Cabronero et al 2010				
Agosta et al 2011				
Salat et al 2010				
Gold et al 2010				
Bosch et al 2010				
Di Paola et al 2010				
O'Dwyer et al 2011				
Alves et al 2012				
Shu et al 2011				

CC-mid

	FA	DR	DA	MD
Vernooij 2008				
Acosta-Cabronero et al 2010				
Agosta et al 2011				
Salat et al 2010				
Gold et al 2010				
Bosch et al 2010				
Di Paola et al 2010				
O'Dwyer et al 2011				
Alves et al 2012				
Shu et al 2011				

CC-post

	FA	DR	DA	MD
Vernooij 2008				
Acosta-Cabronero et al 2010				
Agosta et al 2011				
Salat et al 2010				
Gold et al 2010				
Bosch et al 2010				
Di Paola et al 2010				
O'Dwyer et al 2011				
Alves et al 2012				
Shu et al 2011				

Fornix

	FA	DR	DA	MD
Vernooij 2008				
Acosta-Cabronero et al 2010				
Agosta et al 2011				
Salat et al 2010				
Gold et al 2010				
Bosch et al 2010				
Di Paola et al 2010				
O'Dwyer et al 2011				
Alves et al 2012				
Shu et al 2011				

Legend:
- MCI>ctr
- MCI<ctr
- AD>ctr
- AD<ctr
- MCI>ctr and AD>ctr
- MCI<ctr and AD<ctr
- AD>ctr and AD>MCI

Uncinate

	FA	DR	DA	MD
Vernooij 2008				
Acosta-Cabronero et al 2010				
Agosta et al 2011				
Salat et al 2010				
Gold et al 2010				
Bosch et al 2010				
Di Paola et al 2010				
O'Dwyer et al 2011				
Alves et al 2012				
Shu et al 2011				

Temp

	FA	DR	DA	MD
Vernooij 2008				
Acosta-Cabronero et al 2010				
Agosta et al 2011				
Salat et al 2010				
Gold et al 2010				
Bosch et al 2010				
Di Paola et al 2010				
O'Dwyer et al 2011				
Alves et al 2012				
Shu et al 2011				

Occipital

	FA	DR	DA	MD
Vernooij 2008				
Acosta-Cabronero et al 2010				
Agosta et al 2011				
Salat et al 2010				
Gold et al 2010				
Bosch et al 2010				
Di Paola et al 2010				
O'Dwyer et al 2011				
Alves et al 2012				
Shu et al 2011				

CORPUS CALLOSUM

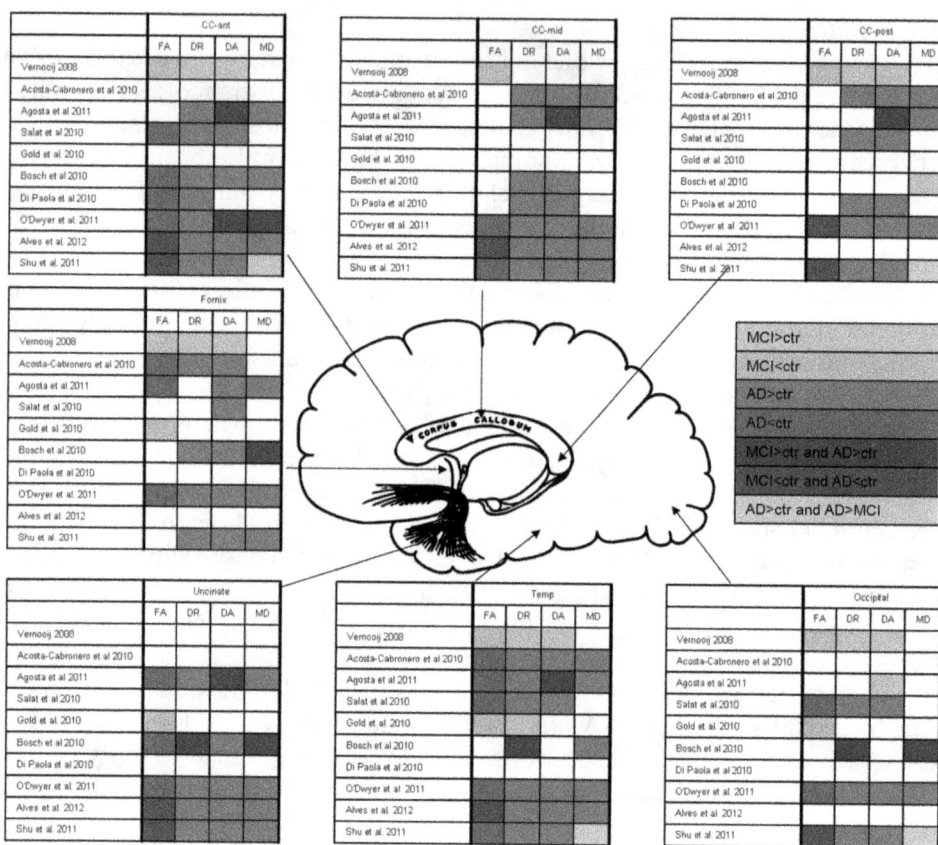

Figure 2.4. Main DTI-TBSS studies carried out with multiple indices: AD and control comparisons.

bodies ultimately lead to degeneration of axons and myelin [96]. Primary damage to WM tracts has been pointed out by recent studies as an alternative explanation for WM disruption (figure 2.4). Interestingly, Aβ deposits around WM vascularity [97] and cellular cytotoxicity provoked by Aβ peptides in oligodendrocytes, the cells responsible for myelin production, have also been reported [98].

The atrophy of the corpus callosum (CC) has been considered the anatomical correlate of Wallerian degeneration of commissural nerve fibers [89]. Indeed, the impact of CC atrophy, as a predictor of cognitive decline, has been demonstrated in a three-year follow up of elderly people with age-related WM leukoaraiosis [99]. Based on the Wallerian degeneration hypothesis and on the AD neuronal degeneration pattern [89], the earlier stages of WM degeneration should be associated with the involvement of the posterior CC sub-region, while in the later stages the anterior segment of the CC would exhibit atrophic changes [89].

The pattern of neuronal disruption in CC has been discussed in a few studies, but the unclear results may be due to the different methods of anatomical parcellation used for investigation. Moreover, the corpus callosum may be susceptible to AD and, depending on its anatomical localization, DTI changes could be associated either with retrogenesis or Wallerian degeneration [99]. Neuroimaging findings favoring Wallerian degeneration or retrogenesis remains disputed [100], sometimes with puzzling results. The CC comprises late-myelinating fibers in the genu [88, 89] and early-myelinating fiber in the splenium. The posterior CC sub-region receives axons directly from the temporo-parietal lobe, which are the same brain regions primarily affected by AD pathology [88]. Conversely, late-myelinating fibers connect the frontal lobes to the limbic system [90]. One study [91] reported lower FA and higher RD in the body of the CC. An increasing body of evidence has described early DTI changes in the genu of the CC in the early (i.e. preclinical) stages of AD. One investigation showed a similar FA profile in the CC between AD and MCI participants who later converted to clinical AD [92]. The CC also exhibited large clusters of voxels with significant differences between MCI converters and non-converters, in particular a decreased FA and an accompanying increase in RD [92].

On the other hand, whether a gradient of posterior–anterior changes or anterior–posterior changes predominates, or even a combination of these patterns occurring simultaneously, is still under intense debate [61, 74, 90, 99–102].

2.3.4.3 Tau deposition hypothesis

Finally, a third mechanism of WM degeneration involving tau has more recently been proposed. The tau protein seems to participate in the integrity and stabilization of the axonal cytoskeleton by binding to microtubules [103]. Axonal extensions may become swollen [104] and axonal transport may be disrupted with the functional failure of tau [87, 105]. An important point to be discussed is the relationship of tau production to neuroplasticity. Initial changes in AD may be identified in the entorhinal transitional neuronal networks, which project through the perforant path to the dental gyrus [13, 106, 107]. Recent studies have demonstrated the mechanism through which tau pathology initially progresses from distal axons to proximal dendrites. Only at later stages may the basal trunk of the dendrite tree and the body of the neuronal cell be damaged by hyperphosphorylated tau [107]. These events are most likely to be involved in the AD pathophysiological cascade.

In summary, the progression of amyloid deposition and hyperphosphorylated tau may hypothetically be linked to the synaptic disconnection of late myelination fibers [107]. The evidence favoring Wallerian degeneration versus retrogenesis remains disputed [100]. According to the retrogenesis model, small-diameter late-myelinating axons of cortical areas would be the earliest and most affected in AD, thereby increasing the susceptibility to amyloid accumulation and hyperphosphorylated tau; conversely, heavily myelinating axons would be less susceptible to AD pathology [73].

2.3.5 Understanding the biophysical properties of DTI

The DTI signal may be regarded as an indirect measure of various aspects of tissue integrity, thereby being influenced by distinct fiber components, including membrane intactness and myelin density [86, 108]. Diffusivity represented by the motion of water in a particular region can thus be altered by ordered structures such as axonal tracts in nervous tissues [109]. Diffusivity oriented by the fiber direction, so-called anisotropic diffusion, is largely restricted in the GM; an increase in anisotropic diffusion may correlate with myelin sheath content, being a valuable parameter for the investigation of WM microstructure integrity [110]. A representation of the ellipsoid can be computed by sampling the diffusivity along multiple directions spaced on a sphere [111, 112]. DTI uses measures derived from the eigenvectors, represented by eigenvalues, which define the diffusion ellipsoid in every voxel [112]. Axial diffusivity (DA) reflects the diffusion coefficient along the principal eigenvector ($\lambda 1$), whereas radial diffusivity (RD) indicates the average diffusion coefficients along the two axes perpendicular to $\lambda 1$. Mean diffusivity (MD) is a measure of the total amount of diffusion within a voxel and is computed as an average of all three diffusion axes [112]. Finally, FA is a scalar value between zero and one and it is calculated from the eigenvalues ($\lambda 1$, $\lambda 2$, $\lambda 3$) of the diffusion tensor [112]. FA measures the overall directionality of water diffusion and reflects the complexity of cytoskeleton architecture, which restricts the intra- and extracellular water movement [112]. The relationship between FA and WM microstructure changes considerably through the lifespan [87]. Finally, RD measures diffusion perpendicular to the WM fibers while diffusion parallel to the fibers is estimated by DA. MD is considered a non-specific marker of degeneration which reflects a decrease in membrane or other barriers to free water diffusion.

2.3.5.1 Interpretation of DTI diffusion indices

The evidence base gathered by DTI investigations in the last decade has helped to better define the pathological cascade underlying AD (figure 2.5) [113]. Diffusion studies of AD were primarily focused on the pattern of lesion distribution, the localization of DTI changes, the distribution of disrupted networks, and the nature of microstructural pathology. Regardless of DTI's sensitivity in assessing WM microstructural changes, differences in diffusion patterns across clinical groups may be challenging to interpret [52]. Several studies reported DTI changes in the parahippocampus, hippocampus, posterior cingulum, and splenium even at the MCI stage [94, 114–117]. Widespread areas of DTI abnormalities may also be observed in AD. It has been estimated that the whole brain may present a mass reduction of nearly 3%–4% per year [118].

The predictive value for conversion to dementia was investigated in a few studies [92]. Von Bruggen and colleagues reported higher parameters of receiver operator characteristic curve (ROC) for RD (0.94) and FA (0.94) in both the corpus callosum and left cingulum, while RD and DA in the fornix showed only fair (0.78) indices [92].

Only a few reports investigated regions of overlap between indices [61, 90, 100, 101, 118, 119]. Overall, there is still considerable variation among studies in the interpretation of multiple indices [90]. Animal models have proposed that increased MD would be more suggestive of myelin breakdown [120], while an increase in RD or DA is more associated with axonal damage [121]. Most authors describe an increase in DA not accompanied by FA changes as gross tissue loss, widespread tissue damage, and increase of extracellular space [100], which in turn may be a consequence of axonal atrophy secondary to Wallerian degeneration [99, 100]. Conversely, significantly reduced RD without differences in DA has been

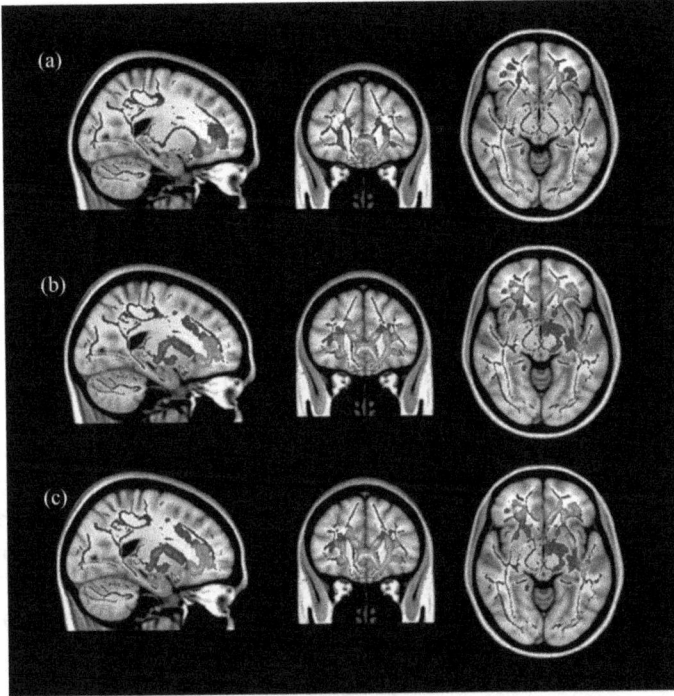

Figure 2.5. DTI changes are evident in Alzheimer's subjects when compared to healthy controls. Overlapping areas of decreased FA/increased RD are indicative of increased diffusion perpendicular to fiber orientation, possibly due to myelin breakdown (yellow–red). These areas can be observed in the corpus callosum (anterior and middle segments), anterior cingulum, and uncinate fasciculus (anterior portion) and remain when controlling for group differences in gray matter atrophy (a) and white matter burden volume (b) and (c). (Adapted from Alves *et al* [71], © 2015 Gilberto Sousa Alves *et al*).

interpreted as a disruption of myelin integrity in the absence of axonal structural irregularities [74, 99]. These changes would indicate specific damage of the myelin sheaths that restrict RD [99].

Another caveat that restricts the interpretation of diffusion indices is the discrepancy of anatomical findings among DTI investigations. Such a limitation may be associated with the different levels of AD severity between participants, which may be responsible for the diverse patterns of distribution of DTI changes. While most studies investigated mild to severe cases [83, 122–124] and mild to moderate cases [101, 102, 118], only mild AD cases were investigated by others [89].

2.3.5.2 Critical aspects of DTI

Notwithstanding the increasing evidence based on multiple indix studies and the possibility of hypothesizing different underlying mechanisms, DTI proxies are not suitable for directly determining the histological background of brain pathology [125, 126]. Hence, multimodal studies incorporating voxel-based morphometry (VBM) and PET techniques, as well as conventional neuropathological studies, may be necessary to clearly validate DTI parameters. Another awaited achievement is the use of high resolution techniques to assess difficult areas such as the fornix and hippocampus. High resolution DTI is based on optimized sequences for the medial temporal lobe and enables a detailed investigation of each individual fiber bundle to image voxel. Despite this, DTI alterations in multiple indices may help to elucidate early pathological changes in the preclinical stages of AD. The accurate prediction of a cognitively healthy individual converting to clinical AD still remains a significant research challenge. Thus, the idea of a biomarker profile, rather than the single use one of these techniques, may offer more robust predictive power to determine who is going to convert to AD with acceptable reliability. Another promising use of DTI is the support vector machine approaches, which consist in the statistical analysis of sensitivity and specificity of DTI indices in the differential diagnosis between groups. One study reported a sensitivity of 93% and a specificity of 92.8% in the discrimination between controls and MCI individuals [128], while in other investigations this discrimination yielded a sensitivity of 90.32% and a specificity of 90.41% [129]. One question raised by support vector investigations is the search for the most accurate DTI indices in voxelwise-analysis. Statistically significant FA differences between controls and MCI were described by some [90, 100, 119] but not all investigations [99, 101, 130]. Nevertheless, non-FA indices (DA, RD, MD), failed to show significant results for MCI–control discrimination [90, 99, 119]. Discrepant results may partially be explained by the fact that some studies [90, 100, 101] employed threshold-free cluster enhancement, which is the most conservative statistical method [131].

Most studies found DA and RD to be more accurate in indicating WM disruption in comparison to FA [100]. However, whether increases in DA or RD, but not decreases of FA, should be highlighted in the interpretation of DTI findings is still under debate. One has to take into account that DA and RD change in the same direction [119]. As a result, FA changes along fibers may be modified by increases of DA or RD, which may potentially suppress the effect of the changed diffusivity of FA. Such a characteristic may also explain major widespread changes in MD and RD, which are absolute diffusion metrics, in comparison to FA [90]. Accordingly, the relation between FA and WM changes presents considerable variations over the disease course, apparently becoming less pronounced in later stages for some WM tracts. A few tracts, such as the internal capsule, may exhibit DA increase with no significant change in RD [119].

Another important constraint of DTI studies is the interpretation of multiple indices, based mostly on animal model studies, which lack a consistent pathological validity [100]. For instance, the proper interpretation of DA may be a controversial issue, since both increases and decreases have been reported in the literature [92, 101, 118, 132]. Possibly, the lack of such association might be associated with axonal fiber organization or, alternatively, with DTI calculation in crossing fiber zones, as reported by some studies [100, 133].

Finally, one aspect that remains relevant is the discrepancy between studies concerning techniques of DTI acquisition and processing: the anatomical segmentation for the extraction of DTI values of regional tracts. For instance, the segmentation of CC has generated some controversy regarding the assumed topography of callosal fibers [74]. The other concern is related to partial volume effects, which may underestimate thinner anatomical regions such as the fornix and some limbic structures [92]. Future studies shall incorporate higher resolution MRI and apply automated VBM to overcome these limitations.

Notwithstanding these limitations, a multiple diffusion indices approach may be employed as one useful tool for the preclinical diagnosis of dementia. Other biomarkers of dementia risk, such as the presence of ApoE4 and inflammatory markers (IL6, CRP) may be associated with a steeper decline of cognitive status or greater neuronal loss [134]. Neuroplasticity refers to compensatory and neuro-protective mechanisms which maintain brain structure and activity [135]. The increase in neuronal activity, one of the variables that induces myelination, has been shown to be modulated by plasticity mechanisms which may be extended into old age [134]. Future research on DTI will need to explore how DTI changes are related to the mechanisms of brain atrophy, neuronal compensation, and plasticity.

2.4 Conclusions

Since its initial pathological description by Alois Alzheimer, outstanding results have been achieved in terms of understanding the neuropathological underpinnings of Alzheimer's disease. Much of the progress in the identification of structural and metabolic changes in AD was achieved thanks to the advent of neuroimaging biomarkers, allowing diagnosis and assessing the progression of the disease. The new neurobiological data have made it possible to move the concept of AD from a clinical–pathological entity toward a laboratorial–biological one. Recent molecular and MRI studies have attempted to look beyond the classical amyloid hypothesis, trying to relate the patterns of AD pathological progression with specific network disconnection syndromes and underlying genetic characteristics. Following these achievements, structural MRI has enabled the understanding of WM disconnection patterns and how they interact with GM atrophy. Molecular and genetic mecha-nisms involving the gradient of myelination seem to play an important role in the degree of vulnerability to AD pathology, with neocortical fibers being the most vulnerable to neurofibrillary and neurotubular changes secondary to hyperphos-phorylation. In this sense, DTI studies with multiple indices seem to offer an advantageous model to understand *in vivo* the cortical gradients of neurodegenera-tion and neuronal compensation observed in AD, and provide an in depth look at the putative molecular mechanisms leading to myelin breakdown and axonal damage. The adoption of multimodal neuroimaging offers thus an exciting opportunity for clarifying the importance of the current models of retrogenesis, Wallerian degeneration, and tau deposition, establishing its clinical importance in the field of age-related cognitive disorders.

References

[1] Caixeta L, Costa J N L, Vilela A C M and da Nóbrega M 2014 The development of the dementia concept in 19th century *Arq. Neuropsiquiatr.* **72** 564–7
[2] Dillmann R J and Stam F C 1992 Alzheimer's disease: variety without unity *Tijdschr. Gerontol. Geriatr.* **23** 41–7

[3] Dubois B 2000 'Prodromal Alzheimer's disease': a more useful concept than mild cognitive impairment? *Curr. Opin. Neurol.* **13** 367–9

[4] Molinuevo J L, Minguillon C, Rami L and Gispert J D 2018 The rationale behind the new Alzheimer's disease conceptualization: lessons learned during the last decades *J. Alzheimers Dis.* **62** 1067–77

[5] Caixeta L 2012 Doença de Alzheimer http://site.ebrary.com/id/10839784 (Accessed: 9 August 2018)

[6] Degenszajn J, Caramelli P, Caixeta L and Nitrini R 2001 Encoding process in delayed recall impairment and rate of forgetting in Alzheimer's disease *Arq. Neuropsiquiatr.* **59** 171–4

[7] Braak H and Braak E 1991 Neuropathological stageing of Alzheimer-related changes *Acta Neuropathol.* **82** 239–59

[8] Jorm A F 1985 Subtypes of Alzheimer's dementia: a conceptual analysis and critical review *Psychol. Med.* **15** 543–53

[9] McKhann G M *et al* 2011 The diagnosis of dementia due to Alzheimer's disease: recommendations from the National Institute on Aging–Alzheimer's Association workgroups on diagnostic guidelines for Alzheimer's disease *Alzheimers Dement.* **7** 263–9

[10] Alladi S *et al* 2007 Focal cortical presentations of Alzheimer's disease *Brain* **130** 2636–45

[11] Cummings J L and Pillai J (ed) 2017 *Neurodegenerative Diseases: Unifying Principles* (Oxford: Oxford University Press)

[12] Rosenberg R N 2006 Time will be of the essence in treating Alzheimer disease *J. Am. Med. Assoc.* **296** 327–9

[13] Braak E, Griffing K, Arai K, Bohl J, Bratzke H and Braak H 1999 Neuropathology of Alzheimer's disease: what is new since A Alzheimer? *Eur. Arch. Psychiatry Clin. Neurosci.* **249** 14–22

[14] Di Luca M, Colciaghi F, Pastorino L, Borroni B, Padovani A and Cattabeni F 2000 Platelets as a peripheral district where to study pathogenetic mechanisms of Alzheimer disease: the case of amyloid precursor protein *Eur. J. Pharmacol.* **405** 277–83

[15] Recuero M, Serrano E, Bullido M J and Valdivieso F 2004 Aβ production as consequence of cellular death of a human neuroblastoma overexpressing APP *FEBS Lett.* **570** 114–8

[16] Schmitt H P 2006 Protein ubiquitination, degradation and the proteasome in neurodegenerative disorders: no clear evidence for a significant pathogenetic role of proteasome failure in Alzheimer disease and related disorders *Med. Hypotheses* **67** 311–7

[17] Sanz-Blasco S, Valero R A, Rodríguez-Crespo I, Villalobos C and Núñez L 2008 Mitochondrial Ca^{2+} overload underlies Aβ oligomers neurotoxicity providing an unexpected mechanism of neuroprotection by NSAIDs *PLoS One* **3** e2718

[18] Rhein V and Eckert A 2007 Effects of Alzheimer's amyloid-beta and tau protein on mitochondrial function—role of glucose metabolism and insulin signalling *Arch. Physiol. Biochem.* **113** 131–41

[19] Vetrivel K S and Thinakaran G 2006 Amyloidogenic processing of β-amyloid precursor protein in intracellular compartments *Neurology* **66** S69–73

[20] Kunjathoor V V, Tseng A A, Medeiros L A, Khan T and Moore K J 2004 Beta-amyloid promotes accumulation of lipid peroxides by inhibiting CD36-mediated clearance of oxidized lipoproteins *J. Neuroinflammation* **1** 23

[21] Nakamura T and Lipton S A 2017 Neurodegenerative diseases as protein misfolding disorders *Neurodegenerative Diseases: Unifying Principles* ed J L Cummings and J Pillai (Oxford: Oxford University Press)

[22] Mazanetz M P and Fischer P M 2007 Untangling tau hyperphosphorylation in drug design for neurodegenerative diseases *Nat. Rev. Drug Discov.* **6** 464–79

[23] Johnson K A *et al* 2007 Imaging of amyloid burden and distribution in cerebral amyloid angiopathy *Ann. Neurol.* **62** 229–34

[24] Wang Z, Zhao C, Yu L, Zhou W and Li K 2009 Regional metabolic changes in the hippocampus and posterior cingulate area detected with 3-Tesla magnetic resonance spectroscopy in patients with mild cognitive impairment and Alzheimer disease *Acta Radiol.* **50** 312–9

[25] Apostolova L G *et al* 2007 Structural correlates of apathy in Alzheimer's disease *Dement. Geriatr. Cogn. Disord.* **24** 91–7

[26] Arendt T, Stieler J and Ueberham U 2017 Is sporadic Alzheimer's disease a developmental disorder? *J. Neurochem.* **143** 396–408

[27] Humpel C 2011 Identifying and validating biomarkers for Alzheimer's disease *Trends Biotechnol.* **29** 26–32

[28] Blennow K 2005 CSF biomarkers for Alzheimer's disease: use in early diagnosis and evaluation of drug treatment *Expert Rev. Mol. Diagn.* **5** 661–72

[29] Blennow K, Hampel H, Weiner M and Zetterberg H 2010 Cerebrospinal fluid and plasma biomarkers in Alzheimer disease *Nat. Rev. Neurol.* **6** 131–44

[30] Marksteiner J, Hinterhuber H and Hinterhuber C 2007 Cerebrospinal fluid biomarkers for diagnosis of Alzheimer's disease: beta-amyloid(1-42), tau, phospho-tau-181 and total protein *Drugs Today* **43** 423

[31] Seab J P, Jagust W J, Wong S T, Roos M S, Reed B R and Budinger T F 1988 Quantitative NMR measurements of hippocampal atrophy in Alzheimer's disease *Magn. Reson. Med.* **8** 200–8

[32] Dickerson B C *et al* 2009 The cortical signature of Alzheimer's disease: regionally specific cortical thinning relates to symptom severity in very mild to mild AD dementia and is detectable in asymptomatic amyloid-positive individuals *Cereb. Cortex* **19** 497–510

[33] Schuff N *et al* 2009 MRI of hippocampal volume loss in early Alzheimer's disease in relation to ApoE genotype and biomarkers *Brain* **132** 1067–77

[34] Seo E H, Park W Y and Choo I H 2017 Structural MRI and amyloid PET imaging for prediction of conversion to Alzheimer's disease in patients with mild cognitive impairment: a meta-analysis *Psychiatry Invest.* **14** 205

[35] Klunk W E *et al* 2004 Imaging brain amyloid in Alzheimer's disease with Pittsburgh compound-B *Ann. Neurol.* **55** 306–19

[36] Buckner R L 2005 Molecular, structural, and functional characterization of Alzheimer's disease: evidence for a relationship between default activity, amyloid, and memory *J. Neurosci.* **25** 7709–17

[37] Schilling L P *et al* 2016 Imaging Alzheimer's disease pathophysiology with PET *Dement. Neuropsychol.* **10** 79–90

[38] Petrella J R 2013 Neuroimaging and the search for a cure for Alzheimer disease *Radiology* **269** 671–91

[39] Jack C R *et al* 2017 Defining imaging biomarker cut points for brain aging and Alzheimer's disease *Alzheimers Dement.* **13** 205–16

[40] Okamura N *et al* 2016 Advances in the development of tau PET radiotracers and their clinical applications *Ageing Res. Rev.* **30** 107–13

[41] Chien D T *et al* 2013 Early clinical PET imaging results with the novel PHF-tau radioligand [F18]-T808 *J. Alzheimers Dis.* **38** 171–84

[42] Harada R *et al* 2016 Characteristics of tau and its ligands in PET imaging *Biomolecules* **6** 7

[43] Sokoloff L *et al* 1977 The [14C]deoxyglucose method for the measurement of local cerebral glucose utilization: theory, procedure, and normal values in the conscious and anesthetized albino rat *J. Neurochem.* **28** 897–916

[44] Rocher A B, Chapon F, Blaizot X, Baron J-C and Chavoix C 2003 Resting-state brain glucose utilization as measured by PET is directly related to regional synaptophysin levels: a study in baboons *NeuroImage* **20** 1894–8

[45] Schöll M, Damián A and Engler H 2014 Fluorodeoxyglucose PET in neurology and psychiatry *PET Clin.* **9** 371–90

[46] Schilling L P, Leuzy A, Zimmer E R, Gauthier S, Rosa-Neto P and Nonamyloid P E T 2014 Biomarkers and Alzheimer's disease: current and future perspectives *Future Neurol.* **9** 597–613

[47] Sperling R A *et al* 2011 Toward defining the preclinical stages of Alzheimer's disease: recommendations from the National Institute on Aging-Alzheimer's Association workgroups on diagnostic guidelines for Alzheimer's disease *Alzheimers Dement.* **7** 280–92

[48] Dubois B *et al* 2007 Research criteria for the diagnosis of Alzheimer's disease: revising the NINCDS–ADRDA criteria *Lancet Neurol.* **6** 734–46

[49] Jack C R *et al* 2018 NIA-AA research framework: toward a biological definition of Alzheimer's disease *Alzheimers Dement.* **14** 535–62

[50] Rosa-Neto P, Hsiung G-Y, Masellis M and on behalf of the CCDTD4 participants 2013 Fluid biomarkers for diagnosing dementia: rationale and the Canadian Consensus on Diagnosis and Treatment of Dementia recommendations for Canadian physicians *Alzheimers Res. Ther.* **5** S8

[51] Simonsen A H *et al* 2017 Recommendations for CSF AD biomarkers in the diagnostic evaluation of dementia *Alzheimers Dement.* **13** 274–84

[52] Matthews P M, Filippini N and Douaud G 2013 Brain structural and functional connectivity and the progression of neuropathology in Alzheimer's disease *J. Alzheimers Dis.* **33** S163–72

[53] Mesulam M M 1998 From sensation to cognition *Brain J. Neurol.* **121** 1013–52

[54] Friese U, Meindl T, Herpertz S C, Reiser M F, Hampel H and Teipel S J 2010 Diagnostic utility of novel MRI-based biomarkers for Alzheimer's disease: diffusion tensor imaging and deformation-based morphometry *J. Alzheimers Dis.* **20** 477–90

[55] Herholz K, Westwood S, Haense C and Dunn G 2011 Evaluation of a calibrated 18F-FDG PET score as a biomarker for progression in Alzheimer disease and mild cognitive impairment *J. Nucl. Med.* **52** 1218–26

[56] Risacher S L *et al* 2009 Baseline MRI predictors of conversion from MCI to probable AD in the ADNI cohort *Curr. Alzheimer Res.* **6** 347–61

[57] Tang Y, Nyengaard J R, Pakkenberg B and Gundersen H J 1997 Age-induced white matter changes in the human brain: a stereological investigation *Neurobiol. Aging* **18** 609–15

[58] Aung W Y, Mar S and Benzinger T L 2013 Diffusion tensor MRI as a biomarker in axonal and myelin damage *Imaging Med.* **5** 427–40

[59] Head D *et al* 2004 Differential vulnerability of anterior white matter in nondemented aging with minimal acceleration in dementia of the Alzheimer type: evidence from diffusion tensor imaging *Cereb. Cortex* **14** 410–23

[60] Sullivan E V *et al* 2001 Equivalent disruption of regional white matter microstructure in ageing healthy men and women *Neuroreport* **12** 99–104

[61] Salat D H *et al* 2010 White matter pathology isolates the hippocampal formation in Alzheimer's disease *Neurobiol. Aging* **31** 244–56

[62] Bendlin B B *et al* 2010 White matter in aging and cognition: a cross-sectional study of microstructure in adults aged eighteen to eighty-three *Dev. Neuropsychol.* **35** 257–77

[63] Brickman A M *et al* 2012 Testing the white matter retrogenesis hypothesis of cognitive aging *Neurobiol. Aging* **33** 1699–715

[64] Braak H and Del Tredici K 2015 *Neuroanatomy and Pathology of Sporadic Alzheimer's Disease* (New York: Springer)

[65] Arendt T 2003 Synaptic plasticity and cell cycle activation in neurons are alternative effector pathways: the 'Dr Jekyll and Mr Hyde concept' of Alzheimer's disease or the yin and yang of neuroplasticity *Prog. Neurobiol.* **71** 83–248

[66] Reisberg B *et al* 1999 Retrogenesis: clinical, physiologic, and pathologic mechanisms in brain aging, Alzheimer's and other dementing processes *Eur. Arch. Psychiatry Clin. Neurosci.* **249** 28–36

[67] Lee M S, Kwon Y T, Li M, Peng J, Friedlander R M and Tsai L H 2000 Neurotoxicity induces cleavage of p35 to p25 by calpain *Nature* **405** 360–4

[68] Clarke R, Smith A D, Jobst K A, Refsum H, Sutton L and Ueland P M 1998 Folate, vitamin B12, and serum total homocysteine levels in confirmed Alzheimer disease *Arch. Neurol.* **55** 1449–55

[69] Diaz-Arrastia R 1998 Hyperhomocysteinemia: a new risk factor for Alzheimer disease? *Arch. Neurol.* **55** 1407–8

[70] Reisberg B, Franssen E H, Souren L E M, Auer S R, Akram I and Kenowsky S 2002 Evidence and mechanisms of retrogenesis in Alzheimer's and other dementias: management and treatment import *Am. J. Alzheimers Dis. Other Demen.* **17** 202–12

[71] Alves G S *et al* 2014 Integrating retrogenesis theory to Alzheimer's disease pathology: insight from DTI-TBSS investigation of the white matter microstructural integrity *BioMed. Res. Int.* **2014** e291658

[72] Bartzokis G 2004 Quadratic trajectories of brain myelin content: unifying construct for neuropsychiatric disorders *Neurobiol. Aging* **25** 49–62

[73] Bartzokis G 2011 Alzheimer's disease as homeostatic responses to age-related myelin breakdown *Neurobiol. Aging* **32** 1341–71

[74] Di Paola M *et al* 2010 Callosal atrophy in mild cognitive impairment and Alzheimer's disease: different effects in different stages *NeuroImage* **49** 141–9

[75] Pillai J 2017 Predementia disorders: neurodegenerative disorder predecessor states as unifying clinical features *Neurodegenerative Diseases: Unifying Principles* ed J L Cummings and J Pillai (Oxford: Oxford University Press)

[76] Zhou J and Seeley W 2017 Brain circuits: neurodegenerative diseases *Neurodegenerative Diseases: Unifying Principles* (Oxford: Oxford University Press)

[77] Zhou J, Gennatas E D, Kramer J H, Miller B L and Seeley W W 2012 Predicting regional neurodegeneration from the healthy brain functional connectome *Neuron* **73** 1216–27

[78] Arendt T 2000 Alzheimer's disease as a loss of differentiation control in a subset of neurons that retain immature features in the adult brain *Neurobiol. Aging* **21** 783–96

[79] Nave K-A 2010 Myelination and support of axonal integrity by glia *Nature* **468** 244

[80] Lee Y *et al* 2012 Oligodendroglia metabolically support axons and contribute to neurodegeneration *Nature* **487** 443–8

[81] Pohanka M 2014 Alzheimer's disease and oxidative stress: a review *Curr. Med. Chem.* **21** 356–64

[82] Yan M H, Wang X and Zhu X 2013 Mitochondrial defects and oxidative stress in Alzheimer disease and Parkinson disease *Free Radic. Biol. Med.* **62** 90–101

[83] Pievani M *et al* 2010 Assessment of white matter tract damage in mild cognitive impairment and Alzheimer's disease *Hum. Brain Mapp.* **31** 1862–75

[84] Thal D R, Rüb U, Orantes M and Braak H 2002 Phases of Aβ-deposition in the human brain and its relevance for the development of AD *Neurology* **58** 1791–800

[85] Beaulieu C, Does M D, Snyder R E and Allen P S 1996 Changes in water diffusion due to Wallerian degeneration in peripheral nerve *Magn. Reson. Med.* **36** 627–31

[86] Beaulieu C 2002 The basis of anisotropic water diffusion in the nervous system—a technical review *NMR Biomed.* **15** 435–55

[87] Canu E *et al* 2011 Mapping the structural brain changes in Alzheimer's disease: the independent contribution of two imaging modalities *J. Alzheimers Dis.* **26** 263–74

[88] Aboitiz F, Scheibel A B, Fisher R S and Zaidel E 1992 Fiber composition of the human corpus callosum *Brain Res.* **598** 143–53

[89] Di Paola M, Spalletta G and Caltagirone C 2010 *In vivo* structural neuroanatomy of corpus callosum in Alzheimer's disease and mild cognitive impairment using different MRI techniques: a review *J. Alzheimers Dis.* **20** 67–95

[90] Alves G S *et al* 2012 Different patterns of white matter degeneration using multiple diffusion indices and volumetric data in mild cognitive impairment and Alzheimer patients *PLoS One* **7** e52859

[91] Huang H *et al* 2012 Distinctive disruption patterns of white matter tracts in Alzheimer's disease with full diffusion tensor characterization *Neurobiol. Aging* **33** 2029–45

[92] van Bruggen T, Stieltjes B, Thomann P A, Parzer P, Meinzer H-P and Fritzsche K H 2012 Do Alzheimer-specific microstructural changes in mild cognitive impairment predict conversion? *Psychiatry Res.* **203** 184–93

[93] Huang J, Friedland R P and Auchus A P 2007 Diffusion tensor imaging of normal-appearing white matter in mild cognitive impairment and early Alzheimer disease: preliminary evidence of axonal degeneration in the temporal lobe *Am. J. Neuroradiol.* **28** 1943–8

[94] Zhang Y *et al* 2007 Diffusion tensor imaging of cingulum fibers in mild cognitive impairment and Alzheimer disease *Neurology* **68** 13–9

[95] Brun A, Gustafson L and Englund E 1990 Subcortical pathology of Alzheimer's disease *Adv. Neurol.* **51** 73–7

[96] Englund E and Brun A 1990 White matter changes in dementia of Alzheimer's type: the difference in vulnerability between cell compartments *Histopathology* **16** 433–9

[97] Iwamoto N, Nishiyama E, Ohwada J and Arai H 1997 Distribution of amyloid deposits in the cerebral white matter of the Alzheimer's disease brain: relationship to blood vessels *Acta Neuropathol.* **93** 334–40

[98] Xu J *et al* 2001 Amyloid-beta peptides are cytotoxic to oligodendrocytes *J. Neurosci.* **21** RC118

[99] Di Paola M *et al* 2010 When, where, and how the corpus callosum changes in MCI and AD: a multimodal MRI study *Neurology* **74** 1136–42

[100] O'Dwyer L *et al* 2011 Multiple indices of diffusion identifies white matter damage in mild cognitive impairment and Alzheimer's disease *PLoS One* **6**

[101] Bosch B *et al* 2012 Multiple DTI index analysis in normal aging, amnestic MCI and AD. Relationship with neuropsychological performance *Neurobiol. Aging* **33** 61–74

[102] Stricker N H *et al* 2009 Decreased white matter integrity in late-myelinating fiber pathways in Alzheimer's disease supports retrogenesis *NeuroImage* **45** 10–6

[103] Braak H and Del Tredici K 2013 Evolutional aspects of Alzheimer's disease pathogenesis *J. Alzheimers Dis.* **33** S155–61

[104] Coleman M 2005 Axon degeneration mechanisms: commonality amid diversity *Nat. Rev. Neurosci.* **6** 889–98

[105] Higuchi M, Lee V M Y and Trojanowski J Q 2002 Tau and axonopathy in neuro-degenerative disorders *Neuromol. Med.* **2** 131–50

[106] Braak H and Braak E 1997 Frequency of stages of Alzheimer-related lesions in different age categories *Neurobiol. Aging* **18** 351–7

[107] Ashford J W and Bayley P J 2013 Retrogenesis: a model of dementia progression in Alzheimer's disease related to neuroplasticity *J. Alzheimers Dis.* **33** 1191–3

[108] Beaulieu C 2009 The biological basis of diffusion anisotropy *Diffusion MRI: From Quantitative Measurement to In vivo Neuroanatomy* (Amsterdam: Elsevier) pp 105–26

[109] Graña M *et al* 2011 Computer aided diagnosis system for Alzheimer disease using brain diffusion tensor imaging features selected by Pearson's correlation *Neurosci. Lett.* **502** 225–9

[110] Johansen-Berg H and Behrens T E J 2009 *Diffusion MRI: From Quantitative Measurement to In-vivo Neuroanatomy* (New York: Academic)

[111] Pierpaoli C, Jezzard P, Basser P J, Barnett A and Di Chiro G 1996 Diffusion tensor MR imaging of the human brain *Radiology* **201** 637–48

[112] Pierpaoli C and Basser P J 1996 Toward a quantitative assessment of diffusion anisotropy *Magn. Reson. Med.* **36** 893–906

[113] Sexton C E, Kalu U G, Filippini N, Mackay C E and Ebmeier K P 2011 A meta-analysis of diffusion tensor imaging in mild cognitive impairment and Alzheimer's disease *Neurobiol. Aging* **32** 2322.e5–18

[114] Chua T C, Wen W, Slavin M J and Sachdev P S 2008 Diffusion tensor imaging in mild cognitive impairment and Alzheimer's disease: a review *Curr. Opin. Neurol.* **21** 83–92

[115] Takahashi S, Yonezawa H, Takahashi J, Kudo M, Inoue T and Tohgi H 2002 Selective reduction of diffusion anisotropy in white matter of Alzheimer disease brains measured by 3.0 Tesla magnetic resonance imaging *Neurosci. Lett.* **332** 45–8

[116] Zhuang L *et al* 2010 White matter integrity in mild cognitive impairment: a tract-based spatial statistics study *NeuroImage* **53** 16–25

[117] Ewers M *et al* 2011 Staging Alzheimer's disease progression with multimodality neuro-imaging *Prog. Neurobiol.* **95** 535–46

[118] Acosta-Cabronero J, Williams G B, Pengas G and Nestor P J 2010 Absolute diffusivities define the landscape of white matter degeneration in Alzheimer's disease *Brain* **133** 529–39

[119] Shu N, Wang Z, Qi Z, Li K and He Y 2011 Multiple diffusion indices reveals white matter degeneration in Alzheimer's disease and mild cognitive impairment: a tract-based spatial statistics study *J. Alzheimers Dis.* **26** 275–85

[120] Harsan L A *et al* 2006 Brain dysmyelination and recovery assessment by noninvasive *in vivo* diffusion tensor magnetic resonance imaging *J. Neurosci. Res.* **83** 392–402

[121] Sun S-W, Liang H-F, Trinkaus K, Cross A H, Armstrong R C and Song S-K 2006 Noninvasive detection of cuprizone induced axonal damage and demyelination in the mouse corpus callosum *Magn. Reson. Med.* **55** 302–8

[122] Zhang Y *et al* 2009 White matter damage in frontotemporal dementia and Alzheimer's disease measured by diffusion MRI *Brain J Neurol.* **132** 2579–92

[123] Jhoo J H *et al* 2010 Discrimination of normal aging, MCI and AD with multimodal imaging measures on the medial temporal lobe *Psychiatry Res.* **183** 237–43

[124] Parente D B *et al* 2008 Potential role of diffusion tensor MRI in the differential diagnosis of mild cognitive impairment and Alzheimer's disease *Am. J. Roentgenol.* **190** 1369–74

[125] Radanovic M *et al* 2013 White matter abnormalities associated with Alzheimer's disease and mild cognitive impairment: a critical review of MRI studies *Expert Rev. Neurother.* **13** 483–93

[126] Jones D K, Knösche T R and Turner R 2013 White matter integrity, fiber count, and other fallacies: the do's and don'ts of diffusion MRI *NeuroImage* **73** 239–54

[127] Oishi K, Mielke M M, Albert M, Lyketsos C G and Mori S 2011 DTI analyses and clinical applications in Alzheimer's disease *J. Alzheimers Dis.* **26** 287–96

[128] O'Dwyer L *et al* 2012 Using support vector machines with multiple indices of diffusion for automated classification of mild cognitive impairment *PLoS One* **7** e32441

[129] Haller S *et al* 2010 Individual prediction of cognitive decline in mild cognitive impairment using support vector machine-based analysis of diffusion tensor imaging data *J. Alzheimers Dis.* **22** 315–27

[130] Agosta F *et al* 2011 White matter damage in Alzheimer disease and its relationship to gray matter atrophy *Radiology* **258** 853–63

[131] Smith S M and Nichols T E 2009 Threshold-free cluster enhancement: addressing problems of smoothing, threshold dependence and localisation in cluster inference *NeuroImage* **44** 83–98

[132] Bennett I J, Madden D J, Vaidya C J, Howard D V and Howard J H Jr 2010 Age-related differences in multiple measures of white matter integrity: a diffusion tensor imaging study of healthy aging *Hum. Brain Mapp.* **31** 378–90

[133] Douaud G *et al* 2011 DTI measures in crossing-fibre areas: increased diffusion anisotropy reveals early white matter alteration in MCI and mild Alzheimer's disease *NeuroImage* **55** 880–90

[134] Lopez O L, Becker J T and Kuller L H 2013 Patterns of compensation and vulnerability in normal subjects at risk of Alzheimer's disease *J. Alzheimers Dis.* **33** S427–38

[135] Barulli D and Stern Y 2013 Efficiency, capacity, compensation, maintenance, plasticity: emerging concepts in cognitive reserve *Trends Cogn. Sci.* **17** 502–9

IOP Publishing

Neurological Disorders and Imaging Physics, Volume 3
Application to autism spectrum disorders and Alzheimer's
Ayman El-Baz and Jasjit S Suri

Chapter 3

Retinal imaging in Alzheimer's disease

Umur Kayabasi

Recent studies suggest that β-amyloid (Aβ) induced tau pathology is responsible for the severe outcomes of the Alzheimer's disease (AD) process. Data from different models support the thesis in which Aβ accumulation acts as a triggering event in the pathogenetic process by accelerating antecedent tau and abnormal aggregation of tau protein ultimately leads to the formation of tangles within nerve cells. Once initiated, the tau aggregation process continues and spreads into previously healthy cells. There are three main characteristics of a taupathy: (a) an increase in tau levels; (b) a modification, such as hyperphosphorylation, sometimes related to other post-translational modifications such as truncation or acetylation; and (c) abnormal tau aggregation. It is possible to detect abnormal protein aggregation in patients with early or advanced AD using spectral domain optical scanning tomography (SD-OCT) and fundus autofluorescein (FAF) tests. Retinal regions with hyper- or hypofluorescence can be inspected using OCT and neurofibrillary filaments or advanced tau tangles can be observed by experienced posterior segment clinicians. In patients with early AD or mild cognitive impairment (MCI), mostly thin filaments are visible in OCT. However, in patients with positron emission tomography (PET)-proven advanced AD, in addition to Aβ plaques, thick tangles can be detected. Some of the thick tangles have a reverse E or number 3 shape. Retinal examination for Aβ is important, but may not be enough to diagnose AD since it can be found in other diseases or simply as a result of ageing. Detecting Aβ triggered tau aggregates may be more specific, and staging of AD may be possible by retinal examination and detection of tau in different development stages. Retinal examination using OCT and FAF is safe, non-invasive, and cheap. In addition OCT and FAF are valuable and reliable biomarkers in the diagnosis of AD.

3.1 Introduction

Alzheimer's disease (AD), first described by German psychiatrist and neuropathologist Alois Alzheimer in 1906, is a chronic neurodegenerative disease characterized

doi:10.1088/978-0-7503-1793-1ch3

by loss of memory and cognitive decline, and is neuropathologically associated with an increase in β-amyloid (Aβ) plaque deposition, neurofibrillary tangle (NFT) formation, neuronal loss, and inflammation [1, 2]. Aβ peptides, which are the dominant component of plaques, are the result of sequential cleavage of the amyloid precursor protein (APP). In addition to cognitive impairment, people with AD often develop visual anomalies in color discrimination, stereoacuity, and contrast sensitivity [3–5]. These visual abnormalities have been attributed, in part, to AD pathology in central visual pathways plus to retinal dysfunction, such as ganglion cell loss, reduction in the thickness of the retinal nerve fiber layer, and optic nerve degeneration [6]. In the past few years, transgenic mouse models have been engineered to explore different aspects of AD neurodegeneration.

These findings have indicated that elevated levels of Aβ peptides are associated with dysfunctional neuronal networks both in the brain and eye [1, 2]. Scientists recognize Alzheimer's as a disease process that begins many years before the symptoms of dementia become evident. New research has found changes in the brain and body up to 20 years before Alzheimer's symptoms arise. This research, published in the journal *Lancet Neurology*, examined a large extended family in Colombia that carried a gene for the early onset form of Alzheimer's, which typically arises before the age of 60. Approximately 30 percent of the 5000 family members carries the defective gene. Inheriting the gene, called presenelin 1, guarantees that the patient will have Alzheimer's at a relative young age [7, 8]. If physicians could reliably identify signs that someone is likely to get the disease, they could begin treatment earlier, possibly warding off the onset of Alzheimer's symptoms years down the road. Someone who has no obvious memory problems might remain problem-free if they take a drug that targets Aβ build-up.

3.2 Lipofuscin hypothesis of AD

Although Aβ protein has been identified as the main component of senile plaques, the triggering factors for the accumulation of Aβ protein have not been fully understood. Following the death of neurons during aging, lipofuscin is released and cannot be rapidly degraded. It may become harmful when it is released into the extracellular space. Lipofuscin contains Aβ and its precursor [9, 10]. Its translocation from the intra- to the extracellular compartment (together with themicroglia, astrocytes, and a neuroinflammatory response) could greatly change its biochemical characteristics. The hydrophobic and insoluble characteristics of lipofuscin may induce an immuneresponse [9]. Mitochondrial autophagocytosis is also a major contributor to lipofuscin formation [10] and finally, the rate of lipofuscin formation is also closely related to oxidative stress [9, 10]. Lipofuscin may therefore be the missing link in the pathogenesis of AD (such as oxidative stress, mitochondrial dysfunction, and the activation of immune responses) [11].

Recent evidence which indicates that vascular factors play an important role in the genesis of AD may be interpreted in this context, because hypoperfusion is a potential cause of damage to neurons and may initiate lipofuscin release [11]. Lipofuscin means 'dark fat' and is also known as 'age pigment' and its distribution is consistent with the release of debris or waste products. Lipofuscin plays a major role in another neuronal degenerative process that is very common in elderly people: age-related macular degeneration (AMD) [11]. Lipofuscin accumulation in the retinal pigment epithelium (RPE) is involved in the pathogenesis of AMD, together with the formation of abnormal extracellular deposits (or drusens). It has been shown that drusens contain a number of molecules and, most significantly, $A\beta$ protein. Drusens arise from material released by lipofuscin-rich retinal pigmented epithelial cells and contain proteins that are also components of senile plaques. This may suggest an unexpected similarity between the pathogenetic mechanisms of AMD and AD [11].

3.3 OCT and FAF in retinal diseases

Fundus autofluorescein (FAF) detects lipofuscin in the retina. The autofluoresence signal from RPE cells is very much correlated with lipofuscin content and accumulation. FAF is increased with RPE dysfunction due to the accumulation, impaired processing, and clearing of lipofuscin. Conversely, the FAF signal may be decreased in the setting of RPE or photoreceptor loss—if there are no photoreceptor outer segments, the source of lipofuscin formation may be lost [12]. The advent of confocal scanning laser ophthalmoscopic (SLO) technology was an important advance in making FAF approaches clinically relevant. Confocal SLO utilizes a focused, low-power laser that is swept across the fundus in a fast pattern. The confocal nature ensures that the reflectance and fluoresence are 'conjugate' (from the same optical plane) [12].

3.3.1 SD-OCT and FAF in diagnosing AD

During the evaluation of patients with MCI or patients with a family history of AD, the FAF test becomes very important. Both hyperfluorescent (regions with excessive lipofuscin) or hypofluorescent (atrophic retina) areas are of concern. To better understand the layer of the defect, spectral domain optical scanning tomography (SD-OCT) can be performed on the abnormal areas of FAF. Some experience is necessary for the differential diagnosis of the lesions [12, 13]. To better understand the nature of the lesion, curcumin may be orally or intravenously administered. It binds to β-amyloid plaques in the retina which are most likely related to AD [14].

3.4 Misfolded proteins in the retina

Tau protein plays a crucial role in many neurodegenerative diseases including AD. Tau dysfunction includes abnormal tau phosphorylation, protein aggregation, neurofibrillary tangle formation, and neurotoxicity [15]. Similar neurodegenerative processes affect the retina, causing impaired contrast sensitivity, reduced visual acuity, and abnormal motion perception. The retina is integrated with the central nervous system (CNS) and has been considered a window to the brain. Approximately 50% of AD patients present with visual deficits that include retinal ganglion cell loss, thinning of the retinal nerve fiber layer, abnormal electroretino-gram response, and reduced blood flow [16]. Tau inclusions and β-amyloid ($A\beta$)

deposition have been described in the post-mortem retina exams of AD patients. Phosphorylated and misfolded tau accumulation has also been observed in retinal ganglion cell soma, dendrites, and transretinal axons in animal models. These pathological changes may cause retinal neurondysfunction and subsequent death and suggest a prominent role for abnormal tau in visual deficits [16].

3.5 Cryo-electron microscopy

Cryogenic-electron microscopy (cryo-EM) was recently used to detect the detailed structure of tau filaments [17]. The researchers discovered C-shaped paired helical filaments (PHFs) composed of tau-subunits in the brains of live AD patients and they presented the C-shape in striking detail (figure 3.1). Tau aggregation consisted of a mixture of PHFs and straight filaments (SFs), the former making up about 90% of the total [3]. A C-shaped core defined the common protofilament of PHFs and SFs, although the two Cs could be arranged differently (figure 3.2).

3.6 Retinal imaging of misfolded proteins

When the retinas of PET-proven live AD patients are examined using SD-OCT and FAF, the hyper- or hypofluorescent lesions in the retina can be scanned using OCT and images that completely correspond to the histopathological and cryo-EM shapes of tau filaments may be observed. Neurofibrillary tangles that are identical to the cryo-EM images in terms of shape and diameter are detected (figures 3.3–3.5). When curcumin is given to the patients, it stains the Aβ lesions. FAF provides dot-like hyperfluorescent images and SD-OCT performed on these lesions reveals the exact location of the accumulations in the layers of the retina (figures 3.6–3.10).

It can be seen that these images of tau are also identical to the histopathologic images (figure 3.11).

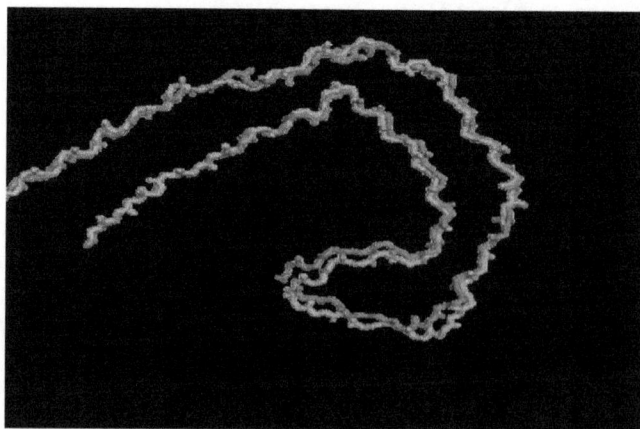

Figure 3.1. Illustration of Tau Imaging in the Brain.

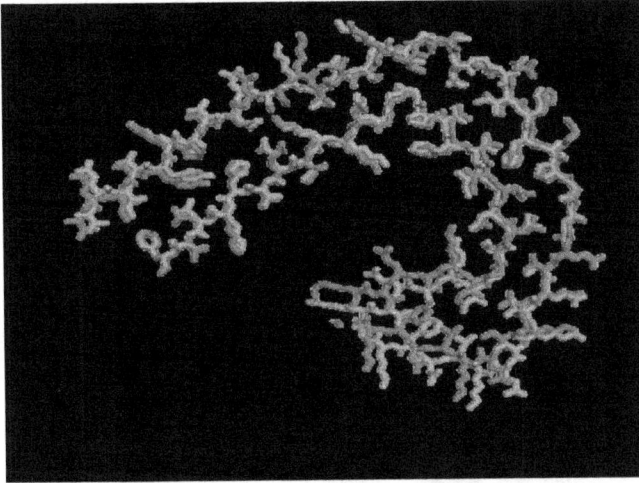

Figure 3.2. Illustration of Tau Imaging in the Brain.

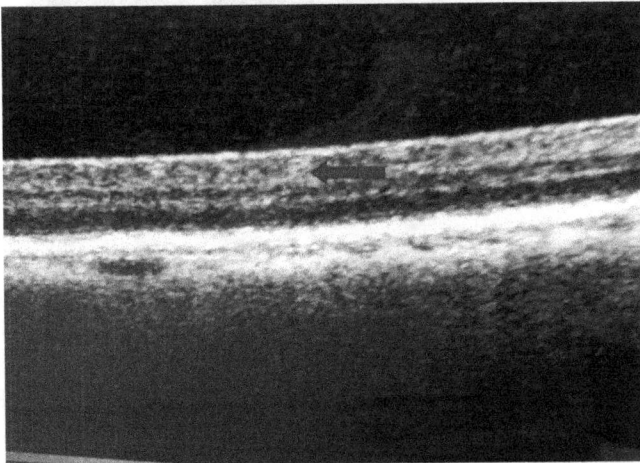

Figure 3.3. Tau image on SD-OCT.

3.7 Curcumin

Curcumin is the major constituent of the Asian spice turmeric, isolated from the rhizome of *Curcuma longa* [18]. Curcumin was isolated in 1815 as a yellow coloring-agent [18]. It has been used historically in Ayurvedic medicine and is extensively used for medicinal purposes in Asia and other parts of the world. Curcumin is used in foods because of its color and flavor. It is also used as cosmetic product, particularly for the skin. Several natural products have been used to delay the progression of disease in the elderly and AD patients. Several groups have studied the efficacy of natural products and antioxidants, including vitamin E, curcumin, *Ginkgo biloba*, and melatonin, to determine if antioxidants reduce $A\beta$ and tau pathologies and enhance cognitive function in mouse models of AD [18].

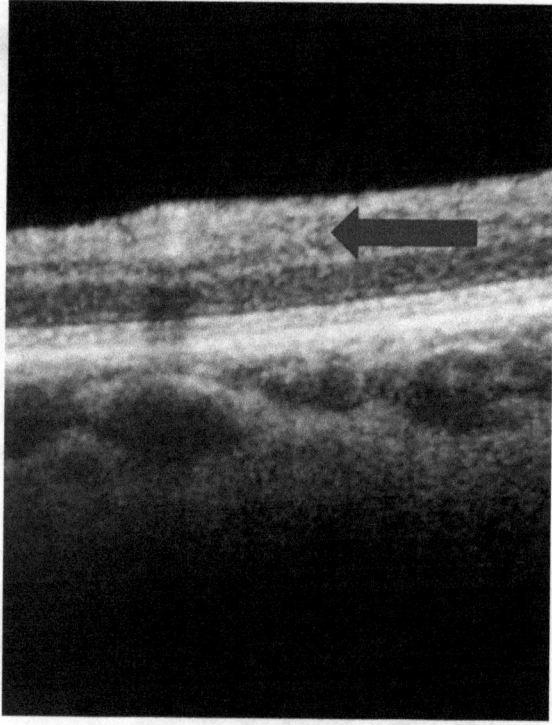

Figure 3.4. Tau on SD-OCT.

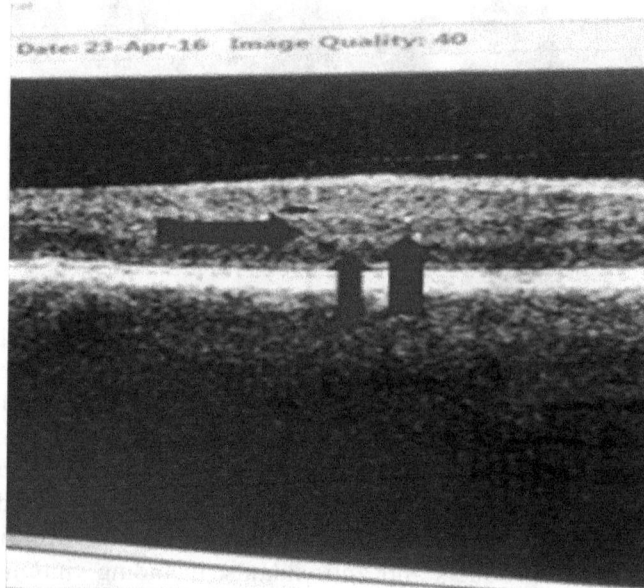

Figure 3.5. Fibrillary tau (early stage) on SD-OCT.

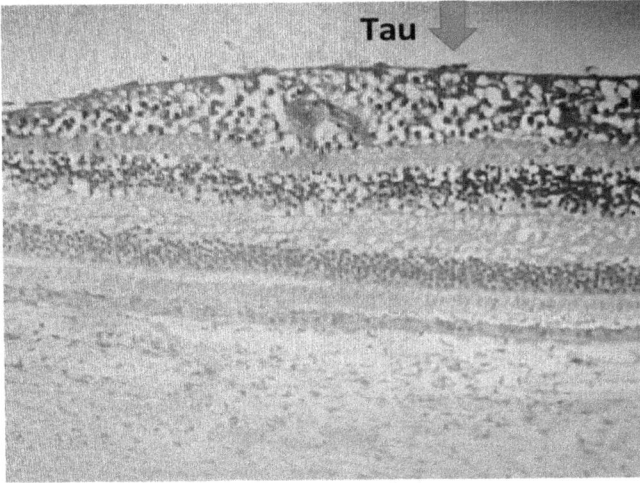

Figure 3.6. Histopathologic image of retinal misfolded proteins.

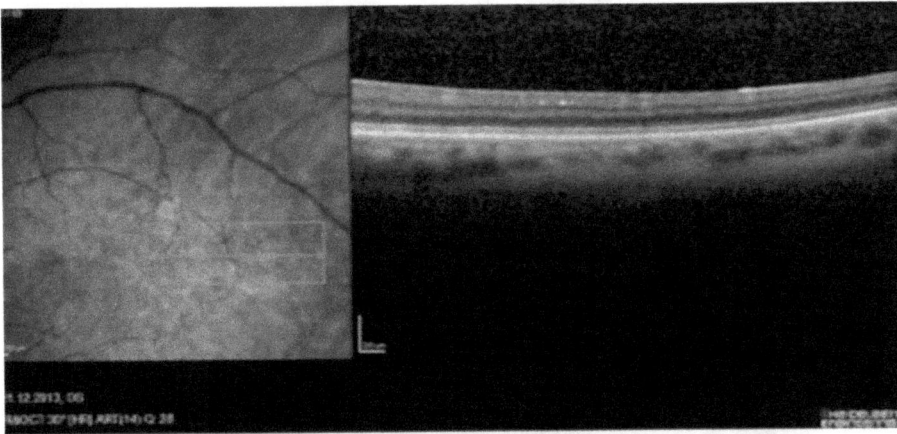

Figure 3.7. β-amyloid plaques in SD-OCT images.

In a recent review article [19], the researchers nicely summarized the latest developments in curcumin research on AD. They focused on the mechanisms of the action of curcumin in AD, including curcumin's effect on the inhibition of $A\beta$ and tau, its copper binding ability, cholesterol lowering ability, anti-inflammatory effects, modulation of microglia, AChE inhibition, antioxidant properties, and modification of insulin signaling pathways. They also covered the bioavailability of curcumin, and the current challenges of curcumin therapy for AD patients [19].

Apart from being used for therapeutic purposes, curcumin's ability to bind with $A\beta$ also makes it suitable for use for diagnostic purposes. After binding with $A\beta$, it becomes possible to image the toxic protein accumulation in the retina (figure 3.11).

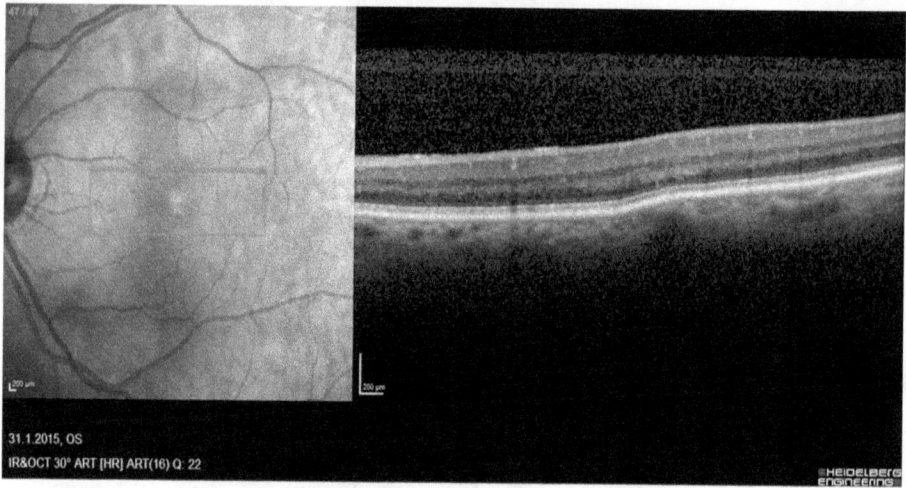

Figure 3.8. Alpha-synuclein in SD-OCT images.

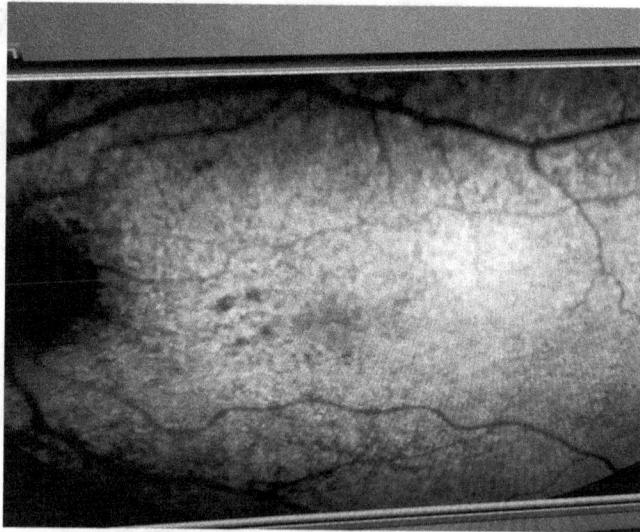

Figure 3.9. Patchy lesions in FAF images.

3.8 AMD and AD

AD and AMD are both neurodegenerative diseases that share common risk factors, such as increased age, systemic hypertension, diabetes, and hypercholesterolemia, and also histopathological features such as the deposition of $A\beta$ in retinal drusens and senile plaques. In addition, oxidative stress and inflammation are thought to contribute to both diseases [20]. However, the genetic risk factors for AD and AMD seem to be different [3]. There is some controversy about the potential associations between AD and AMD. Some studies have reported an association between AMD

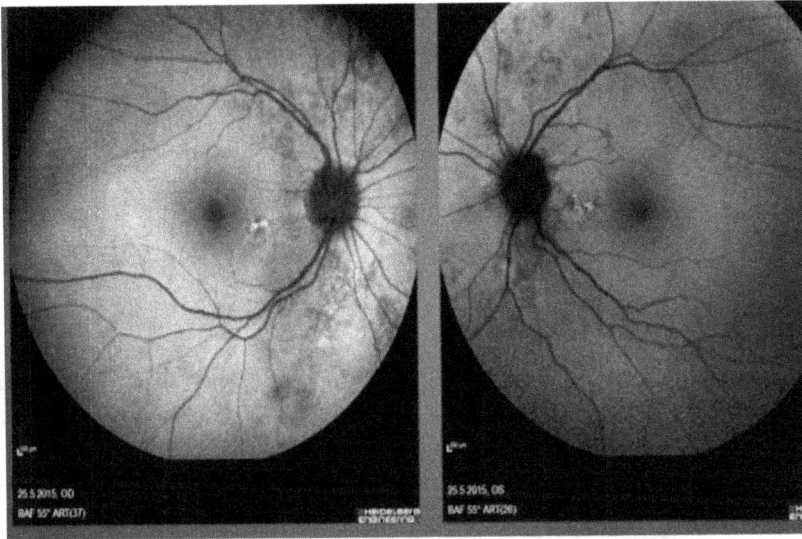

Figure 3.10. Hypo- and hyperfluorescent lesions in FAF images.

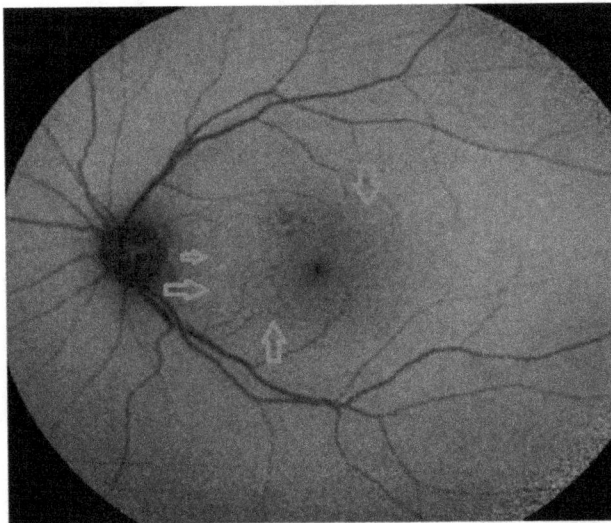

Figure 3.11. Curcumin stained lesions in FAF images.

and cognitive impairment, based on mental state examination, while other epidemiological studies found no significant association between AD and AMD [20]. There is no definitive answer to whether AD is associated with AMD. Finding the answer to this is important, because the treatment for one disease may protect against or exacerbate the other disease. In addition, patients with one disease may need examination for the other disease.

The case report shown here clearly displays AMD in a patient diagnosed with AD. The curcumin stained regions within the degenerated macula show $A\beta$ inside drusen-like accumulations (figures 3.12–3.15). Curcumin's ability to bind to Aβ is known and is also proof of the nature of these accumulations. Aβ was shown in drusens in previous studies [21]. The results that suggest no association between AD and AMD may be due to the fact that AD patients are more reluctant to attend regular ophthalmologic examinations [21]. In fact, if an association between the two diseases is established in multi-centered trials, a novel drug may treat the two devastating illnesses simultaneously. This report may be a step towards the realization of this conflicting situation [22].

3.9 Glaucoma and AD

Recent research conducted in France showed that glaucoma patients were four times more likely to develop dementia. This finding was not associated with high eye pressures or glaucoma medication usage, suggesting that the patients who were most vulnerable might be those with normal eye pressure or low-tension glaucoma [23].

It has already been shown that Alzheimer's patients have retinal nerve fiber layer thinning and loss of the retinal ganglion cells that compose the optic nerve, both hallmarks of glaucoma. One of the other pathologic hallmarks of AD, abnormal tau proteins, has been found in the vitreous jelly (the clear jelly-like substance that fills the space from the lens to the retina) of the eye in glaucoma patients [24]. Animal models of glaucoma have demonstrated that there is Aβ protein exposure in the retina which results in increased retinal ganglion cell loss, while treatment of these animals with drugs that are commonly used in the treatment of AD resulted in decreased death of the retinal ganglion cells lost due to glaucoma. [25].

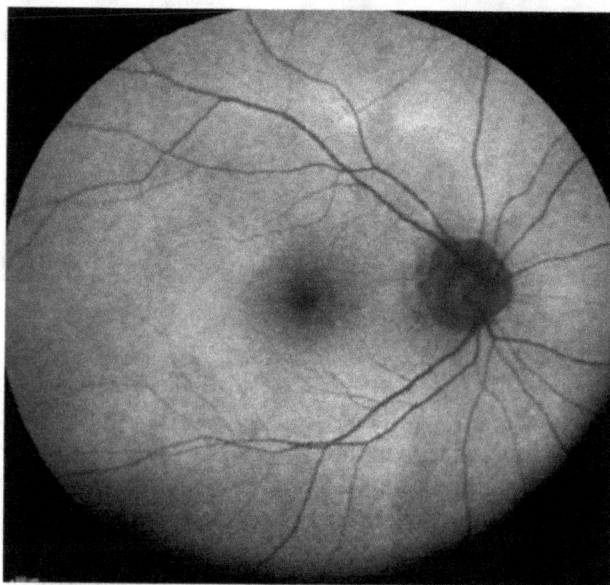

Figure 3.12. FAF before curcumin.

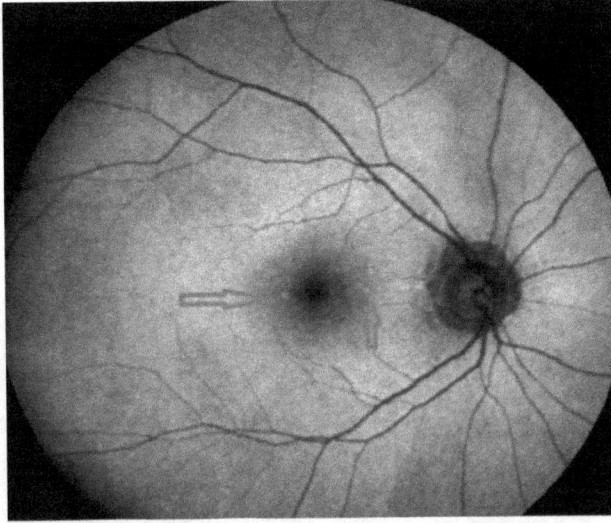

Figure 3.13. FAF after curcumin revealing hyperfluorescent spots.

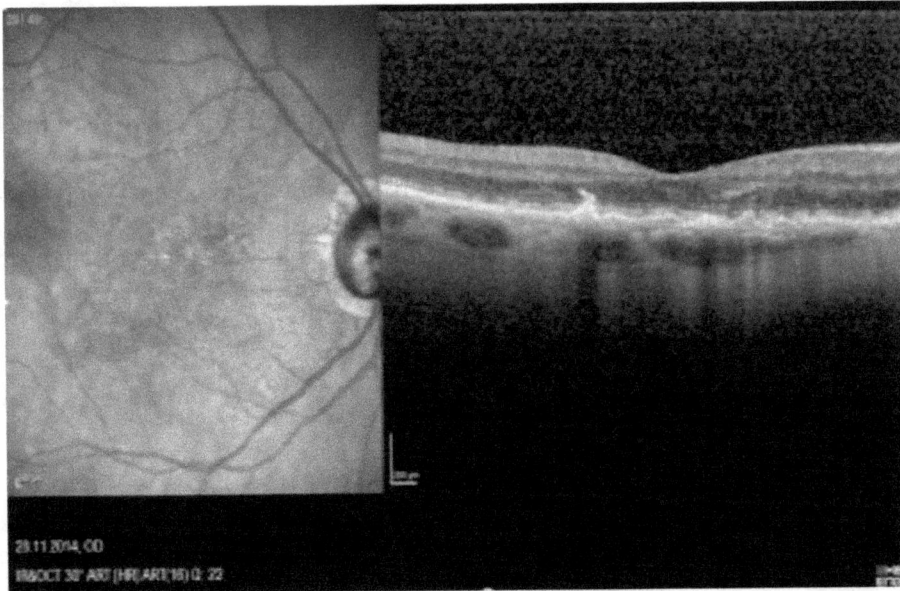

Figure 3.14. SD-OCT revealing dry AMD and $A\beta$ in the lesion.

3.10 Alpha-synuclein in AD

AD and Parkinson's disease (PD) affect different regions of the brain and have distinct genetic and environmental risk factors. However, at the biochemical level these two neurodegenerative diseases look similar. In both AD and PD, a sticky protein forms toxic clumps in brain cells. In AD, the troublemaker inside the cells is

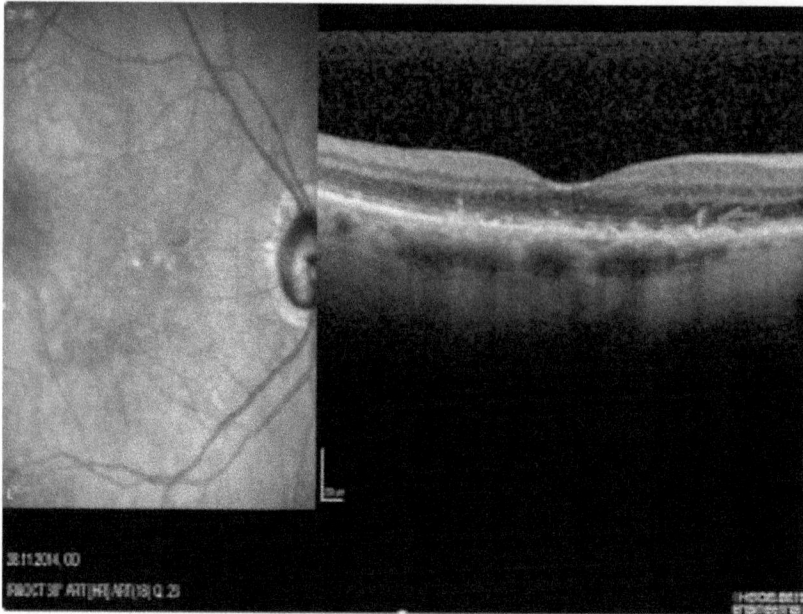

Figure 3.15. Aβ in dry AMD.

tau, creating neurofibrillary tangles, while in PD the sticky protein is alpha-synuclein, forming Lewy bodies [26]. Retinal examination may also disclose alpha-synuclein accumulation (figure 3.8).

The abnormal proteins found in neurodegenerative diseases such as AD, PD, and Huntington's seem to depend on the same mechanism to spread within the brain. Recognizing the involvement of these proteins and the underlying mechanism of aggregate protein assembly and spreading are key factors to understanding the progressive nature of neurodegenerative diseases [27].

3.11 Early diagnosis of AD

Treatments for AD that failed in clinical trials may have been started too late. Given the uncertainty around definite diagnostic tests for AD, treatment for the disease typically begins after the first signs of memory loss become evident. Unfortunately, this may be too late. To date, clinical trials assessing agents targeting either tau or Aβ have yielded mixed results. That is why new findings could be a game changer in the development of new treatments for AD. Current research suggests that the time window to intervene to slow AD progression may need to be earlier than thought [28].

Historically, deciding when to start treatment has been the biggest clinical challenge in the management of AD. Since the mid-2000s, research has focused on the role of amyloid and tau in the development of the condition, with the idea of using either or both as a means of assessing individual risk. According to the working theory, those with elevated levels of the proteins would be at risk of developing AD. So-called 'anti-amyloid' therapy would be prescribed even before obvious cognitive decline has been observed.

AD begins decades before symptom onset, but it is unknown at what point it starts to exert harmful neurological effects. Researchers are continuing to collect longitudinal data in order to follow up on any changes in amyloid accumulation, tau accumulation, and how these pathologies are interacting to influence memory loss [28].

Certain elements of brain structure may make people less likely to develop AD. Brain samples from patients at memory clinics were analyzed and it was found that the presence of healthy dendritic spines, which are connections between neurons, provide protection against AD in people whose brains have misfolded proteins associated with the disease. About 30% of the aging population have amyloid and tau build-up but never develop dementia. These individuals have larger, more numerous dendritic spines than those with dementia, indicating that spine health plays a major role in the onset of disease [29]. One obvious culprit in AD is the loss of dendritic spines and thus the loss of synapses.

Healthy dendritic spines could be genetic or the result of healthy lifestyle habits—such as good diet and exercise—which are known to reduce the risk of dementia. This findings provide a new target for drugs that would be designed to support and maintain dendritic spine health. These data suggest that rebuilding neurons may be possible. In the case where an increase of amyloid and tau is identified early in the disease, even before symptoms arise, a new medication that can contribute to maintaining healthy dendritic spines may be offered [29].

3.12 Biomarkers in AD

Neuroimaging and a lumbar puncture for the collection of cerebrospinal fluid are the tests generally used to determine a diagnosis. However, given the invasiveness and cost, it is not feasible to screen everyone in the general population who is worried about AD. New biomarkers in blood are being investigated, such as plasma phospho-tau181 and the apolipoprotein E (APOE) gene [30], which would allow wider screeening. Retinal examination also has a place as a novel biomarker.

3.13 Discussion

A recent article authored by an experienced team of authors suggested that retinal amyloid burden is correlated with brain amyloid burden in AD [31]. Moreover, it was stated that retinal amyloid plaques appear before they start to invade the hippocampus [31]. Screening for $A\beta$ in the retina is definitely a major breakthrough in AD. Comparing retinal amyloid burden with brain amyloid is going one step further in the trials for the early diagnosis of AD. Imaging the retina is easy, non-invasive, cheap, and does not involve radiation. However, there are two important issues that come to mind when considering the recent developments in AD studies:

1. The Mayo Clinic study of thousands of brains revealed tau as the driver of AD [2]. So, do we need to show tau protein alongside $A\beta$ in the retina to be able to make correct assumptions about the disease process [32]?

2. Post-mortem studies have demonstrated that people with AD have far more Aβ plaques in their brains than healthy people. However, roughly 30% of people without any signs of dementia have brains 'chock-full' of Aβ at autopsy [32]. So, the presence of Aβ may not be enough to prove that any patient will develop the disease in the future. Increased deposition of Aβ isoforms has been described on photoreceptor outer segments and along the RPE–Bruch's membrane interface in the aging human and mouse retina [33]. Another study using immunostaining revealed Aβ deposition on photoreceptor outer segments all throughout the retina in humans as an ageing process [33]. This deposition process may well be only a sign of ageing and not related to AD. Analyses of drusen components have shown deposition of Aβ within vesicles in the eyes of AMD patients [22]. A better approach may be to demonstrate and quantify both proteins in the retina [34, 35]. Tau aggregates have also been detected in the retina of live patients. In the brain, tau deposits have been found in the temporal lobe early in the disease, which proves that tau PET imaging may also be necessary to be able to correlate the abnormal protein burden with brain plaques. Recent studies have confirmed that a combination of tau and Aβ aggregates combined with neuroinflammation causes the disease. Another proof of this theory is the unsuccessful results of Aβ aimed therapies in AD. However, new treatments are on the horizon.

In retinal examinations, hypopigmented areas in FAF show atrophic retinal changes which may indicate an advanced stage of the disease. Hyperpigmented lesions contain more lipofuscin, but the retina may be less compromised and not atrophic. In patients who are given bioavailable curcumin capsules, the reason for the increased hypofluorescence instead of hyperfluorescence may be the thinning in the choroid and degeneration of RPE [21]. Thinning of the choroid and changes in RPE were recently demonstrated by clinical trials. Hyperintense dots in FAF which become noticeable after curcumin use prove that it has affinity for Aβ. Examination of the retina using FAF and OCT gives us valuable information about neuro-degeneration and these tests may be reliable biomarkers for AD. It may be possible to detect the disease and also examine the progression of lesions [36]. The early detection of AD is of paramount importance. The retina is an excellent model for neurodegenerative diseases. If the plaques begin first in the retina, as has been reported [31], this will give us time to halt the disease. Supplements, regular exercise, and a healthy diet can contribute to the slowing (or reversing) of the disease. Maybe in the near future new drugs will halt the disease before the plaques appear in the brain. Much more needs to be done to fight AD. Detecting the disease early is the most important goal; people who have relatives with AD in their families or with mild cognitive impairment should have a thorough medical examination. Detailed retinal examination using FAF and OCT will certainly contribute to neurological evaluation in neurodegenerative diseases.

References

[1] Prince M, Bryce R and Ferri C 2011 *World Alzheimer Report 2011: the Benefits of Early Diagnosis and Intervention* (London: Alzheimer's Disease International)

[2] Hyman B T *et al* 2012 National Institute on Aging–Alzheimer's Association guidelines on neuropathologic assessment of Alzheimer's disease *Alzheimers Dement.* **8** 1–13

[3] Bloudek L M, Spackman D E, Blankenburg M and Sullivan S D 2011 Review and meta-analysis of biomarkers and diagnostic imaging in Alzheimer's disease *J. Alzheimers Dis.* **26** 627–45

[4] Bateman R J *et al* 2012 Clinical and biomarker changes in dominantly inherited Alzheimer's disease *N. Engl. J. Med.* **367** 795–804

[5] Roberts R O *et al* 2008 The Mayo Clinic Study of Aging: design and sampling, participation, baseline measures and sample characteristics *Neuroepidemiology* **30** 58–69

[6] Simao L M 2013 The contribution of optical coherence tomography in neurodegenerative diseases *Curr. Opin. Ophthalmol.* **24** 521–7

[7] Barnes D E and Yaffe K 2011 The projected effect of risk factor reduction on Alzheimer's disease prevalence *Lancet Neurol.* **10** 819–28

[8] Blennow K and Hampel H 2003 CSF markers for incipient Alzheimer's disease *Lancet Neurol.* **2** 605–13

[9] Terman A and Brunk U T 2004 Lipofuscin *Int. J. Biochem. Cell Biol.* **36** 1400–4

[10] Seehafer S S and Pearce D A 2006 You say lipofuscin, we say ceroid: defining autofluorescent storage material *Neurobiol. Aging* **27** 576–88

[11] Giaccone G, Orsi L, Cupidi C and Tagliavini F 2011 Lipofuscin hypothesis of Alzheimer's disease *Dement. Geriatr. Cogn. Dis. Extra* **1** 292–6

[12] Olsen T W 2008 The Minnesota Grading System using fundus autofluorescence of eye bank eyes: a correlation to age-related macular degeneration (an AOS thesis) *Trans. Am. Ophthalmol. Soc.* **106** 383–401

[13] Huang D *et al* 1991 Optical coherence tomography *Science* **254** 1178–81

[14] Koronyo-Hamaoui M *et al* 2011 Identification of amyloid plaques in retinas from Alzheimer's patients and noninvasive *in vivo* optical imaging of retinal plaques in a mouse model *Neuroimage* **54** S204–17

[15] Crowther R A and Goedert M 2000 Abnormal tau-containing filaments in neuro-degenerative diseases *J. Struct. Biol.* **130** 271–9

[16] Chiasseu M *et al* 2017 Tau accumulation in the retina promotes early neuronal dysfunction and precedes brain pathology in a mouse model of Alzheimer's disease *Mol. Neurodegener.* **12** 58

[17] Fitzpatrick A W P *et al* 2017 Cryo-EM structures of tau filaments from Alzheimer's disease *Nature* **547** 185–90

[18] Ringman J M, Frautschy S A and Cole G M 2005 A potential role of the curry spice curcumin in Alzheimer's disease *Curr. Alzheimer Res.* **2** 131–6

[19] Small G W *et al* 2017 Memory and brain amyloid and tau effects of a bioavailable form of curcumin in non-demented adults: a double-blind, placebo-controlled 18-month trial *Am. J. Geriatr. Psychiatry* **26** 266–77

[20] Cerman E, Eraslan M and Cekic O 2015 Age-related macular degeneration and Alzheimer disease *Turk. J. Med. Sci.* **45** 1004–9

[21] Kaarniranta K *et al* 2011 Age-related macular degeneration (AMD): Alzheimer's disease in the eye? *J. Alzheimer's Dis.* **24** 615–31

[22] Kayabasi U and Sahbaz I 2017 Beta amyloid in age-related macular degeneration lesions in a patient with Alzheimer's disease *EC Ophthalmol.* **7** 18–21

[23] Ishikawa M, Yoshitomi T and Covey D F 2018 Neurosteroids and oxysterols as potential therapeutic agents for glaucoma and Alzheimer's disease *Neuropsychiatry* **8** 344–59

[24] Yoneda S *et al* 2005 Vitreous fluid levels of beta-amyloid and tau in patients with retinal diseases *Jpn. J. Ophthalmol.* **49** 106–8

[25] Nucci C *et al* 2018 Neuroprotective agents in the management of glaucoma *Eye* **32** 938–45

[26] Suzuki M, Sango K and Wada K 2018 Pathological role of lipid interaction with α-synuclein in Parkinson's disease *Neurochem. Int.* **119** 96–106

[27] Santiago J A, Bottero V and Potashkin J A 2017 Dissecting the molecular mechanisms of neurodegenerative diseases through network biology *Front. Aging Neurosci.* **9** 166

[28] Crous-Bou M, Minguillón C and Gramunt N 2017 Alzheimer's disease prevention: from risk factors to early intervention *Alzheimers Res. Ther.* **9** 71

[29] Boros B D *et al* 2017 Dendritic spines provide cognitive resilience against Alzheimer's disease *Ann. Neurol.* **82** 602–14

[30] Ghidoni R *et al* 2018 Innovative biomarkers for Alzheimer's disease: focus on the hidden disease biomarkers *J. Alzheimers Dis.* **62** 1507–18

[31] Koronyo Y *et al* 2017 Retinal amyloid pathology and proof-of-concept imaging trial in Alzheimer's disease *JCI Insight* **2** pii: 93621

[32] Kayabasi U 2017 Retinal examination for plaque burden *Clin. Surg.* **2** 1736

[33] Hoh Kam J, Lenassi E and Jeffery G 2010 Viewing ageing eyes: diverse sites of amyloid beta accumulation in the ageing mouse retina and the upregulation of macrophages *PLoS One* **5**

[34] Kayabasi U 2016 Tau in the retina *EC Neurol.* **3** 493–9

[35] Kayabasi U 2017 Retinal examination by OCT to reveal neurodegeneration in the brain *EC Ophthalmol.* **6** 139–40

[36] Kayabasi U, Sergott R C and Rispoli M 2014 Retinal examination for the diagnosis of Alzheimer's disease *Int. J. Ophthalmol. Clin. Res.* **1** 002

IOP Publishing

Neurological Disorders and Imaging Physics, Volume 3
Application to autism spectrum disorders and Alzheimer's
Ayman El-Baz and Jasjit S Suri

Chapter 4

Clinically relevant depression and risk of Alzheimer's disease in the elderly: meta-analysis of cohort studies

Javier Santabárbara, Patricia Gracia-García, Anais Sevil-Pérez, Beatriz Villagrasa and Raúl López-Antón

A meta-analysis of population-based cohort studies is carried out to investigate the risk of Alzheimer's disease (AD) according to clinically relevant depression, assessed with standardized instruments. A rapid systematic literature search using the search engine PubMed of studies published up to September 2018 was performed to identify all longitudinal studies on the association between clinically relevant depression (diagnosed using the GMS, DSM, or ICD criteria) and AD risk in the elderly. Data extraction and methodological quality assessment were conducted independently by two authors. We calculated the pooled relative risks (RR) to examine depression as a possible risk factor for AD in community studies, as well as to compute the population attributable fraction (PAF) of AD for clinically significant depression.

Eight studies met inclusion criteria for the systematic review. All of these provided enough information to perform a meta-analysis. Participants with clinically relevant depression had a two-fold higher risk of Alzheimer's disease (pooled RR = 2.01 (95% CI: 1.70–2.39); $p < 0.001$). The Alzheimer's disease PAF attributable to clinically relevant depression was 16.3% (95% CI: 10.9%–19.6%).

Thus clinically relevant depression appears to be associated with an increased risk of AD in the elderly community, with a potential impact on AD risk which is higher than other known/recognized risk factors. Future studies should explore the mechanisms linking depression and AD as well as whether an effective treatment of clinically significant depression could prevent AD development.

4.1 Introduction

Dementia has now been recognized as a public health priority by the World Health Organization (WHO) [1]. Indeed, an estimated 47 million people worldwide were living with dementia in 2015, and this number is projected to triple by 2050 [2]. Alzheimer's disease (AD) is the most common form of dementia and may contribute to 60%–70% of cases [3].

Dementia affects memory, thinking, behavior, and the ability to perform every-day activities. This is overwhelming not only for the patient but above all for their caregivers and relatives [1]. It also has significant social and economic implications because it is one of the major causes of disability and dependence among older people worldwide [3].

Reducing the risk of dementia is one of the prioritised areas in the 'Global action plan on the public health response to dementia 2017–25' [3]. Although dementia is not an inevitable consequence of ageing, the strongest known risk factor is age [3]. However, up-to-date scientific evidence has shown that regular physical activity and management of cardiovascular risk factors (diabetes, obesity, smoking, and hyper-tension) may reduce the risk of cognitive decline and dementia, and a healthy diet and lifelong learning/cognitive training may also reduce the risk of cognitive decline [4].

AD, like other common chronic diseases, most likely develops as a result of multiple factors rather than a single cause, with the exception of those cases presenting genetic abnormalities and symptoms before the age of 65 [2]. Prevalence of AD increases continuously and steeply with age, with rates doubling approximately every five years, and it is reported at higher rates in women than in men [5]. Some risk factors such as age and family history cannot be modified. In the case of gender, a survival bias might explain the disproportionate prevalence of AD among women, that is, they have a longer life-expectancy than men, while men may die of competing causes of death earlier in life (e.g. cardiovascular causes) leaving only the most resilient and healthy men to survive to old-age. Both biological and socio-cultural differences between men and women (e.g. access to education and occupation) will also need to be examined to determine which account for gender differences in AD prevalence [6]. Other risk factors can be modified to reduce the risk of AD. Evidence regarding seven potentially modifiable risk factors for AD, namely cognitive inactivity or low educational attainment, smoking, physical inactivity, depression, diabetes, midlife hypertension, and midlife obesity, has been reported [7, 8]. After accounting for non-independence between risk factors, these potentially modifiable risk factors were associated with around a third [7] of AD cases worldwide, but when taken together amounted up to half [8] of cases. A 10%–25% reduction in all seven risk factors was likely to prevent as many as 1.1–3.0 million AD cases worldwide [8]. It was also hypothesized that relative reductions of 10% per decade in the prevalence of each of the seven risk factors could reduce the prevalence of AD in 2050 by 8.3% worldwide.

Several studies on the association between depression and AD have suggested that depression is a risk factor for AD. However, individual studies have yielded inconsistent results which could be linked to variation in several factors, such as length of observation, rate of follow-up participation, potentially modifying the variables controlled for (e.g. gender, education level), and/or the approach to the assessment of depression and AD. To assess depression, symptom-based scales are frequently used as a continuous variable or with a cut-off score below the threshold reported as optimum for screening for major depression (e.g. the Centre for Epidemiological Studies Depression scale (CES-D)). Therefore, these studies do not provide enough information on a relevant clinical subject such as treatable depression [9, 10].

Previous meta-analyses of epidemiological studies found that subjects with late-life depression have a 60%–65% increased risk of incident AD [11, 12]. However, these meta-analyses have also indicated important limitations, such as heterogeneity in the classification of cases and non-cases of depression resulting from the inclusion of studies that assess depressive symptoms through rating scales, using pre-established cut-off scores instead of structured interviews for diagnosis of depressive disorders [12]. Moreover, these results have been obtained by combining findings based on different instruments and using both continuous measures of depressive symptomatology and categorical classification of major or minor depression, including findings based on self-reports [10]. These methodological limitations are likely to have led to less reliable estimates of risk which are not clearly attributable to clinically relevant depression [10]. Cherbuin et al [10] argue that reliable measures of risk exposure based on specific, validated instruments are needed for the identification of risk factors and estimation of the magnitude of their effect. As a result, individuals at higher risk can be identified and interventions and/or treatment can be calibrated at the individual level. Therefore, evidence-based clinical practice requires that clinical advice and decision-making be based on objective measures of risk [10]. This level of evidence was not available in previous meta-analyses due to the above-mentioned methodological limitations. To produce precise dementia risk estimates associated with clinically significant late-life depression which could be related to specific thresholds on validated instruments, Cherbuin et al [10] conducted a meta-analysis pooling separately compatible findings from population studies using specified cut-offs using validated instruments (e.g. CES-D). This meta-analysis aimed to assess depression status and it was based only on those studies which relied on widely accepted clinical criteria for dementia (e.g. the *Diagnostic and Statistical Manual of Mental Disorders* (DSM)-IV). Analyses of studies using a cut-off score consistent with major depression and previously validated against clinical criteria (CED-D > 20) demonstrated somewhat higher risk estimates than those using a more lenient cut-off (CES-D > 16). These findings suggest that clinical depression is probably associated with an approximately 80%–100% increased risk of dementia and AD, and that mild depression is still associated with an approximately 60%–70% increased risk of dementia and AD [10]. Previous studies of large population samples that assess depression with the Geriatric Examination for Computer Assisted Taxonomy (GMS-AGECAT) have found even higher risk estimates than those using symptomatic rating scales for clinically significant severe depression and dementia [13] and specifically AD [14].

A valid approach for the detection of 'depression requiring clinical attention' in community samples is based on GMS-AGECAT [15]. After symptom assessment in stage I, stage II allows the diagnosis of depression. At this stage, a computer program compares syndrome clusters (dementia, depression, anxiety, etc) to reach a final diagnosis, recorded as either a diagnostic 'subsyndromal' (confidence levels 1 and 2) or a diagnostic 'case' (confidence levels $\geqslant$ 3). AGECAT 'caseness' implies the 'desirability of intervention' [15, 16]. Agreement is only moderate between the AGECAT and DSM criteria for depression [16], while cases of depression considered to be 'severe' using AGECAT criteria may be similar to those considered 'major' by other investigators [17].

To produce precise AD risk estimates associated with clinically relevant depression, in this paper we will review published studies based on cohort design that assess clinically relevant depression by standardized clinical diagnostic criteria (International Classification of Diseases (ICD) or DSM) and/or GMS-AGECAT criteria (confidence levels $\geqslant$ 3), and we will pool their risk estimates through a meta-analysis. If the association between clinically relevant depression and AD is confirmed, and assuming a causal relationship between both conditions, we will calculate the population attributable fraction (PAF) of AD due to clinically relevant depression.

4.2 Methods

We followed the PRISMA guidelines for reporting a systematic review and meta-analysis [18].

4.2.1 Search strategy

A rapid systematic review [19] of all cohort studies investigating an association between clinically relevant depression (diagnosed by standardized clinical criteria or with GMS-AGECAT system) and AD risk was undertaken in September 2018 on PubMed by one researcher (ASP). Another researcher (JS) reviewed a random sample of 10% of the studies to assess agreement on inclusion and exclusion criteria and to approve the studies meeting the final eligibility criteria.

The PubMed search strategy was: (depression [Title]) AND (Alzheimer [Title] OR dementia [Title]) AND (risk [Title] OR incidence [Title] OR cohort studies [Mesh]).

In addition, the reference lists of selected publications were also screened for potentially eligible studies. Authors of studies were contacted directly when insufficient data were available in articles meeting the inclusion criteria.

4.2.2 Study selection

Studies selected for analysis had to meet the following requirements: (1) the identification of depression 'caseness' (dichotomous variable) was based on GMS-AGECAT (cut-off 3), DSM diagnosis, or ICD diagnosis; (2) the study design was a prospective or retrospective cohort; (3) it investigated the association between depression and incidence of AD; (4) there was an absence of dementia in the baseline assessment; and (5) it included a summary estimate (relative risk, odds ratio, or hazard ratio) with reported confidence intervals.

Studies that focused on MCI samples, as well as review articles and meta-analyses, were excluded. Studies not reporting original, published peer-reviewed results were also excluded to ensure only high quality research was included in the analyses.

4.2.3 Data extraction

Two reviewers (JS and PGG) independently extracted data for the included studies. We used a predesigned data extraction form to obtain information on country, sample size, number of prevalent cases of anxiety, number of incident cases of dementia, percentage of females, average age, dementia assessment and clinical criteria, covariates adjusted for in the analysis, adjusted RR estimates, and duration of follow-up.

4.2.4 Quality assessment

Study quality was assessed with the Newcastle–Ottawa scale (NOS) for cohort studies [20]. The NOS is a nine-point scale to assess the quality of nonrandomized studies in terms of design and content. It measures exposure (0–4 points), the comparability of cohorts (0–2 points), and the identification of the outcome and adequacy of follow-up (0–3 points). We assigned scores of 0–3, 4–6, and 7–9 to indicate low, moderate, and high quality studies, respectively. One researcher (JS) assessed the quality of all included studies.

4.2.5 Statistical analysis

We used relative risks (RRs) as the common measure of association across studies, and considered hazard ratios (HRs) and odds ratios (ORs) as equivalents, as considered appropriate when the outcome condition is relatively rare (prevalence < 15%) [21]. We preferentially pooled risk estimates from fully adjusted models. We conducted a random-effects model that allows for HRs and ORs to be incorporated into the same MA, as well as accounting for heterogeneity between studies [22].

The Hedges Q statistic was used to describe the heterogeneity (the statistical significance was set at $p < 0.10$). Additionally, to quantify heterogeneity we report the I^2 statistic [23], with its 95% confidence interval, as recommended when the number of studies is small [24]. We assigned low heterogeneity for I^2 values between 25%–50%, moderate for 50%–75%, and high for $\geqslant 75\%$. We performed subgroup and meta-regression analyses [25] to explore the sources of the heterogeneity expected in meta-analyses of observational studies [26].

Visual inspection of funnel plots and Begg's and Egger tests were used to assess publication bias [27, 28]. A sensitivity analysis was performed to assess the influence of each individual study on the overall results, by omitting studies in turn one by one. All statistical analyses were performed with STATA statistical software (version 10.0; College Station, TX, USA), and p values are reported as two-sided, with 0.05 accepted as statistically significant except where otherwise indicated.

4.3 Results

4.3.1 Study selection

Figure 4.1 presents the results of the literature search and study selection process. The primary search yielded 227 potential records, of which 209 articles were excluded as their title/abstract did not meet the selection criteria. The full-text of the 18 remaining articles was read, after which 13 were excluded and three were

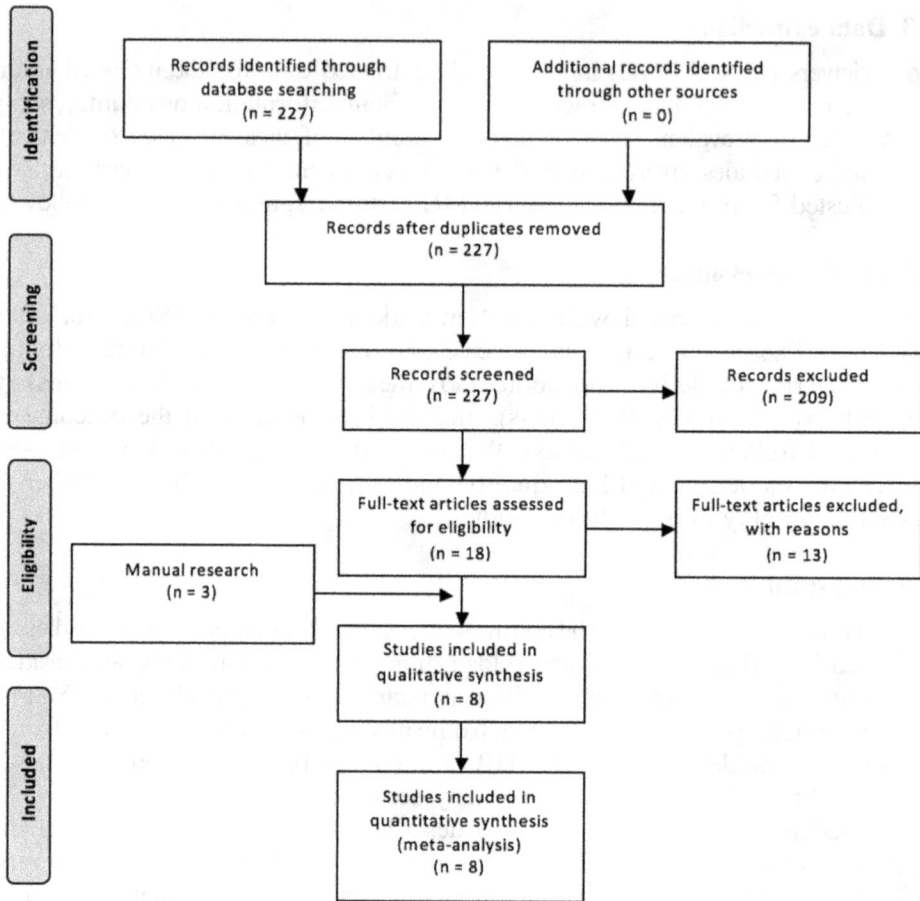

Figure 4.1. Flow-chart of meta-analysis of depression and AD risk.

included in the final review of this report by means of manual searches in the reference sections of the articles selected. Finally, eight articles were included for the present meta-analysis.

4.3.2 Description of included studies

The eight included articles were published between 2000 and 2013 [14, 29–35]. They reported a total of 24 044 participants and 1459 cases of incident AD.

Study details are presented in tables 4.1 and 4.2. Six studies were conducted in Europe [14, 29, 31–34] and the others in South Korea [30] and the United States of America [35].

The studies differed in the criteria used for the diagnosis of clinically relevant depression: three used GMS-AGECAT [14, 29, 30], four DSM [31–34], and one ICD [35]. AD diagnosis was most frequently established based on the National Institute of Neurological and Communicative Disorders and Stroke and the Alzheimer's Disease and Related Disorders Association (NINCDS-ADRDA) or

Table 4.1. Characteristics of included studies in meta-analysis ($n = 8$).

Author, year (no. of participants)	Study	Country	Follow-up (years)	Age at baseline (years), mean (SD)	Female (%)	Drop-out rate (%)	Depression measure	No. AD incident cases (criteria)
Barnes et al 2012 ($n = 13\,535$)	MHC	USA	6	81.1 (4.5)	57.9	55.5	Late-life (previous 3-13 years) depression. #ICD-9	$n = 1051$ (ICD-9)
Geerlings et al 2000 ($n = 1911$)	AMSTEL	The Netherlands	4	73.1 (5.5)	62.3	39.3	GMS-AGECAT	$n = 54$ (DSM-IV)
Gracia-Garcia et al 2013 ($n = 3864$)	ZARADEMP	Spain	4.5	71.8 (8.9)	54.4	37.9	GMS-AGECAT	$n = 70$ (DSM-IV, NINCDS-ADRDA)
Heser et al 2013 ($n = 2663$)	AgeCoDe	Germany	6	81.3 (3.4)	65.3	11.2	Lifetime prevalence. Age of onset. Major depression DSM-IV	$n = 152$ (DSM-IV)
Heun et al 2006 ($n = 615$)	NR	Germany	10	68.1 (8.2)	59.8	18.8	Lifetime prevalence. Late-onset (>60 y). Major depression DSM-IIIR	$n = 38$ (DSM-IIIR)
Jungwirth et al 2009 ($n = 487$)	VITA Study	Austria	5	75.8 (0.5)	60.9	16.8	History of depressive disorder (NS[a]). DSM-IV	$n = 30$ (NINCDS-ADRDA)
Kim et al 2011 ($n = 518$)	NR	South Korea	2.4	71.8 (5.1)	54.4	7.1	GMS	$n = 34$ (NINCDS-ADRDA)
Vilalta-Franch et al 2012 ($n = 451$)	Girona cohort study	Spain	5	76.9 (5.5)	65.4	5.1	Current. Major depression[b] DSM-IV	$n = 30$ (DSM-IV)

SD: standard deviation; NR: not reported; GMS: Geriatric Mental State; AGECAT: Geriatric Examination for Computer Assisted Taxonomy; MMSE: Mini-Mental State Examination; NINCDS-ADRDA: National Institute of Neurological and Communicative Disorders and Stroke and the Alzheimer's Disease and Related Disorders Association; ICD: International Classification of Diseases. AMSTEL: Amsterdam Study of the Elderly; AgeCoDe: Study on Ageing, Cognition, and Dementia in Primary Care Patients; MHC: Multiphasic Health Checkup; VITA: Vienna Transdanube Aging Study; ZARADEMP: Zaragoza Dementia Depression Study.
codes: 296.2 (major depressive disorder), 296.3 (recurrent major depressive disorder), 298.0 (depressive type psychosis), 300.4 (dysthymic disorder), and 311.0 (depressive disorder not elsewhere classified).
[a] Not specified.
[b] They also provide data of minor depression/dysthymic disorder, but the results are not included in our analysis.

Table 4.2. Results of the studies included in the meta-analysis (n = 8).

Author, year (no. of participants)	Statistical analysis	Effect measure (95% CI)	Covariates	Quality assessment[a]
Barnes et al 2012 (n = 13 535)	Cox (proportional hazards) regression	Hazard ratio (HR)—Late-life depression: 2.06 (1.67–2.55)	Education, sex, race, and number of comorbilities[b].	8
Geerlings et al 2000 (n = 1911)	Logistic regression	Odds ratio (OR)—Depression (AGECAT ⩾ 3): 1.67 (0.76–3.63)	Age, gender, education, memory complaints, and psychiatric history.	9
Gracia-Garcia et al 2013 (n = 3864)	Fine and Gray (competing risk) regression	Hazard ratio (HR)—Depression (AGECAT ⩾ 3): 1.23 (0.64–2.35)[c]	Age, gender, education, cognitive status at baseline (MMSE score), functional disability, vascular risk factors, and diseases.	9
Heser et al 2013 (n = 2663)	Cox (proportional hazards) regression	Hazard ratio (HR)—Very late-onset depression (⩾ 70 years): 1.31 (0.58–2.95)	Age, sex, education, and ApoE-e4.	9
Heun et al 2006 (n = 615)	Logistic regression	Odds ratio (OR)—Lifetime diagnosis of depression (after the age of 60 y): 2.37 (1.07–5.23)	Age, gender, MCI, subjective memory impairment, family history of AD, and ApoE-e4.	9
Jungwirth et al 2009 (n = 487)	Logistic regression	Odds ratio (OR)—History of depressive disorder: 2.70 (1.28–5.70)	—	7
Kim et al 2011 (n = 518)	Logistic regression	Odds ratio (OR)—Depression (AGECAT ⩾ 3): 2.41 (1.05–5.59)[d]	—	8
Vilalta-Franch et al 2012 (n = 451)	Cox (proportional hazards) regression	Hazard ratio (HR)—Late-onset major depression: 3.94 (1.31–11.84)	Age, sex, education, executive dysfunction, civil status, MMSE, and stroke	9

AD: Alzheimer's disease; CVD: cardiovascular disease; MCI: mild cognitive impairment; MMSE: mini-mental state examination; TMT-A: trail making test-A.
[a] Rating based on the Newcastle–Ottawa scale for cohort studies.
[b] Age was used as the time-scale variable.
[c] Data provided by authors.
[d] Pooled OR estimation obtained by meta-analysis of reported ORs for AGECAT = 0 and AGECAT = 1–2.

DSM-III/IV criteria. The duration of follow-up ranged from 2.4 [30] to 10 years [32], with a median follow-up of 5.0 years (IQR: 4.2–6.0).

The level of adjustment for covariates differed across the studies (table 4.2), and we used the risk estimates from the most fully adjusted models in estimating the pooled RR. The adjusted RR varied between 1.23 (95% CI: 0.64–2.36) [14] and 3.94 (95% CI: 1.31–11.84) [34].

Quality assessments for the cohort studies are shown in table 4.2. Seven studies had a quality score of 8–9, indicating a low risk of bias [14, 29–32, 34, 35] while one study [33] had a quality score of 7, indicating a medium risk of bias.

4.3.3 Effect estimation of AD risk based on depression

Individual study estimates as well as the overall estimate for incident dementia according to depression status are shown in figure 4.2. All RR estimates were above unity (significant in five studies [30, 32, 34, 35]) resulting in a pooled RR of 2.01 (95% CI: 1.70–2.39; $p < 0.001$).

Overall, compared with the reference group (non-depression), participants with depression had a two-fold higher AD risk, with an absence of between-study heterogeneity ($I^2 = 0\%$; 95% CI: 0%–68%; $p = 0.548$) (figure 4.2).

4.3.4 The risk of publication bias

Visual inspection of the funnel plot (figure 4.3) could suggest a small publication bias, but Egger ($p = 0.989$) and Begg ($p = 0.536$) tests indicated no publication bias.

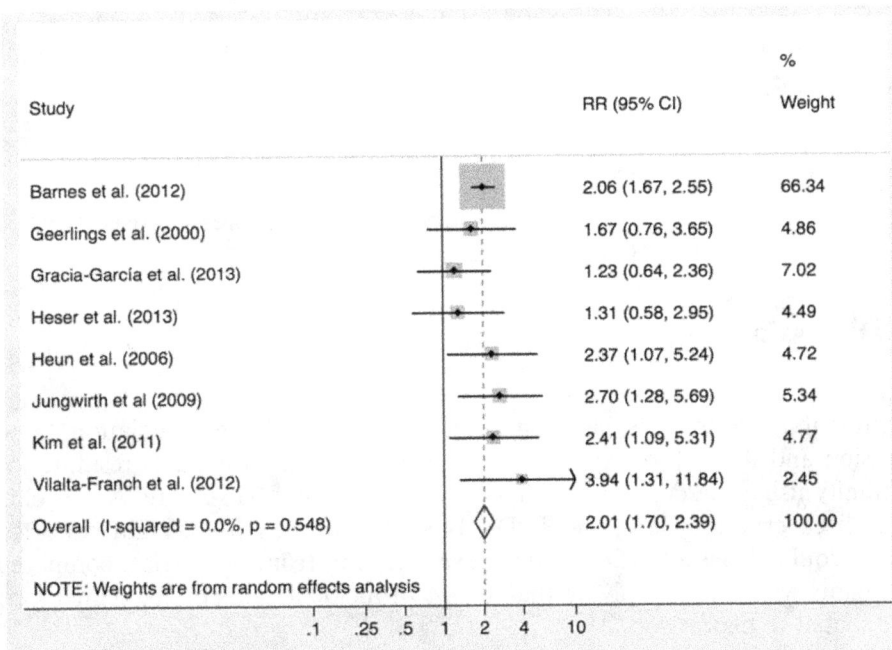

Study	RR (95% CI)	% Weight
Barnes et al. (2012)	2.06 (1.67, 2.55)	66.34
Geerlings et al. (2000)	1.67 (0.76, 3.65)	4.86
Gracia-García et al. (2013)	1.23 (0.64, 2.36)	7.02
Heser et al. (2013)	1.31 (0.58, 2.95)	4.49
Heun et al. (2006)	2.37 (1.07, 5.24)	4.72
Jungwirth et al (2009)	2.70 (1.28, 5.69)	5.34
Kim et al. (2011)	2.41 (1.09, 5.31)	4.77
Vilalta-Franch et al. (2012)	3.94 (1.31, 11.84)	2.45
Overall (I-squared = 0.0%, p = 0.548)	2.01 (1.70, 2.39)	100.00

NOTE: Weights are from random effects analysis

Figure 4.2. Forest plot of meta-analysis of depression and AD risk.

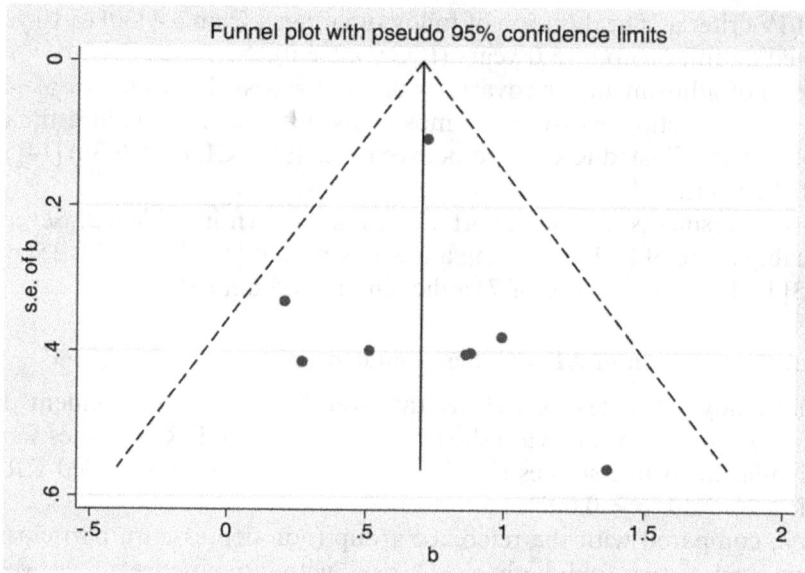

Figure 4.3. Funnel plot.

4.3.5 Influence analysis

Sensitivity analyses that excluded each study in turn showed moderate robustness, since the overall combined RR did not change substantially, with a range from 1.93 (95% CI: 1.43–2.59) [35] to 2.09 (95% CI: 1.75–2.50) [14]. This clearly shows no major impact of any single study on the overall combined and statistically significant RR (figure 4.4).

4.3.6 Population attributable fraction

We estimated that the proportion of the population with depression was 19.3%. This yields a PAF for AD of 16.3% (95% CI: 10.9%–19.6%).

4.4 Discussion

4.4.1 Main results

The present meta-analysis has quantitatively evaluated the association between depression and the risk of Alzheimer's disease (AD) in individuals residing in the community using the depression diagnostic instrument GMS-AGECAT or clinical standardized criteria (DSM and ICD). Based on the results of eight cohorts, we found a doubled risk of AD among those suffering from depression compared to participants without it, implying that 16.3% of incident cases of Alzheimer's could be attributed to depression.

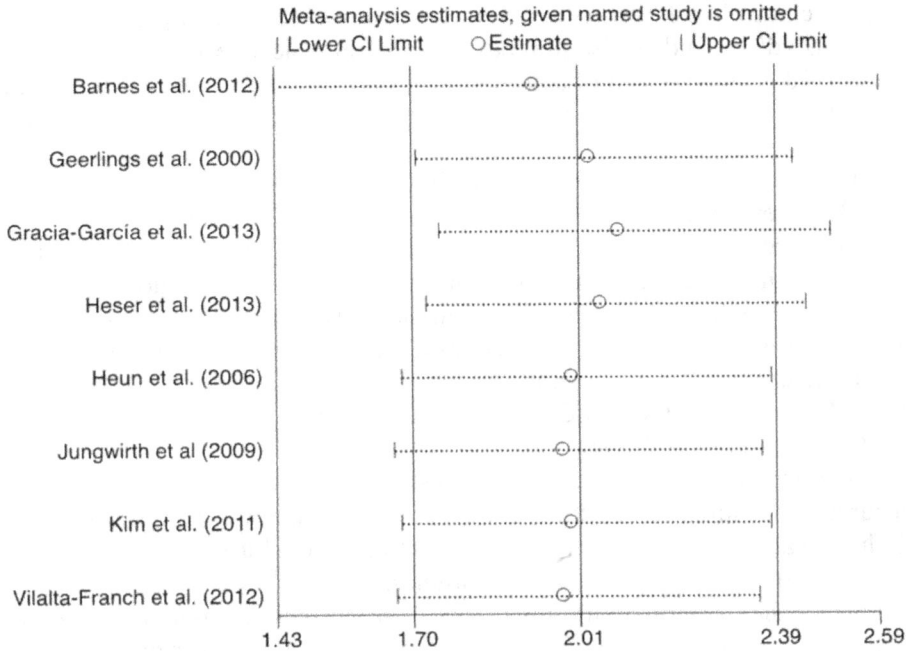

Figure 4.4. Influence plot.

4.4.2 Comparison with previous studies

After a first meta-analysis by Ownby *et al* [11] in which it was found that depression increases the risk of AD by 90%, Diniz *et al* [12] showed that depression increases the risk of AD by 65%. In a subsequent study, Cherbuin *et al* [10] found that depression doubles the risk of Alzheimer's dementia. To date, the present study is the fourth meta-analysis on the effect of depression on the risk of AD in the elderly. Our investigation complements the previous ones by examining eight prospective cohort studies in the elderly population and reporting results regarding the risk of Alzheimer-type dementia associated with 'clinically relevant' depression.

Although two previous meta-analyses [10, 12] also included studies conducted in the community, our result was slightly lower and this is linked to two fundamental issues: (i) we only included cohort studies since, in observational studies in the elderly, the strongest evidence of causal links between exposure and outcome came from cohort studies [36]; and (ii) we included studies with a clinical diagnosis of depression such as GMS-AGECAT. This is the first time that this is reported in the literature.

4.4.3 Strengths and limitations

This study has multiple strengths. First, it conducts a meta-analysis of all available studies of depression and risk of AD which results in a greater power to detect an effect than any of the included individual studies. Through inclusion of only studies that assess clinically relevant depression, evidence-based clinical practice mandates and clinical advice and decision-making based on objective measures of risk are reinforced [10]. The selection of cohort studies that avoided the influence of memory and selection biases makes it a robust study. Previously, only one study used a retrospective cohort [35] and determined depression through retrospective data of a comprehensive electronic medical record database to ensure that memory bias was not a risk. Likewise, the inclusion of large sample studies in our study reduced the risk of the effects of small studies and contributed a significant number of incident cases. In addition, each included study had a sufficiently long follow-up period to observe the potential association between depression and the risk of dementia. Moreover, given that case–control or cross-sectional studies were excluded from this analysis, the exclusive use of cohort studies provided more evidence regarding the efficacy of establishing the cause and effect relationship according to the criteria of causality proposed by Hill [37].

The requirement of the relative risk adjusted by the relevant covariables yielded an accurate risk estimate. In addition, the combined relative risks were consistent in the influence analysis. Likewise, the random-effects model used in our analysis took into account the heterogeneity between studies.

This study has several limitations. First, the current meta-analysis contains a limited number of effect quantities which could affect the power of the tests used. However, research has shown that a meta-analysis of only a few studies could continue to provide important information [38, 39]. In this sense, we have used the confidence interval of I^2 for the evaluation of heterogeneity instead of using only the Q contrast statistic recommended in meta-analyses of a few individual studies [24, 40]. Second, although some studies [31, 33] included current depressive symptomatology (assessed through symptom-based scales) as a possible confounding variable in their analysis, half of the prospective studies [31–33] in this meta-analysis diagnosed lifetime depression and not current depression.

A limitation found in previous meta-analyses [11, 12], as in ours, is that the data did not distinguish between the risk for AD in persons who may have had one episode of depression compared with persons who may have had several episodes or chronic minor depression. Although some individual studies included in our revision referred to characteristics of depression such as a four-times increased risk of AD in severe depression [14], we calculated the HR for clinically relevant depression to homogenize the definition of depression between studies. However, some individual data in our meta-analysis specifically addressed late-onset depression (defined from 60 to 70 years old in different studies) [31, 32, 34, 35].

4.4.4 Pathogenic hypotheses

The specificity of the association between late-onset depression and incident AD found in some studies has led some authors [31, 34] to suggest that depression is a prodromal feature of AD rather than a risk factor for the disease. Moreover, this hypothesis is reinforced by the finding that late-onset depression in combination with current depressive symptoms [31] and executive dysfunction [34] was more predictive of AD than late-onset depression without current depressive symptoms and without executive dysfunction. However, this is a controversial issue because other studies support the risk factor hypothesis. For example, in the meta-analysis carried out by Ownby *et al* [11] it was found that the interval between the diagnosis of depression and AD was positively related to an increased risk of developing AD.

To explain the findings which show that clinically relevant depression might increase the risk of AD, we tend to favor the 'brain reserve' hypothesis as the final common pathway. Depression could diminish brain reserves leading to an early expression of cognitive impairment in subjects who do not suffer from depression [41] and might injure the brain through different pathways such as an increased incidence of vascular disease [42], neurodegenerative changes as result of chronic inflammation [43], and an impaired capacity for neuroplasticity as result of chronic stress [44]. Some studies also suggest that depression may act in synergy with other risk factors, such as ApoE-ε4, to accelerate the deposition of amyloid in the brain [30]. Of note, physical and cognitive inactivity, often present in depression, have also been identified as modifiable risk factors of dementia [7, 8].

4.4.5 Clinical implications

We agree with Cherbuin *et al* [10] that evidence-based clinical practice mandates that clinical advice and decision-making based on objective measures of risk should be reinforced. We add that these measures should be based on methods of assessment and diagnosis used in regular clinical practice. This is why we focus in this meta-analysis on studies based on clinically relevant depression.

Standardized international criteria such as the American Psychiatric Association's (APA) *Diagnostic and Statistical Manual of Mental Disorders* (DSM) were intended to improve the reliability and validity of routine psychiatric diagnosis [45]. We included in this meta-analysis four studies that assessed depression according to the DSM criteria. Most of these studies (three) diagnosed major depression cases. We also included one study that assessed depression (various clinical categories) according to the ICD-9-CM criteria, that is, the official system of assigning codes to diagnoses and procedures associated with hospital use in the United States and which is based on ICD criteria developed by the World Health Organization [46]. Given that AGECAT 'caseness' is a valid approach for the detection of 'depression requiring clinical attention' in community samples and that it implies the 'desirability of intervention' [15, 16], studies that assessed depression using GMS-AGECAT were also considered.

Our study found a doubled risk of AD for subjects with clinically relevant depression. This highlights, once again, the importance of treatment and long-term follow-up of subjects with this clinical condition, particularly in cases of onset in old-age, to eventually prevent development of AD or, in any case, to allow an early diagnosis of AD.

4.4.6 Public health implications

The present meta-analysis shows a two-fold increase in the risk of Alzheimer's in individuals with clinically relevant depression, this effect being greater than that of any other well-established dementia risk factor, such as education (HR = 1.27), ApoE-ε4 (HR = 1.47), ischemic heart disease (HR = 1.39), or diabetes (HR = 1.85) [47]. In addition, we documented a population fraction of dementia attributable to depression (FAP) of 16.3%. This indicates a significant contribution of depression to AD, as suggested by the fact that this fraction is higher than that of other classic dementia risk factors such as diabetes (4.9%) or the allele ApoE-ε4 (7.1%) [47]. Therefore, our findings underscore the need for future work to develop preventive strategies in this risk group.

4.5 Conclusion

In conclusion, this meta-analysis of cohort studies suggests that clinically relevant depression significantly increases the risk of AD. Taking into account the high prevalence of depression worldwide and the important burden of AD, our findings provide evidence for the prevention of AD by identifying modifiable risk factors such as depression.

References

[1] World Health Organization 2012 *Dementia, a Public Health Priority* https://who.int/mental_health/publications/dementia_report_2012/en/

[2] Alzheimer's Association 2017 Alzheimer's disease: facts and figures *Alzheimers Dement.* **13** 325–73

[3] World Health Organization 2017 *Fact-sheets: Dementia* http://who.int/news-room/fact-sheets/detail/dementia

[4] Baumghart M 2015 Summary of the evidence on modifiable risk factors for cognitive decline and dementia: a population-based perspective *Alzheimers Dement.* **11** 718–26

[5] Lobo S *et al* 2000 Prevalence of dementia and major subtypes in Europe: a collaborative study of population-based cohorts. Neurologic Diseases in the Elderly Research Group *Neurology* **54** S4–9

[6] Mielke M M, Vemuri P and Rocca W A 2014 Clinical epidemiology of Alzheimer's disease: assessing sex and gender differences *Clin. Epidemiol.* **6** 37–48

[7] Norton S, Matthews F, Barnes D, Yaffe K and Brayne C 2014 Potential for primary prevention of Alzheimer's disease: an analysis population-based data *Lancet Neurol.* **13** 788–94

[8] Barnes D E and Yaffe K 2011 The projected effect of risk factor reduction on Alzheimer's disease prevalence *Lancet Neurol.* **10** 819–28

[9] Gracia-García P, De la Cámara C, López R, Roy J F and Lobo A 2015 Depresión y riesgo de demencia. Una revisión actualizada de la bibliografía *Psicogeriatría* **5** 15–22

[10] Cherbuin N, Kim S and Anstey K J 2015 Dementia risk estimates associated with measures of depression: a systematic review and meta-analysis *BMJ Open* **5** e008853

[11] Ownby R L, Crocco E, Acevedo A, John V and Loewestein D 2006 Depression and risk for Alzheimer disease: systematic review, meta-analysis, and metaregression analysis *Arch. Gen. Psychiatry* **63** 530–8

[12] Diniz B S, Butters M A, Albert S M, Dew M A and Reynolds C F 2013 Late-life depression and risk of vascular dementia and Alzheimer's disease: systematic review and meta-analysis of community-based cohort studies *Br. J. Psychiatry* **202** 329–35

[13] Chen R, Hu Z, Wei L, Qin X, McCraken C and Copeland J R 2008 Severity of depression and risk for subsequent dementia: cohort studies in China and the UK *Br. J. Psychiatry* **193** 373–7

[14] Gracia-García P *et al* 2015 Depression and incident Alzheimer disease: the impact of disease severity *Am. J. Geriatr. Psychiatry* **23** 119–29

[15] Copeland J R *et al* 2004 Depression among older people in Europe: the EURODEP studies *World Psychiatry* **3** 45–9

[16] Schaub R T, Linden M and Copeland J R 2003 A comparison of GMS-A/AGECAT, DSM-III-R for dementia and depression, including subthreshold depression (SD)--results from the Berlin Aging Study (BASE) *Int. J. Geriatr. Psychiatry* **18** 109–17

[17] Penninx B W, Geerlings S W, Deeg D J, van Eijk J T, van Tilburg W and Beekman A T 1999 Minor and major depression and the risk of death in older persons *Arch. Gen. Psychiatry* **56** 889–95

[18] Moher D, Liberati A, Tetzlaff J and Altman D GPRISMA Group 2009 Preferred reporting items for systematic reviews and meta-analyses: the PRISMA statement *PLoS Med.* **6** e1000097

[19] Ganann R, Ciliska D and Thomas H 2010 Expediting systematic reviews: methods and implications of rapid reviews *Implement. Sci.* **5** 56

[20] Wells G A, Shea B, O'Connell D, Peterson J, Welch V, Losos M and Tugwell P 2016 *The Newcastle-Ottawa Scale (NOS) for assessing the quality of non randomised studies in meta-analyses.* http://ohri.ca/programs/clinical_epidemiology/oxford.asp

[21] Shrier I and Steele R 2006 Understanding the relationship between risks and odds ratios *Clin. J. Sport Med.* **16** 107–10

[22] DerSimonian R and Laird N 1986 Meta-analysis in clinical trials *Control Clin. Trials* **7** 177–88

[23] Higgins J P and Thompson S G 2002 Quantifying heterogeneity in a meta-analysis *Stat. Med.* **21** 1539–58

[24] Thorlund K, Imberger G, Johnston B C, Walsh M, Awad T, Thabane L, Gluud C, Devereaux P J and Wetterslev J 2012 Evolution of heterogeneity (I^2) estimates and their 95% confidence intervals in large meta-analyses *PLoS One* **7** e39471

[25] Thompson S G and Higgins J P 2002 How should meta-regression analyses be undertaken and interpreted? *Stat. Med.* **21** 1559–73

[26] Egger M, Schneider M and Smith D 1998 Spurious precision? Meta-analysis of observational studies *Br. Med. J.* **316** 140–4

[27] Begg C B and Mazumdar M 1994 Operating characteristics of a rank correlation test for publication bias *Biometrics* **50** 1088–101

[28] Egger M, Smith G D, Schneider M and Minder C 1997 Bias in meta-analysis detected by a simple, graphical test *Br. Med. J.* **315** 629–34

[29] Geerlings M I, Schmand B, Braam A W, Jonker C, Bouter L M and Van Tilburg W 2000 Depressive symptoms and risk of Alzheimer's disease in more highly educated older people *J. Am. Geriatr. Soc.* **48** 1092–7

[30] Kim J M, Stewart R, Kim S Y, Kim S W, Bae K Y, Yang S J, Shin I S and Yoon J S 2011 Synergistic associations of depression and apolipoprotein E genotype with incidence of dementia *Int. J. Geriatr. Psychiatry* **26** 893–8

[31] Heser K *et al* Age CoDe Study Group 2013 Age of major depression onset, depressive symptoms, and risk for subsequent dementia: results of the German Study on Ageing, Cognition, and Dementia in Primary Care Patients (AgeCoDe) *Psychol. Med.* **43** 1597–610

[32] Heun R, Kölsch H and Jessen F 2006 Risk factors and early signs of Alzheimer's disease in a family study sample *Eur. Arch. Psychiatry Clin. Neurosci.* **256** 28–36

[33] Jungwirth S, Zehetmayer S, Bauer P, Weissgram S, Tragl K H and Fischer P 2009 Prediction of Alzheimer dementia with short neuropsychological instruments *J. Neural Transm.* **116** 1513–21

[34] Vilalta-Franch J, Lopez-Pousa S, Llinas-Regla J, Calvó-Perxas L, Merino-Aguado J and Garre-Olmo J 2013 Depression subtypes and 5-year risk of dementia and Alzheimer's disease in patients aged 70 years *Int. J. Geriatr. Psychiatry* **28** 341–50

[35] Barnes D E, Yaffe K, Byers A L, McCormick M, Schaefer C and Whitmer R A 2012 Midlife vs late-life depressive symptoms and risk of dementia: differential effects for Alzheimer disease and vascular dementia *Arch. Gen. Psychiatry* **69** 493–8

[36] Kingston A and Jagger C 2018 Review of methodologies of cohort studies of older people *Age Ageing* **47** 215–9

[37] Hill A B 1965 The environment and disease: association or causation? *Proc. R. Soc. Med.* **58** 295–300

[38] Goh J X, Hall J A and Rosenthal R 2016 Mini meta-analysis of your own studies: some arguments on why and a primer on how *Soc. Pers. Psychol. Compass* **10** 535–49

[39] Valentine J C, Pigott T D and Rothstein H R 2010 How many studies do you need? A primer on statistical power of meta-analysis *J. Educ. Behav. Stat.* **35** 215–47

[40] Von Hippel P T 2015 The heterogeneity statistic I^2 can be biased in small meta-analyses *BMC Med. Res. Methodol.* **15** 35

[41] Butters M A, Young J B, López O, Aizenstein H J, Mulsant B H, Reynolds C F, DeKosky S T and Becker J T 2008 Pathways linking late-life depression to persistent cognitive impairment and dementia *Dialogues Clin. Neurosci.* **10** 345–57

[42] Camus V, Kraehenbuhl H, Preisig M, Bula C J and Waeber G 2004 Geriatric depression and vascular diseases: what are the links? *J. Affect. Disord.* **81** 1–16

[43] Hurley L and Tizabi Y 2013 Neuroinflammation, neurodegeneration and depression *Neurotox. Res.* **23** 131–44

[44] Pittenger C and Duncan R S 2008 Stress, depression, and neuroplasticity: a convergence of mechanisms *Neuropsychopharmacology* **33** 88–109

[45] Spitzer R L, Endicott J and Robins E 1975 Clinical criteria for psychiatric diagnosis and DSM-III *Am. J. Psychiatry* **132** 1187–92

[46] Centers for Disease Control and Prevention (CDC) National Center for Health Statistics 2011 *International Classification of Diseases, Ninth Revision, Clinical Modification (ICD-9-CM)* https://cdc.gov/nchs/icd/icd9cm.htm

[47] Ritchie K, Carriere I, Ritchie C W, Berr C, Artero S and Ancelin M L 2010 Designing prevention programmes to reduce incidence of dementia: prospective cohort study of modifiable risk factors *Br. Med. J.* **341** c3885

IOP Publishing

Neurological Disorders and Imaging Physics, Volume 3
Application to autism spectrum disorders and Alzheimer's
Ayman El-Baz and Jasjit S Suri

Chapter 5

The implications of genetic factors in autism spectrum disorder and Alzheimer's disease

Ola M Eid and Maha M Eid

Autism spectrum disorder (ASD) and Alzheimer's disease (AD) are neurodevelopmental and neurodegenerative disorders, respectively. These disorders have distressing effects on both the affected individuals and society. Both disorders are expressed at different stages of life. Certain susceptible genes are implicated in these disorders. Recent advances in research technology have revealed many new genes that make one susceptible to these disorders. Subsequent advances in the psychological diagnosis and pharmacological/nutritional aspects of these disorders have created new therapeutic approaches [1]. This chapter focuses on the genetic background of these two disorders.

5.1 Autism spectrum disorder

Autism spectrum disorder (ASD) is a group of neurodevelopmental and neuropsychiatric disorders that manifest in early childhood. They share features of impaired communication, lack of social interaction and responsiveness, repetitive stereotyped behavior, and restricted interest. Currently there are very few medication options that efficiently treat ASD [2–7].

The prevalence is approximately 60 ASD patients per 10 000 individuals. ASD shows a profound sex-bias. Males are four times more likely to be affected than females, while the recurrence risk for female siblings is larger than for male siblings. The average age of diagnosis is 4.5 years [3, 5, 8, 9].

ASD can be classified according to cognitive function into high and low functioning ASD subclasses, according to the developmental course into ASD with or without regression subclasses, and into syndromic and non-syndromic ASD subclasses. Syndromic ASD is secondary to a genetic syndrome, such as Angelman syndrome, fragile X syndrome (FXS), or Rett syndrome. Non-syndromic ASD is diagnosed when ASD is the main diagnosis and is not a part of a disorder [7].

doi:10.1088/978-0-7503-1793-1ch5

From twin and family studies, the solid heritability of ASD is well recognized and estimated to be approximately 90%. However, a recognized genetic cause is only detected in around 10% of ASD patients. More than 800 genes and genomic loci among patients with ASD were identified using methodologies for genome-wide scans of copy number variations (CNVs), such as array-based comparative genomic hybridization [2, 5, 7]. These genes and their respective proteins are involved in various biological processes. The development of therapeutic interventions is probably dependent on the targetable biological pathways, not the genetic insults identified. Most molecular or cellular mechanisms of ASD pathophysiology are individually quite broad and are not totally independent, as the same genes or molecular pathways contribute to several of these processes at various phases during development. Moreover, evidence from known mutations proposes considerable convergence in the pathways where the mutations are present. However, there is restricted information on the precise genetic contributions to ASD predisposition and the full extent of convergence will be better identified after more elucidation of the genetic background [7, 10].

In the following section, we will briefly discuss the different underlying genetic contributions to the possible pathophysiology and biological pathways of ASD.

5.1.1.1 Altered fetal cortical development

Neuropathological studies have identified a number of cortical abnormalities in ASD individuals that could be due to dysregulation of fetal cortical development. Many studies have recognized that certain genetic mutations related to fetal brain development are associated with syndromic ASD, such as mutations in fragile X mental retardation 1 (FMR1), tuberous sclerosis 1 and 2 (TSC1 and TSC2), phosphatase and tensin homolog (PTEN), contactin associated protein-like 2 (CNTNAP2), and chromodomain helicase DNA binding protein 7 (CHD7). The genetic mutations of some of these syndromes may converge in the rapamycin (mTOR) pathway. This pathway regulates cell proliferation, growth, and neuronal morphogenesis. Moreover, PTEN mutations are consistent with defects in corticogenesis [10–12].

Furthermore, a number of ASD associated genetic variants are involved in the Wnt pathway. The Wnt pathway is a regulator of the balance between radial glia self-renewal and neuronal differentiation, in addition to dorsoventral patterning in the brain. Many of these variants that may disrupt gene function are in genes that regulate canonical Wnt signaling and that are involved in chromatin modification and regulation of gene expression, for example CHD8, T-box brain 1 (TBR1), and members of the Brg1-associated factors (BAF) and mixed-lineage leukemia (MLL) complexes. Mutations in these genes affect cortical development, as expected, as they are highly coexpressed in the human fetal brain during the period of neurogenesis as well as being expressed in both neural progenitors and newly born neurons [10, 13, 14].

5.1.1.2 Synaptic dysfunction

Several targeted sequencing studies and neuropathological studies of ASD demonstrated the association of ASD and mutations in genes encoding excitatory and inhibitory synaptic cell-adhesion molecules (such as neurexins (NRXNs) and neuroligins (NLGNs)), excitatory synaptic scaffolding molecules (such as the SH3 and multiple ankyrin-repeat domain (SHANK) proteins), the excitatory glutamatergic receptor GRIN2B and inhibitory GABAergic receptor (GABAR) subunits, as well as exonic deletions in the gene encoding the inhibitory synaptic scaffolding molecule gephyrin (GPHN). Moreover, there is less statistical evidence of the involvement of synaptotagmin and synapsin mutations, neurotransmitter release regulators, in ASD [10].

5.1.1.3 *Activity-dependent transcription and translation*
ASD is associated with disruptions in the neuronal activity-dependent transcriptional regulators or their targets. These disruptions may be caused by mutations in methyl-CpG-binding protein 2 (MECP2) and calcium channel, voltage-dependent L-type, alpha 1C subunit (CACNA1C), *de novo* mutations in the neuronal activity-induced transcription factor myocyte enhancer factor 2C (MEF2C) and *de novo* mutations in TBR1, whose product is required for activity-dependent Grin2b expression. Also, these disruptions may be caused by abnormal imprinting and microdeletion of MEF2-regulated ubiquitin-protein ligase E3A (UBE3A) or duplication of 15q11-q13 (that involves UBE3A) [10].

Moreover, mutations in TSC1 and TSC2, which encode canonical components of the mTOR pathway, FMR1, which encodes fragile X mental retardation protein (FMRP), and dup15q11-q13, which contains the FMRP interactor and translational repressor cytoplasmic FMR1-interacting protein 1 (CYFIP1), suggest convergence in neuronal translational regulation in individuals with ASD. Furthermore, convergence between synaptic function and translational regulation is suggested in individuals with ASD, probably due to alteration of translation of neuroligin 2 (NLGN2), neurexin 1 (NRXN1), and SHANK3 through FMRP, disrupting mTOR signaling or translation initiation [10].

5.1.1.4 *Altered neural circuitry*
Neuroimaging and neuropathology studies in humans suggest that alterations occur in resting state network activity and in macro-circuit connectivity within the cortex and in corticostriatal circuits. To date, the brain circuitry underlying social behavior has not been fully clarified. Moreover, cerebellar function has been implicated in social behavior. Experimental knockout of TSC1 in cerebellar Purkinje cells provided evidence that cerebellar dysfunction may lead to ASD-like social deficits. Moreover ASD risk may increase many-fold through developmental injury in the cerebellar circuitry, whereas social behavior is not impaired by adult injury. This finding suggests that cerebellar injury during early development may lead to behavioral deficits through a cascade of long-term deficits in cerebellar-associated targets [10, 15, 16].

However, the neuroanatomical substrate for repetitive behavior is better understood. Substantial evidence suggests that striatal dysfunction is the underlying factor for repetitive behavior and motor routine learning. Striatal dysfunction was present in mice lacking SHANK3B and stereotyped motor routines were manifested in mice lacking NLGN3 [17–19].

5.1.1.5 Dysregulated neuron–glia signaling and neuroinflammation

Neuropathological and positron emission tomography (PET) imaging studies of ASD individuals have identified the existence of activated microglia and astrocytosis in multiple brain regions. Moreover, transcriptomic studies of the post-mortem brains of idiopathic ASD individuals detected upregulation of genes enriched in activated microglia and astrocytes in their cortex, and to a lesser extent in their cerebellum. However, no genetic variants of microglia- or astrocyte-specific genes were recognized in ASD individuals. This evidence suggests that this finding is probably secondary to underlying synaptic dysfunction. However, the absence of genetic evidence for association with ASD does not diminish the importance of these microglial or astrocyte pathways as promising therapeutic possibilities to investigate [10].

In the next section, we will present the current knowledge on the roles of some well-established candidate genes, classifying them according to their biological role, keeping in mind that a single gene can have several functions.

5.1.2.1 Chromatin remodelers

5.1.2.1.1 MECP2

The MECP2 gene is an X-linked gene that is located on the q arm of the X chromosome (Xq28). MECP2 encodes methyl-CpG-binding protein 2. MECP2 acts as both a global gene repressor and activator, with mixed transcriptional profiling. MECP2 plays an important role in neuronal differentiation, neural development, and synaptic plasticity. Loss of function mutations of MECP2 cause Rett syndrome, which predominantly affects girls, while they are generally lethal in boys. Rett syndrome is a severe neurodevelopmental disorder where ASD is present in more than 50% of affected individuals. However, ASD individuals show significant reduction in MECP2 expression, even in the absence of MECP2 mutations. Occasionally, *de novo* MECP2 mutations can lead to relatively asymptomatic phenotypes, depending mainly on the X-inactivation pattern, which tends to be greatly skewed in the co-existence of mutations affecting X-linked genes. Moreover, MECP2 duplication is associated with a high prevalence of ASD [4, 7, 20].

5.1.2.1.2 CHD8

The chromodomain helicase DNA binding protein 8 gene (CHD8) is located at 14q11.2. It encodes an ATP-dependent chromatin remodeler that binds to trimethy-lated histone H3 lysine 4, a posttranslational histone modification present at active promoters. Mutations in CHD8 are implicated in ASD, particularly among individuals with macrocephaly. Experimental suppression of CHD8 in neural progenitor cells indirectly down-regulates numerous ASD candidate genes [7, 21, 22].

5.1.2.1.3 EHMT1

The euchromatic histone-lysine *N*-methyltransferase 1 gene (EHMT1) is located on chromosome 9q34.3. EHMT1 encodes a histone methyltransferase which regulates gene expression by modification of chromatin structure and by interactions with other transcription factors. Its deletion or mutation causes Kleefstra syndrome, sufferers of which are frequently diagnosed with ASD [7, 23].

5.1.2.1.4 ANKRD2

The ankyrin-repeat domain 2 (ANKRD2) gene is located at 16q24.3. ASD is associated with CNVs and mutations disrupting ANKRD2. ANKRD2 plays an important role in neural development because it is vital for proliferation of radial precursors and correct localization of neurons [7].

5.1.2.2 Transcription factors

5.1.2.2.1 FOXP1

The FOXP1 gene is located at 3p13. FOXP1 is transcription factor. It regulates the expression of ASD relevant genes that are involved in axon guidance, neuronal development, and neuronal differentiation. Loss of function variants of the FOXP1 gene have been detected in ASD individuals [24–26].

5.1.2.2.2 TBR1

TBR1 is a T-box transcription factor that is essentially expressed in postmitotic projection neurons. It regulates axon guidance, expression of an *N*-methyl-D-aspartate receptor subunit, as well as neuronal differentiation and migration. The TBR1 gene is located at 2q24. Loss of function variants of TBR1 and microdeletions involving TBR1 have been identified in ASD individuals [27–29].

5.1.2.2.3 RAI1

Retinoic acid induced 1 (RAI1) regulated genes are involved in many biological processes, such as cell cycle regulation and neuronal differentiation. Complete deficiency of Rai1 causes severe neurobehavioral abnormalities. The RAI1 gene is located at 17p11.2. Duplication of the 17p11.2 region causes Potocki Lupski syndrome, which presents intellectual disability and ASD, while its deletion leads to Smith Magenis syndrome which is accompanied by behavioral abnormalities [7, 30].

5.1.2.3 RNA binding and regulation

5.1.2.3.1 FMR1

Fragile X mental retardation gene 1 (FMR1) is located on the X chromosome at Xq27.3. Unstable expansion of a CGG trinucleotide repeat in the 5′ untranslated region of the FMR1 gene produces abnormal methylation, FMR1 transcription silencing, as well as decreased FMRP protein levels in the brain. This leads to fragile X syndrome (FXS). It is the most common single-gene disorder that causes ASD. Variants in the intronic region or 5′ regulatory sequences of FMR1, in addition to variants of other genes in the FMR1 pathway, such as CYFIP1 and CAMK4, may modify ASD risk [4, 7].

5.1.2.4 Cell growth and proliferation

5.1.2.4.1 TSC1 and TSC2
The TSC1 and TSC2 genes are located at 9q34 and 16p13.3, respectively. Mutation in either TSC1 or TSC2 causes tuberous sclerosis (TSC). TSC is an autosomal dominant neurocutaneous syndrome. ASD is present in about 25% of TSC individuals. A variety of nonsense, missense, insertion, and deletion mutations involve nearly all exons of TSC1 and TSC2. Individuals with TSC2 mutations present with a significantly more severe phenotype compared to those with TSC1 mutations [4, 7].

5.1.2.4.2 PTEN
PTEN (phosphatase and tensin homolog) is a tumor suppressor located at 10q23. It act as a dual-specificity phosphatase active in various pathways involved in cellular growth. As a lipid phosphatase, it down-regulates the phosphoinositol 3-kinase/ AKT pathway that is involved in cell proliferation. Dysregulation of this pathway has been associated with overgrowth syndromes. PTEN point mutations and deletions have been reported with autism, typically associated with macrocephaly [7, 31].

5.1.2.4.3 NF1
Neurofibromatosis type 1 (NF1) is a common autosomal dominant genetic disorder. Its prevalence is one per 4560 individuals. The NF1 gene is located on chromosome 17 (17q11.2). Half of cases are inherited; the other half are sporadic due to spontaneous mutation of the NF1 gene. NF1 gene encodes for neurofibromin, which negatively regulates RAS GTPase activation, key in the regulation of cell differentiation, growth, and apoptosis. The prevalence of ASD and other behavioral abnormalities associated with NF1 disorder has been increasingly documented recently. Moreover, NF1 single-nucleotide polymorphism has been described in ASD. NF1 is considered as an important single-gene disorder model for studying ASD [7, 32–34].

5.1.2.4.4 SYNGAP1
The synaptic Ras GTPase-activating protein 1 (SYNGAP1) gene is located at chromosome 6p21.31. Its protein is a major signaling protein that plays a critical role in regulating essential molecular changes in dendritic spine synaptic morpho-logical and functional modifications. Moreover, mutations of the SYNGAP1 gene have frequently been detected in ASD cases [7, 35].

5.1.2.4.5 MET
The MET gene is a well-established risk factor for ASD. A single-nucleotide variant at the MET 5′ promoter region was reported among ASD families. Moreover, a non-

coding promoter variant of the MET gene that reduces MET transcription and protein translation is associated with increased ASD risk. Moreover, MET expression was found to be reduced in the temporal lobe of ASD subjects. MET gene transcription can be regulated by FOXP2 and MECP2, ASD-related factors. MET transcription is repressed by direct FOXP2 binding to the 5′ regulatory region of MET and the risk of language disorders is increased with FOXP2 mutations, while abnormalities in MECP2 lead to Rett syndrome and ASD as mentioned before [36, 37].

5.1.2.4.6 CDKL5
The cyclin-dependent kinase-like 5 (CDKL5) gene is located on the X chromosome at Xp22. It encodes a proline-directed kinase that belongs to the CDC-like family of serine/threonine kinases. Mutations in the CDKL5 gene are associated with ASD, intellectual disability, and early-onset epilepsy. Moreover, it was proposed that an increased dosage of CDKL5 might cause interactions of this kinase with its substrates. These interactions lead to perturbation of synaptic plasticity and manifest in autistic behavior, developmental and speech delay, hyperactivity, and macrocephaly [7, 38].

5.1.2.4.7 DYRK1A
The dual-specificity tyrosine phosphorylation-regulated kinase 1A (DYRK1A) gene lies within the Down syndrome critical region on chromosome 21. It plays a major role in brain development, particularly neurogenesis, neural plasticity, and cellular death. Recurrent disruptions of DYRK1A have been found in 0.5% of cases of ASD. Moreover, ASD is diagnosed in up to 40% of cases with DYRK1A mutations. However, most of the remaining cases have features consistent with ASD [7, 39].

5.1.2.5 Protein ubiquitination

5.1.2.5.1 UBE3A
The UBE3A gene is located on chromosome 15 at 15q11.2. It encodes E3 ubiquitin ligase protein 3a. The loss of paternally imprinted, maternally expressed sequences on 15q11-13 leads to Angelman syndrome. Angelman syndrome individuals show several symptoms and signs of ASD (such as language impairment), but only a subgroup of them meet the diagnostic criteria for ASD. Moreover, duplications, typically of maternal origin, were found in 1%–3% of cases of ASD. It was suggested that UBE3A plays an important role in the early stages of neuronal development [7, 40].

5.1.2.5.2 MAGEL2 and USP7
MAGEL2 (melanoma antigen family L2) is one of five protein-coding genes in the Prader–Willi syndrome (PWS) critical region that is located on chromosome 15q11-q13. They are maternally imprinted, paternally expressed genes. MAGEL2 paternal allele truncating point mutations lead to Schaaf–Yang syndrome. The phenotype of this syndrome overlaps with PWS. However, it is clinically different with a higher prevalence of ASD (up to 75% of affected individuals). The MAGEL2 protein forms a protein complex with TRIM27 (ring ubiquitin ligase) and USP7 (ubiquitin specific peptidase) that helps WASH-dependent endosomal protein recycling. *De novo* loss

of function mutations and haploinsufficiency of USP7 have been reported in intellectual disability and ASD individuals [7, 41].

5.1.2.6 Synaptic organization and activity

5.1.2.6.1 SHANK family

Shank/ProSAP proteins are scaffold proteins. They are critical for synaptic development and functions. They are present at the postsynaptic density of glutamatergic synapses. Shank proteins are encoded by the three genes SHANK1, SHANK2, and SHANK3, which are SHANK family genes. Genetic abnormalities in the SHANK family genes have been linked to neurodevelopmental and neuropsychiatric disorders (Shankopathies), such as ASD, schizophrenia, and Alzheimer's disease. However, the involvement of SHANK family genes in ASD affects patients differently [42, 43].

The SHANK1 gene is located on chromosome 19q13.33. It comprises 23 exons and two different promoters. The two different promoters lead to two different isoforms. It was proposed that SHANK1 deletion could be associated with mild autism in males. However, for an unknown reason, females are protected. Nevertheless, the relation of SHANK1 and the sex of ASD patients should be further evaluated. A meta-analysis study presented that deletions disrupting SHANK1 were detected in 0.04% of ASD individuals. However, no SHANK1 duplications were reported in ASD patients [43–45].

The SHANK2 gene is located on chromosome 11q13.3. It comprises 25 exons. This gene contains three alternative promoters and one stop codon that lead to four different isoforms. *De novo* heterozygous deletions and mutations of SHANK2 were identified in ASD and ID patients by several studies and revealed the contribution of SHANK2 to ASD [46–51]. Moreover, *de novo* SHANK2 deletions were associated with inherited 15q11-q13 CNVs in the form of nicotinic receptor CHRNA7 duplication and synaptic translation repressor CYFIP1 deletion in two ASD patients. These finding suggest an oligogenic 'multiple hit' underlying genetic causes in ASD. However, no SHANK2 duplications were reported in ASD individuals [43, 49, 52].

The SHANK3 gene is located on chromosome 22q13.3. It comprises 22 exons. This gene contains multiple intragenic promoters and alternative splicing of coding exons. Consequently it codifies a large number of mRNA and protein isoforms. The SHANK3 gene has been widely studied as it is the causative gene of Phelan–McDermid syndrome (PMS). SHANK3 haploinsufficiency is considered the main cause of the neurological symptoms of this syndrome [43, 53]. SHANK3 *de novo* mutations, interstitial, or terminal deletion have been identified in several studies of ASD patients [45, 54, 55]. Moreover, it was reported that a PMS patient with 22q13 deletion has an Asperger's syndrome brother with reciprocal 22qter duplication. Also, a boy affected by ASD with a SHANK3 deletion was reported to have a sister affected by attention deficit and hyperactivity disorder with reciprocal duplication. These findings suggest that SHANK3 levels are crucial in ASD pathogenesis [54, 56]. Furthermore, the high level of SHANK3 gene methylation detected in a cohort of ASD individuals suggested that epigenetic dysregulation of the SHANK3 gene may have a role in ASD pathogenesis [57].

5.1.2.6.2 NRXNs and NLGNs

Neurexins (NRXNs) and neuroligins (NLGNs) are synaptic transmembrane proteins. Both of them form the transsynaptic NRXN/NLGN complex. This complex is critical to synaptic function, but not synaptic formation. The NRXN/NLGN complex and its downstream cascades have been proposed in the underlying processes of the specific synaptic pathway for ASD etiology [7, 58].

NLGNs are postsynaptic cell-adhesion molecules. They comprise five family members (NLGN1, 2, 3, 4X, and 4Y). In the extracellular domain, NLGN proteins and presynaptic NRXN form a transsynaptic complex, whereas the cytoplasmic domain interacts with postsynaptic molecules such as SHANK. Numerous variants in NLGN3 and NLGN4 have been detected in non-syndromic ASD individuals. These variants result in loss- or gain-of-functions. Moreover, CNVs and SNVs involving NLGN1 have been reported in ASD [58].

NRXNs is the third crucial player in the autism-related synaptic network. NRXNs are a family of presynaptic single-pass transmembrane proteins. They act as synaptic organizers. The NRXNs comprise three genes (NRXN1, NRXN2, and NRXN3). Each gene encodes longer α- and shorter β-forms. NRXNs have five canonical alternative splice sites in the extracellular region. Moreover, a sixth splice site has been recognized in NRXN1 and NRXN3. More than 4000 structural variants of the NRXN proteins are generated by this combination of exon-insertion splice sites. This structural variety defines the NRXNs' binding specificities to extracellular ligands, such as NLGNs. Moreover, NRXNs are essential mediators for neurotransmitter release through connecting calcium (Ca^{2+}) channels to synaptic vesicle exocytosis. NRXN genomic alterations are frequently identified in ASD. NRXN1 CNVs and NRXN1 and NRXN2 truncating mutations were detected in ASD individuals. NRXN3 deletions have been reported in four ASD individuals, three of them were inherited from either subclinical or apparently healthy parents. This observation emphasizes the issues of penetrance and expressivity at this locus. Moreover, NRXN1 bi-allelic deficiency is associated with Pitt–Hopkins syndrome, wherein autism is a primary feature [4, 59, 60].

5.1.2.6.3 GABRB3

The 15q11-q13 locus harbors a cluster of GABA receptor subunit genes such as GABRB3. The GABRB3 gene encodes the $\beta3$ subunit, a component of the GABAergic signaling pathway. One of the most common chromosomal anomalies in idiopathic ASD is 15q11-q13 duplication. Gene expression analysis suggested a causative role for the GABAergic system in autism. SNPs occurring at the promoter of the GABRB3 might alter the transcription factor binding sites and consequently reduce its expression [61].

5.1.2.6.4 SCN1A

Neuronal voltage gated sodium channel genes such as SCN1A, SCN2A, and SCN3A are implicated in psychiatric disorders and neurological diseases. These genes are clustered within 600 kB on chromosome 2q24. They encode distinct α-subunit isoforms that are highly expressed in neurons and glia throughout the nervous system. Locus 2q, including the SCN1A, SCN2A, and SCN3A genes, was reported as a potential susceptibility region for autism. Loss of function mutations in the SCN1A gene cause Dravet syndrome (DS). DS is a diverse seizure disorder, occasionally manifested by autistic behaviors. About 70%–80% of DS patients have SCN1A mutations, whereas some patients display variants in other genes such as SCN2A [62].

5.1.2.6.5 STXBP1
Syntaxin-binding protein 1 (STXBP1) is involved in the synaptic release of neuro-transmitters. It binds to the SNARE complexes. These complexes control synaptic vesicle priming. Heterozygous variants in STXBP1 have been described in ASD individuals with epilepsy and intellectual disability [63].

5.1.2.6.6 CTNND2
Delta 2 catenin gene (CTNND2) encodes the delta catenin protein. CTNND2 is highly expressed in the fetal brain and is strongly linked to other ASD genes. CNVs and point mutations in CTNND2 have been identified in ASD individuals. These variants are loss of function abnormalities and affect Wnt signaling [7, 64].

5.1.2.7 Cell adhesion

5.1.2.7.1 CNTNAP2
The Contactin associated protein-like 2 (CNTNAP2) gene is located on chromosome 7q35-36.1. CNTNAP2 encodes CASPR2, a member of the neurexin family. Its expression is restricted to neurons. CASPRs function as cell-adhesion molecules mainly between neuronal and glial cells. CNTNAP2 variants have been found in association with ASD. Its loss of function is responsible for the behavioral deficits linked to ASD [7, 65–67].

5.2 Alzheimer's disease

5.2.1 Introduction

Alzheimer's disease (AD) is an irreversible and progressive neurodegenerative disease that ultimately leads to death. It is the most common cause of age-related dementia, and affects people worldwide with dementia. AD is defined on a pathological basis by severe neuronal loss, precipitation of amyloid β ($A\beta$) in extracellular plaques, and hyperphosphorylated tau protein that leads to the formation of intraneuronal neurofibrillary tangles [68].

It is clinically characterized by progressive deterioration of memory, and as the disease progresses the patient may become aggressive, irritable, and confused [69], and finally full-time care becomes necessary. Currently no curative therapy is available, and the available therapies provide short term symptom relief without slowing the progress of the disease [69].

There are two major types of the disease: early onset <60 y.o. (familial type) and late onset (sporadic type). Three genes are responsible for the familial type (APP, PSEN1 and PSEN2). Many genes have been implicated in the development of the sporadic form, however the APOE gene is considered as a high risk factor for late onset familial type and sporadic type AD [70].

5.2.2 Clinical assessment

5.2.2.1 Symptoms of Alzheimer's disease

AD clinically presents as dementia that begins with a gradual memory loss which slowly progresses until it eventually become severe. The first warning sign is the inability to remember recent events or information. However, memories of distant events are spared until later stages of the disease. Also, as the disease progresses, impairments in other areas of cognition develop, such as language, executive function, and decision making. The severity of the cognitive impairment is variable, but is usually associated with difficulties at work or with household activities or social communications. Mood instability usually accompanies the decline in memory. Delusions and hallucinations may develop at any time through the course of the illness. Less common symptoms that may occur at later stages of the disease include seizures, hypertonia, incontinence, and mutism. Death commonly occurs from malnutrition and pneumonia [71].

5.2.2.2 Types of Alzheimer's disease

According to the age of onset, the disease is divided into two categories: early-onset AD (EOAD) and late-onset AD (LOAD). The rate of the early-onset subtype is approximately 1%–6% of all cases and the range of age of onset is roughly estimated to be from 30 to 60 years. On the other hand, the late-onset subtype, which is by far the most common subtype, has been recorded to develop after the age of 65. A positive family history of AD may be present in both categories [70].

About 60% of EOAD cases have more than one affected patient within their families, and approximately 13% are inherited as autosomal dominant with at least three generations affected [72].

Apart from the autosomal dominant families, most AD cases appear to be a complex disorder that is likely to involve multiple susceptibility genes and environmental factors [73].

5.2.3 Risk and protective factors

In addition to the genetic factors that play a major role in late-onset AD, other non-genetic factors have been implicated in the development of the disease, such as diabetes, cardiovascular conditions, obesity, and brain and systemic inflammation [69].

5.2.3.1 Personal habits

5.2.3.1.1 Diet

The traditional Mediterranean diet is acknowledged as optimal for a healthy life, as it is associated with reduced mortality and morbidity. The Mediterranean diet is characterized by: high consumption of vegetables, fruits, nuts, and cereals; moderate consumption of fish; moderate intake of dairy products; and low consumption of meat. Also, olive oil is a major source of fat. Thus it was found that adapting the Mediterranean pattern of diet is strongly associated with a lower risk of AD and better cognitive function [74].

5.2.3.1.2 Sleep

A reduction of the total nocturnal sleeping time and excessive daytime sleepiness are common in patients with AD [75]. Sleep disturbances can result from neuro-degenerative changes that can increase the risk of AD. Sleep disturbance has been related to lower baseline cognitive function and progressive cognitive function decline [76].

Animal studies have indicate the potential impact of sleep disturbance on AD. While sleeping, as a result of the increased interstitial space that surrounds brain cells, more exchanges of CSF and interstitial fluid in the brain occurs which leads to increased clearance of Aβ peptides and other toxic substances in comparison to the situation during wakefulness. Deficient sleep may decrease the rate of the removal of the toxic materials from the brain, which leads to an increased propensity for AD. This mutual relationship between sleep and AD forms a vicious cycle that probably will lead to further neuropathological changes in patients [77].

5.2.3.1.3 Physical activity

Researchers have suggested a positive reciprocal relationship between physical activity and cognitive function. Exercise decreases the risk of cerebrovascular diseases and also improves attention and memory, and all other brain functions such as processing speed and executive function. Moreover, aerobic exercise reversed age-related volume loss of the hippocampus in older adults without dementia [78]. Exercise exerts its impact on the brain by improving cerebral metabolism, circulation, and endurance towards oxidative stress. All of these properties are important for brain plasticity and thus potentially could prevent AD [79].

5.2.3.1.4 Cognitive reserve

Education is very crucial for cognitive reserve. High educational levels have been recorded to decrease the risk of AD and protect the brain against more deterioration of cognitive function for years after the appearance of the first signs. It can be explained by the development of a cognitive reserve that helps the brain to endure more pathological insults and become more resilient to cognitive damage, or may allow the brain to evolve compensatory pathways to deal with the pathology, hence delaying the onset of AD. However, at the time of AD onset, these same individuals may show more rapid cognitive decline because excessive pathological changes have been accumulated over the years. Also, occupational involvement and engaging in activities also reduce the risk of developing dementia [80, 81].

5.2.3.2 Diseases

5.2.3.2.1 Diabetes mellitus

The incidence of AD is very high in patients with type 2 diabetes mellitus (T2DM), reaching up to 50% to 100% more than in those without. Impairment of insulin signaling may be the underlying common pathological change in both diseases. Also, the pathological changes of T2DM and AD are similar in many other ways; T2DM is characterized by aggregation of islet amyloid polypeptide in the pancreas and loss of β-cells, and in AD there is Aβ plaque aggregation and loss of neuronal function in the brain [82].

5.2.3.2.2 Accumulation of neurotoxic metabolites

Ammonia is known to be toxic to the CNS. According to the duration and level of exposure, high brain ammonia has known neurological symptoms which also develop in AD, among these are memory loss, disturbance of cognitive and spatial learning, and pathophysiological aberrations include the following: mitochondrial dysfunction, and disruption of cellular glucose metabolism, neurotransmission, and regulation of inflammatory responses [83].

Reactive oxygen species (ROS) play a key role in the pathogenesis of AD. Altered cellular oxygen metabolism will result in molecules containing oxygen that play important roles in the development of multiple diseases, including neurodegenerative diseases [84]. Under normal situations, the ROS participate in redox reactions, and assist as second messengers for regulatory functions. The main change in AD is the abnormal accumulation of Aβ which is modified by the ROS into toxic products, which progressively aggregate into senile plaques signaling p53 mediated apoptosis. Eventually this leads to loss of neuronal functions [85].

In addition, ROS cause tau protein hyperphosphorylation which might be related to impaired memory [86]. Tau phosphorylation inhibits its microtubule-association action and supports its self-aggregation [87]. Aggregation of hyperphosphorylated tau is the main element of the degenerating neurons in AD, it also may directly contribute to neurodegeneration [88].

5.2.4 Neuropathological changes

5.2.4.1 Amyloidopathy

The deposition of amyloid beta (Aβ) peptide, the main component of senile plaques in neuronal cells, is known as 'amyloidopathy'. It is becoming evident that the extracellular aggregation of Aβ plaques is largely associated with neuronal atrophy and the concomitant damage of synapses in different brain regions that results in gradual neuronal death and memory loss [89].

The amyloid cascade hypothesis, originally proposed by Hardy and Higgins in 1992, is the most accepted and well understood mechanism of deposition of Aβ protein in Alzheimer's pathogenesis. Amyloid precursor protein (APP) is a large transmembrane protein that produces Aβ which consists of either 40 or 42 amino acids. Aβ40 is a commonly found protein molecule whereas Aβ42 is highly toxic to neuronal cells. In normal physiological conditions, membrane APP plays a critical role in synapse formation and neuronal activity, and transmission via two proteolytic pathways: the 'nonamyloidogenic' and 'amyloidogenic' pathways [90].

In the 'nonamyloidogenic' pathway, soluble APPα is released to the extracellular compartment through APP cleavage via α- and γ-secretases. The 'amyloidogenic' pathway is triggered the by β-secretase 1 (β-site amyloid precursor protein cleaving enzyme 1, BACE-1) mediated cleavage pathway, resulting in soluble extracellular APPβ, which is subsequently cleaved by a γ-secretase complex, leading to the release of Aβ into the extracellular space. Then, Aβ is transported and degraded by the APOE2 and APOE3 isoform and cleared completely through the blood–brain barrier, together with insulin degrading enzyme (IDE) or with the neprilysin degradation pathway (NDP) [91].

On the other hand, the binding of Aβ to the APOE4 isoform leads largely to Aβ aggregation in the extracellular region of the central nervous system. Under normal conditions, during Aβ generation there is a feedback mechanism that releases the APP intracellular C-terminal domain and increases the level of neprilysin, which promotes Aβ turnover. The disturbance of the normal physiological pathway may develop through mutations and changes in the expression of APP, BACE-1, IDE, APOE, and neprilysin, which mitigate the accumulation of Aβ and the manifestation of AD [92].

5.2.4.2 Hyperphosphorylation of tau protein

Tau protein is a microtubule binding protein that is found largely in neurons, specifically in axons and dendrites. Tauopathies are diseases associated with the intracellular pathological aggregation of tau protein in neurofibrillary tangles, which is observed in many neurodegenerative diseases, including Alzheimer's disease. In AD, the paired helical filament type of tau aggregates and forms tangles in neuronal cells which is neurotoxic and leads to neurodegeneration [93]. The microtubule-associated protein tau (MAPT) gene located on chromosome 17 regulates the expression of six isoforms of the tau protein in human neurons. The six isoforms are 4R2N, 4R1N, 3R2N, 4R0N, 3R1N, and 3R0N, and they are produced as a result of alternative mRNA splicing of the MAPT gene. The shortest form, the 3R0N isoform, is expressed predominantly in the fetal brain [94]. The loss of functional tau due to the hyperphosphorylation state of paired helical filaments (PHFs) of all six isoforms of tau proteins is the pathological hallmark of AD which leads to disruption in microtubule assembly, decreased axonal transport, and neurodegeneration in areas such as the frontotemporal lobe of the human brain [95].

Many scientific studies have clearly indicated that an increased generation of reactive oxygen species (ROS) that is not eliminated by an internal defense mechanism results in oxidative stress and manifestations of neurodegenerative diseases. Also, studies have demonstrated a relation between mitochondrial oxidative stress and the hyperphosphorylation of tau protein. Accordingly, tau phosphorylation was halted by the administration of a high dose of antioxidant, which indicates that an efficient antioxidant defense system prevents tau hyperphosphorylation and delays the manifestation of AD [96].

5.2.4.3 Synaptic damage

A plethora of studies has determined that the key problems of AD are the fewer synapses, the alteration in synaptic plasticity, and the severity of synaptic decline at specific locations in the brain, such as the hippocampus which has great relevance for the cognitive decline in AD patients [97].

5.2.4.4 Other pathological changes

The other pathological changes that play role in AD are neuroinflammation, mitochondrial dysfunction with excessive ROS production, protein oxidation, lipid peroxidation, and depletion of endogenous antioxidant enzymes [89].

5.2.5 Genetics of AD

Although studies support the genetic involvement in LOAD, in more than 90% of patients their AD was found to be sporadic and have a late age of onset at over 65 years of age. No causative gene has been yet identified. However, the only gene that has been consistently found to be associated with sporadic LOAD, is the apolipoprotein E (APOE) gene. However, many carriers of the APOE risk allele ($\varepsilon 4$) may live into their 90s, which suggests the presence of other factors which still needed to be investigated.

5.2.5.1 Genes associated with early-onset AD (EOAD)

Only hundreds of families carry one of the following mutations and account for less than 1% of cases.

5.2.5.1.1 Amyloid precursor protein (APP)

APP is an integral membrane glycoprotein from which Aβ is derived by a proteolytic process involving secretase enzymes [98]. The function of APP is not fully understood. However, it has been implicated in neuronal development and synaptic formation and repair, which is upregulated after exposure to neuronal injury [69]. Mutations of the APP gene account for 10%–15% of early-onset familial AD and appear to be family specific, and were not found in the majority of sporadic cases of AD [71].

The encoding gene of APP is mapped to chromosome 21q21.3, which consists of 19 exons and is about 240 kb in size. The resulting protein is expressed in many tissues, and as a result of the alternative splicing sites different isoforms are created. APP695 (exons 1–6, 9–18), APP751 (exons 1–7, 9–18), and APP770 (exons 1–18) are the most abundant isoforms. All of these transcripts produce a multidomain protein. APP751 and APP770 are different from APP695 in that they possess exon seven, which encodes a serine protease inhibitor domain. APP695 is the predominant isoform in neuronal tissues [99, 100].

APP is further proteolytically processed by α-, β-, and γ-secretases following two pathways: nonamyloidogenic or amyloidogenic. The proteolysis of APP by the nonamyloidogenic pathway is carried out by α- and γ-secretases and results in nonpathogenic fragments (sAPPα and α-C-terminal fragment). On the other hand, the proteolysis of APP by the amyloidogenic pathway, which is enriched in neurons, is carried out by β-secretase and γ-secretase gives rise to a mixture of Aβ peptides with different amino-acid number. There are two major Aβ species: Aβ1–40 (90%)

and $A\beta1$–42 (10%). It is the $A\beta1$–42 fragments that possess a greater aggregation tendency which are the main constituent of the extracellular amyloid plaques in the brains of AD patients that are considered the hallmarks of Alzheimer's disease [101, 102]. About forty APP mutations are reported to affect proteolysis of APP in favor of $A\beta1$–42 [103, 104]. Interestingly, a rare protective variant of APP has been identified and located near the β proteolytic cleavage site which results in failure of the cleavage of APP and a reduced level of $A\beta1$–40 and $A\beta1$–42 followed by a decreased rate of AD [105]. Also, another mutation at the same site has been identified (p.A673V) as pathogenic in the homozygous state and protective in the heterozygous state, which suggests that a complex of wild-type and mutant APP alleles may affect the aggregation properties of $A\beta$ peptides in a good way [106].

5.2.5.1.2 Presenilin 1 (PSEN1)

Presenilins are essential components of the γ-secretase complex, which catalyzes the cleavage of membrane proteins, including APP [107, 108]. The encoding gene of PSEN1 is located on chromosome 14q24.2. It consists of 12 exons that encode a 467 amino-acid protein [109, 110]. PSEN1 mutations resulted in an increase of the $A\beta42$ to $A\beta40$ ratio, by altering the functions that lead to impairment of the γ-secretase mediated cleavage of APP in $A\beta$ fragments. It has been proposed that deposition of $A\beta42$ may be an early preclinical event in the carrier of PSEN1 mutation [111, 112].

Mutations in PSEN1 are the most frequent causes of EOAD. PSEN1 missense mutations account for 18% to 50% of autosomal dominant EOAD cases with complete penetrance, and are responsible for the severe form of AD that can occur as early as 25 years of age [113, 114]. The age of onset is similar among affected family members and among people from different families carrying the same mutation. However, there is a great variability in age of onset of AD with different mutations of PSEN1 having an average age of onset >58 years [115].

About 176 PSEN1 mutations have been reported in 390 families and the majority of these mutations are missense [70]. These missense mutations cause amino acid substitutions throughout the PSEN1 protein and result in an increase in the ratio of $A\beta42$ to $A\beta40$ peptides either through increased $A\beta42$ production, decreased $A\beta40$ production, or a combination of both [116].

5.2.5.1.3 Presenilin 2 (PSEN2)

The PSEN2 gene is located on chromosome 1 (1q42.13) and has 12 exons that encode a 448 amino-acid peptide [117, 118]. Missense mutations in the PSEN2 gene are a rare cause of EOAD. Like the PSEN1 gene, the PSEN2-associated mutations have been reported to increase the ratio of $A\beta42$ to $A\beta40$ in humans [116]. However, the clinical features of PSEN2-affected families generally have an older age of onset (45–88 years) and the age of onset is highly variable among PSEN2-affected members of the same family [109, 119]. Also, missense mutations in the PSEN2 gene have a lower penetrance than the PSEN1 gene and therefore it has been suggested to be subject to modifying action by other genes or environmental factors [109, 120].

5.2.5.2 Genes associated with risk in late-onset AD (familial and sporadic)
Unlike early-onset familial AD which is inherited in a Mendelian dominant pattern, late-onset AD (LOAD) is considered to be multifactorial with strong evidence of genetic predisposition [121]. However, the genetic component itself is complex and heterogeneous [122]. Although APOE was for a long time considered to be the major gene known to increase AD risk, an additional large number of gene polymorphisms was suggested to be associated with increased risk of LOAD. However, the recurrence of these associations was not consistent. These polymorphisms affect genes regulating inflammation, oxidative stress, and vascular biology, and other functions affecting brain pathogenesis [123, 124].

5.2.5.2.1 Apolipoprotein E (APOE)
The APOE gene is associated with late-onset AD whether familial or sporadic. It encodes a polymorphic glycoprotein expressed in different body cells such as the liver, brain, and macrophages. APOE protein participates in the transport of cholesterol and other lipids. It is also involved in neuronal growth and regeneration, repair response to tissue injury, immunoregulation, and activation of lipolytic enzymes [122, 125]. APOE binds to $A\beta$ and participate in the clearance of soluble $A\beta$ from extracellular space and reduction of $A\beta$ aggregation [128].

The APOE gene is located on chromosome 19q13.2 and consists of four exons that encode a 299 amino-acid protein. The APOE gene is in a cluster with other apolipoprotein genes: APOC1, APOC2, and APOC4 [126]. It has three major allelic variants ($\varepsilon 2$, $\varepsilon 3$, and $\varepsilon 4$), encoding different isoforms (APOE2, APOE3, and APOE4) that differ in two sites of the amino-acid sequence [127]. The APOE $\varepsilon 4$ allele was found to be the one involved in the increased risk of AD in familial and sporadic early-onset and late-onset type AD, but its presence alone is not sufficient to develop the disease [125, 127]. Each APOE isoform exerts a different effect, while APOE $\varepsilon 4$ is thought to be less efficient in mediating $A\beta$ clearance from the brain extracellular tissue, the APOE $\varepsilon 3$ isoform has a higher $A\beta$-binding capacity that facilitates clearance of $A\beta$, and the APOE $\varepsilon 2$ allele is thought to have a protective effect and hence delay AD onset age [128–130]. The gene dose of APOE $\varepsilon 4$ is a major risk factor for AD, with many studies reporting an association between gene dose, age at onset, and cognitive decline [131].

The mechanisms that control APOE toxicity in brain tissue are not fully understood. Some proposed pathways include isoform-specific toxicity, such as APOE $\varepsilon 4$ mediated amyloid aggregation, and APOE $\varepsilon 4$ mediated tau hyper-phosphorylation [129].

5.2.5.2.2 Clusterin (CLU)
Clusterin is a 75 kDa apolipoprotein that is widely expressed throughout the body, particularly in the brain [132]. The CLU gene is located on chromosome 8 (8p21), it is composed of nine exons, covering 16 kb of DNA [133]. The name clusterin is derived from the ability to cluster together cells of various types [134]. It has the ability to interact with many different molecules [135]. Both APOE and clusterin have been involved in the clearance of beta-amyloid from neural tissue and CSF [136]. It is also

similar to APOE through its abundance in the hippocampus region that is affected by the pathology of Alzheimer's disease and cerebrospinal fluid (CSF), as well as being present in amyloid plaques and binding to beta-amyloid [137, 138].

5.2.5.2.3 *ATP-binding cassette transporter 7 (ABCA7)*

ATP-binding cassette subfamily A member 7 (ABCA7) was considered as a risk factor for AD, the function of its product is to transport substrates across cell membranes [139]. The ABCA7 gene is located on chromosome 19p13.3 and contains 46 exons spans about 32 kb, which encodes a polypeptide of 2149 amino acids [140]. It can undergo alternative splicing to produce two transcripts, both of which are expressed in the brain [141]. Polymorphisms in this region increase LOAD risk, however, its impact in AD is poorly understood [142, 143]. ABCA7 is expressed in the hippocampal region and at ten-fold higher levels in microglia [144] and associated with plaque aggregation in AD brains [145]. ABCA7 mRNA expression in brain tissue is also associated with advanced cognitive decline [142, 143]. ABCA7 is involved in the efflux of lipids from cells into lipoprotein particles. *In vitro*, ABCA7 stimulates cholesterol efflux and inhibits $A\beta$ secretion [146]. Researchers have suggested that ABCA7 may influence AD risk via cholesterol transfer to APOE or by clearing $A\beta$ aggregates [146–148].

Recently, three novel ABCA7 alternative splicing events were identified. One isoform in particular—which is formed through exon 19 skipping—lacks the first nucleotide binding domain of ABCA7 and is abundant in brain tissue [149].

The above three genes APOE, CLU, and ABCA7 are implicated in the pathogenesis of AD through cholesterol metabolism, however, a plethora of other genes have been described to be associated with AD risks involving other pathways [150].

Neuroinflammation and dysregulation of the immune response are the main features of AD [151]. Common variants have been identified to be associated with LOAD which involve neuroinflammation and dysregulation of the immune response, these include CR1, CD33, MS4A, and CLU [152–154]. Also a rare coding variant was described in TREM2 [155, 156].

Endocytosis is critical for normal processing of APP, which is crucial in AD pathogenesis. Furthermore, synaptic activity and neurotransmitter release is disrupted in AD [152]. Genes associated with endocytosis and synaptic function were identified in LOAD, such as BIN1, PICALM, CD2AP, EPHA1, and SORL1 [152, 157].

Additional loci were identified in the largest LOAD genome-wide association studies (GWAS). Less is known of the role of these genes in AD, however, many of these genes fit into known pathways that are altered in AD, genes involved in either the immune response, synaptic function, cytoskeletal function, axonal transport, or tau metabolism [152]. However, several of these susceptibility genes occur in gene-dense regions, so, it remains unclear which gene is responsible for the association [150].

References

[1] Khan S A, Khan S A, Narendra A R, Mushtaq G, Zahran S A, Khan S and Kamal M A 2016 Alzheimer's disease and autistic spectrum disorder: is there any association? *CNS Neurol. Disord. Drug Targets* **15** 390–402

[2] Bremer A, Giacobini M, Nordenskjöld M, Brøndum-Nielsen K, Mansouri M, Dahl N, Anderlid B and Schoumans J 2010 Screening for copy number alterations in loci associated with autism spectrum disorders by two-color multiplex ligation-dependent probe amplification *Am. J. Med. Genet.* B **153B** 280–5

[3] Huguet G, Ey E and Bourgeron T 2013 The genetic landscapes of autism spectrum disorders *Annu. Rev. Genomics Hum. Genet.* **14** 191–213

[4] Persico A M and Napolioni V 2013 Autism genetics *Behav. Brain Res.* **251** 95–112

[5] Chung B H, Tao V Q and Tso W W 2014 Copy number variation and autism: new insights and clinical implications *J. Formos. Med. Assoc.* **113** 400–8

[6] Elsheikh S, Kuusikko-Gauffin S, Mattila M L, Jussila K, Ebeling H, Loukusa S, Omar M, Riad G, Rautio A and Moilanen I 2016 Neuropsychological performance of Finnish and Egyptian children with autism spectrum disorder *Int. J. Circumpolar Health* **75** 29681

[7] Yin J and Schaaf C P 2017 Autism genetics—an overview *Prenat. Diagn.* **37** 14–30

[8] Jorde L B, Hasstedt S J, Ritvo E R, Mason-Brothers A, Freeman B J, Pingree C, McMahon W M, Petersen B, Jenson W R and Mo A 1991 Complex segregation analysis of autism *Am. J. Hum. Genet.* **49** 932–8

[9] Rosti R O, Sadek A A, Vaux K K and Gleeson J G 2014 The genetic landscape of autism spectrum disorders *Dev. Med. Child Neurol.* **56** 12–8

[10] de la Torre-Ubieta L, Won H, Stein J L and Geschwind D H 2016 Advancing the understanding of autism disease mechanisms through genetics *Nat. Med.* **22** 345–61

[11] Magri L *et al* 2011 Sustained activation of mTOR pathway in embryonic neural stem cells leads to development of tuberous sclerosis complex-associated lesions *Cell Stem Cell* **9** 447–62

[12] O'Roak B J *et al* 2012 Multiplex targeted sequencing identifies recurrently mutated genes in autism spectrum disorders *Science* **338** 1619–22

[13] Sowers L P *et al* 2013 Disruption of the non-canonical Wnt gene PRICKLE2 leads to autism-like behaviors with evidence for hippocampal synaptic dysfunction *Mol. Psychiatry* **18** 1077–89

[14] Miller J A *et al* 2014 Transcriptional landscape of the prenatal human brain *Nature* **508** 199–206

[15] Tsai P T, Hull C, Chu Y, Greene-Colozzi E, Sadowski A R, Leech J M, Steinberg J, Crawley J N, Regehr W G and Sahin M 2012 Autistic-like behaviour and cerebellar dysfunction in Purkinje cell Tsc1 mutant mice *Nature* **488** 647–51

[16] Wang S S, Kloth A D and Badura A 2014 The cerebellum, sensitive periods, and autism *Neuron* **83** 518–32

[17] Langen M, Durston S, Kas M J, van Engeland H and Staal W G 2011 The neurobiology of repetitive behavior: …and men *Neurosci. Biobehav. Rev.* **35** 356–65

[18] Peça J, Feliciano C, Ting J T, Wang W, Wells M F, Venkatraman T N, Lascola C D, Fu Z and Feng G 2011 SHANK3 mutant mice display autistic-like behaviours and striatal dysfunction *Nature* **472** 437–42

[19] Rothwell P E, Fuccillo M V, Maxeiner S, Hayton S J, Gokce O, Lim B K, Fowler S C, Malenka R C and Südhof T C 2014 Autism-associated neuroligin-3 mutations commonly impair striatal circuits to boost repetitive behaviors *Cell* **158** 198–212

[20] Liyanage V R and Rastegar M 2014 Rett syndrome and MeCP2 *Neuromol. Med.* **16** 231–64

[21] Sugathan A *et al* 2014 CHD8 regulates neurodevelopmental pathways associated with autism spectrum disorder in neural progenitors *Proc. Natl Acad. Sci. USA* **111** E4468–77

[22] Stolerman E S, Smith B, Chaubey A and Jones J R 2016 CHD8 intragenic deletion associated with autism spectrum disorder *Eur. J. Med. Genet.* **59** 189–94

[23] Koemans T S *et al* 2017 Functional convergence of histone methyltransferases EHMT1 and KMT2C involved in intellectual disability and autism spectrum disorder *PLoS Genet.* **13** e1006864

[24] Le Fevre A K *et al* 2013 FOXP1 mutations cause intellectual disability and a recognizable phenotype *Am. J. Med. Genet.* A **161A** 3166–75

[25] Iossifov I *et al* 2014 The contribution of de novo coding mutations to autism spectrum disorder *Nature* **515** 216–21

[26] Araujo D J *et al* 2015 FOXP1 orchestration of ASD-relevant signaling pathways in the striatum *Genes Dev.* **29** 2081–96

[27] Huang T N, Chuang H C, Chou W H, Chen C Y, Wang H F, Chou S J and Hsueh Y P 2014 TBR1 haploinsufficiency impairs amygdalar axonal projections and results in cognitive abnormality *Nat. Neurosci.* **17** 240–7

[28] Deriziotis P, O'Roak B J, Graham S A, Estruch S B, Dimitropoulou D, Bernier R A, Gerdts J, Shendure J, Eichler E E and Fisher S E 2014 *De novo* TBR1 mutations in sporadic autism disrupt protein functions *Nat. Commun.* **5** 4954

[29] Notwell J H *et al* 2016 TBR1 regulates autism risk genes in the developing neocortex *Genome Res.* **26** 1013–22

[30] Carmona-Mora P and Walz K 2010 Retinoic acid induced 1, RAI1: a dosage sensitive gene related to neurobehavioral alterations including autistic behavior *Curr. Genomics* **11** 607–17

[31] Varga E A, Pastore M, Prior T, Herman G E and McBride K L 2009 The prevalence of PTEN mutations in a clinical pediatric cohort with autism spectrum disorders, developmental delay, and macrocephaly *Genet. Med.* **11** 111–7

[32] Garg S, Green J, Leadbitter K, Emsley R, Lehtonen A, Evans D G and Huson S M 2013 Neurofibromatosis type 1 and autism spectrum disorder *Pediatrics* **132** e1642–8

[33] Garg S, Lehtonen A, Huson S M, Emsley R, Trump D, Evans D G and Green J 2013 Autism and other psychiatric comorbidity in neurofibromatosis type 1: evidence from a population-based study *Dev. Med. Child Neurol.* **55** 139–45

[34] Garg S, Plasschaert E, Descheemaeker M J, Huson S, Borghgraef M, Vogels A, Evans D G, Legius E and Green J 2015 Autism spectrum disorder profile in neurofibromatosis type I *J. Autism Dev. Disord.* **45** 1649–57

[35] Jeyabalan N and Clement J P 2016 SYNGAP1: Mind the gap *Front. Cell Neurosci.* **10** 32

[36] Plummer J T, Evgrafov O V, Bergman M Y, Friez M, Haiman C A, Levitt P and Aldinger K A 2013 Transcriptional regulation of the MET receptor tyrosine kinase gene by MeCP2 and sex-specific expression in autism and Rett syndrome *Transl. Psychiatry* **3** e316

[37] Peng Y, Lu Z, Li G, Piechowicz M, Anderson M, Uddin Y, Wu J and Qiu S 2016 The autism-associated MET receptor tyrosine kinase engages early neuronal growth mechanism and controls glutamatergic circuits development in the forebrain *Mol. Psychiatry* **21** 925–35

[38] Szafranski P *et al* 2015 Neurodevelopmental and neurobehavioral characteristics in males and females with CDKL5 duplications *Eur. J. Hum. Genet.* **23** 915–21

[39] Earl R K, Turner T N, Mefford H C, Hudac C M, Gerdts J, Eichler E E and Bernier R A 2017 Clinical phenotype of ASD-associated DYRK1A haploinsufficiency *Mol. Autism* **8** 54

[40] Guffanti G, Strik Lievers L, Bonati M T, Marchi M, Geronazzo L, Nardocci N, Estienne M, Larizza L, Macciardi F and Russo S 2011 Role of UBE3A and ATP10A genes in autism

susceptibility region 15q11-q13 in an Italian population: a positive replication for UBE3A *Psychiatry Res.* **185** 33–8

[41] Fountain M D and Schaaf C P 2016 Prader–Willi syndrome and Schaaf–Yang syndrome: neurodevelopmental diseases intersecting at the MAGEL2 gene *Diseases* **4** 2

[42] Wang X, Bey A L, Chung L, Krystal A D and Jiang Y H 2014 Therapeutic approaches for shankopathies *Dev. Neurobiol.* **74** 123–35

[43] Sala C, Vicidomini C, Bigi I, Mossa A and Verpelli C 2015 Shank synaptic scaffold proteins: keys to understanding the pathogenesis of autism and other synaptic disorders *J. Neurochem.* **135** 849–58

[44] Sato D *et al* 2012 SHANK1 deletions in males with autism spectrum disorder *Am. J. Hum. Genet.* **90** 879–87

[45] Leblond C S *et al* 2014 Meta-analysis of SHANK mutations in autism spectrum disorders: a gradient of severity in cognitive impairments *PLoS Genet.* **10** e1004580

[46] Berkel S *et al* 2010 Mutations in the SHANK2 synaptic scaffolding gene in autism spectrum disorder and mental retardation *Nat. Genet.* **42** 489–91

[47] Pinto D *et al* 2010 Functional impact of global rare copy number variation in autism spectrum disorders *Nature* **466** 368–72

[48] Wischmeijer A *et al* 2011 Olfactory receptor-related duplicons mediate a microdeletion at 11q13.2q13.4 associated with a syndromic phenotype *Mol. Syndromol.* **1** 176–84

[49] Leblond C S *et al* 2012 Genetic and functional analyses of SHANK2 mutations suggest a multiple hit model of autism spectrum disorders *PLoS Genet.* **8** e1002521

[50] Chilian B, Abdollahpour H, Bierhals T, Haltrich I, Fekete G, Nagel I, Rosenberger G and Kutsche K 2013 Dysfunction of SHANK2 and CHRNA7 in a patient with intellectual disability and language impairment supports genetic epistasis of the two loci *Clin. Genet.* **84** 560–5

[51] Schluth-Bolard C *et al* 2013 Breakpoint mapping by next generation sequencing reveals causative gene disruption in patients carrying apparently balanced chromosome rearrangements with intellectual deficiency and/or congenital malformations *J. Med. Genet.* **50** 144–50

[52] Guilmatre A, Huguet G, Delorme R and Bourgeron T 2014 The emerging role of SHANK genes in neuropsychiatric disorders *Dev. Neurobiol.* **74** 113–22

[53] Phelan K and McDermid H E 2012 The 22q13.3 deletion syndrome (Phelan–McDermid syndrome) *Mol. Syndromol.* **2** 186–201

[54] Durand C M *et al* 2007 Mutations in the gene encoding the synaptic scaffolding protein SHANK3 are associated with autism spectrum disorders *Nat. Genet.* **39** 25–7

[55] Boccuto L *et al* 2013 Prevalence of SHANK3 variants in patients with different subtypes of autism spectrum disorders *Eur. J. Hum. Genet.* **21** 310–6

[56] Moessner R *et al* 2007 Contribution of SHANK3 mutations to autism spectrum disorder *Am. J. Hum. Genet.* **81** 1289–97

[57] Zhu L *et al* 2014 Epigenetic dysregulation of SHANK3 in brain tissues from individuals with autism spectrum disorders *Hum. Mol. Genet.* **23** 1563–78

[58] Nakanishi M, Nomura J, Ji X, Tamada K, Arai T, Takahashi E, Bućan M and Takumi T 2017 Functional significance of rare neuroligin 1 variants found in autism *PLoS Genet.* **13** e1006940

[59] Vaags A K *et al* 2012 Rare deletions at the neurexin 3 locus in autism spectrum disorder *Am. J. Hum. Genet.* **90** 133–41

[60] Kasem E, Kurihara T and Tabuchi K 2018 Neurexins and neuropsychiatric disorders *Neurosci. Res.* **127** 53–60

[61] Noroozi R, Taheri M, Ghafouri-Fard S, Bidel Z, Omrani M D, Moghaddam A S, Sarabi P and Jarahi A M 2018 Meta-analysis of GABRB3 gene polymorphisms and susceptibility to autism spectrum disorder *J. Mol. Neurosci.* **65** 432–7

[62] Nickel K, Tebartz van Elst L, Domschke K, Gläser B, Stock F, Endres D, Maier S and Riedel A 2018 Heterozygous deletion of SCN2A and SCN3A in a patient with autism spectrum disorder and Tourette syndrome: a case report *BMC Psychiatry* **18** 248

[63] Uddin M *et al* 2017 Germline and somatic mutations in STXBP1 with diverse neuro-developmental phenotypes *Neurol. Genet.* **3** e199

[64] Turner T N *et al* 2015 Loss of δ-catenin function in severe autism *Nature* **520** 51–6

[65] Peñagarikano O *et al* 2011 Absence of CNTNAP2 leads to epilepsy, neuronal migration abnormalities, and core autism-related deficits *Cell* **147** 235–46

[66] Peñagarikano O and Geschwind D H 2012 What does CNTNAP2 reveal about autism spectrum disorder? *Trends Mol. Med.* **18** 156–63

[67] Werling A M, Bobrowski E, Taurines R, Gundelfinger R, Romanos M, Grünblatt E and Walitza S 2016 CNTNAP2 gene in high functioning autism: no association according to family and meta-analysis approaches *J. Neural Transm.* **123** 353–63

[68] Van Cauwenberghe C, Van Broeckhoven C and Sleegers K 2016 The genetic landscape of Alzheimer disease: clinical implications and perspectives *Genet. Med.* **18** 421–30

[69] Barber R C 2012 The genetics of Alzheimer disease *Scientifica* **2012** 246210

[70] Bekris L M, Yu C, Bird T D and Tsuang D W 2010 Genetics of Alzheimer disease *J. Geriatr. Psychiatry Neurol.* **23** 213–27

[71] Bird T D 2008 Genetic aspects of Alzheimer disease *Genet. Med.* **10** 231–9

[72] Brickell K L, Steinbart E J, Rumbaugh M, Payami H, Schellenberg G D, Van Deerlin V, Yuan W and Bird T D 2006 Early-onset Alzheimer disease in families with late onset Alzheimer disease: a potential important subtype of familial Alzheimer disease *Arch. Neurol.* **63** 1307–11

[73] Roses A D 2006 On the discovery of the genetic association of apolipoprotein E genotypes and common late-onset Alzheimer disease *J. Alzheimers Dis.* **9** 361–6

[74] Lourida I, Soni M, Thompson-Coon J, Purandare N, Lang I A, Ukoumunne O C and Llewellyn D J 2013 Mediterranean diet, cognitive function, and dementia: a systematic review *Epidemiology* **24** 479–89

[75] Peter-Derex L, Yammine P, Bastuji H and Croisile B 2015 Sleep and Alzheimer's disease *Sleep Med. Rev.* **19** 29–38

[76] Lim A S, Kowgier M, Yu L, Buchman A S and Bennett D A 2013 Sleep fragmentation and the risk of incident Alzheimer's disease and cognitive decline in older persons *Sleep* **36** 102732

[77] Xie L *et al* 2013 Sleep drives metabolite clearance from the adult brain *Science* **342** 373–7

[78] Erickson K I *et al* 2011 Kramer AF.Exercise training increases size of hippocampus and improves memory *Proc. Natl Acad. Sci. USA* **108** 3017–22

[79] Yee A *et al* 2018 Alzheimer's disease: insights for risk evaluation and prevention in the Chinese population and the need for a comprehensive programme in Hong Kong/China *Hong Kong Med. J.* **24** 492–500

[80] Amieva H, Mokri H, Le Goff M, Meillon C, Jacqmin-Gadda H, Foubert-Samier A, Orgogozo J M, Stern Y and Dartigues J F 2014 Compensatory mechanisms in higher-educated subjects with Alzheimer's disease: a study of 20 years of cognitive decline *Brain* **137** 1167–75

[81] Stern Y 2012 Cognitive reserve in ageing and Alzheimer's disease *Lancet Neurol.* **11** 1006–12

[82] De Felice F G and Ferreira S T 2014 Inflammation, defective insulin signaling, and mitochondrial dysfunction as common molecular denominators connecting type 2 diabetes to Alzheimer disease *Diabetes* **63** 2262–72

[83] Jin Y Y, Singh P, Chung H J and Hong S T 2018 Blood ammonia as a possible etiological agent for Alzheimer's disease *Nutrients* **10** 564

[84] Poprac P, Jomova K, Simunkova M, Kollar V, Rhodes C J and Valko M 2017 Targeting free radicals in oxidative stress-related human diseases *Trends Pharmacol. Sci.* **38** 592–607

[85] Shafi O 2016 Inverse relationship between Alzheimer's disease and cancer, and other factors contributing to Alzheimer's disease: a systematic review *BMC Neurol.* **16** 236

[86] Kang S W, Kim S J and Kim M S 2017 Oxidative stress with tau hyperphosphorylation in memory impaired 1,2-diacetylbenzene-treated mice *Toxicol. Lett.* **279** 53–9

[87] Ando K, Maruko-Otake A, Ohtake Y, Hayashishita M, Sekiya M and Iijima K M 2016 Stabilization of microtubule-unbound tau via tau phosphorylation at Ser262/356 by Par-1/MARK contributes to augmentation of AD-related phosphorylation and Aβ42-induced tau toxicity *PLoS Genet.* **12** e1005917

[88] Guo J, Cheng J, North B J and Wei W 2017 Functional analyses of major cancer-related signaling pathways in Alzheimer's disease etiology *Biochim. Biophys. Acta* **1868** 341–58

[89] Magalingam K B, Radhakrishnan A, Ping N S and Haleagrahara N 2018 Current concepts of neurodegenerative mechanisms in Alzheimer's disease *Biomed. Res. Int.* **2018** 3740461

[90] Priller C, Bauer T, Mitteregger G, Krebs B, Kretzschmar H A and Herms J 2006 Synapse formation and function is modulated by the amyloid precursor protein *J. Neurosci.* **26** 7212–21

[91] Huifang M, Lesn'e S and Kotilineketal L 2007 Involvement of β-site APP cleaving enzyme1 (BACE1) in amyloid precursor protein mediated enhancement of memory and activity-dependent synaptic plasticity *Proc. Natl Acad. Sci. USA* **104** 8167–72

[92] Kim J, Basak J M and Holtzman D M 2009 The role of apolipoprotein E in Alzheimer's disease *Neuron* **63** 287–303

[93] Goedert M and Spillantini M G 2006 A century of Alzheimer's disease *Science* **314** 777–81

[94] Andreadis A, Brown W M and Kosik K S 1992 Structure and novel exons of the human tau gene *Biochemistry* **31** 10626–33

[95] Köpke E, Tung Y C, Shaikh S, ADC A, Iqbal K and Grundke-Iqbal I 1993 Microtubule-associated protein tau: abnormal phosphorylation of a non-paired helical filament pool in Alzheimer disease *J. Biol. Chem.* **268** 24374–84

[96] Melov S *et al* 2007 Mitochondrial oxidative stress causes hyperphosphorylation of tau *PLoS One* **2** e536

[97] Chen Y G 2018 Research progress in the pathogenesis of Alzheimer's disease *Chin. Med. J.* **131** 1618–25

[98] Giaccone G, Tagliavini F, Linoli G, Bouras C, Frigerio L, Frangione B and Bugiani O 1989 Down patients: extracellular preamyloid deposits precede neuritic degeneration and senile plaques *Neurosci. Lett.* **97** 232–8

[99] Kang J, Lemaire H G, Unterbeck A, Salbaum J M, Masters C L, Grzeschik K H, Multhaup G, Beyreuther K and Müller-Hill B 1987 The precursor of Alzheimer's disease amyloid A4 protein resembles a cell-surface receptor *Nature* **325** 733–6

[100] Yoshikai S, Sasaki H, Doh-ura K, Furuya H and Sakaki Y 1990 Genomic organization of the human amyloid beta-protein precursor gene *Gene* **87** 257–63

[101] Suzuki N, Cheung T T, Cai X D, Odaka A, Otvos L Jr, Eckman C, Golde T E and Younkin S G 1994 An increased percentage of long amyloid beta protein secreted by familial amyloid beta protein precursor (beta APP717) mutants *Science* **264** 1336–40

[102] Esler W P and Wolfe M S 2001 A portrait of Alzheimer secretases—new features and familiar faces *Science* **293** 1449–54

[103] Goate A *et al* 1991 Segregation of a missense mutation in the amyloid precursor protein gene with familial Alzheimer's disease *Nature* **349** 704–6

[104] St George-Hyslop P H *et al* 1987 The genetic defect causing familial Alzheimer's disease maps on chromosome 21 *Science* **235** 885–90

[105] Jonsson T *et al* 2012 A mutation in APP protects against Alzheimer's disease and age-related cognitive decline *Nature* **488** 96–9

[106] Di Fede G *et al* 2009 A recessive mutation in the APP gene with dominant-negative effect on amyloidogenesis *Science* **323** 1473–7

[107] De Strooper B, Saftig P, Craessaerts K, Vanderstichele H, Guhde G, Annaert W, Von Figura K and Van Leuven F 1998 Deficiency of presenilin-1 inhibits the normal cleavage of amyloid precursor protein *Nature* **391** 387–90

[108] Wolfe M S, Xia W, Ostaszewski B L, Diehl T S, Kimberly W T and Selkoe D J 1999 Two transmembrane aspartates in presenilin-1 required for presenilin endoproteolysis and gamma-secretase activity *Nature* **398** 513–7

[109] Sherrington R *et al* 1995 Cloning of a gene bearing missense mutations in early onset familial Alzheimer's disease *Nature* **375** 754–60

[110] Hutton M and Hardy J 1997 The presenilins and Alzheimer's disease *Hum. Mol. Genet.* **6** 1639–46

[111] Cruts M and Van Broeckhoven C 1998 Presenilin mutations in Alzheimer's disease *Hum. Mutat.* **11** 183–90

[112] Lippa C F, Nee L E, Mori H and St George-Hyslop P 1998 Aβ-42 deposition precedes other changes in PS-1 Alzheimer's disease *Lancet* **352** 1117–8

[113] Theuns J, Del-Favero J, Dermaut B, van Duijn C M, Backhovens H, Van den Broeck M V, Serneels S, Corsmit E, Van Broeckhoven C V and Cruts M 2000 Genetic variability in the regulatory region of presenilin 1 associated with risk for Alzheimer's disease and variable expression *Hum. Mol. Genet.* **9** 325–31

[114] Cruts M, Theuns J and Van Broeckhoven C 2012 Locus-specific mutation databases for neurodegenerative brain diseases *Hum. Mutat.* **33** 1340–4

[115] Wolfe M S 2007 When loss is gain: reduced presenilin proteolytic function leads to increased Aβ42/Aβ40. Talking Point on the role of presenilin mutations in Alzheimer disease *EMBO Rep.* **8** 136–40

[116] Scheuner D *et al* 1996 Secreted amyloid beta-protein similar to that in the senile plaques of Alzheimer's disease is increased *in vivo* by the presenilin 1 and 2 and APP mutations linked to familial Alzheimer's disease *Nat. Med.* **2** 864–70

[117] Rogaev E I *et al* 1995 Familial Alzheimer's disease in kindreds with missense mutations in a gene on chromosome 1 related to the Alzheimer's disease type 3 gene *Nature* **376** 775–8

[118] Levy-Lahad E, Wijsman E M, Nemens E, Anderson L, Goddard K A, Weber J L, Bird T D and Schellenberg G D 1995 A familial Alzheimer's disease locus on chromosome 1 *Science* **269** 970–3

[119] Sherrington R *et al* 1996 Alzheimer's disease associated with mutations in presenilin 2 is rare and variably penetrant *Hum. Mol. Genet.* **5** 985–8

[120] Tandon A and Fraser P 2002 The presenilins *Genome Biol.* **3** reviews 3014

[121] Gatz M 1, Reynolds C A, Fratiglioni L, Johansson B, Mortimer J A, Berg S, Fiske A and Pedersen N L 2006 Role of genes and environments for explaining Alzheimer disease *Arch. Gen. Psychiatry* **63** 168–74

[122] Saunders A M *et al* 1993 Association of apolipoprotein E allele epsilon 4 with late-onset familial and sporadic Alzheimer's disease *Neurology* **43** 1467–72

[123] Colhoun H M, McKeigue P M and Smith G D 2003 Problems of reporting genetic associations with complex outcomes *Lancet* **361** 865–72

[124] Ioannidis J P A, Ntzani E E, Trikalinos T A and Contopoulos-Ioannidis D G 2001 Replication validity of genetic association studies *Nat. Genet.* **29** 306–9

[125] Farrer L A, Cupples L A, Haines J L, Hyman B, Kukull W A, Mayeux R, Myers R H, Pericak-Vance M A, Risch N and van Duijn C M 1997 Effects of age, sex, and ethnicity on the association between apolipoprotein E genotype and Alzheimer disease. A meta-analysis *J. Am. Med. Assoc.* **278** 1349–56

[126] Mahley R W, Weisgraber K H and Huang Y 2006 Apolipoprotein E4: a causative factor and therapeutic target in neuropathology, including Alzheimer's disease *Proc. Natl Acad. Sci. USA* **103** 5644–51

[127] Corder E H, Saunders A M, Strittmatter W J, Schmechel D E, Gaskell P C, Small G W, Roses A D, Haines J L and Pericak-Vance M A 1993 Gene dose of apolipoprotein E type 4 allele and the risk of Alzheimer's disease in late onset families *Science* **261** 921–3

[128] Deane R, Sagare A, Hamm K, Parisi M, Lane S, Finn M B, Holtzman D M and Zlokovic B V 2008 ApoE isoform-specific disruption of amyloid beta peptide clearance from mouse brain *J. Clin. Invest.* **118** 4002–13

[129] Huang Y 2006 Molecular and cellular mechanisms of apolipoprotein E4 neurotoxicity and potential therapeutic strategies *Curr. Opin. Drug Discov. Devel.* **9** 627–41

[130] Corder E H *et al* 1994 Protective effect of apolipoprotein E type 2 allele for late onset Alzheimer disease *Nat. Genet.* **7** 180–4

[131] Martins C A, Oulhaj A, de Jager C A and Williams J H 2005 Predict APOE alleles the rate of cognitive decline in Alzheimer disease: a nonlinear model *Neurology* **65** 1888–93

[132] Jones S E and Jomary C 2002 Clusterin *Int. J. Biochem. Cell Biol.* **34** 427–31

[133] Wong P, Taillefer D, Lakins J, Pineault J, Chader G and Tenniswood M 1994 Molecular characterization of human TRPM2/clusterin, a gene associated with spermmaturation, apoptosis and neurodegeneration *Eur. J. Biochem.* **221** 917–25

[134] Fritz I B, Burdzy K, Setchell B and Blaschuk O 1983 Ram rete testis fluid contains a protein (clusterin) which influences cell–cell interactions *in vitro Biol. Reprod.* **28** 1173–88

[135] Calero M, Rostagno A, Matsubara E, Zlokovic B, Frangione B and Ghiso J 2000 Apolipoprotein J (clusterin) and Alzheimer's disease *Microsc. Res. Tech.* **50** 305–15

[136] Bell R D, Sagare A P and Friedmanetal A E 2007 Transport pathways for clearance of human Alzheimer's amyloidβ-peptide and apolipoproteins E and J in the mouse central nervous system *J. Cereb. Blood Flow Metab.* **27** 909–18

[137] Liang W S *et al* 2008 Altered neuronal gene expression in brain regions deferentially affected by Alzheimer's disease: a reference data set *Physiol. Genomics* **33** 240–56

[138] Nuutinen T, Suuronen T, Kauppinen A and Salminen A 2009 Clusterin: a forgotten player in Alzheimer's disease *Brain Res. Rev.* **61** 89–104

[139] Kim W S, Weickert C S and Garner B 2008 Role of ATP-binding cassette transporters in brain lipid transport and neurological disease *J. Neurochem.* **104** 1145–66

[140] Kaminski W E, Orsó E, Diederich W, Klucken J, Drobnik W and Schmitz G 2000 Identification of a novel human sterol-sensitive ATP-binding cassette transporter (ABCA7) *Biochem. Biophys. Res. Commun.* **273** 532–8

[141] Ikeda Y *et al* 2003 Posttranscriptional regulation of human ABCA7 and its function for the apoA-I-dependent lipid release *Biochem. Biophys. Res. Commun.* **311** 313–8

[142] Karch C M, Jeng A T, Nowotny P, Cady J, Cruchaga C and Goate A M 2012 Expression of novel Alzheimer's disease risk genes in control and Alzheimer's disease brains *PLoS One* **7** e50976

[143] Vasquez J B, Fardo D W and Estus S 2013 ABCA7 expression is associated with Alzheimer's disease polymorphism and disease status *Neurosci. Lett.* **556** 58–62

[144] Kim W S, Guillemin G J, Glaros E N, Lim C K and Garner B 2006 Quantitation of ATP-binding cassette subfamily-A transporter gene expression in primary human brain cells *Neuroreport* **17** 891–6

[145] Shulman J M 1 *et al* 2013 Genetic susceptibility for Alzheimer disease neuritic plaque pathology *JAMA Neurol.* **70** 1150–7

[146] Chan S L, Kim W S, Kwok J B, Hill A F, Cappai R, Rye K A and Garner B 2008 ATP-binding cassette transporter A7 regulates processing of amyloid precursor protein *in vitro J. Neurochem.* **106** 793–804

[147] Kim W S, Li H, Ruberu K, Chan S, Elliott D A, Low J K, Cheng D, Karl T and Garner B 2013 Deletion of ABCA7 increases cerebral amyloid-beta accumulation in the J20 mouse model of Alzheimer's disease *J. Neurosci.* **33** 4387–94

[148] Wildsmith K R, Holley M, Savage J C, Skerrett R and Landreth G E 2013 Evidence for impaired amyloid beta clearance in Alzheimer's disease *Alzheimers Res. Ther.* **5** 33

[149] De Roeck A *et al* 2018 An intronic VNTR affects splicing of ABCA7 and increases risk of Alzheimer's disease *Acta Neuropathol.* **135** 827–37

[150] Karch C M and Goate A M 2015 Alzheimer's disease risk genes and mechanisms of disease pathogenesis *Biol. Psychiatry* **77** 43–51

[151] Holtzman D M, Morris J C and Goate A M 2011 Alzheimer's disease: the challenge of the second century *Sci. Transl. Med.* **3** 77sr71

[152] Lambert J C *et al* 2013 Meta-analysis of 74,046 individuals identifies 11 new susceptibility loci for Alzheimer's disease *Nat. Genet.* **45** 1452–8

[153] Bertram L *et al* 2008 Genome-wide association analysis reveals putative Alzheimer's disease susceptibility loci in addition to APOE *Am. J. Hum. Genet.* **83** 623–32

[154] Lambert J C *et al* 2009 Genome-wide association study identifies variants at CLU and CR1 associated with Alzheimer's disease *Nat. Genet.* **41** 1094–9

[155] Guerreiro R 1 *et al* Alzheimer Genetic Analysis Group 2013 TREM2 variants in Alzheimer's disease *N. Engl. J. Med.* **368** 117–27

[156] Jonsson T *et al* 2013 Variant of TREM2 associated with the risk of Alzheimer's disease *N. Engl. J. Med.* **368** 107–16

[157] Harold D *et al* 2009 Genome-wide association study identifies variants at CLU and PICALM associated with Alzheimer's disease *Nat. Genet.* **41** 1088–93

IOP Publishing

Neurological Disorders and Imaging Physics, Volume 3
Application to autism spectrum disorders and Alzheimer's
Ayman El-Baz and Jasjit S Suri

Chapter 6

Nuclear neurology of autism spectrum disorder

Michelle Hartley-McAndrew, Robert Miletich and Osman Farooq

Autism spectrum disorder (ASD) is a neurodevelopmental disorder with multi-factorial etiology and varied presentation. It is characterized by impairments in social interaction and communication with a restricted range of interests and repetitive behaviors. Individuals with ASD tend to have altered neurotransmitter functions, including dopamine, GABA, serotonin, etc. The nuclear medicine techniques of positron emission tomography (PET) and single-photon emission computed tomography (SPECT) have shown unique findings in ASD. Although nuclear imaging is currently not indicated in the evaluation of ASD, characteristic imaging patterns seen on such imaging modalities can provide corroborative information and increase the diagnostic yield. Molecular imaging can therefore provide the preliminary data for promising therapeutic interventions. Herein, we describe the differences in neurochemical and molecular features in individuals with ASD.

6.1 Introduction

Autism spectrum disorder (ASD) has a constellation and spectrum of symptoms, signs, history, and course. It is a disorder characterized by impairments in social interaction and communication along with a restricted range of interests and repetitive behaviors [1]. The terms autism, pervasive developmental disorder and autism spectrum disorder have been used synonymously to describe these features. The fact that there have been changes in diagnostic criteria indicates the hetero-geneity of this disorder, making abnormalities in patterns of neuronal networking a challenge to elucidate. It is likely that multiple genetic, polygenic, and environmental influences play roles in the pathogenesis of ASD. Therefore, there are probably multiple etiologies for autism. The objective diagnosis of ASD can be a challenge and is based on history and clinical observation with the assistance of objective measures such as the Autism Diagnostic Observation Schedule (ADOS), or the Autism Diagnostic Interview-Revised (ADI-R). These are structured

doi:10.1088/978-0-7503-1793-1ch6

instruments that require training in order to provide valid and reliable administration and measure. Early establishment of the diagnosis of ASD is optimal for the best outcome. Numerous studies have been performed that explore the genetic, anatomic, electrophysiological, and pathological features of ASD, aiming to clarify etiology and possible avenues of treatment. The nuclear neurology techniques of positron emission tomography (PET) and single-photon emission computed tomography (SPECT) have shown unique findings in autism spectrum disorder.

6.1.1 Imaging modalities

Nuclear neurology studies the physiology of the central nervous system using *in vivo* imaging. There are two classes of physiology measured: basal physiology and neurochemical specific physiology. Both types of physiology can be examined with both PET and SPECT. Basal physiology refers to tissue properties, such as regional glucose metabolism or cerebral perfusion or blood flow. Specific neurochemical physiology relates to identified entities, such as neurotransmitters, enzymes, transporters, etc. Nuclear neurology typically involves the intravenous administration of a drug which has had a radioisotope incorporated into it. Although ultimately both detect gamma radiation, PET and SPECT involve two different kinds of radiation emission and thus must be specifically designed to detect that radiation. PET radioisotopes emit positrons. Positrons are anti-electrons. Those emitted positrons travel a short distance before colliding with an electron, with the pair annihilating each other, transforming into two high-energy photons traveling at 180 degrees, i.e. opposite, to each other. PET cameras are typically designed for coincidence detection of these two photons on opposite sides of the camera gantry opening (where the patient is lying). SPECT, on the other hand, measures single photons. Once the geometric grid of emitted radioactivity is captured, and various correction factors are incorporated, the camera software can reconstruct images of the spatially determined radioactivity. These images can be projections, i.e. whole organ views, or tomographs, i.e. slices through the organ. The distribution of the radioactivity in the body is based on the tracer principle, wherein low concentrations of radioactive drug trace a physiologic process. Low concentration is defined relative to the endogenous molecule or process that the drug traces or follows. Measuring the radioactivity provides a measure of the process or molecule.

The image reconstruction corrections available for PET are physically and mathematically precise, allowing high fidelity to the actual tissue radioactivity. PET typically involves radioisotopes of high photon flux and short half-life. High count rates dramatically improve image quality, while short half-lives diminish the radiation dose to the patient to the range typical for all nuclear medicine procedures. Typical radioisotopes used in PET are ^{11}C, ^{18}F, ^{13}N, and ^{15}O. These are the same atoms from which the molecules of life are made (F acts as a hydroxyl moiety analog). Thus, PET can conceivably measure anything in the body if you can tag the appropriate molecule. Typical SPECT radioisotopes include ^{99m}Tc, ^{201}Tl, ^{123}I, ^{131}I, ^{111}In, and ^{67}Ga. Many of these are not found in biological systems and are bulky atoms requiring different radiochemistry in order to get them into a radiotracer of interest. The rate of emission is lower and their half-lives are longer, resulting in a radiation dose to the patient similar to that of PET. The image quality is typically inferior to PET, but the remarkable aspect of SPECT is that it can be performed on practically all standard nuclear medicine cameras resulting in a much wider potential availability than PET.

6.2 Specific neurochemical physiology

Neurotransmitter system function can be measured through assays of neurotransmitter concentrations, its metabolite concentrations, its synthesis rates, or the amount of tracer binding to its receptors or transporters. These are dynamic systems and unfortunately, measuring only one dimension of the system provides only a very limited view of the system's overall function. This is further complicated by the fact that changes in the amount of binding of a tracer to a receptor or transporter may be secondary to changes in the amount of that binding protein or to changes in the amount of endogenous neurotransmitter competitively binding to that protein. The influence of endogenous neurotransmitter on measured binding is tracer specific and has not been worked out for all tracers mentioned in this chapter. As such, we present the literature results merely as changes in the amount of binding or binding protein.

6.2.1 Dopaminergic neurotransmission

Dopamine is a neurotransmitter that is involved in multiple processes including social reward and social motivation. A limited number of studies have investigated dopamine *in vivo* in individuals with ASD to determine the baseline dopamine transporter, dopamine synthesis, or effects of treatment on the dopamine system. In 1997, Ernst studied 14 children with autism and 10 healthy children with ^{18}F-labeled fluorodopa (FDOPA) using PET and found a 39% reduction in the anterior medial prefrontal cortex/occipital cortex ratio in the group with autism, but there were no significant differences in any other region measured. The authors postulate that decreased dopaminergic function in the prefrontal cortex in individuals with autism may contribute to the cognitive impairments seen [2]. In 2005, the dopamine transporter was studied in children with autism using tracer Tc 99m TRODAT-1 (a tropane derivative) imaged by SPECT. They reported a whole brain increase in dopamine transporter binding in the autism group, whereas the striatum/cerebellum ratio showed no differences between the groups [3]. In one study by Makkonen *et al*, SPECT with *N*-(2-fluoroethyl)-2beta-carbomethoxy-3beta(40[^{123}I]iodophenyl)-nortropane([^{123}I]nor-b-CIT), which is a specific, non-selective radiotracer for the dopamine transporter (DAT) and the serotonin transporter (SERT), was administered to 15 children ages 5–16 with ASD and 10 age-matched control participants. They found that there was no difference between uptake in the striatum of participants with ASD compared to those without ASD [4].

In another study with 20 high functioning individuals with ASD ages 18–26, Nakamura measured dopamine transporter binding in adults with autism using PET [^{11}C]-labeled 2 beta-carbomethyoxy-3beta-(4-fluorophenyl) tropane ([^{11}C]WIN35,428).

When compared to age and IQ matched controls, they found that dopamine transporter binding was significantly higher in the orbital frontal cortex in the autism group and was inversely correlated with binding of the serotonin transporter (SERT) [5]. These studies support the presence of altered dopaminergic function in the frontal cortical regions but not in the striatum in individuals with autism [6].

6.2.2 Serotoninergic neurotransmission

Dysfunction in serotonin, 5-hydroxytryptamine (5-HT) metabolism, was evidenced with increased levels in whole blood measured in 23 children with ASD and severe intellectual disability [7]. In the same study, a lesser elevation of whole-blood serotonin was observed in seven individuals with intellectual disability without ASD. Comparatively, four typically developing children without intellectual disability and without ASD, as well as 12 children with borderline or mild intellectual disability, showed normal concentrations of serotonin. The same study indicated that 12 children with ASD showed higher urinary concentrations of 5-hydroxyindole-acetic acid (5-HIAA), a urinary metabolite of serotonin, than 6 children with mild intellectual disability and without ASD. Other studies have shown elevated whole-blood serotonin in young adults with ASD [8, 9]. Studies also indicate that parents and siblings of individuals with ASD have elevated levels of serotonin. This finding led to further studies of serotonin metabolism in the brain, as follows [10].

6.2.3 Serotonin synthesis

The radiotracer alpha[^{11}C]methyl-L-tryptophan ([^{11}C]AMT) has been used to measure serotonin synthesis. Typically developing children demonstrate high serotonin synthesis, decreasing to adult levels at around 5 years of age. However, children with ASD show reduced serotonin synthesis. In children with ASD, the rate of serotonin synthesis increased between 2 and 15 years of age. Studies of adults with ASD indicate 1.5 times the synthesis of serotonin compared to the levels of typical adults [11, 12].

In a study using [^{11}C]AMT with PET scanning, in eight children with ASD (7 male, 1 female), abnormalities in serotonin metabolism were observed in the seven males, but not in the female. Five of the seven males showed reduced uptake of [^{11}C] AMT in the left frontal cortex and the left thalamus along with increased [^{11}C]AMT in the right dentate nucleus of the cerebellum. The other two males demonstrated the opposite pattern. They showed reduced uptake of [^{11}C]AMT in the right frontal cortex, the right thalamus, and increased uptake in the left dentate nucleus of the cerebellum [13]. These findings provide evidence for abnormalities in the serotonin neurotransmission through the dentatothalamocortical pathway. The abnormalities in the frontal lobe provide anatomical evidence for challenges with social behavior and communication in males with ASD. The frontal lobe abnormalities indicate the basis for difficulties with executive functioning while the abnormalities in the dentate nucleus of the cerebellum provide reasons for difficulties with coordination, as exhibited by many males with ASD [13].

In a study by Chandana in 2005, a sample of 117 children with ASD showed the absence of cortical symmetry of [^{11}C]AMT accumulation in 55% of children with ASD on PET scanning. In 31 children with reduced left cortex accumulation, this is correlated with severe language impairment while the children with reduced right cortical accumulation exhibited more left and mixed handedness [14].

6.2.4 Serotonin transporter (SERT)

Serotonin plays a regulatory role in brain development, modulating processes such as neuronal migration and synaptogenesis [15]. There are multiple lines of evidence including elevation in whole blood levels of serotonin in individuals with ASD. Several PET/SPECT studies have focused on this neurotransmitter, making it the most common neurotransmitter system investigated *in vivo* in ASD [15]. In a study of 15 children with ASD between 5–16 years of age who were administered *N*-(2-fluoroethyl)-2beta-carbomethoxy-3beta-(40[^{123}I] iodophenyl)-nortropane ([^{123}I] nor-β-CIT) for SPECT scanning, found significantly lower SERT binding in the medial frontal cortex, midbrain, and both the left and right temporal lobe than in ten age-matched controls [4]. In a study of 20 men between 18–26 years of age with high functioning ASD, abnormal SERT metabolism was again supported using [^{11}C](+) McN5652 PET scanning, which found that a reduction of SERT density in the anterior and posterior cingulate cortices correlated with impairments of social cognition. In addition, among men with ASD, repetitive and obsessive behaviors and interests correlated with SERT binding in the thalamus. In individuals with ASD, SERT binding in the orbitofrontal cortex was inversely correlated with binding of the dopamine transporter [5].

6.2.5 Serotonin (5-HT 2A) receptor

In a study of 17 individuals with Asperger's syndrome and 17 typical individuals without ASD using [^{11}C]MDL100907, which is a radioligand selective for serotonin receptors, showed no significant differences in SERT binding in any regions of interest [16]. However [^{18}F] setoperone, a reliable radioligand that is highly specific and selective for serotonin receptors in the frontal cortex, PET demonstrated less thalamic binding in six adults with high functioning ASD compared to typical adults. Cortical 5-HT2 receptors (using [^{18}F] setoperone) were shown to be reduced in the brains of 19 parents of families with ASD compared to 17 typical adults. The parents also showed an inverse correlation between platelet 5-HT2 receptor levels and cortical 5-HT2 levels [17]. A study in 2012 by Beversdorf using [^{18}F] setoperone PET showed that there was significantly lower 5-HT2 binding in the thalamus in adults with autism but did not reach significance in other brain regions. It appeared that in these adults with ASD, communication problems were correlated with reduced thalamic 5-HT2 binding [18].

In essence, PET has provided evidence that typically developing children synthesize serotonin at twice the rate of adults and gradually decrease to adult rates between 2–15 years of age. However, children with ASD increase serotonin synthesis gradually to 1.5 the rate of typical adults. Also, SERT is abnormal in ASD. There is reduced SERT binding throughout the brain in adult men with ASD. This seems to correlate with impaired social cognition when reductions of SERT are in the cingulate cortex. There is increased orbitofrontal DAT that is inversely correlated with SERT and reductions in thalamic serotonin receptor binding associated with communication difficulties in adults with ASD.

6.2.6 GABA

Gamma-aminobutyric acid (GABA) plays a central role in synaptic pruning and early developmental processes. Evidence that suggests a potential link between GABA dysfunction and ASD includes data from genetic studies, postmortem studies, as well as animal models [15]. One pilot PET study of three individuals with ASD and three matched controls and one SPECT study have been performed to investigate GABA in ASD. Both studies showed a reduction in GABAa receptor in brain regions including those associated with reward, emotional, and social processing [19, 20]. Mori compared nine children with ASD (Asperger's) to age-matched controls with partial epilepsy. When comparing the ASD group to the focal epilepsy group, there were significant decreases in binding in the left superior frontal gyrus, the right superior frontal gyrus, and the left medial frontal gyrus. When dividing the autism group based on intellectual impairment, only the group with intellectual impairment showed significantly decreased binding. Further, they examined the ASD group based on the presence or absence of focal epileptiform discharges and found that both groups had decreased frontal lobe binding, although decreases were larger in the group with focal epileptiform discharges [19]. In the pilot PET study, Mendez used the tracer [^{11}C]Ro15-4513 that binds to the alpha1 and alpha5 subunits of the GABAa receptor. In the ASD group, 83 different brain regions were studied and significant reductions in the right and left nucleus accumbens and right and left subcallosal areas were found [20].

6.3 Basal physiology

6.3.1 Protein synthesis

Protein synthesis is an integral step for regulating diverse cellular functions. Therefore, determining how a cell allocates its synthesis capacity for each protein provides foundational information for systems biology. In 2011, Shandal examined the cerebral protein synthesis rate in the language regions of children with developmental delay with and without pervasive developmental disorder. Using L-[1-^{11}C]-Leucine PET, children with pervasive developmental disorder with developmental delay were compared to children with developmental delay. They found that there was a higher protein synthesis rate in children with pervasive developmental disorder in the left posterior middle temporal region. Also, there was significant asymmetric protein synthesis (right greater than left) observed in the developmental delay group without pervasive developmental disorder in the middle frontal and posterior middle temporal regions [21]. This may indicate that abnormal language area protein synthesis in children with developmental delays is related to pervasive developmental disorder.

6.3.2 Glucose metabolism

The brain utilizes glucose as a primary energy source. Glucose metabolism has been investigated in individuals with ASD under different states, including awake, asleep, and anesthetized. The radioligand 2-[^{18}F] fluoro-2-deoxy-D-glucose ([^{18}F]FDG) is used in the measurement of glucose metabolism. In most situations, there is a nearly stoichiometric relationship between cerebral glucose metabolism and synaptic activity. Glucose metabolism represents a summation measure of multiple variables, including neuron number, synapse number, and magnitude of unit work of those elements [55].

Studies comparing brain glucose metabolism in individuals with autism and co-morbid conditions such as seizure, infantile spasm, tuberous sclerosis, etc, are not included in this discussion. As far back as 1985, it was used in PET studies and demonstrated widespread hypermetabolism of glucose in ten individuals with ASD compared to individuals without ASD [22]. Further studies, however, have not been consistent with these initial findings. Some studies have found no significant metabolic differences [23, 24] and others have found hypometabolism [25]. There have been technological advancements over time and different studies utilize different measurement techniques. As such, changes in global glucose metabolism cannot be entirely discounted [6].

Focal abnormalities of glucose metabolism have been reported in several other studies in individuals with autism in which global brain glucose metabolism was not addressed. In one study that examined adults 19–36 years in age with age-matched controls, Heh *et al* examined glucose metabolism in the cerebellum and demonstrated no significant differences in mean glucose metabolic rates for cerebellar hemispheres or vermal lobes VI and VII in autistic individuals [26]. Cerebral glucose metabolic rate studies of adults have found decreased glucose metabolism in the whole parietal lobe, the frontal premotor/supplementary motor and eye-field areas, rostral orbito-frontal cortex, and amygdala, and increased glucose metabolism in the posterior cingulate and occipital visual cortices, orbitofrontal area 12, medial temporal cortex, hippocampus, and basal ganglia in individuals with ASD compared to typical subjects without ASD [27]. That study did not find any metabolic differences in the cerebellum. During memory tasks, atypical glucose metabolism was observed in the cingulate, occipital, and parietal cortices in individuals with ASD [28]. In a study using [^{18}F] FDG PET, involving ten individuals with ASD ages 11–19 and 15 individuals without ASD ages 8–18, showed abnormalities in four individuals with ASD, and none of the individuals without ASD. These abnormalities included bilateral cerebellar hypometabolism in two, bilateral anterior temporal cortical hypometabolism and bilateral cerebellar hypometabolism in one, and bilateral frontal cortical hypermetabolism with bilateral medial occipital hypermetabolism in one [29]. Hybrid PET/CT was used in the evaluation of glucose metabolism in 23 individuals with ASD using 259 MBq (7 mCi) [^{18}F] FDG and 15 individuals with ASD using 5.55 MBq (0.15 mCi) [^{18}F] FDG. It was found that there was hypometabolism in the temporal lobes in both groups. The findings were, however, inconsistent in other regions [30].

Recently Sharma *et al* performed PET/CT scans with $[^{18}F]FDG$ and investigated age related developmental changes on brain metabolism. They found that individuals with autism below 5 years of age had evidence of greater metabolism in all the regions of the brain except the hippocampus compared to healthy control data, where the hippocampal area showed lower metabolic uptake. In the 5–10 year age group, individuals with autism showed lower metabolic uptake in all brain regions compared to controls. In the 10–15 year age group, the autism group also showed lower metabolic uptake in all regions compared to healthy control data with the exception of the insula metabolism which was slightly higher than healthy controls. Overall, this study seemed to support the presence of a linear decrease in glucose metabolism in the brain of an autistic individual as they grow older [31].

6.4 Regional cerebral blood flow

6.4.1 PET investigation

Similar to glucose metabolism, there is a coupling of regional synaptic activity and regional cerebral perfusion. Hence, measuring blood flow gives an indication of neuronal function [55]. In contrast to imaging glucose metabolism, imaging blood flow shows two phenomena simultaneously: neuronal function and the adequacy of the vascular plumbing, in other words, the presence of any arterial tree restriction of blood delivery. Interpreting blood flow imaging requires distinguishing these two phenomena.

PET has been used to measure cerebral blood flow and may be measured using $[^{15}O]$ CO_2. One initial study demonstrated no differences in six males with ASD compared to eight males without ASD in a study using PET techniques [24]. Since then studies have supported the evidence of distinctive cerebral hypoperfusion in individuals with ASD. Many studies using PET have been conducted during sleep, rest, and socio-cognitive tasks such as emotion processing [32], theory of mind [33], and language [34]. Therefore, it is difficult to compare data across studies and across age groups due to differences in study design. In 1995, Zilbovicius examined five children diagnosed with ASD at ages 3–4 years and again 3 years later. Initially they found frontal hypoperfusion in the children, but it had normalized by 6–7 years of age [35]. Later, in 2000, Zilbovicius measured regional cerebral blood flow using $[^{15}O]H_2O(22)$ in 21 children diagnosed with ASD of ages 8–13 years and compared them to ten individuals without ASD but with intellectual disability, and found significant hypoperfusion in both temporal lobes centered in the associative auditory and adjacent multimodal cortex in children with ASD [36]. Zilbovicius suggests that this dysfunction of the auditory cortex may explain why children with autism often have the appearance of being deaf and often have severe communication impairments. Dysfunction of the superior temporal sulcus may explain the emotional and cognitive components of autism since this multimodal association region is connected with frontoparietal and limbic regions.

Pagani in 2012 showed with PET that 13 adults with ASD and normal intelligence aged 20–48 years, compared to ten IQ and gender matched controls, had significant cerebral blood flow increases in the right para-hippocampal, posterior cingulate, primarily visual and temporal cortex, putamen, caudatus, substantia nigra, and cerebellum compared to controls. No correlation between cerebral blood flow and IQ was found [37].

6.4.2 SPECT investigation

Many studies involving subjects with autism measuring cerebral blood flow have been performed using SPECT and demonstrate a variety of global and focal abnormalities [38–40]. In 1992, George *et al* found global hypoperfusion in the resting state in adult males with autism compared to controls. In particular, hypoperfusion was found in the frontotemporal cortices [41]. Zilbovicius, in 1992, studied regional cerebral blood flow with SPECT. When the results from the group with primary autism were compared to an age-matched group of nonautistic children with slight to moderate language disorders ($N = 14$), no cortical regional abnormalities were found. However, methodological limitations such as sedation of the study group and not the control group were cited [42]. Another study in 1995 localized hypoperfusion to the vermis and right cerebellar hemisphere [43]. In a study by Starkstein using SPECT scanning, 30 patients diagnosed with ASD ages 4–18 and 14 patients with intellectual disability but without ASD ages 7–15 demonstrated that individuals with autism have significantly lower cerebral blood perfusion in the right temporal lobe in the basal and inferior areas, occipital lobes, thalami, and left basal ganglia [44].

Using SPECT scanning, Kaya studied 18 individuals with ASD compared to 11 controls without ASD and demonstrated cerebral hypoperfusion in the bilateral frontal, frontotemporal, temporal, and temporo-occipital regions [45]. Wilcox in a study in 2002 explored cerebral blood flow using SPECT scanning in 14 patients with ASD ages 3–37 years and found significant hypoperfusion in the prefrontal areas of individuals with ASD compared to controls [46]. In a further SPECT study involving ten children with ASD and intellectual disability ages 4–8 years, generalized hypoperfusion was demonstrated in all ten cases as compared to age-matched controls [47]. Frontal and prefrontal regions revealed maximum hypo-perfusion. Subcortical areas also indicated hypoperfusion. In a study involving high functioning individuals with ASD, Ito demonstrated hypoperfusion in the left temporal region compared to controls [48].

In a study of ten children with ASD, Degirmenci found significant hypoperfusion in the right inferior and superior frontal, left superior frontal, right parietal, right mesial temporal, and right caudate nucleus compared to five age-matched controls [49]. In 2011, a study with SPECT scanning showed that 23 children with ASD had significant reduction in bilateral frontal lobe and bilateral basal ganglia cerebral blood flow. In a lesser affected Asperger's syndrome group, a decrease in the bilateral frontal, temporal, and parietal lobes as well as the cerebellum was seen but with pronounced asymmetry in the hemispheric hypoperfusion compared to the more severely affected group [50].

Studies have revealed that regional cerebral blood flow is decreased in ASD and seems to be related to symptomatology. Using SPECT scans in individuals with

ASD, a correlation between syndrome scores and regional cerebral blood flow was observed. Ohnishi reported that each syndrome was associated with a particular pattern of perfusion in the limbic system and in the medial prefrontal cortex [51]. Perfusion abnormalities seemed to be related to cognitive dysfunction in ASD, such as abnormal responses to sensory stimuli and obsessive desire for sameness. The underlying abnormal behavioral patterns seem to correlate with the location and level of perfusion patterns. Cerebral blood flow in the thalamus in individuals with ASD showed associated repetitive, self-stimulatory behaviors and unusual sensory interest. A decreased IQ in individuals with ASD showed hypoperfusion of the temporal and frontal lobes [52]. Differences in processing facial expression and emotion correlated with decreased blood flow to the temporal lobes and amygdala, and difficulties recognizing familiar faces correlated with a decrease in blood flow to the fusiform gyrus [53, 54]. In addition, a decrease in cerebral blood flow to Wernicke's and Brodmann's areas correlated to impairments in language development and auditory processing [39, 46].

6.5 Conclusion

There has been significant variability observed among the characteristics of individuals with ASD. This may be, in part, due to the differences in clinical findings and etiologies among individuals on the autism spectrum. The old adage stating 'when you have met one individual with autism, you have met one individual with autism' comes into play. This implies that each individual has their own levels of strengths and challenges along the spectrum of social, communicative, and behavioral features. In addition, the studies described above have differences in techniques that are employed to assess the individuals, including differences in scanners, radiotracers, and analytic methods. Other sources of variability include differences in sample age, gender, ethnic background, social status, exposure to environmental substances, and potential toxins.

The overall evidence seems to support the existence of altered neurotransmitter functions in individuals with ASD. Decreased dopaminergic function in the frontal cortices and decreased GABA function in brain regions, including those associated with reward, emotional, and social processing, are examples. PET gives evidence that young children synthesize serotonin at twice the rate of adults and attain the serotonin synthesis rates of adults by their adolescent years, and that children with ASD develop a rate of serotonin synthesis 1.5 times the rate of typical adults. Dysfunction of serotonergic neurotransmission has consistently been observed in ASD. There are decreased levels in SERT in children, adolescents, and adults with ASD. Reduced levels of SERT in the cingulate correlate with impaired social cognition. Communication difficulties are seen with reduced serotonin receptor levels in the thalamus. Decreased glucose metabolism along with an overall decrease in regional cerebral blood flow has often been reported in the brains of individuals with ASD. In addition, it appears that regional cerebral blood flow is decreased in ASD and seems to be related to symptomatology.

The studies described herein explore the differences in neurochemical and molecular features in individuals with ASD. Further research that utilizes nuclear neurology measures comprising of multi-center studies with more uniform procedures are warranted to elucidate potential biomarkers and pharmacological interventions for this population. Future directions will be aided by advancements in identifying the genetic causes of ASD, therefore, decreasing the heterogeneity in samples. Clarifying the characteristics pertaining to the features of ASD provide potential avenues for arresting pathophysiological deterioration and improvement of symptomatology.

References

[1] American Psychiatric Association 2013 *Diagnostic and Statistical Manual of Mental Disorders (DSM-5)* 5th edn (Arlington, VA: American Psychiatric Publishing)

[2] Ernst M, Zametkin A, Matochik J A, Pascualvaca D and Cohen R M 1997 Low medial prefrontal dopaminergic activity in autistic children *Lancet* **350** 638–70

[3] Xiao-Mian S, Jing Y, Chongxuna Z, Min L and Hui-Xing D 2005 Study of 99mTc-TRODAT-1 imaging on human brain with children autism by single photon emission computer tomography *Conf. Proc. IEEE Eng. Med. Biol. Soc.* **5** 5328–30

[4] Makkonen I, Riikonen R, Kokki H, Airaksinen M M and Kuikka J T 2008 Serotonin and dopamine transporter binding in children with autism determined by SPECT *Dev. Med. Child Neurol.* **50** 593–7

[5] Nakamura K *et al* 2010 Brain serotonin and dopamine transporter bindings in adults with high-functioning autism *Arch. Gen. Psychiatry* **67** 59–68

[6] Chugani D 2014 Application of PET and SPECT to the study of autism spectrum disorders *PET and SPECT in Psychiatry* (New York: Springer) pp 691–707

[7] Schain R J and Freedman D X 1961 Studies on 5-hydroxyindole metabolism in autistic and other mentally retarded children *J. Pediatr.* **58** 315–20

[8] McBride P A *et al* 1989 Serotonergic responsivity in male young adults with autistic disorder: results of a pilot study *Arch. Gen. Psychiatry* **46** 213–21

[9] Brasic J R 2017 *Autism. Medscape Drugs and Diseases* 21 April http://emedicine.medscape.com/article/912781-overview (Accessed 9 September 2018)

[10] Chugani D C 2012 Neuroimaging and neurochemistry of autism *Pediatr. Clin. N. Am.* **59** 63–73

[11] Brasic J R and Mohamed M 2014 Human brain imaging of autism spectrum disorders *Imaging of Human Brain in Health and Disease* ed P Seeman and B Madras (Oxford: Academic, Elsevier) pp 373–406

[12] Chugani D C *et al* 1999 Developmental changes in brain serotonin synthesis capacity in autistic and non autistic children *Ann. Neurol.* **45** 287–95

[13] Brasic J R and Wong D F 2015 *PET scanning in autism spectrum disorders. Medscape Drugs & Diseases* 14 Oct http://emedicine.medscape.com/article/1155568-overview (Accessed: 1 September 2018)

[14] Chandana S R *et al* 2005 Significance of abnormalities in developmental trajectory and asymmetry of cortical serotonin synthesis in autism *Int. J. Dev. Neurosci.* **23** 171–82

[15] Zurcher N R, Bhanot A, McDougle C J and Hooker J M 2015 A systematic review of molecular imaging (PET and SPECT) in autism spectrum disorder: current state and future research opportunities *Neurosci. Biobehav. Rev.* **52** 56–73

[16] Girgis R R *et al* 2011 The 5-HT2A receptor and serotonin transporter in Asperger's disorder: a PET study with [^{11}C]MDL 100907 and [^{11}C]DASB *Psychiatry Res.* **194** 230–4

[17] Goldberg J *et al* 2009 Cortical serotonin type-2 receptor density in parents of children with autism spectrum disorders *J. Autism Dev. Disord.* **39** 97–104

[18] Beversdorf D Q *et al* 2012 5-HT2 receptor distribution shown by [18F] setoperone PET in high-functioning autistic adults *J. Neuropsychiatry Clin. Neurosci.* **24** 191–7

[19] Mori T *et al* 2012 Evaluation of the GABAergic nervous system in autistic brain: (123) I-iomazenil SPECT study *Brain Dev.* **34** 648–54

[20] Mendez M A *et al* 2013 The brain GABA-benzodiazepine receptor alpha-5 subtype in autism spectrum disorder; a pilot [(11)C]Ro 15-4513 positron emission tomography study *Neuropharmacology* **68** 195–201

[21] Shandal V, Sundaram S K, Chugani D C, Kumar A, Behen M E and Chugani H T 2011 Abnormal brain protein synthesis in language areas of children with pervasive developmental disorder: a L-[1-11 C]-leucine PET study *J. Child Neurol.* **26** 1347–54

[22] Rumsey R M *et al* 1985 Brain metabolism in autism *Arch. Gen. Psychiatry* **42** 448–55

[23] De Volder A, Bol A, Michel C, Cogneau M and Goffinet A M 1987 Brain glucose metabolism in children with the autistic syndrome: positron tomography analysis *Brain Dev.* **9** 581–7

[24] Herold S, Frackowiak R S, Le Couteur A, Rutter M and Howlin P 1988 Cerebral blood flow and metabolism of oxygen and glucose in young autistic adults *Psychol. Med.* **18** 823–31

[25] Hwang B J, Mohamed M A and Brasic J B 2017 Molecular imaging of autism spectrum disorder *Int. Rev. Psychiatry* **29** 530–54

[26] Heh C *et al* 1989 Positron emission tomography of the cerebellum in autism *Am. J. Psychiatry* **146** 242–5

[27] Mitelman S A *et al* 2018 Positron emission tomography assessment of cerebral glucose metabolic rates in autism spectrum disorder and schizophrenia *Brain Imaging Behav.* **12** 532–46

[28] Hazlett E A *et al* 2004 Regional glucose metabolism within cortical Brodmann areas in healthy individuals and autistic patients *Neuropsychobiology* **49** 115–25

[29] Anil Kumar B N, Malhotra S, Bhattacharya A, Grover S and Batra Y K 2017 Regional cerebral glucose metabolism and its association with phenotype and cognitive functioning in patients with autism *Indian J. Psychol. Med.* **39** 262–70

[30] LeBlanc H 2017 Brain abnormality findings in F18-FDG PET/CT imaging and its role in the clinical diagnosis of autism *J. Nucl. Med.* **58** 828

[31] Sharma A, Gokulchandran N, Sane H, Nivins S, Paranjape A and Badhe P 2018 The baseline pattern and age related developmental metabolic changes in the brain of children with autism as measured on positron emission tomography/computer tomography scan *World J. Nucl. Med.* **17** 94–101

[32] Hall G B, Szechtman H and Nahmias C 2003 Enhanced salience and emotion recognition in autism: a PET study *Am. J. Psychiatry* **160** 1439–41

[33] Happe F *et al* 1996 Theory of mind in the brain: evidence from a PET scan study of Asperger syndrome *Neuroreport* **8** 197–201

[34] Muller R A *et al* 1998 Impairment of dentate-thalamo-cortical pathway in autistic men: language activation data from positron emission tomography *Neurosci. Lett.* **245** 1–4

[35] Zilbovicius M *et al* 1995 Delayed maturation of the frontal cortex in childhood autism *Am. J. Psychiatry* **152** 248–52

[36] Zilbovicius M *et al* 2000 Temporal lobe dysfunction in childhood autism: a PET study *Am. J. Psychiatry* **157** 1988–93

[37] Pagani M *et al* 2012 Brief report: alterations in cerebral blood flow as assessed by PET/CT in adults with autism spectrum disorder with normal IQ *J. Autism Dev. Disord.* **42** 313–8

[38] Bjørklund G *et al* 2018 Cerebral hypoperfusion in autism spectrum disorder *Acta Neurobiol. Exp.* **78** 21–9

[39] Boddaert N and Zilbovicius M 2002 Functional neuroimaging and childhood autism *Pediatr. Radiol.* **32** 1–7

[40] Galuska L *et al* 2002 PET and SPECT scans in autistic children *Orv. Hetil.* **143** 1302–4

[41] George M, Costa D, Kouris K, Ring H A and Ell P J 1992 Cerebral blood flow abnormalities in adults with infantile autism *J. Nerv. Ment. Dis.* **180** 413–7

[42] Zilbovicius M *et al* 1992 Regional cerebral blood flow in childhood autism: a SPECT study *Am. J. Psychiatry* **149** 924–30

[43] McKelvey J R, Lambert R, Mottron L and Shevell M 1995 Right-hemisphere dysfunction in Asperger's syndrome *J. Child Neurol.* **10** 310–4

[44] Starkstein S E *et al* 2000 SPECT findings in mentally retarded autistic individuals *J. Neuropsychiatry Clin. Neurosci.* **12** 370–5

[45] Kaya M *et al* 2002 The relationship between 99mTc-HMPAO brain SPECT and the scores of real life rating scale in autistic children *Brain Dev.* **24** 77–81

[46] Wilcox J, Tsuang M T, Ledger E, Algeo J and Schnurr T 2002 Brain perfusion in autism varies with age *Neuropsychobiology* **46** 13–6

[47] Gupta S K and Ratnam B V 2009 Cerebral perfusion abnormalities in children with autism and mental retardation: a segmental quantitative SPECT study *Indian Pediatr.* **46** 161–4

[48] Ito H *et al* 2005 Findings of brain 99mTc-ECD SPECT in high-functioning autism— 3-dimensional stereotactic ROI template analysis of brain SPECT *J. Med. Invest.* **52** 49–56

[49] Degirmenci B *et al* 2008 Technetium-99m HMPAO brain SPECT in autistic children and their families *Psychiatry Res.* **162** 236–43

[50] Yang W H *et al* 2011 Regional cerebral blood flow in children with autism spectrum disorders: a quantitative ^{99}mTc-ECD brain SPECT study with statistical parametric mapping evaluation *Chin. Med. J.* **124** 1362–6

[51] Ohnishi T *et al* 2000 Abnormal regional cerebral blood flow in childhood autism *Brain* **123** 1838–44

[52] Hashimoto T, Sasaki M, Fukumizu M, Hanaoka S, Sugai K and Matsuda H 2000 Single-photon emission computed tomography of the brain in autism: effect of the developmental level *Pediatr. Neurol.* **23** 416–20

[53] Critchley H D *et al* 2000 The functional neuroanatomy of social behaviour: changes in cerebral blood flow when people with autistic disorder process facial expressions *Brain* **123** 2203–12

[54] Pierce K, Haist F, Sedaghat F and Courchesne E 2004 The brain response to personally familiar faces in autism: findings of fusiform activity and beyond *Brain* **127** 2703–16

[55] Clarke D D and Sokoloff L 1999 Circulation and energy metabolism of the brain *Basic Neurochemistry: Molecular, Cellular, and Medical Aspects* ed G J Siegel (Philadelphia, PA: Lippincott, Williams and Wilkins) pp 637–69

IOP Publishing

Neurological Disorders and Imaging Physics, Volume 3
Application to autism spectrum disorders and Alzheimer's
Ayman El-Baz and Jasjit S Suri

Chapter 7

Ethylene and ammonia in neurobehavioral disorders

Ana-Maria Bratu, Cristina Popa (Achim), Mioara Petrus and Dan C Dumitras

Oxidative stress in biological molecules occurs when the balance between the reactive oxygen species (ROS) produced and the cell's antioxidant defense mechanisms is modified and the function of free radical inactivation is affected. Reactive oxygen species include ions, free radicals, and peroxides. During times of environmental stress ROS levels can increase dramatically which can result in significant damage to cell structures, particularly in the absence of antioxidant defenses, such as the enzymes superoxide dismutase, catalase, glutathione peroxidase, and glutathione reductase, or the antioxidant vitamins A, C, and E, and polyphenol antioxidants. Although the brain is known to be vulnerable to the effects of ROS because of its high oxygen consumption and high content of polyunsaturated fatty acids, it is still not quite clear how the oxidative stress itself contributes to neurological disorders. In this study, we will discuss the role of oxidative stress in autism and schizophrenia, which provides ample opportunities and hope for a better understanding of their pathophysiology, which may lead to new therapeutic strategies.

7.1 Introduction

The fight against neurological disorders has led to research innovations and important developments in this field. Neurological disorders are diseases of the central and peripheral nervous system. Problems with part of your nervous system can lead to incorrect movement, speech, swallowing, breathing, or learning. You can also have problems with your memory, senses, or mood [1].

The nervous system is anatomically and functionally divided into the central and peripheral subsystems. The central nervous system (CNS) includes the brain and spinal cord, and CNS dysfunction can be subdivided into two general categories, namely neurobehavioral and motor/sensory. Neurobehavioral difficulties involve two primary categories: cognitive decline, including memory problems and

doi:10.1088/978-0-7503-1793-1ch7

dementia; and neuropsychiatric disorders, including neurasthenia (a collection of symptoms including difficulty concentrating, headache, insomnia, and fatigue), depression, posttraumatic stress disorder (PTSD), and suicide. Other CNS problems can be associated with motor difficulties, characterized by problems such as weakness, tremors, involuntary movements, poor coordination, and gait/walking abnormalities. These are usually associated with subcortical or cerebellar system dysfunction. The anatomic elements of the peripheral nervous system (PNS) include the spinal rootlets that exit the spinal cord, the brachial and lumbar plexus, and the peripheral nerves that innervate the muscles of the body. PNS dysfunctions, involving either the somatic nerves or the autonomic system, are known as neuropathies.

Neurological dysfunction can be further classified as either global or focal. For example, in neurobehavioral disorders, global dysfunction can involve altered levels of consciousness or agitated behavior, whereas focal changes give rise to isolated signs of cortical dysfunction such as aphasia or apraxia. Likewise, global neuropathies could affect all peripheral nerves of the body, whereas a focal lesion would damage only a single nerve [2].

Some physical brain disorders, such as schizophrenia, schizoaffective disorder, bipolar and major depressive disorders, autism, pervasive developmental disorders, obsessive compulsive disorder, Tourette's disorder, anxiety and panic disorders, and attention deficit hyperactivity disorder, are estimated to affect hundreds of millions of people worldwide [3]. Scientifically verifiable abnormalities in brain anatomy, brain chemistry, and brain function help explain these disorders. Schizophrenia is a chronic and severe mental disorder that affects how a person thinks, feels, and behaves. People with schizophrenia may seem like they have lost touch with reality. Although schizophrenia is not as common as other mental disorders, the symptoms can be very disabling. Schizophrenia affects about one in 100 people and is difficult to diagnose and treat because it manifests differently in different people [4].

Autism has often been described as a neurological disorder that affects how an individual behaves, communicates, and interacts with others [5]. It is known as a 'spectrum' disorder because there are a wide range of symptoms which can vary in severity, and a person's diagnosis can fall anywhere along that spectrum. Since the exact causes of autism are unknown, diagnosing autism can be difficult [6].

Most previous studies in neurobehavioral disorders have been invasive [7, 8], requiring samples of blood or cerebrospinal fluid, or indirect measures of antioxidant enzyme levels have been used. A new way to noninvasively evaluate neurological disorders in humans could be the measurement of volatile organic compounds in the breath. Breath analysis is an emerging methodology which, being noninvasive and rapid, is ideally suited to clinical monitoring.

Breath analysis is a potentially powerful tool for the diagnosis and study of medical diseases [9]. In breath analysis, biomarkers give information very quickly compared to blood or urine tests. Breath is a complex mixture of gases, vapors, and aerosols. The bulk matrix of breath is a mixture of nitrogen (78.6%), oxygen (16%), carbon dioxide (0.9%), water vapor, and gases: inorganic gases (e.g. NO, CO_2, and CO), volatile organic compounds (e.g. isoprene, ethane, pentane, and acetone), and other typically non-volatile substances (isoprostanes, peroxynitrite, cytokines, and nitrogen). More than 3500 different components have been identified in exhaled breath, among these more than 1000 volatile organic compounds (35 breath biomarkers). Breath analysis is a method for obtaining noninvasive information on the clinical state of an individual by monitoring volatile organic compounds present in the exhaled breath and this method has great clinical potential [10–12].

Breath analysis continues to be an attractive field for noninvasive diagnosis of serious illnesses. Biomarker analysis in exhaled breath may be the most simple, rapid, and safe way to accurately determine the stage or the severity of a disease. Although numerous biomarkers have been identified so far, very little is known about their origin, if they are metabolic or not [13–18]. The field of analysis of volatile organic compounds has attracted a considerable amount of scientific interest during the last decade, but one of the challenges in the field of analysis of the volatile trace gases in exhaled breath is to be able to relate their concentrations to corresponding plasma levels [13–15]. In contrast to NO, which is predominantly generated in the bronchial system, volatile organic compounds are mainly blood borne and therefore enable monitoring of different processes in the body. Efforts have been made to detect and quantify ammonia and ethylene present in the breath, which are now known to be indirect biomarkers for various diseases and malfunctions of the human body [13–15, 17–28]. Elevation of breath levels of some of these gases (ammonia, ethane, isoprene, butane, nitric oxide, ethanol, carbon monoxide, methane, hydrogen sulfide, and acetone) is associated with a variety of metabolic and pathologic conditions [21–35]. The level of compounds in human breath can change slightly depending on individual oral hygiene (if these persons used care products before breath collection or not), and also on the time of day that the breath sample is taken. Therefore, each person has a unique 'breathprint' (or chemical composition) that remains distinct and relatively stable.

Ethylene from human breath is a marker of oxidant stress and can be directly attributed to the biochemical events surrounding lipid peroxidation. In the human body, the fatty acids inside the membrane lipids are mainly linoleic acid and arachidonic acid. The peroxidation of these fatty acids produces two volatile alkanes: ethylene and pentane. Both of them are considered in the literature to be good biomarkers of free-radical-induced lipid peroxidation in humans [36]. The fact that ethylene is highly volatile, not significantly metabolized by the body, and not soluble in body fat means that it diffuses rapidly into the bloodstream after generation and it is transported to the lungs. In the lungs, the gas is excreted in the expired breath and then is collected. A normal concentration of ethylene in the breath is 20–100 ppb (parts per billion).

Elevated breath ammonia could be due to liver disease, kidney disease, genetic diseases, periodontal disease, or physical exercises. The normal concentration of ammonia in the breath is 50–2000 ppb (parts per billion) [37, 38]. Since ammonia passively diffuses from the blood to both the salivary and sweat glands, it can be detected in oral fluid and sweat [39, 40].

Breath ethylene and ammonia have been detected in exhaled breath using various methods, including selected ion flow tube mass spectrometry [41], ion mobility spectrometry [42], cavity ring-down spectroscopy [43], and photoacoustic spectroscopy [44]. In this chapter, we utilize the CO_2 laser photoacoustic spectroscopy method to compare ethylene and ammonia exhalations from individuals with a healthy physiological state with ethylene and ammonia exhalations from schizophrenia patients with a pathological state (before and after treatment with Levomepromazine), thereby allowing for the identification of schizophrenia-related breath biomarkers in the exhaled air.

Oxidative stress, in general, occurs when oxidants exceed the antioxidant defense with damage to biomolecules and functional impairment. Autism is a behavioral disorder, with hallmark communication and social deficits. It has been suggested that oxidative stress may play a role in the pathophysiology underlying the behaviors that define autism. Another serious behavioral disorder, schizophrenia, features high oxidative biomarkers and there is documentation of a clinical response to antioxidants. Many antipsychotic medications used in the treatment of schizophrenia are, in fact, potent antioxidants.

7.2 Method

Laser photoacoustic spectroscopy (LPAS) is a sensitive technique for the detection and monitoring of trace gases at very low concentrations. The application of LPAS for rapid measurement of breath biomarkers has emerged in recent years as a very powerful investigation technique for monitoring and diagnosis, as it is able to measure trace-gas concentrations at sub-parts per billion (ppb) levels. The technique operates on the principle that the amount of light absorbed by a sample is related to the concentration of the target species in the sample. Light of known intensity is directed through a gas sample cell and the amount of light absorbed by the sample is measured as a sound intensity by a detector, usually a sensitive microphone. During the last few years the LPAS technique has been developed to a high degree of perfection.

Applications of LPAS include concentration measurements and trace-gas analysis, accurate determination of thermophysical properties, and detection of dynamic processes such as mixing of gases or chemical reactions, relaxation processes, spectroscopic experiments, and measurement of aerosols. Trace-gas detection techniques are important for applications such as breath diagnostics, security, workplace surveillance, air-quality measurements, and atmospheric monitoring. The laser-based instruments can also be used for the detection of a wide variety of industrial gases, a broad range of chemical warfare agents, blistering agents, poisonous gases, or explosives [45, 46]. Considering the wide gamut of application areas, the requirements for LPAS are various and the development and implementation of versatile analytical tools is challenging. Important features are multi-component capability, high sensitivity and selectivity (being immune to interference), high accuracy and precision, large dynamic range (usually larger than six orders of magnitude, from 100 ppt to 100 parts per million (ppm)), no or only minor sample preparation, good temporal resolution, ease of use, versatility, reliability, robustness, and a relative low cost per unit [47, 48].

CO_2 LPAS is a relatively accurate and reliable method for detecting breath biomarkers from the exhaled breath of schizophrenia patients, which could represent an effective and convenient screening method for this neurological disability. Spectroscopic systems include differential optical absorption spectroscopy, Fourier transform infrared spectrometers, and light detection and ranging systems. Although there is no ideal instrument that would fulfill all the requirements mentioned above, the sensing techniques based on LPAS principles offer some important advantages in breath monitoring [48]. The success of the photoacoustic based trace-gas sensing techniques crucially depends on the availability and the performance of the tunable laser source (accessible wavelengths, tuning characteristics, typical power range) and of the detection scheme employed. Lasers offer the advantage of high spectral power density owing to their intrinsic narrow linewidth in the range of megahertz. Since the laser linewidth is usually much smaller than the molecular absorption linewidth (the gigahertz region at atmospheric pressure), it is not an important issue in most measurements. The most widely used sources are CO and CO_2 lasers, lead salt diode lasers, quantum cascade lasers, and nonlinear optical devices such as optical parametric oscillators and difference frequency generation. Because the spectrum of a CO_2 laser overlaps, at room temperature and normal atmospheric pressure, the absorption spectra of numerous gases (volatile organic compounds), a good choice is to use a frequency stabilized CO_2 laser and a photoacoustic cell (PA cell) in performing the measurements on the patients' exhaled breath. The kind and number of detectable substances is related to the spectral overlapping of the laser emission with the absorption bands of the trace-gas molecules. The number of detectable compounds is first limited by the laser wavelength range, which should overlap the absorption spectrum of each individual gaseous compound, and second by the fact that the laser source (CO_2 laser) enables only discrete wavelength tuning. In addition, a partial overlapping of the individual absorption spectra of several compounds existing in the sample could happen, making it difficult to distinguish between them. This issue could be overcome by looking for a specific wavelength placed at a reasonable distance in the spectrum at which one of the compounds has a strong absorption, while the other one is transparent and vice versa. A generally applicable method to limit the gases' interference is to separate gases using gas chromatographic methods, selective trapping inside a cold trap, or a specific chemical reaction (e.g. CO_2 using $KOH \Rightarrow K_2CO_3$ and water).

To distinguish trace gases, mouth-exhaled breath and nose-exhaled breath have been investigated using CO_2 laser photoacoustic spectroscopy. This method is very sensitive and selective and is well known in the field of trace-gas detection (at the ppb level). It was used in our study for quantitative determination, in real time, of carbon dioxide, ethylene, methanol, ethanol, and ammonia, with operational simplicity, easy calibration, and no need for sample preparation. Laser photoacoustic spectroscopy is essentially based on the generation of an acoustic wave in a gas excited by a modulated laser beam. This is a technique where laser radiation is modulated at a wavelength that overlaps with the spectral feature of the target species. A fraction of the ground-state molecular population of the target molecule is excited by absorption of the incident laser radiation (see figure 7.1). Between vibrational levels

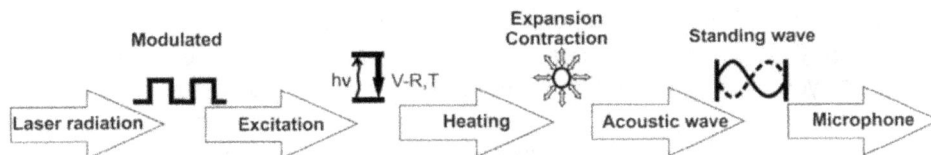

Figure 7.1. Schematic of the physical processes during optical excitation of molecules in photoacoustic spectroscopy.

and from vibrational states to rotational degrees of freedom, energy exchange processes occur. The energy absorbed by a vibrational–rotational transition is almost completely converted to kinetic energy by collisional de-excitation of the excited state. At the modulation frequency, the kinetic energy is converted into a periodic local heating. During expansion and contraction of the gas in a closed volume, a pressure variation leads to the formation of a standing acoustic wave in the resonator. These acoustic waves are measurable with sensitive microphones [49–56].

The main components of the LPAS detector are a frequency stabilized line-tunable CO_2 laser and a resonant photoacoustic (PA) cell in which the gas concentration is measured. Multicomponent mixtures can be measured with high sensitivity and necessary selectivity using this system for the molecules that possess high absorption strengths and a characteristic absorption pattern in the wavelength range of the CO_2 laser [57].

Key indicators of the quality of a measuring instrument are the reliability and validity of the measures. The process of developing and validating an instrument is in large part focused on reducing error in the measurement process. Reliability estimates evaluate the stability of measures, the internal consistency of measurement instruments, and the inter-rater reliability of instrument scores.

In this way, the emitted gas concentration measurements analyzed using the LPAS technique offer a high sensitivity that makes it possible to evaluate absorption coefficients on the order of 10^{-8} cm^{-1} [50–55]. In order to analyze the gases from respiration, we operate a sensitive instrument and we report the gases from respiration measured at the CO_2 laser wavelengths in precisely controlled experimental conditions that include a detector running in a resonant regime, a certified calibrated mixture of gases, a continuous wave (CW) line-tunable CO_2 frequency stabilized laser (with a discrete spectrum consisting of 60 distinct spectral lines as a result of the vibrational–rotational transitions of the CO_2 within the 9.2–10.8 μm region, grouped in four branches: 9R, 9P, 10R, and 10P) together with a full test point data acquisition and real-time processing system. The CO_2 laser is frequency stabilized to the center of the curve representing its output power versus frequency. The measurements are made with the laser operating in the TEM00 mode for all wavelengths. The CW tunable CO_2-laser beam is chopped, focused by a ZnSe lens, and introduced in the cell. The light beam is modulated by a high chopper model DigiRad (30 aperture blade), operated at the appropriate resonant frequency of the cell. The laser beam diameter is typically 6.2 mm at the point of insertion of the chopper blade and is nearly equal to the width of the chopper aperture. An approximately square waveform is produced with a modulation depth of 100% and a duty cycle of 50% so that the average power measured by the power meter at the exit of the cell is half the CW value. Enclosing the chopper wheel in a housing with a small hole (10 mm) for the laser beam to enter and exit, reduces chopper-induced sound vibrations in air that can be transmitted to the microphone detector as noise interference. A compatible phase reference signal is provided for use with a lock-in amplifier (time constant 1 s, sensitivity 1–100 mV) [50–57].

The PA cell consists of an acoustic resonator tube, windows, gas inlets and outlets, microphones, and an acoustic filter to suppress the window noise. The PA cell windows are made of ZnSe and positioned at the Brewster angle to their mounts. The resonant conditions are obtained as longitudinal standing waves in an open tube (excited in its first longitudinal mode). To achieve an optimum signal, we chose a long absorption path length of 300 mm and an inner diameter of the pipe of 7 mm. The fundamental longitudinal wave, therefore, has a nominal wavelength of 600 mm and a resonance frequency of 564 Hz.

The two buffer volumes placed near the Brewster windows have a length of 75 mm and a diameter of 57 mm. The inner wall of the stainless steel resonator tube is highly polished. It is centered inside the outer stainless steel tube with Teflon spacers. A massive spacer is positioned at one end to prevent bypassing of gas in the flow system; another is partially open to avoid the formation of closed volumes. Gas is admitted and exhausted through two ports located near the ends of the resonator tube. The perturbation of the acoustic resonator amplitude by the gas flow noise is thus minimized. The acoustic waves generated in the PA cell are detected by four Knowles electrets miniature microphones (sensitivity 20 mV Pa^{-1} each) in series, mounted flush with the wall. They are situated at the loops of the standing wave pattern at an angle of $90°$ to one another. The electrical output from these microphones is summed and the signal is selectively amplified by the lock-in amplifier [50].

The detection sensitivity is highly improved when the cell is built as an acoustic resonator working in a resonant regime which is achieved for a certain modulation frequency of the laser beam, known as the cell resonance frequency. This sensitivity brought by the resonant regime is described by a quality factor, Q, representing the amount of signal enhancement that occurs when the laser is modulated at the resonance frequency [50, 51]. At resonance, the amplitude of the photoacoustic signal is Q times larger than the amplitude far from the resonance frequency, i.e. the amplification is equal to the value of the Q factor. The quality factor of the system Q (experimentally determined value: 16.1) is the ratio between the energy stored in a specific mode and the energy losses per cycle of this acoustic wave [50–53].

To summarize, a gas sample is introduced into the PA cell by using a vacuum/gas handling system. As breath gases entered the PA cell of the system, the tunable CO_2 laser excites the gas molecules. The light energy used for vibrational excitation of molecules is then converted to heat that gives rise to a pressure variation in a closed volume (an acoustic wave) that can be detected with sensitive miniature microphones placed inside the PA cell [50–56]. Once the gas is analyzed by the lock-in amplifier, the digital data are introduced in the acquisition board and all the experimental data are processed in real time and stored in a computer (see figure 7.2).

Our measurement procedure to determine the concentrations of gases involves the following basic steps: cleaning the cell, calibration of the cell or measurement of the cell responsivity, and acquiring spectra of ethylene and ammonia vapors. For the measurement and detection of the gases from breath, the laser was kept tuned to where ethylene and ammonia exhibit the strongest and most characteristic peaks [50–53]. The cleaning of the cell was carefully performed by successively flushing out with nitrogen of purity 6.0 (99.9999%) at atmospheric pressure, executed each time when the cell contents was changed. An appropriate degree of cell cleaning was considered to be achieved if the photoacoustic signal measured in the nitrogen atmosphere had a fair low level, typically 30 μV. The PA cell responsivity was measured using a calibrated mixture of 0.996 ppmV ethylene in nitrogen 6.0 (99.9999%), supplied by Linde Gas, at atmospheric pressure. The 10P (14) line from the CO_2 laser spectrum was selected for this task due to the fact that ethylene presents here the highest absorption of 30.4 atm^{-1} cm^{-1}. The measured responsivity of the cell was 320 cm V W^{-1}. The absorption spectra for all gases were built recording the values measured for one cell filling by successively tuning the laser on the available lines [50–57]. As sampling gases we used certified gases supplied by Linde Gas with the following concentrations: for ethylene a calibrated mixture of 9.88 ppmV ethylene diluted in nitrogen 6.0, and for ammonia a calibrated mixture of 10 ppmV diluted in nitrogen.

The gas concentrations in the mixtures were chosen to be low enough to avoid the saturation effect and high enough to produce a significant value of the signal and the choice of nitrogen complies with two requirements: a buffer gas that is non-absorbing in the IR region and a gas close to the real air composition. The absorption measurements were always performed at room temperature (20 °C–22 °C). To collect a clean breath air sample [54] we used aluminized multi-patient collection bags (750 ml aluminum-coated bags), composed of a disposable mouthpiece, a tee-mouthpiece assembly (including a plastic tee and a removable one-way flutter valve), and a discard

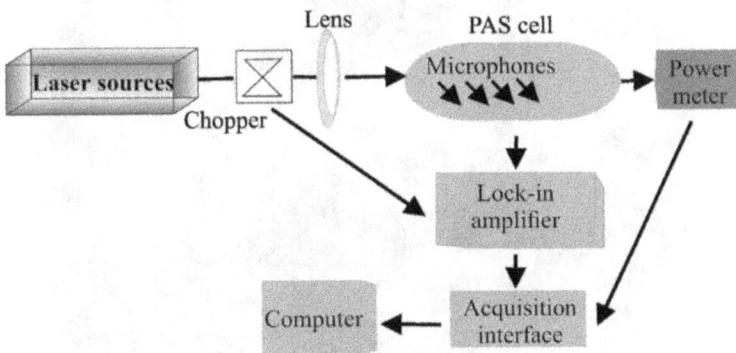

Figure 7.2. Schematic of the CO_2 laser photoacoustic technique.

(figure 7.3). Multi-patient collection bags are designed to collect multiple samples from patients and hold a sample for a maximum of 6 h.

After an approximately normal inspiration, the subject places the mouthpiece in his/her mouth, forming a tight seal around it with the lips. A normal expiration is then made through the mouth/nose, in order to empty the lungs of as much air as required to provide the breath sample [50–54]. For the mouth-exhaled breath sample, the first portion of the expired air is directed into the discard bag (with the role of collecting the 'dead-space' air: the first portion of an expired breath), while the alveolar air is diverted to the collection bag. When an adequate sample is collected, the subject stops exhaling and removes the mouthpiece/tube.

After the volunteer has exhaled via the mouth and the sample is collected, the gas from the sample is transferred into the PA cell and can be analyzed immediately or later. In either case, it is recommendable to seal the large port with the collection bag port cap furnished with the collection bag. The use of the port cap assures that the sample volume will not be lost due to a leak. Its use also avoids the contamination of the sample by gas diffusion through the one-way valve in the large port, if the sample is stored for a long period of time prior to its analysis.

The experiments were conducted using a gas handling and vacuum system used to ensure gas purity in the PA cell, to pump out the cell, introduce the sample gas in the PA cell at a controlled flow rate, and monitor the total and partial pressures of the gas mixtures. The gas handling system also includes two gas flow controllers, the sample bag, and the potassium hydroxide (KOH) scrubber. The breath samples were collected in special bags equipped with valves that sealed them after filling. The stored breath samples are transferred from bags into the measuring cell by the gas flow controller. Before entering the PA cell, the gas mixture passes through a KOH scrubber, which retains most of the interfering carbon dioxide and water vapor [53].

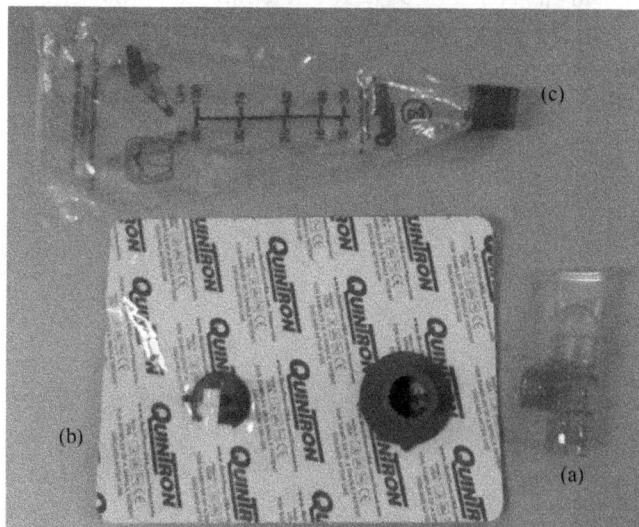

Figure 7.3. (a) Mouthpiece and tee-connector. (b) 0.75 l aluminum-coated bag. (c) 0.40 l discard bag.

Some means of selective spectral discrimination is required if ethylene is to be detected interference free in the matrix of absorbing gases. There are several ways to overcome this problem. One way is to remove CO_2 from the flowing sample by absorption on a KOH-based scrubber inserted between the sampling bag and the PA cell. Taking into account the nature of the specific chemical reactions involved in the CO_2 removal by KOH, a certain amount of water is also absorbed from the sample passing the scrubber. In this way, concentrations below 1 volume parts per million (ppmV) CO_2 (equivalent to a concentration of 0.07 ppbV of C_2H_4) can be achieved without influencing the C_2H_4 concentration. The pressure of the gas inside the cell influences the responsivity R (cm V W^{-1}) of the PA cell. The initial pressures in the sample bags filled by healthy humans differ from subjects with different disorders and it is necessary to know the pressure dependence of the PA cell responsivity. The exhaled air sample is transferred to the PA cell at a controlled flow rate of $36 \, l \, h^{-1}$ (600 standard cubic centimeters per minute (sccm)), and the total pressure of the gas in the cell is measured, applying then the correction factor for the responsivity according to the calibration curve (see figure 7.4) [53]. Our laser-based spectrometer for optical trace detection is one of the most sensitive instruments, with a limiting sensitivity 5.9×10^{-9} cm^{-1} that corresponds to a mixture of 0.96 ppmV of ethylene in nitrogen. Real-time detection of exhaled biomarkers using CO_2 LPAS provides access for research into breath testing for clinical diagnosis.

CO_2 LPAS performs well in terms of sensitive and selective detection of trace gases and it allows near on-line measurements. The calibration measurements (concentration-dependent response) for both ammonia and ethylene (figure 7.4) were experimentally determined using commercially prepared, certified gas mixtures containing 0.96 ppmV ethylene diluted in pure nitrogen and 10 ppmV ammonia diluted in pure nitrogen [53].

For calibration, we examined this reference mixture at a total pressure of ~1013 mbar and a temperature of 23 °C, using the commonly accepted values 30.4 cm^{-1} atm^{-1} (for ethylene) and 57 cm^{-1} atm^{-1} (for ammonia). To analyze the gas from the bags, we evacuated the extra gas and then we flushed the system with

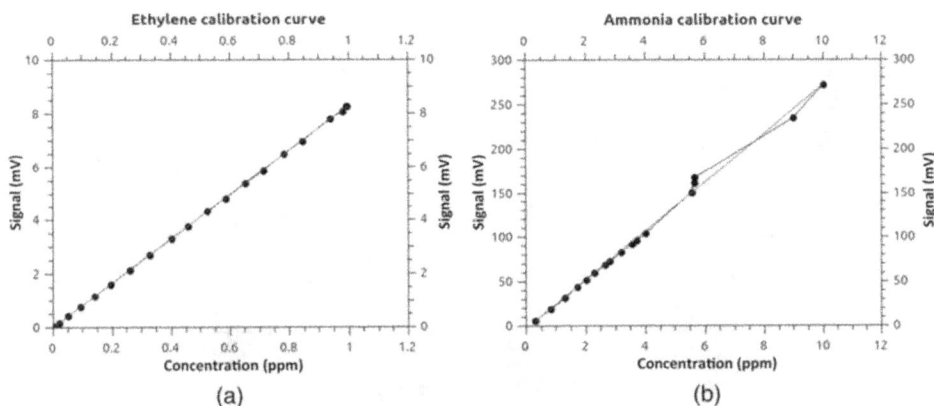

Figure 7.4. The concentration-dependent response for (a) ethylene and (b) ammonia.

pure nitrogen at atmospheric pressure for few minutes, then the exhaled air sample can be transferred to the cell using a controlled flow rate.

Because ammonia is a highly adsorbing compound and the results of successive measurements are often altered by the molecules previously adsorbed on the pathway and cell wall, an intensive cycle of N_2 washing was performed between samples in order to have a maximum increase of 10% for the background PA signal (to ensure the quality of each measurement). It has to be emphasized that the measured PA signal is due mainly to the absorption of ammonia and ethylene, but some traces of CO_2, H_2O, ethanol, etc, influence the measurements (the overall contribution is <10%). The response to all absorbing species at a given laser wavelength (PA signal) decreased considerably when we inserted a KOH trap (with a volume > 100 cm^3), proving that the amounts of CO_2 and H_2O vapors in the breath can significantly alter the results, thus making their removal compulsory [53].

An important parameter in the measurements is the responsivity R (cm V W^{-1}) of the PA cell, which depends on the pressure of the gas inside the cell. Taking into account the fact that the initial pressures in the sample bags filled by the healthy humans and by the subjects with different disorders differ from one case to another, it is necessary to know the pressure dependence of the PA cell responsivity (figure 7.5).

The exhaled air sample was transferred to the PA cell at 600 standard cubic centimeters per minute, and the total pressure of the gas in the PA cell was measured, then the correction factor for the responsivity according to the calibration curve from figure 7.4 was applied.

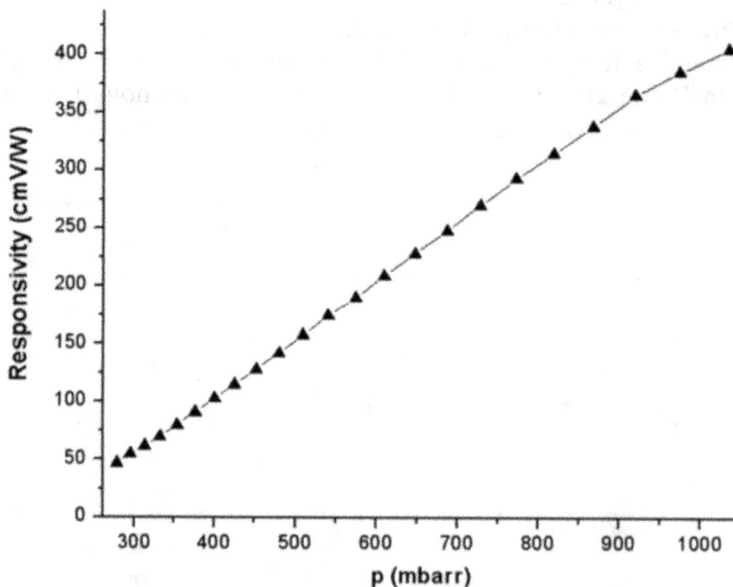

Figure 7.5. The responsivity of the photoacoustic cell against pressure.

The responsivity of the PA cell was determined using a calibrated mixture (Linde Gas) of 0.96 ppmV (~2%) C_2H_4 diluted in nitrogen 6.0 (purity 99.9999%) and of 10 ppmV (~5%) NH_3 diluted in nitrogen 5.0 (purity 99.999%). The pressure dependence of the responsivity was always measured at the center of the CO_2 laser line by using a frequency stabilized laser (instability 3×10^{-8}).

The absorption coefficients of ethylene and ammonia at different CO_2 laser wavelengths had been precisely measured previously [58] and the CO_2 laser was kept tuned at the 10P(14) line (10.53 μm) where ethylene exhibits a strong peak, corresponding to an absorption coefficient of 30.4 cm^{-1} atm^{-1} and at the 9R(30) CO_2 laser line (9.22 μm), where the ammonia absorption coefficient has a maximum value of 57 cm^{-1} atm^{-1}.

In this context, we utilized the CO_2 LPAS method to compare ethylene and ammonia exhalations from individuals with a healthy physiological state to ethylene and ammonia exhalations from schizophrenic patients with a pathological state (before and after treatment with Levomepromazine), thereby allowing for the identification of schizophrenia-related breath biomarkers in exhaled air. The levels of oxidative stress (given by the exhaled ethylene) in young adults with autism were also compared to the level of oxidative stress in individuals with a healthy physiological state.

7.3 Volatile organic compounds in autism and schizophrenia

7.3.1 Schizophrenia

Schizophrenia is a common psychiatric disorder, marked by gross distortion from reality and disturbances in thinking, feeling, and behavior. It has a life-time prevalence of ~1% of the world's population [59] and affects how a person thinks, feels, and acts. People with schizophrenia may have difficulty distinguishing between what is real and what is imaginary and may have difficulty expressing normal emotions in social situations. Most schizophrenic individuals are not violent and do not pose a danger to others. Schizophrenia has no symptoms that are identical for each person and is not caused by childhood experiences, poor parenting, or a lack of willpower. The cause of schizophrenia is still unclear. Investigations in this field include genetics (heredity), biology (abnormalities in the brain's chemistry or structure), and/or possible viral infections and immune disorders. Schizophrenia is usually treated with a combination of medication and therapy tailored to each individual. Usually antipsychotic medicines and cognitive behavioral therapy (CBT) are used.

People with schizophrenia should receive help from a community mental health team, which offers day-to-day support and treatment. Many people recover from schizophrenia, although they may have periods when symptoms return (relapses). Support and treatment can help reduce the impact the condition has on daily life.

It is believed that increased oxidative stress may be relevant to the pathophysiology of schizophrenia, but most of the results regarding this subject are contrasting [60–63]. Behavior disorders in the absence of mental health and social problems are best managed with psychological therapies, but the success rate is variable. Some individuals may, therefore, end up being treated with antipsychotic medications along with other approaches, despite the lack of a clear evidence base for drug use in this area [64], with the exception of Risperidone, which, in small doses, has been found to be beneficial for a subgroup of patients with behavior disorders [65–68].

Levomepromazine (Methotrimeprazine) is a phenothiazine that was first introduced (in treatments) in 1956. It is structurally similar to Chlorpromazine and Clozapine [69]. The contraindications, cautions, and side effects of Levomepromazine listed in the British National Formulary [70] are essentially the same as those for other typical antipsychotics, such as Haloperidol and Chlorpromazine. It is also known to cause hypothermia [71] and postural hypotension in ambulant patients over the age of 50 years.

7.3.2 Autism

Autism is a spectrum condition, a behavioral disorder, with hallmarks of communication and social deficits [72]. The routes to obtain a precise diagnosis are very difficult because the characteristics of autism may vary from one person to another [73]. Autism spectrum disorder (ASD) and autism are both general terms for a group of complex disorders of brain development. These disorders are characterized, in varying degrees, by difficulties in social interaction, verbal and nonverbal communication, and repetitive behaviors. There are different ways a person can be affected by ASD and thus making a diagnostic decision is a complex process [74].

Autism appears in early brain development and symptoms tend to emerge between 2 and 3 years of age, or some children with autism develop normally until toddlerhood when they stop acquiring or lose previously gained skills. Autistic people have difficulties with understanding language, and may think people always mean exactly what they say. They may find it difficult to use or understand facial expressions, tone of voice, jokes, and sarcasm [75].

Autism is a lifelong condition and has been observed that is more common in boys than in girls, with the latter manifesting less obvious signs compared to boys. However, many children diagnosed with ASD go on to live independent, productive, and fulfilling lives [76]. Autism differs from person to person in severity and combination of symptoms. There is a great range of abilities and characteristics of children with autism spectrum disorder—no two children appear or behave the same way. Symptoms can range from mild to severe and often change over time [75]. Early intervention is important to help a child reach their full potential. A child's treating team or specialist will help to develop an action plan for the family that can include information resources, parent training, strategies for family support, and an action plan for the child.

There is no 'cure' for autism and the exact cause of autism is still being investigated. A combination of factors—genetic and environmental—may account for differences in development. Conclusive diagnosis of ASD can be made by a team of professionals experienced in the field, such as a psychologist, psychiatrist, neurologist, developmental pediatrician, or similar qualified medical professional

[77]. Many autism biomarkers have been proposed for diagnostic of ADS [78] but at present there are no validated biomarkers for use in clinical practice [79].

7.3.3 Ethylene in mental disorders

In a normal healthy human body, the generation of pro-oxidants in the form of reactive oxygen species and reactive nitrogen species is effectively kept in check by the various levels of antioxidant defense [80]. When the body is exposed to adverse physicochemical, environmental, or pathological agents, such as cigarette smoking, atmospheric pollutants, ultraviolet rays, radiation, toxic chemicals, over nutrition, and advanced glycation end-products in diabetes, this delicately maintained balance is shifted in favor of pro-oxidants, resulting in oxidative stress. This has been implicated in the etiology of several (>100) human diseases and in the process of aging. All the biological molecules present in our body are at risk of being attacked by free radicals. Such damaged molecules can impair cell functions and even lead to cell death, eventually resulting in diseased states. The membrane lipids present in subcellular organelles are highly susceptible to free radical damage. When lipids react with free radicals, they can undergo the highly damaging chain reaction of lipid peroxidation, leading to both direct and indirect effects [81]. During lipid peroxidation, a large number of toxic by-products are also formed that can have effects at a site away from the area of generation, behaving as 'secondary messengers'. The damage caused by lipid peroxidation is highly detrimental to the functioning of the cell [82]. Lipid peroxidation is a free-radical-mediated process. Initiation of a peroxidative sequence is due to an attack by any species which can abstract a hydrogen atom from a methylene group (CH_2), leaving behind an unpaired electron on the carbon atom ($\bullet CH$). The resultant carbon radical is stabilized by molecular rearrangement to produce a conjugated diene, which can then react with an oxygen molecule to produce a lipid peroxyl radical ($LOO\bullet$). These radicals can further abstract hydrogen atoms from other lipid molecules to form lipid hydroperoxides (LOOH) and at the same time further propagate lipid peroxidation. The process of lipid peroxidation gives rise to many products of toxicological interest such as malondialdehyde, 4-hydroxynonenal, and a variety of hydrocarbons including pentane, ethane, and ethylene [83].

Oxidative stress has been associated with the pathophysiology of mental disorders. In contrast to other organs in the body, brain tissues exhibit high vulnerability to oxidative stress because of their high oxygen consumption, high content of polyunsaturated fatty acids (PUFA), and low level of antioxidant defenses in addition to a high metal content, which can catalyze the formation of ROS/RNS. Under physiological conditions, the potential for free radical mediated damage is counteracted by the antioxidant defense system, which is composed of a series of enzymatic and nonenzymatic components. The critical antioxidant enzymes include superoxide dismutase, catalase, and glutathione peroxidase. In mental disorders, the antioxidant defense is considered to be weak and oxidative stress to be present. Superoxide dismutase converts free radicals into hydrogen peroxide, which is then decomposed into water and oxygen by catalase, thereby preventing the formation of hydroxy radicals that initiate lipid peroxidation [84].

The process of lipid peroxidation ends in many products including ethylene, an important biomarker to monitor oxidative stress. A recent study [85] has correlated the systemic oxidative stress with changes in brain metabolism, defining ethane as a terminal product of the oxidation of omega-3 PUFA. Ethylene is a product of the lipid peroxidation of linoleic acid and can assess free radical damage [86, 87]. Given the correlation of breath ethylene with brain metabolism [84, 85, 88], measuring the breath concentration of this compound may represent a useful means to examine oxidative stress in schizophrenia and autism.

7.3.4 Ammonia in mental disorders

Changes in protein metabolism in neurological disorders have not been extensively studied. Dysfunctions of ammonia metabolism are associated with severe neurological impairments [89]. Medication works by reducing the psychotic symptoms. Understanding any relation between schizophrenia, antipsychotic treatment, and changes in metabolic variables in people with schizophrenia is important to managing the incidence of these events.

Ammonia is an important component in the human body, being a major by-product of systematic and cerebral nitrogen metabolism [90]. It is a source of nitrogen supply and helps in the synthesis of amino acids, which are considered to be the building blocks of protein in the body. As a by-product of protein metabolism, a large portion of ammonia is generated by the gastrointestinal tract. Bacteria present in the gastrointestinal tract digest protein into polypeptides, amino acids, and ammonia. Ammonia is converted into urea by the liver, and finally excreted by the body in the form of urine through the kidneys. In the human body urea is produced in the liver by a complex cyclical series of reactions known as the urea cycle. Ammonia is generated in the liver from glutamate and in the kidney by deamidation of glutamine. The glutamine derived from brain, muscle, and other tissues acts as an energy source and releases ammonia for urea synthesis. In too-high concentrations ammonia becomes toxic to the human body [91, 92]. Thus, for a healthy person, blood ammonia is tightly regulated via the urea cycle, with excess ammonia being converted to urea and excreted in urine.

Ammonia is toxic to the central nervous system when it reacts with α-ketoglutarate to form glutamate. The metabolite α-ketoglutarate impairs the function of the citric acid cycle in neurons, depriving them of energy production. Because glutamate is a potent neurotransmitter, any significant increase in the concentration of glutamate could have abnormal effects on synaptic transmission [89, 93, 94]. The neurotransmitter glutamate system shows strong involvement in the pathogenesis of schizophrenia [95]. The brain shows changes in the levels of glutamate and in the function and expression of its transporters and receptors [96].

Abnormal body concentrations of ammonia can now be studied using breath analysis. This analysis offers a unique and noninvasive method for compounds that circulate in the blood, rapidly diffuse across the pulmonary alveolar membrane, and appear in the exhaled breath to be detected [97, 98]. The level of ammonia in human breath has been measured as being between 50 and 2000 ppb (where 1 ppbV of ammonia in human breath is approximately 0.67 μg m^{-3}) and is dependent on a range of factors including the health status of the patient, the route of sampling (nasal or oral), contribution from oral bacteria, as well as diet, pharmaceutical use and levels of metabolic activity [93, 99].

7.4 Results and discussion

7.4.1 Protocol for breath gas sampling from schizophrenic patients

Schizophrenic patients were recruited and informed consent was obtained from staff at the CSCCHS Center, Calarasi, Romania (all gave their informed consent to participate in this research, which was approved by the institutional review boards of both institutions). The trial protocol was reviewed by ward consultants at the CSCCHS Center, and patients were matched for age, gender, and smoking status. The diagnosis of schizophrenia was made based on the criteria of schizophrenia disorders as evaluated in the Complex Evaluation Service of the CSCCHS Center.

A total of 15 subjects (6 male and 9 female, age range from 20–23 years, mean ± SD: 21.46 ± 1.45) who had previously been diagnosed as suffering from schizophrenia and 19 subjects without any history of psychiatric illness or other diseases and nonsmokers were selected as a control group (15 males and 4 females, age range from 25–33 years, mean ± SD: 30.05 ± 1.96) and included in the study. The control subjects were non- or ex-smokers, nonalcoholic, nonrenal, nondiabetic, and free from psychiatric disorders, somatic diseases, or brain tumors, and had never been treated with antidepressants or antipsychotic medications.

The schizophrenia group, comprising 15 patients, were nonsmokers, nonalcoholic, nonrenal, nondiabetic, and taking a range of drug therapies including an antipsychotic and anxiolytic treatment: Levomepromazine. Some studies [100–102] indicate that Levomepromazine may be useful in a small number of patients with severe aggression; the drug appears to be efficacious not only in controlling aggression but also lethargy, stereotypy, irritability, and hyperactivity symptoms.

Prior to the analysis of breath, the subjects were asked to avoid for at least 6 h, before or at any time during the breath sample collection, alcohol and coffee, or any other food or beverages, and to refrain from exercise in the morning. On the day prior to the test, products such as onions, leeks, eggs, and garlic were to be avoided.

7.4.2 Ethylene and ammonia assessment using LPAS from schizophrenic patients

In this study, ethylene and ammonia concentrations from breath samples were measured before/after treatment with Levomepromazine in schizophrenia patients, and the results were compared to healthy controls using CO_2 LPAS. Figure 7.6 shows the average concentrations of breath ethylene for schizophrenic patients, before and 30 min after ingestion of Levomepromazine treatment compared to the ethylene concentrations of a healthy group control.

As an observation of our primary result of interest, we see that the mean ethylene level of schizophrenic patients is higher (0.07 ppm) compared to the mean ethylene

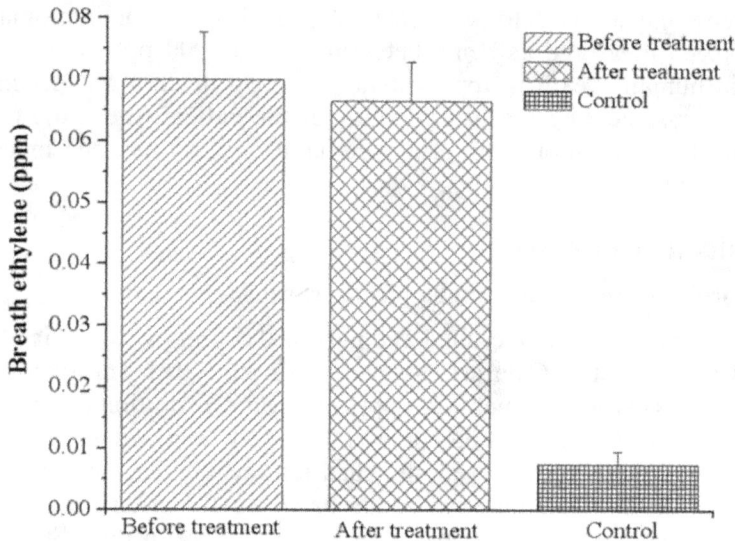

Figure 7.6. Breath ethylene biomarker in 15 patients with schizophrenia and 19 age-matching control individuals.

level of healthy subjects (0.008 ppm). In addition, at 30 min after the start of the treatment with Levomepromazine, the mean ethylene level of schizophrenic patients is smaller (0.066 ppm) than before treatment (but still high compared to the control subjects).

Oxidative stress seems to be a key piece in the schizophrenia pathophysiology. When oxidants exceed the antioxidant defense, biological systems suffer oxidative stress with damage to biomolecules and functional impairment. The possible responsible factor for the differences between the concentrations of breath ethylene before and after the treatment with Levomepromazine could be explained by the difference between untreated and treated schizophrenic patients. Most invasive measurements of oxidative stress in patients with schizophrenia have been made on peripheral tissues [105–110]. There is a lack of information on oxidative processes in the cerebrospinal fluid and brain. It must be mentioned that traces of oxidative damage may originate from various sources in the body and, consequently, such a peripheral indicator may not necessarily reflect the conditions of the oxidative stress parameters in the brain [109].

Using gas chromatography and mass spectrometry, previous studies [103, 104] reported an increase in exhaled ethane (like ethylene, ethane is also a hydrocarbon derived from n-3 PUFA) of patients with schizophrenia (e.g. 5.15 ppb or 8 ppbV) compared with those of the healthy controls (e.g. 2.63 ppb or 2.5 ppbV). It is important to mention that the schizophrenic patients in the previous studies [103, 104] had not been in receipt of psychotropic medication for three weeks prior to participating in the study but had received medication for the purpose of the study.

Thus our findings confirm previous determinations that oxidative stress is increased in schizophrenia and that this is unlikely to be a consequence of the

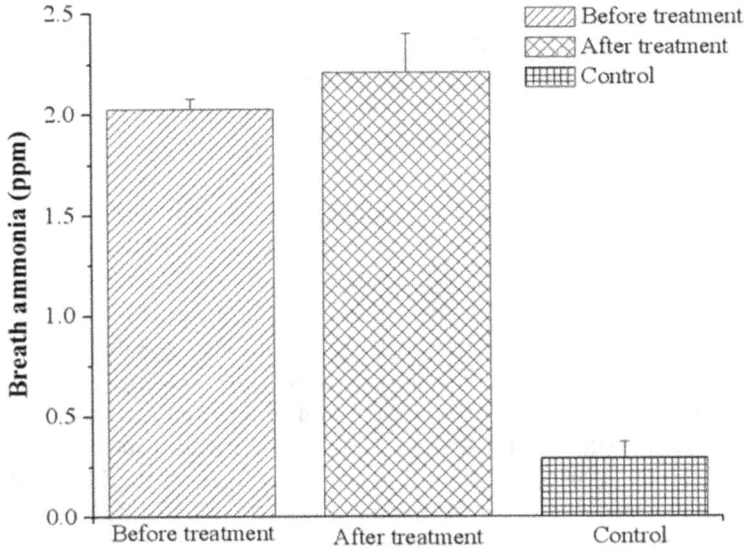

Figure 7.7. Breath ammonia biomarker in 15 schizophrenic patients and 19 age matched control individuals.

antipsychotic medications because the breath biomarkers after treatment were not significantly increased.

Whereas ethylene is produced as a by-product of oxidative stress, ammonia is produced as a by-product of amino acids and protein ingestion. Figure 7.7 shows the average concentrations of breath ammonia for schizophrenic patients, before and 30 min after ingestion of Levomepromazine compared to the ammonia concentrations of a healthy control group.

It should be pointed out that the mean ammonia level of schizophrenic patients is higher (2.02 ppm) compared to the mean ammonia level of healthy subjects (0.29 ppm). At 30 min after the start of the treatment with Levomepromazine, the mean ammonia level is even higher (2.2 ppm). Other possible confounding variables, such as age or sex, showed no statistically significant differences between the two groups.

Our measurements are based on the detection of biomarkers from breath and are in good agreement with those (based on oxidative stress analysis) reported in the literature [111–120]. While the majority of invasive studies have reported decreased antioxidant defense in patients with schizophrenia, there are also some studies where the opposite has been reported [121–128]. Several factors, such as the differences in measuring techniques, differences in material tested, exposure to antipsychotic treatment, sampling of patients at different stages of disease, lifestyle, and dietary patterns, may be responsible for this discrepancy.

Our study also reviewed the efficacy of Levomepromazine in patients with schizophrenia, and the findings indicate that breath ethylene decreases after treatment and breath ammonia increases after treatment (but not significantly). So, while the oxidative stress is mildly reduced after treatment, a mild impairment of metabolic liver function will produce increased blood (breath) ammonia.

Taking into consideration that Levomepromazine effectiveness is achieved in 2–3 h depending on the route of administration [129], at 30 min after the administration there is no significant change in the chemical levels in the breath of patients. The physiological basis of these findings is still speculative and future studies are needed that would clearly identify the etiologic relation between breath biomarkers and treatment with Levomepromazine.

The relation between the level of ammonia in the exhaled breath and schizophrenia could be explained by the treatment with Levomepromazine that can lead to a deficiency of amino acids which are required to detoxify toxins in the liver [130]. Along with their useful effects, most medicines can cause unwanted side effects, although not everyone experiences them. It seems that in schizophrenic patients Levomepromazine mildly reduces kidney function resulting in an insufficient detoxification pathway with a very small accumulation of ammonia in the breath [131]. The most important route for ammonia is the formation of urea in the liver. The urea is then transported to the blood from the liver to the kidneys and finally appears in the exhaled breath of schizophrenic patients. From the results of this study, the ammonia in the breath of schizophrenic patients was identified in higher concentrations (in treated patients) when compared to the healthy group.

Our data support a dysregulation of energy metabolism in schizophrenia and suggest new markers that may contribute to a better understanding of this disease. Both the feasibility and the importance of monitoring exhaled ammonia and exhaled ethylene from different subjects have been shown.

7.4.3 Protocol for breath gas sampling from autistic young adults

Breath samples were investigated at the Optics and Lasers in Life Sciences, Environment and Manufacturing Laboratory of the National Institute for Laser, Plasma and Radiation Physics and collected by a team with experience at the CSCCHS Center, Calarasi, Romania. The young people with autism were recruited and informed consent was obtained (all gave their informed consent to participate in this research, which was approved by the institutional review boards of both institutions). The trial protocol was reviewed by ward consultants at the CSCCHS Center, and patients matched for age and gender. The diagnosis of autism was made based on the criteria of autism disorders as evaluated in the Complex Evaluation Service of the CSCCHS Center.

A total of nine young adults (7 males and 2 females, age range from 23–26 years) who had been previously diagnosed as suffering from autism were included in the study and nine subjects without any history of psychiatric illness or other diseases were selected as a control group (9 males, age range from 25–33 years). It is important to mention that the adults with autism took some vitamin complex one month prior to the study. The non-autistic subjects were non- or ex-smokers, nonalcoholic, nonrenal, nondiabetic and free from psychiatric disorders, somatic diseases, and brain tumors and had never been treated with antidepressant or antipsychotic medications.

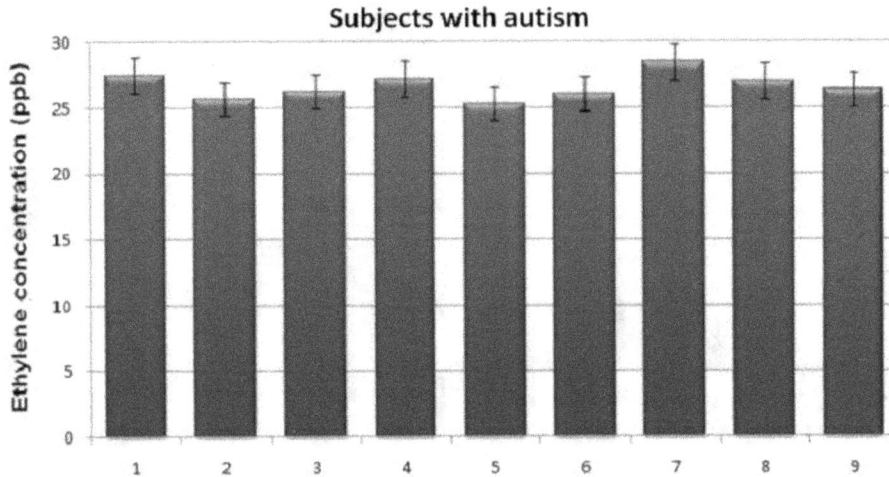

Figure 7.8. Ethylene gas average levels for adults with autism (error bars included).

Prior to the analysis of breath, the subjects avoided for at least 3 h (and at any time during the breath sample collection) alcohol and coffee, and other food or beverages, and refrained from exercise in the morning. On the day prior to the test, they avoided all consumption of products such as onions, leeks, eggs, and garlic. The collection of breath samples was made in 0.75 l aluminum-coated bags, designed to collect multiple samples from the subject and hold a sample for a maximum of 6 h. All of the collected samples were analyzed within 3 h after sampling over a period of 1 month. A total of ten samples per subject were collected over this period.

7.4.4 Ethylene and ammonia assessment using LPAS in autistic young adults

Most studies on oxidative stress in autism have been invasive, requiring samples of blood or indirect measures of antioxidant enzyme levels. A new way to measure noninvasive lipid peroxidation in humans is to measure free radical damage by analyzing the early products of oxidation, such as exhaled hydrocarbons. In this study, ethylene concentrations from breath samples were measured in young adults with autism and the results were compared to healthy controls using CO_2 laser photoacoustic spectroscopy. The sample bags that were utilized to collect exhaled air from the autistic patients and healthy subjects did not release contaminants at room temperature; moreover, the bags underwent standard washing and evacuation procedures prior use to exclude gas contamination from the external environment.

Figure 7.8 shows the average concentrations of breath ethylene for subjects with autism, compared to the ethylene concentrations of a healthy group control (figure 7.9). As an observation of our primary result of interest, we see that the mean ethylene levels of adults with autism were not significantly increased compared to the mean ethylene level for healthy adults.

In addition to applied behavior analysis (ABA) therapy, pharmaceutical treatment can be effective to reduce medical interference or disruptive behaviors. People

Healthy subjects

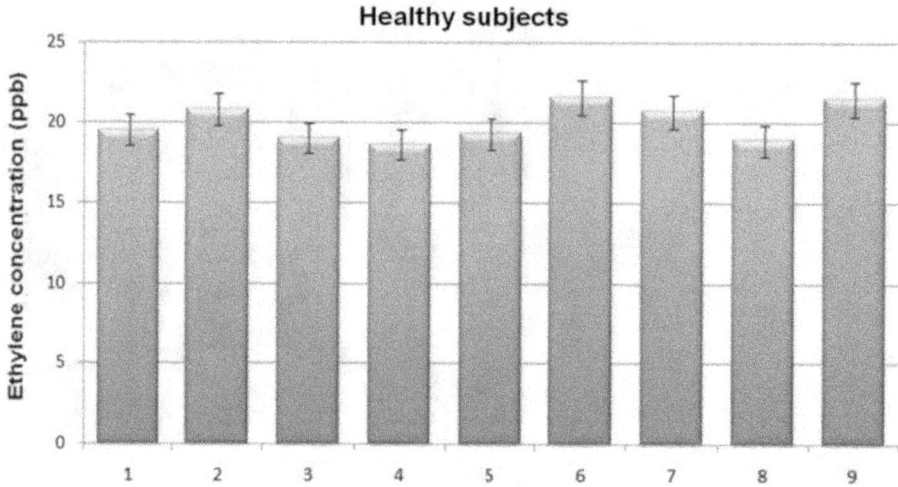

Figure 7.9. Ethylene gas average levels for healthy adults (error bars included).

with autism generally receive antioxidants because they cross the blood–brain barrier and combat oxidative stress in the brain cells and provide neurostimulative treatment. Pharmaceutical treatments can help to ameliorate some of the behavioral symptoms of ASD, including irritability, aggression, and self-injurious behavior. Medications should be prescribed and monitored by a qualified physician. The subjects of our studies received supplements with vitamin B-complex, Tonotil-N, Neuro Optimizer® 60cps, and cod liver oil for three consecutive months together with therapies. Cod liver oil was administrated every day and the others only for 10 days/month. Essential fatty acids such as cod liver oil are necessary fats that children and adults cannot synthesize, and must be obtained through the diet or supplements. Essential fatty acids are long-chain polyunsaturated fatty acids derived from linolenic, linoleic, and oleic acids. Essential fatty acids are found in cod liver oil and are an absolute requirement for proper brain functioning and allow decreased hyperactivity in autism or improvement in other symptoms of autism. The subjects took daily doses of 1000 mg of cod liver oil. The vitamin B-complex includes thiamine (vitamin B1), riboflavin (vitamin B2), niacin (vitamin B3), pantothenic acid (vitamin B5), pyridoxine (vitamin B6), biotin, folic acid, and the cobalamins (vitamin B12). These nutrients play an important role in the formation of serotonin and other neurotransmitters, and deficiencies—which are linked to high levels of homocysteine—are linked with mood disorders. Vitamin B-complex helps to support the nervous system and prevent imbalances from occurring. The subjects used vitamin B-complex (100 mg) 10 days per month. Tonotil-N stimulates brain activity, energy production, and cellular oxygen use. Tonotil-N complements deficient cellular components without generating degradation products. The subjects took 1 bottle in the morning before breakfast for 10 days per month. The natural Neuro Optimizer® product helps to improve the synthesis of adrenaline neuro-transmitters and dopamine and maintains the balance between sodium, potassium, calcium, and magnesium concentrations in the body. It provides antioxidant protection against neuronal destruction caused by free radicals and contributes to stimulating memory, concentration, and adaptability to intellectual effort. The subjects took 1 capsule of Neuro Optimizer® for 10 days per month. Subjects who took part on our study were administrated this treatment 1 month prior to the study and this together with the therapeutic lifestyle and dietary patterns may be an explanation for the normality between the concentrations of breath ethylene in adults with autism and the control adults.

How oxidative stress contributes to autism is not clearly understood and implications for future research and interventions are currently underway. Despite the amount of research and attention on ADS we still know little about this disorder, and its treatment and prevention. To date, we still have no effective medication for this disorder. A number of previous studies have reported evidence of immune dysregulation and/or inflammation in individuals with autism and support the notion that oxidative stress is associated with autism in children [132–136], including gene changes pertaining to the immune system [137–140].

In the literature, there are studies on lipid peroxides, 4-hydroxynonenal, malondialdehyde (MDA), isoprostanes, levuglandin adducts, nitrotyrosine, oxidized nucleic acids, protein carbonyls, advanced glycation end-products, cellular apoptosis, nutrient and antioxidant enzyme concentrations, total nitrite + nitrate, enzyme-binding affinities, and luminal NO• by rectal catheter. One study evaluated the levels of antioxidant enzymes, superoxide dismutase (SOD), and glutathione peroxidase (GSH-Px), and levels of malondialdehyde (MDA), a marker of lipid peroxidation [141]. They compared levels of SOD, GSH-Px, and MDA in children with autism and controls. In children older than 6 years, there was no significant difference in any of these values between cases and controls. They concluded that children with autism are more vulnerable to oxidative stress in the form of increased lipid peroxidation and deficient antioxidant defense mechanism, in particular younger children.

It is important to mention that the subjects with autism in the previous studies were all children, whereas the present study analyzed young adults. Further effort is needed to actually measure the process of oxidative stress in ADS subjects, in particular without an invasive procedure.

7.5 Conclusions and future directions

The use of related markers in exhaled breath air for neurological disorder analysis is theoretically reasonable. Metabolic changes are assumed to occur in patients that inevitably lead to the production of certain abnormal metabolites. These metabolites are transported through the blood to the alveoli of the lungs, through alveolar gas exchange, and volatile metabolites will then be discharged into the air as components of each exhaled breath.

We analyzed breath ethylene in young adults with autism and we compared the results to the exhaled breath of normal controls. The results indicated that in our case the relationship between autism and oxidative stress-induced damage is insufficient to recommend ethylene as a biomarker for this disorder.

It must be mentioned that traces of oxidative damage may originate from various sources in the body, our measurements being based on the detection of ethylene gas from breath. While the majority of noninvasive studies have reported increased oxidative stress in children with autism, our measurements on young adults reported the opposite. Several factors, such the differences in measuring techniques, differences in material tested, exposure to antioxidant supplements, sampling of patients at different stages of the disease, lifestyle, and dietary patterns, may be responsible for this discrepancy.

The physiological basis of these findings is still speculative and future studies that would clearly identify the etiologic relation between breath ethylene biomarkers and oxidative stress in adults with autism need to be carefully determined. Although photoacoustic spectroscopy is a sensitive, noninvasive, and real-time method to accurately analyze breathing gas concentrations, finding a sensitive, specific, and noninvasive biomarker of autism which could be measured in alveolar air still remains an important task.

Breath biomarkers in schizophrenic patients were also analyzed. The breath ethylene and breath ammonia of schizophrenic patients were measured before and after treatment with Levomepromazine, and we compared the results with the exhaled breath of normal controls. The sample bags that were utilized to collect exhaled air from the schizophrenic patients and healthy subjects did not release contaminants at room temperature. Moreover, the bags underwent standard washing and evacuation procedures prior to use to exclude gas contamination from the external environment.

In the results of this study, ethylene and ammonia were identified in higher concentrations in the breaths of schizophrenic patients when compared to the healthy group. The results also revealed that the ethylene levels can be considered as a measure of oxidative stress index in schizophrenic people. In conclusion, the data from this study support the hypothesis of the oxidant/antioxidant balance as a key component that may contribute to schizophrenia pathology. Based on a noninvasive sampling method, that is stable in biological materials and easy to measure, we conclude that CO_2 LPAS analyses of breath ethylene/ammonia in alveolar air appeared to distinguish patients with schizophrenia from non-schizophrenic controls. Although CO_2 LPAS is a sensitive, noninvasive, and real-time method to accurately analyze breathing gas concentrations, finding a sensitive, specific, and noninvasive biomarker of schizophrenia, which could be measured in alveolar air, still remains an important task.

Important conclusions resulted from our studies. In the case of schizophrenia, the ethylene concentration is increased compared to the control group, whereas in the case of autistic people, the ethylene concentration shows no different values when compared to the control group. Considering that oxidative stress is a factor that can be corrected, future studies that would clearly identify the etiologic relation between antioxidant deficiencies and schizophrenia may provide prophylactic treatments, as well as new treatment schemes in addition to available antipsychotic schemes. From these results, we can conclude that clear differences exist between people diagnosed with schizophrenia and autism. Autism spectrum disorder and schizophrenia are distinct disorders with unique characteristics, but they share similarities in social dysfunction. Studying the differences between the social impairments found in autism and schizophrenia will help develop better treatments for people with both disorders.

Further studies with a larger number of subjects will be necessary to carefully determine which antioxidants, and what dosages/in what combinations, will have the greatest therapeutic benefit, considering the importance of OS in many biological reactions. With improved sensitivity and specificity, CO_2 LPAS analyses of alveolar air might offer a new approach for the evaluation and a better understanding of neurological disorders.

References

[1] Medline Plus 2019 Neurological diseases https://medlineplus.gov/neurologicdiseases.html

[2] Institute of Medicine (US) Committee to Review the Health Effects in Vietnam Veterans of Exposure to Herbicides 1994 *Veterans and Agent Orange: Health Effects of Herbicides Used in Vietnam* (Washington, DC: National Academies Press)

[3] LeClerc S and Easley D 2015 Pharmacological therapies for autism spectrum disorder: a review *Pharm. Ther.* **40** 389–97

[4] National Institute of Mental Health 2016 Schizophrenia https://nimh.nih.gov/health/topics/schizophrenia/index.shtml

[5] National Institute of Child Health and Human Development 2019 Autism Spectrum Disorder (ASD) https://nichd.nih.gov/health/topics/autism

[6] Simple Steps Autism 2019 Diagnosis and coming to terms with autism https://simplestep-sautism.com/About-What-is-Autism

[7] Puri B K, Ross B M and Treasaden I H 2008 Increased levels of ethane, a non-invasive, quantitative, direct marker of n-3 lipid peroxidation, in the breath of patients with schizophrenia *Prog. Neuro-Psychopharmacol. Biol. Psychiatry* **32** 858–62

[8] Mukerjee S *et al* 1996 Impaired antioxidant defense at the onset of psychosis *Schizophr. Res.* **19** 19–26

[9] Lourenço C and Turner C 2014 Breath analysis in disease diagnosis: methodological considerations and applications *Meta* **4** 465–98

[10] Risby T and Tittel F K 2010 Current status of mid-infrared quantum and interband cascade lasers for clinical breath analysis *Opt. Eng.* **49** 111123

[11] Amann A, Miekisch W, Schubert J, Buszewski B, Ligor T, Jezierski T, Pleil J and Risby T 2014 Analysis of exhaled breath for disease detection *Annu. Rev. Anal. Chem.* **7** 455–82

[12] Navas M J, Jimenez A M and Asuero A G 2013 Human biomarkers in breath by photoacoustic spectroscopy *Clin. Chim. Acta* **413** 1171–8

[13] Wang T, Pysanenko A, Dryahina K, Španěl P and Smith D 2008 Analysis of breath, exhaled via the mouth and nose, and the air in the oral cavity *J. Breath Res.* **2** 037013

[14] Miekisch W, Schubert J K and Noeldge-Schomburg G F 2004 Diagnostic potential of breath analysis—focus on volatile organic compounds *Clin. Chim. Acta* **347** 25–39

[15] Lourenço C and Turner C 2014 Breath analysis in disease diagnosis: methodological considerations and applications *Meta* **4** 465–98

[16] Patrick S, Kseniya D and David S 2013 A quantitative study of the influence of inhaled compounds on their concentrations in exhaled breath *J. Breath Res.* **7** 017106

[17] Alyssa M F, Benjamin L M and Ivan R M 2013 Chemical analysis of exhaled human breath using a terahertz spectroscopic approach *Appl. Phys. Lett.* **103** 133703

[18] Khalid T Y, Saliha S, Greenman J, de Lacy Costello B, Probert C S J and Ratcliffe N M 2013 Volatiles from oral anaerobes confounding breath biomarker discovery *J. Breath Res.* **7** 017114

[19] Mazzone P J 2008 Analysis of volatile organic compounds in the exhaled breath for the diagnosis of lung cancer *J. Thorac. Oncol.* **3** 774–80

[20] Sowmya Y 2016 A review on the human oral microflora *Res. Rev.: J. Dent. Sci.* **4** 1–5

[21] Turner C, Spanel P and Smith D 2006 A longitudinal study of methanol in the exhaled breath of 30 healthy volunteers using selected ion flow tube mass spectrometry, SIFT-MS *Physiol. Meas.* **27** 637–48

[22] Simic M, Ajdukovic N, Veselinovic I, Mitrovic M and Djurendic-Brenesel M 2012 Endogenous ethanol production in patients with diabetes mellitus as a medicolegal problem *Forensic Sci. Int.* **216** 97–100

[23] Hibbard T and Killard A J 2011 Breath ammonia levels in a normal human population study as determined by photoacoustic laser spectroscopy *J. Breath Res.* **5** 037101

[24] Karlsson M 2014 Characterization of absorption spectra of molecular constituents in the mid-infrared region and their role as potential markers for breath analysis *Student Thesis* Umea University

[25] Kearney D J, Hubbard T and Putnam D 2002 Breath ammonia measurement in *Helicobacter pylori* infection *Dig. Dis. Sci.* **47** 2523–30

[26] Popa C, Dutu D C A, Cernat R, Matei C, Bratu A M, Banita Ş and Dumitras D C 2011 Ethylene and ammonia traces measurements from the patients' breath with renal failure via LPAS method *Appl. Phys.* B **105** 669–74

[27] Olde Damink S W, Deutz N E, Dejong C H, So eters P B and Jalan R 2002 Interorgan ammonia metabolism in liver failure *Neurochem. Int.* **41** 177–88

[28] David Weiner I, Mitch W E and Sands J M 2014 Urea and ammonia metabolism and the control of renal nitrogen excretion *Clin. J. Am. Soc. Nephrol.* **10** 1444–58

[29] Pleil J D, Stiegel M A and Risby T R 2013 Clinical breath analysis: discriminating between human endogenous compounds and exogenous (environmental) chemical confounders *J. Breath Res.* **7** 017107

[30] Boots A W, van Berkel J J, Dallinga J W, Smolinska A, Wouters E F and van Schooten F J 2012 The versatile use of exhaled volatile organic compounds in human health and disease *J. Breath Res.* **6** 027108

[31] Moeskops B W M, Steeghs M M L, van Swam K, Cristescu S M, Scheepers P T J and Harren F J M 2006 Real-time trace gas sensing of ethylene, propanol and acetaldehyde from human skin *in vivo Physiol. Meas.* **27** 1187–96

[32] Paetznick D J, Reineccius G A, Peppard T L, Herkert J M and Lenton P 2010 Comparison of breath and in-mouth collection for the measurement of oral malodorous compounds by gas chromatography using sulfur chemiluminescence detection *J. Breath Res.* **4** 017106

[33] Popa C, Băniţă S, Bratu A M, Paţachia M, Matei C and Dumitraş D C 2013 The level of ethylene biormarker in renal failure of elderly patients analyzed by photoacoustic spectro-scopy *Laser Phys.* **41** 125701

[34] Schmidt F M, Vaittinen O, Metsälä M, Lehto M, Forsblom C, Groop P H and Halonen L 2013 Ammonia in breath and emitted from skin *J. Breath Res.* **7** 017109

[35] Schubert J K, Miekisch W, Geiger K and Nöldge-Schomburg G F 2004 Breath analysis in critically ill patients: potential and limitations *Expert Rev. Mol. Diagn.* **4** 619–29

[36] Giubileo G 1998 *Proc. SPIE* **3405** 642

[37] Hibbard T and Killard A 2011 *Anal. Chem.* **41** 21

[38] Hibbard T, Crowley K and Killard A J 2013 *Anal. Chim. Acta* **779** 56

[39] Huizenga J R, Vissink A, Kuipers E J and Gips C H 1999 *Clin. Oral Invest.* **3** 84

[40] Czarnowski D, Gorski J, Jozwiuk J and Boron-KaczMarska A 1992 *Eur. J. Appl. Physiol.* **65** 135

[41] Davies S, Spanel P and Smith D 1997 *Kidney Int.* **52** 223

[42] Neri G, Lacquaniti A, Rizzo G, Donato N, Latino M and Buemi M 2012 *Nephrol. Dial. Transplant.* **27** 2945–52

[43] Schmidt F M, Vaittinen O, Metsala M, Lehto M, Forsblom C, Groop P-H and Halonen L 2013 *J. Breath Res.* **7** 017109

[44] Wang J, Zhang W, Li L and Yu Q 2011 *Appl. Phys.* B **103** 263

[45] Lima G R, Sthel M S, da Silva M G, Schramm D U S, de Castro M P P and Vargas H 2011 *J. Phys.: Conf. Ser.* **274** 012086

[46] Stepanov E V 2007 *Physics of Wave Phenomena* **vol 15** 149

[47] Wang C and Sahay P 2009 *Sensors* **9** 8230

[48] Dumitras D C, Dutu D C, Matei C, Magureanu A M, Petrus M and Popa C 2007 Laser photoacoustic spectroscopy: principles, instrumentation, and characterization *J. Optoelectron. Adv. Mater.* **9** 3655–701

[49] Sethi S, Nanda R and Chakraborty T 2013 Clinical application of volatile organic compound analysis for detecting infectious diseases *Clin. Microbiol. Rev.* **26** 462–75

[50] Dumitras D C, Dutu D C, Matei C, Magureanu A M, Petrus M, Popa C and Patachia M 2008 Measurements of ethylene concentration by laser photoacoustic techniques with applications at breath analysis *Rom. Rep. Phys.* **60** 593–602

[51] Dumitras D C, Banita S, Bratu A M, Cernat R, Dutu D C A, Matei C, Patachia M, Petrus M and Popa C 2010 Ultrasensitive CO_2 laser photoacoustic system *J. Infrared Phys. Technol.* **53** 308–14

[52] Bratu A M, Popa C, Matei C, Banita S, Dutu D C A and Dumitras D C 2011 Removal of interfering gases in breath biomarker measurements *J. Optoelectron. Adv. Mater.* **13** 1045–50

[53] Popa C, Patachia M, Banita S and Dumitras D C 2013 Exertion in kangoo jumps aerobic: evaluation and interpretation using spectroscopic technique determinations *J. Spectrosc.* **2013** 602434

[54] Dumitras D C, Banita S, Bratu A M, Cernat R, Dutu D C A, Matei C, Patachia M, Petrus M and Popa C 2010 Ultrasensitive CO_2 laser photoacoustic system *Infrared Phys. Technol. J.* **53** 308–14

[55] Dumitras D C, Dutu D C A, Matei C, Cernat R, Banita S, Patachia M, Bratu A M, Petrus M and Popa C 2011 Evaluation of ammonia absorption coefficients by photoacoustic spectroscopy for detection of ammonia levels in human breath *Laser Phys.* **21** 796–800

[56] Ivascu I R, Matei C E, Patachia M, Bratu A M and Dumitras D C 2015 Multicomponent detection in photoacoustic spectroscopy applied to pollutants in the environmental air *Rom. Rep. Phys.* **67** 1558–64

[57] Ivascu I R, Matei C E, Patachia M, Bratu A M and Dumitras D C 2016 CO_2 laser photoacoustic measurements of ethanol absorption coefficients within infrared region of 9.2–10.8 μm *Spectrochim. Acta* A **163** 115–9

[58] Popa C, Banita S, Patachia M and Dumitras D C 2013 Spectroscopic study of ammonia at subjects with kidney failure: a case control study *Rev. Roum. Chim.* **58**

[59] Schultz S H, North S W and Shields C G 2007 Schizophrenia: a review *Am. Fam. Physician* **75** 1821–9

[60] Kunz M, Gama C S and Andreazza A C 2008 Elevated serum superoxide dismutase and thiobarbituric acid reactive substances in different phases of bipolar disorder and in schizophrenia *Prog. Neuropsychopharmacol. Biol. Psychiatry* **32** 1677–81

[61] Dadheech G *et al* 2008 Evaluation of antioxidant deficit in schizophrenia *Indian J. Psychiatry* **50** 16–20

[62] Wood S J *et al* 2009 Neurobiology of schizophrenia spectrum disorders: the role of oxidative stress *Ann. Acad. Med. Singapore* **38** 396–406

[63] Padurariu M *et al* 2010 Changes of some oxidative stress markers in the serum of patients with mild cognitive impairment and Alzheimer's disease *Neurosci. Lett.* **469** 6–10

[64] Brylewski J and Duggan L 2004 Antipsychotic medication for challenging behaviour in people with learning disability *Cochrane Database Systemat. Rev.* **3**

[65] Bokszanska A *et al* 2003 Olanzapine and Risperidone in adults with learning disability: a clinical naturalistic study *Int. Clin. Psychopharmacol.* **18** 285–91

[66] Cohen S A *et al* 1998 Risperidone for aggression and self-injurious behaviour in adults with mental retardation *J. Autism Dev. Disord.* **28** 229–33

[67] McAdam D B *et al* 2002 Effects of Risperidone on aberrant behaviour in persons with developmental disabilities: II. Social validity measures *Am. J. Ment. Retard.* **107** 261–9

[68] Zarcone J R *et al* 2001 Effects of Risperidone on aberrant behaviour in persons with developmental disabilities: I. A double-blind crossover study using multiple measures *Am. J. Ment. Retard.* **106** 525–38

[69] Devapriam J *et al* 2008 Use of Levomepromazine in the management of aggression in adults with intellectual disability *Br. J. Dev. Disabil* **54** 11–7

[70] British Medical Association and Royal Pharmaceutical Society of Great Britain 2006 *British National Formulary* vol 52 (London: BMJ Group)

[71] Van Marum R J, Jansen S and Ponssen H H 2003 Antipsychotic medication as a cause of deep hypothermia *Ned. Tijdschr. Geneeskd.* **147** 1201–4

[72] National Institute of Neurological Disorders and Stroke 2019 Autism spectrum disorder fact sheet https://www.ninds.nih.gov/Disorders/Patient-Caregiver-Education/Fact-Sheets/Autism-Spectrum-Disorder-Fact-Sheet

[73] https://autismspeaks.org/family-services/tool-kits/100-day-kit/diagnosiscausessymptoms

[74] National Institute of Mental Health 2018 Autism Spectrum Disorder https://nimh.nih.gov/health/topics/autism-spectrum-disorders-asd/index.shtml

[75] https://autism.org.uk/about/what-is/asperger.aspx

[76] Murphy C M, Wilson C E, Robertson D M, Ecker C, Daly E M, Hammond N, Galanopoulos A, Dud I, Murphy D G and McAlonan G M 2016 Autism spectrum disorder in adults: diagnosis, management, and health services development *Neuropsychiatr. Dis. Treat.* **12** 1669–86

[77] Centers for Disease Control and Prevention 2018 Screening and diagnosis of autism spectrum disorder https://cdc.gov/ncbddd/autism/screening.html

[78] Ruggeri B, Sarkans U, Schumann G and Persico A M 2013 Biomarkers in autism spectrum disorder: the old and the new *Psychopharmacology* **231** 1201–16

[79] Momeni N *et al* 2012 A novel blood-based biomarker for detection of autism spectrum disorders *Transl. Psychiatry* **2** e91

[80] Powers S K and Jackson M J 2008 Exercise-induced oxidative stress: cellular mechanisms and impact on muscle force production *Physiol. Rev.* **88** 1243–76

[81] Lobo V, Patil A, Phatak A and Chandra N 2010 Free radicals, antioxidants and functional foods: Impact on human health *Pharmacogn. Rev.* **4** 118–26

[82] Barrera G 2012 Oxidative stress and lipid peroxidation products in cancer progression and therapy *ISRN Oncol.* **2012** 137289

[83] Focke W W and Radusch H-J 2014 *Engineering of Polymers and Chemical Complexity, Volume II: New Approaches, Limitations and Control* (New York: Academic)

[84] Bitanihirwe B K Y and Woo T-U 2011 Oxidative stress in schizophrenia: an integrated approach *Neurosci. Biobehav. Rev.* **35** 878–93

[85] Puri B K *et al* 2008 Evidence from *in vivo* 31-phosphorus magnetic resonance spectroscopy phoshodiesters that exhaled ethane is a biomarker of cerebral n-3 polyunsaturated fatty acid peroxidation in humans *BMC Psychiatry* **8**

[86] Puiu A *et al* 2007 Stress monitoring in a Guinness 10-day scuba dive *Laser Phys.* **17** 772

[87] Kocielnik R *et al* 2013 Smart technologies for long-term stress monitoring at work *IEEE 26th Int. Symp. on Computer-Based Medical Systems* 53–8

[88] Mukerjee S *et al* 1996 Impaired antioxidant defense at the onset of psychosis *Schizophr. Res.* **19** 19–26

[89] Butterworth R F 2001 *Ment. Retard. Dev. Disabil. Res. Rev.* **7** 276–9

[90] Beckett B S 1986 *Biology: A Modern Introduction* (Oxford: Oxford University Press)

[91] Cooper A J L, Lai J C K and Gelbard A S 1989 *Hepatic Encephalopathy Experimental Biology and Medicine* vol 22 (New York: Humana) pp 27–48

[92] Hibbard T and Killard A J 2011 *J. Breath Res.* **5** 037101

[93] Cheeke P R and Dierenfeld E S 2010 *Comparative Animal Nutrition and Metabolism* (Wallingford: CABI)

[94] Murray R K, Bender D and Botham K M 2012 *Harper's Illustrated Biochemistry* 29th edn (New York: McGraw-Hill)

[95] Sh. Burbaeva G, Boksha I S, Turishcheva M S, Savushkina O K, Tereshkina E B and Vorob'eva E A 2001 *Vestn. Ross. Akad. Med. Nauk.* **7** 34–7

[96] Rubio M D, Drummond J B and Meador-Woodruff J H 2012 *Biomol. Ther.* **20** 1–18

[97] Phillips M 1992 *Sci. Am.* **267** 74–9

[98] Hibbard T and Killard A J 2011 *Crit. Rev. Anal. Chem.* **41** 21–35

[99] Aguilar A D *et al* 2008 *IEEE Sens. J.* **8** 269–73

[100] Devapriam J *et al* 2008 Use of Levomepromazine in the management of aggression in adults with intellectual disability *Br. J. Dev. Disabil.* **54** 11–7

[101] Handlogten M E *et al* 2005 Expression of the ammonia transporter proteins, Rh B glycoprotein and Rh C glycoprotein, in the intestinal tract *Am. J. Physiol. Gastrointest. Liver Physiol.* **288** G1036–47

[102] Aukst-Margetic B *et al* 2004 Levomepromazine helps to reduce sleep problems in patients with PTSD *Eur. Psychiatry* **19** 235–6

[103] Puri B K, Ross B M and Treasaden I H 2008 Increased levels of ethane, a non-invasive, quantitative, direct marker of n-3 lipid peroxidation, in the breath of patients with schizophrenia *Prog. Neuro-Psychopharmacol. Biol. Psychiatry* **32** 858–62

[104] Ross B M, Shah S and Peet M 2011 Increased breath ethane and pentane concentrations in currently unmedicated patients with schizophrenia *Open J. Psychiatry* **1** 1–7

[105] Yanik M *et al* 2003 Is the arginine-nitric oxide pathway involved in the pathogenesis of schizophrenia? *Neuropsychobiology* **47** 61–5

[106] Do K Q *et al* 2000 Schizophrenia: glutathione deficit in cerebrospinal fluid and prefrontal cortex *in vivo* Eur. *J. Neurosci.* **12** 3721–8

[107] Uttara B *et al* 2009 Oxidative stress and neurodegenerative diseases: a review of upstream and downstream antioxidant therapeutic options *Curr. Neuropharmacol.* **7** 65–74

[108] Halliwell B 2001 Role of free radicals in the neurodegenerative diseases: therapeutic implications for antioxidant treatment *Drugs Aging* **18** 685–716

[109] Mahadik S P, Evans D and Lal H 2001 Oxidative stress and role of antioxidant and omega-3 essential fatty acid supplementation in schizophrenia *Prog. Neuropsychopharmacol. Biol. Psychiatry* **25** 463–93

[110] Fendri C *et al* 2006 Oxidative stress involvement in schizophrenia pathophysiology: a review *Encephale* **32** 244–52

[111] Michel T M *et al* 2004 Cu, Zn- and Mn-superoxide dismutase levels in brains of patients with schizophrenic psychosis *J. Neural Transm.* **111** 1191–201

[112] Pavlovic D, Tamburic V and Stojanovic I 2002 Oxidative stress as marker of positive symptoms in schizophrenia *Facta Univ.* **9** 157–61

[113] Khan M M *et al* 2002 Reduced erythrocyte membrane essential fatty acids and increased lipid peroxides in schizophrenia at the never-medicated first-episode of psychosis and after years of treatment with antipsychotics *Schizophr. Res.* **58** 1–10

[114] Phillips M *et al* 1995 Volatile organic compounds in the breath of patients with schizophrenia *J. Clin. Pathol.* **48** 466–9

[115] Ranjekar P K *et al* 2003 Decreased antioxidant enzymes and membrane essential poly-unsaturated fatty acids in schizophrenic and bipolar mood disorder patients *Psychiatry Res.* **121** 109–22

[116] Boskovic M *et al* 2011 Oxidative stress in schizophrenia *Curr. Neuropharmacol.* **9** 301–12

[117] Ciobica A *et al* 2011 Oxidative stress in schizophrenia—focusing on the main markers *Psychiatr. Danub.* **23** 237–45

[118] Akyol O *et al* 2002 The indices of endogenous oxidative and antioxidative processes in plasma from schizophrenic patients. The possible role of oxidant/antioxidant imbalance *Prog. Neuropsychopharmacol. Biol. Psychiatry* **26** 995–1005

[119] Dietrich-Muszalska A, Olas B and Rabe-Jablonska J 2005 Oxidative stress in blood platelets from schizophrenic patients *Platelets* **16** 386–91

[120] Othmen L B *et al* 2008 Altered antioxidant defense system in clinically stable patients with schizophrenia and their unaffected siblings *Prog. Neuropsychopharmacol. Biol. Psychiatry* **32** 155–9

[121] Altuntas I *et al* 2000 Erythrocyte superoxide dismutase and glutathione peroxidase activities, and malondialdehyde and reduced glutathione levels in schizophrenic patients *Clin. Chem. Lab. Med.* **38** 1277–81

[122] Dakhale G *et al* 2004 Oxidative damage and schizophrenia: the potential benefit by atypical antipsychotics *Neuropsychobiology* **49** 205–9

[123] D'souza B and D'souza V 2003 Oxidative injury and antioxidant vitamins E and C in schizophrenia *Indian J. Clin. Biochem.* **18** 87–90

[124] Halliwell B and Whiteman M 2004 Measuring reactive species and oxidative damage *in vivo* and in cell culture: how should you do it and what do the results mean? *Br. J. Pharmacol.* **142** 231–55

[125] Kuloglu M *et al* 2002 Lipid peroxidation and antioxidant enzyme levels in patients with schizophrenia and bipolar disorder *Cell Biochem. Funct.* **20** 171–5

[126] Rukmini M, D'souza B and D'souza V 2004 Superoxide dismutase and catalase activities and their correlation with malondialdehyde in schizophrenic patients *Indian J. Clin. Biochem.* **19** 114–8

[127] Surapaneni K 2007 Status of lipid peroxidation, glutathione, ascorbic acid, vitamin E and antioxidant enzymes in schizophrenic patients *J. Clin. Diagn. Res.* **1** 39–44

[128] Zhang X *et al* 2006 The effects of Ginkgo biloba extract added to haloperidol on peripheral T cell subsets in drug-free schizophrenia: a double-blind, placebo-controlled trial *Psychopharmacology* **188** 12–7

[129] Sivaraman P, Rattehalli R D and Jayaram M B 2013 Levomepromazine for schizophrenia. Cochrane Database *Syst Rev.* **2010** CD007779

[130] Hibbard T, Crowley K and Killard A J 2013 Direct measurement of ammonia in simulated human breath using an inkjet-printed polyaniline nanoparticle sensor *Anal. Chim. Acta* **779** 56–63

[131] Lynne R and Andrew M 2010 Palliative and end-of-life care in advanced renal failure *Clin. Med.* **10** 279–81

[132] Gupta S, Wallqvist A, Bondugula R, Ivanic J and Reifman J 2010 Unraveling the conundrum of seemingly discordant protein–protein interaction datasets *Conf. Proc. IEEE Eng. Med. Biol. Soc.* **1** 783–6

[133] Onore C, Careaga M and Ashwood P 2012 The role of immune dysfunction in the pathophysiology of autism *Brain Behav. Immun.* **26** 383–92

[134] Rossignol D A, Frye R E, Rossignol D A and Frye R E 2012 A review of research trends in physiological abnormalities in autism spectrum disorders: immune dysregulation, inflammation, oxidative stress, mitochondrial dysfunction and environmental toxicant exposures *Mol. Psychiatry* **17** 389–401

[135] Depino A M 2013 Peripheral and central inflammation in autism spectrum disorders *Mol. Cell. Neurosci.* **53** 69–76

[136] Goines P E and Ashwood P 2012 Cytokine dysregulation in autism spectrum disorders (ASD): possible role of the environment *Neurotoxicol. Teratol.* **36** 67–81

[137] Michel A M, Choudhury K R, Firth A E, Ingolia N T, Atkins J F and Baranov P V 2012 Observation of dually decoded regions of the human genome using ribosome profiling data *Genome Res.* **22** 2219–29

[138] Poultney C S *et al* 2013 Identification of small exonic CNV from whole-exome sequence data and application to autism spectrum disorder *Am. J. Hum. Genet.* **93** 607–19

[139] mostafa G A and Kitchener N 2009 Serum anti-nuclear antibodies as a marker of autoimmunity in Egyptian autistic children *Pediatr. Neurol.* **40** 107–12

[140] Mostafa G A and Al-Ayadhi L Y 2013 The possible relationship between allergic manifestations and elevated serum levels of brain specific auto-antibodies in autistic children *J. Neuroimmunol.* **261** 77–81

[141] Meguid N A, Dardir A A, Abdel-Raouf E R and Hashish A 2011 Evaluation of oxidative stress in autism: defective antioxidant enzymes and increased lipid peroxidation *Biol. Trace Elem. Res.* **143** 58–65

IOP Publishing

Neurological Disorders and Imaging Physics, Volume 3
Application to autism spectrum disorders and Alzheimer's
Ayman El-Baz and Jasjit S Suri

Chapter 8

The impact of stress on parental behavior following a diagnosis of autism

Diana Delgado, Kimberly N Frame and Laura B Casey

Extensive research shows that parents of children with autism spectrum disorder (ASD) experience depression, anxiety, and symptoms that mirror those of post-traumatic stress more than parents with children that have other developmental delays or parents with typically developing children. Factors such as their unfamiliarity with the child's diagnosis, fear of the unknown once a diagnosis is received, and lack of information regarding evidence-based treatments often result in poor treatment selections. Thus, it is important that parental well-being is assessed at the outset of a diagnosis. When parents are stressed, decision-making skills may be compromised, which may result in selecting non-evidence-based treatments. This chapter discusses the nature of stress and suggests possible ways to curb the symptoms or even prevent their onset. Future research should explore the variables that predict and control parental stress and parental adherence to evidence-based treatments, and investigate the effects of treatment adherence on reducing stress related symptoms in parents of children with ASD.

8.1 Parental stress and the ASD diagnosis

Parenting a child diagnosed with ASD involves negotiating a number of situations that can be categorized as stressors. The initial stressor occurs when a parent becomes aware that the development of their child is delayed. This can occur either from personal observation, concerned friends or family, or a medical professional suggesting that something may be outside of the normal range of typical development. Unfortunately, the time frame from recognition of atypical development to a formal diagnosis can be years. This lapse of time can result in a myriad of problems including the child's behaviors worsening, parents seeking information from a variety of sources of varying quality, loss of employment, and/or a decrease in quality of life [1–3].

doi:10.1088/978-0-7503-1793-1ch8

Although parental reactions to their child's developmental trajectory may be idiosyncratic, research has found that parents of children with a disability experience stress more often than parents with a neurotypical child. Additionally, parents of children with ASD reportedly experience more stress than parents with other types of disabilities. These findings date back over a decade. In 2008, Davis and Carter described that parents of autistic children report acute and chronic stress more often than parents of children with other developmental disabilities. Most of this research on parental stress, whether acute or chronic, is often focused on the anxiety related to the unknown, the child's disruptive behavior and lack of pro-social behavior, and the uncertainties of the future as their child ages [4]. In addition to the research on the daily experiences surrounding living with a child with autism, researchers have attempted to focus on specific triggers that may have a causal relationship with respect to stress related symptoms. Specifically, the delivery of the diagnosis is an area that researchers have begun to investigate as a common trigger for stress.

Current research on receiving the formal diagnosis is mixed. While some researchers report that receiving a formal diagnosis is relieving and akin to receiving a missing piece to a puzzle they were trying to solve, other researchers report up to 20% of parents experiencing symptoms that mirror those of post-traumatic stress following the diagnosis [6]. The nature of the symptoms that are commonly experienced by someone with post-traumatic stress include hyper-arousal, intrusion of thoughts, and flashbacks. These symptoms differ from those of a parent describing their experiences of daily stress as involving general depression or an anxiety related concern.

Research is actively being conducted on ways to mitigate negative parental responses to their child receiving a diagnosis (e.g. how the diagnosis is made or delivered and direct education on bedside manner). However, there are other day-to-day events when raising a child with ASD that, according to parents, are also stress triggers and efforts to support parents in these areas need attention as well. For example, parents may find themselves unable to understand communication attempts by their child [7], have difficulty handling problematic behavior, and struggle with non-socially appropriate bids for attention, access to toys, or avoidance of unwanted activities [8, 9], all of which can affect family routines and disrupt family cohesion. For example, Brobst *et al* (2009) [10] report that relationship satisfaction is lower among parents with children with autism. Of course, all parents experience stress from time to time as they juggle work, school, extracurricular activities, family problems, etc, however, extant research suggests that parents of children with ASD have additional stressors that put them at a higher risk.

Although responses to stressors or stimuli that are potentially threatening constitute adaptive physiological reactions, they can be harmful when maintained for long periods of time. Symptoms associated with prolonged exposure to stress such as depression, anxiety, digestive problems, irritability, insomnia, a higher propensity for illnesses, nightmares, isolation, difficulty planning, impaired decision-making skills, among others are known to be detrimental to the health and well-being of the individual [6]. In addition, individuals who are exposed to stressful situations, and who do not have a supportive social network or adequate coping skills can sometimes handle stress in ways that can have further deleterious effects on their health (e.g. smoking, drinking) [1]. Parents experiencing these symptoms should be treated by trained clinicians so that their health and well-being can improve. If left untreated, stress, anxiety, and depression may also have detrimental effects on the child [11].

8.2 The potential effect of stress on parental treatment choices

Unfortunately, fad treatments for ASD have proliferated on the treatment market with over 400 treatments available for parents and caregivers to choose from, most with little to no evidence base to support their use [12]. These treatments are quite popular among parents and caregivers despite a lack of robust scientific evidence supporting their effectiveness. Some of the most popular fad treatments include: those based on biomedical approaches (gluten-free diets, chelation, vitamin and hormone therapies, secretin therapy, hyperbaric chambers, etc), animal therapies (dolphin-assisted therapy and horseback riding therapies), and many others such as sensory integration, facilitated communication, floor time, holding therapy, music therapy, prism glasses, weighted vests, and essential oils therapy (information on each treatment can be found at asatonline.org).

The American Psychological Association has criteria to evaluate therapies as having an evidence base, therapies that are promising, and those that have no evidence at all [13]. In addition, Horner and Kratcohwill [14] have published as a requirement for a treatment to be considered as having an evidence base, the use of a single subject design, a common research design used in the field of behavior analysis. Unfortunately, knowledge of these guidelines is not the same as being able to evaluate evidence-based research. Most families are not taught how to evaluate social and behavioral research, much less to know the requirements for rigorous experimental, quasi-experimental, or qualitative research. The inability to find reputable information for a treatment may contribute to the stress associated with identifying and selecting treatments. The enormous difficulty of this task is probably best demonstrated in the extant research related to preferred sources for information about ASD in which parents indicate other parents of children with ASD as highly preferred sources of information [15].

In addition to issues regarding the research evidence (or lack thereof) for autism treatment, many parents often use more than one treatment for their child [16, 17]. The reasons for using what in [18] is called the 'buffet approach' to treatment selection vary, and can include distress over lack of progress, being given a 'time frame' for best outcomes (i.e. treatment should begin by a certain age), or being overwhelmed by the sheer number of treatment choices [19, 20]. The fact that there are no consistent recommendations from professionals as to which treatments to complete (e.g. contrary to a medical illness which may have 1–2 typical treatment recommendations) may further complicate this decision. Some parents have reported receiving no recommendations at all from the diagnosing doctors [19].

8.2.1 The prevalence of fad treatments

According to Vyse [21] there are several reasons why fad therapies prevail. One reason he gives is the expectation that most individuals have that a treatment for a health problem should be completely effective. For example, when you go to the doctor for an ear infection, you receive medication, and the infection goes away. This is in stark contrast to treatments for autism, where even the effect of evidence-based therapies may be at times slow or long delayed [19]. With the medication example above the doctor is able to tell you an approximate length of time in which you should see results, however, with autism treatments the length of time to see results may be unknown and the results uncertain. The second reason Vyse gives is that the evidence-based treatment may be 'onerous or distasteful' [21, p 10]. Applied behavior analysis, an evidence-based treatment for ASD, typically has treatment recommendations of up to 40 h per week [22].

The beliefs about the etiology of a disorder are also likely to affect the probability that families will select non-evidence-based treatments. For example, those individuals that believe ASD is caused by a spirit may choose a fortune teller [23] or those that believe ASD is caused by mercury poisoning may choose chelation therapy [24]. Another reason mentioned by Vyse [21] for selecting fad therapies is the promotion of a treatment by a professional group. An example of this is facilitated communication. This is a thoroughly discredited treatment for teaching communication skills to individuals with disabilities [25], the promoters of which are housed at a well-respected university (Syracuse). The final reason Vyse gives for the prevalence of fad treatments is science denial. Science denial is definitely at work for those who rely on facilitated communication, but it is also evidenced in beliefs about the origin of ASD (e.g. vaccines).

When it comes to selecting a treatment for a child, arguments of scientific empirical data supporting treatment effectiveness may not mean much to families that are desperately seeking help, and thus may fall short of their aim. The challenge doctors and clinicians face is how to affect parents' choices with respect to their children and the challenges that an autism diagnosis poses. The high prevalence and demand for fad treatments by consumers of behavioral services, and the findings from studies showing higher levels of stress symptoms in mothers of children with ASD [3, 10], should be a red flag indicating that some aspect of the treatment plan is not being sufficiently covered. We suggest that what may be missing is an analysis of the variables that predict and control parental behavior with respect to the ASD diagnosis and the selection of treatment choices.

Focusing explicitly on parental responses to the child's behavior may need to be part of comprehensive behavioral programs. The contributions of the behavioral literature on parent training may constitute the basis upon which further research in this area may be developed. Increasing effective parenting behaviors that will facilitate parent–child interactions should be one of the primary goals of researchers and clinicians.

8.3 Parents as the agents of behavioral change

Applied behavior analysis (ABA) has effectively improved the lives of children with ASD for over 30 years [26]. Providers of behavioral services and in particular, Board Certified Behavior Analysts (BCBAs), are trained to design and deliver highly effective interventions that result in significant reductions in problem behavior, inclusion of children with an ASD diagnosis in general education classrooms, and acquisition of skills that promote independent functioning and successful integration in society. However, the behavior of the parents, a most influential part of the child's development, has not received equal attention.

There is extensive literature to show that for behavior programs to be effective, parent involvement is necessary [27, 30–32]. When some aspects of the child's environment are changed to produce positive changes in his/her behavior, these will be maintained only to the extent that they also occur with the same level of consistency in all other environments that the child encounters.

Most children with ASD struggle with generalization of learned responses to novel stimuli or environments. In other words, it is insufficient for a therapist to teach a skill such as independently making a request in the clinic and assume that requesting will also occur in other environments (at the restaurant, at the park, at home, during dinner, etc). For this reason, explicitly programming for generalization needs to involve the participation of parents and caregivers as interventionists at home [28]. Because parents are the stimuli that are common across the many different settings in which the child is likely to be, they can quickly acquire discriminative control over behavior change, thus enhancing generalization.

Further, the long-term cost effectiveness of parent training is undeniable. When contingencies at home start to mirror those arranged at clinics, schools, or centers, goals will be reached faster, thus reducing the duration of services. In some cases, portions of the behavioral program may be parent implemented, and in some other cases, the majority or the totality of the program may be implemented by parents. Such is the case when families live in remote areas where access to behavioral services is limited or requires commuting for several hours. For behavioral programs to be successful, it is imperative that parents are in a position to be an active part of the child's therapy.

Among the most effective practices for parent training, behavioral skills training (BST), is one of the most commonly used and empirically researched [33, 35]. BST uses four basic components to train a skill to a mastery criterion, which are instructions, modeling, rehearsal, and feedback. In the instructions component, the trainer provides written and oral instructions describing the steps that the skill requires, and explains its importance and rationale. During the modeling component, the trainer models the skills while the trainee observes. In rehearsal, the trainer allows the trainee to practice the skill, and in the feedback component, the trainer provides positive feedback for correct performances and corrective feedback for incorrect responses.

BST has been used to train parents in a variety of strategies that can be used to improve problem behaviors at home. For example, Mueller *et al* [34] trained mothers in feeding interventions to treat their children's food-related challenging behaviors, Miles and Wilder [35] successfully taught parents how to increase their children's frequency of compliance to demands, and Lafasakis and Sturmey [36] trained parents to teach gross motor imitation skills to children who were non-verbal and engaged in severe disruptive behaviors. Results from these studies show that parents were able to successfully implement the treatment procedures, and as a result, children reduced the frequency of their disruptive behaviors.

In a more recent study, Dogan *et al* [37] trained parents on the use of BST to teach children social skills at home. Using BST, parents taught their children two socially relevant skills: joining a conversation and asking for help. To accomplish this goal, each of these two skills was divided into four simpler steps, which were modeled and practiced using contrived social situations. During the modeling phase, the trainers first demonstrated how to use BST with a graduate assistant playing the role of the child, and later, with the parent playing the role of the child. Finally, during the rehearsal and feedback phases, the parent was given the opportunity to teach the skills to the trainer playing the role of the child. Comparisons of data between conditions and across participants showed increased performance of children's social skills, and increased accuracy of parents' implementation of BST steps in post-training, generalization probes, and in a one-month follow-up. All of these successful empirical demonstrations of the effectiveness of BST as a model for training parents demonstrated high procedural integrity, maintenance at follow-ups, generalization of the trained skills to novel conditions and settings, and high ratings of parent's satisfaction with respect to the procedures and the training outcomes.

8.4 Factors affecting parental involvement

8.4.1 Making parent training accessible

The available data indicate that BST constitutes an effective evidence-based technology for parent training. In most of these studies, parents report that they find the acquired skills useful and that they are likely to continue to use them at home. However, the extent to which parents actually do so outside the context of research studies has not been examined [35, 36].

A lack of support and of continued access to training may be one of the factors contributing to the discontinuation of parent implemented interventions. Gerow *et al* [38] examined the social validity of parent implemented functional communication craining (FCT) across 26 studies. Among several other social validity indicators, the authors examined whether parent training was conducted using resources and conditions that are available for parents outside the context of the study. Their results show that although in 85% of the studies parent training was conducted at home, the trainer was either the experimenter or a person who was not regularly available to the parents.

Despite the recognized importance and well-researched effectiveness of parent training, restricted resources and funding may make it difficult for service providers to incorporate parent training programs regularly. In addition, many small towns and rural areas lack enough trained professionals to fully cover the needs of these families. In these circumstances, the need for services can only be met with the resources made available by service providers such as school administrators, and highly qualified staff members and supervisors in clinics, agencies, and centers.

In the past ten years, several research efforts have evaluated the effectiveness of distance learning programs [39, 40], online parent training options [41, 42], and educator delivered parent training, with very promising outcomes. These resources allow parents who have limited access to behavioral services to learn how to effectively implement basic ABA techniques at home. For example, Heitzman-Powell *et al* [41] evaluated an ABA distance training program with seven families who live in rural areas with no access to behavioral services. The parent training program included a series of interactive training modules, an assessment, and an online coaching session. During coaching sessions parents had the opportunity to discuss the content of the modules, and obtain feedback on their use of incidental teaching strategies. The modules reviewed the basic principles of behavior and other topics including the definition and observation of target behaviors, manipulation of environmental variables, skill acquisition strategies, and basic procedures for the reduction of challenging behaviors. Parents advanced through the modules by obtaining scores of 90% or higher on their corresponding evaluations, and 80% or higher on treatment fidelity measures.

Coaching sessions took place in a videoconference room at the local school or community center where the parents lived. Throughout the training, parents were also encouraged to engage in short free-play sessions with the children. In these sessions, parents were instructed to: place no demands; engage enthusiastically with the child; and provide immediate, frequent, and specific reinforcement for appropriate interactions. Their overall findings show increased measures of parents' performance after completion of this training program. Furthermore, despite the time intensive investment required by the program, all parents indicated that the program was beneficial in improving their child's behavior. Future research is required to support the outcomes of these parent training models, document their impact on the quality of life of families, and to investigate the extent to which parents are able and willing to integrate these skills successfully as part of their parenting repertoires.

Another useful resource for parents with limited access to behavioral services is to participate in parent training programs delivered by trained educators. Unfortunately, teachers and staff in public school special education programs often lack training on evidence-based interventions that reduce disruptive behaviors and facilitate learning in young children with ASD. An example of a successful parent training model designed for early childhood special education teachers in public schools is described in a study by Ingersoll and Dvortcsac [30]. This program included nine weekly group sessions and three individual coaching sessions. The authors conducted the majority of the group and coaching sessions with the parents while the school staff (the special education teachers, autism specialist, speech language pathologist, and occupational therapist) observed and took notes for later discussion with the trainers. During the last few sessions, educators conducted the sessions with the parents and received feedback from the authors.

Group sessions included topics such as selection of behavioral goals, implementing interventions based on incidental teaching, and other techniques such as modeling, shaping, prompting, and using reinforcement based procedures. During these sessions, parents watched and discussed videos showing the implementation of these procedures in naturalistic contexts. In individual coaching sessions, the trainer modeled specific interventions with the child, observed the parents implementing the modeled interventions, provided feedback, and discussed opportunities to use the interventions at home. Increments in the post-training scores of parents' correct implementation of the procedures were observed in all parents who participated in the program.

Coaching sessions were the preferred component of this program as reported by both parents and teachers. In these sessions, parents learned to use daily interactions in the home as opportunities to teach and strengthen communication skills in their children. Teachers and school staff also reported that they would like to continue using the program as part of the special education curriculum and that they would use this as a resource to instruct other educators on the delivery of the training program [30].

One of the main strengths of these programs is their focus on incidental teaching. While learning that occurs during structured training trials such as discrete trial training (DTT) is also recommended, incidental teaching strategies favor generalization and maintenance of treatment gains [28]. In summary, the success of these programs provides evidence to support the development of parent training models that can be adapted to fit the needs of families with restricted access to behavioral services. Their design and implementation benefit a wide range of families in need of behavioral services and also promote the dissemination of empirically validated techniques in families, schools, and communities.

8.4.2 Focusing on the contingencies that maintain parents' behaviors

To date, there is no research documenting the extent to which parents continue to use the skills acquired in parent training programs [32, 38]. However, high levels of stress, anxiety, and depression in parents of children with ASD, and increasing demands for fad treatments, seem to suggest that the effects of parent training programs may be short-lived. Generally, the focus of behavior analytic services is on the children as the direct recipients of those services. However, it is possible that the field has underestimated the extent to which parents require support from behavioral service providers. It is possible that parents' responses to the challenges involved in having a child with ASD need further attention from both researchers and practitioners.

While we have spent an enormous amount of time researching treatment effectiveness, we lack data documenting the extent to which parents continue to use these strategies, or examining the variables that produce their continuous use.

Working towards lasting behavior change may require examining the variables that control parental behavior at home, and designing programs that will increase the sustainability of parent implemented programs over significant periods of time.

A behavior analytic view implies acknowledging that while some of the symptoms of stress experienced by parents are automatic reactions of physiological systems, some of the behaviors involved in what is called stress, i.e. anxiety, depression, difficulty planning, etc, are also maintained by their effect on the environment, and are thus, under the control of operant contingencies. Therefore, what we may be lacking in our parent training models is an operant analysis of parenting behavior that accounts for the factors that contribute to the continued use of treatment recommendations.

One of the factors that is likely to contribute to parental non-adherence is the immediate negative reinforcement effect produced by the termination or avoidance of the child's problem behaviors. For example, if delivering attention reliably follows the occurrence of problem behavior, doing so will momentarily terminate each particular episode of the behavior. This momentary termination (or avoidance) of the problem behavior is a powerful reinforcer for parents and caregivers, that inadvertently reinforces or strengthens the problem behaviors of the child [27].

For this reason, behavioral procedures such as extinction and differential reinforcement are particularly difficult for parents and caregivers: if providing attention, providing access to a preferred activity, or terminating a demand have prevented or terminated the occurrence of challenging behaviors in the past, parents may be likely to continue doing so when other alternatives do not result in the same effect [43]. Studies have demonstrated that parents and caregivers present fewer demands if children engage in problem behavior, and are more likely to deliver attention (whether positive attention or verbal reprimands), when undesirable behaviors occur [32, 44]. In addition, children will engage in problem behaviors when access to the things they want and need is the reliable consequence of doing so.

Further complicating matters for treatment adherence is the fact that the consistency with which these operant relations are established is such that resurgence of the problem behaviors is likely if treatment variables are not as consistently applied. As the literatures in behavioral resurgence and resistance to extinction demonstrate, if reinforcement for the alternative appropriate behavior is withheld, the problem behavior is likely to re-occur [45, 46]. Likewise, if acquired parenting skills fail to produce the reinforcing consequence, e.g. appropriate behavior in the child, the parenting behaviors that prevailed prior to parent training will quickly take the place of the more recently acquired skills.

Because establishing a new behavioral repertoire for parents and children, under the control of new contingencies of reinforcement, takes a considerable amount of discipline and persistence, parent training programs need to include powerful reinforcing contingencies for parental adherence. One very effective and natural reinforcer for parent involvement is the child's success. Small gains that can be attributed to parents' changes in their own behavior can significantly contribute to the parents' continued involvement. Teaching the child to reinforce appropriate parenting behavior [32], and designing self-monitoring programs for parents to keep track of their own behavior as agents of change may also contribute to facilitating long lasting contingency changes in the home.

In accordance with a functional analysis, however, for any of these strategies to be effective, clinicians need to understand the controlling variables of parental behavior. Identifying these variables through a functional assessment may facilitate the design of parent training programs that take into account the needs and motivations of the parents. Future research in this area should evaluate (a) the effectiveness of parent training models that include contingencies for the sustained use of parent implemented procedures, and (b) the extent to which a functional analysis of parents' behavior will increase the effectiveness of these models by identifying the relevant controlling variables of the behavior of the parents [27].

An analysis of the variables maintaining appropriate parental behavior may be a powerful contribution to parent training models and behavioral intervention plans. It will facilitate the design of programs that are effective and also socially significant in terms of their long-term impact on the well-being of families. An account of the social relevance of behavioral goals, procedures, and outcomes of ABA interventions involves, among other indicators, examining the long-term effects of treatment programs and their impact in the quality of life of the families that those programs are designed to serve [38]. If ABA is not producing a lasting improvement in the family's quality of life, or if it is not evaluating what needs to be done to produce this outcome, it may be falling short of its aim.

8.5 Tying it all together: mitigating stress, selecting evidence-based treatments, and increasing parental involvement

Symptoms of stress in parents of children with ASD can occur for a variety of reasons from a poor delivery by a medical professional confirming the diagnosis, to being unable to maintain a cohesive family unit, to ultimately selecting and implementing non-evidence-based treatments. When there is a lack of emphasis on parental well-being or when parental issues are completely neglected by providers, medical staff and therapists could unintentionally increase the parents' symptoms of stress and hinder the child's progress. One recommendation could be that front-line staff delivering the diagnosis be trained to evaluate parent and child, and to ensure that timely services are sought at the outset of the diagnosis or when parental stress symptoms are apparent. In these cases medical staff and behavior analysts should either recommend counseling services to help the parent (for issues related to the parent's mental health), or design strategies that are conducive to teaching the parent how to work more effectively with their child. Empowering parents through education would increase their self-efficacy and possibly in turn reduce stress symptoms.

One way to reduce the onerous nature of a treatment and to increase the probability of treatment gains is effective parent training. Although many clinics and centers offer parent training programs, initial or sustained parent involvement in treatment goals, procedures, and changes in those procedures may not always occur. In an effort to secure parent buy-in, parent training programs should entail programming reinforcement contingencies for parents' behaviors that result in sustained involvement in the child's behavioral program and implementation of the recommended procedures at home.

Generalization and maintenance of positive treatment outcomes may be threatened when treatments are withdrawn, implemented inconsistently, and/or combined with treatments that have no evidence base demonstrating their effectiveness. Therefore, engaging parents continuously in short-term goals towards the improvement of the child's behavior involves their participation in activities such as collaborating in the goal setting process, looking at progress graphs, learning how to implement behavior change procedures, and discussing the cost effectiveness of seeking alternative treatments.

Parental stress as a result of parenting a child with ASD is a critical issue that needs to be addressed by professionals who treat children with ASD, in particular if the upward trend in the prevalence of ASD continues. If neglected, it may increase the number of children who are left untreated or subjected to ineffective treatments. Future research must evaluate the impact of parent training models on parental well-being, determine the success of these models in providing appropriate and timely access to treatment, and examine their overall impact in reducing the stress levels reported by parents of children with ASD.

References

[1] Benson P R 2018 The impact of child and family stressors on the self-rated health of mothers of children with autism spectrum disorder: Associations with depressed mood over a 12-year period *Autism* **22** 489–501

[2] Gibson A N, Kaplan S and Vardell E 2017 A survey of information source preferences of parents of individuals with Autism Spectrum Disorder *J. Autism Dev. Disord.* **47** 2189–204

[3] McStay R L, Trembath D and Dissanayake C 2014 Maternal stress and family quality of life in response to raising a child with autism: From preschool to adolescence *Res. Dev. Disab* **35** 3119–30

[4] Davis N O and Carter A S 2008 Parenting stress in mothers and fathers of toddlers with autism spectrum disorders: Associations with child characteristics *J. Autism Dev. Disord.* **38** 1278

[5] Beck A, Daley D, Hastings R P and Stevenson J 2004 Mothers' expressed emotion towards children with and without intellectual disabilities *J. Intellec. Disab. Res* **48** 628–38

[6] Casey L B, Zanksas S, Meindl J N, Parra G R, Cogdal P and Powell K 2012 Parental symptoms of posttraumatic stress following a child's diagnosis of autism spectrum disorder: a pilot study *Res. Autism Spectrum Disord.* **6** 1186–93

[7] Goin-Kochel R P and Myers B J 2005 Congenital versus regressive onset of autism spectrum disorders: parents' beliefs about causes *Focus Autism Other Dev. Disab.* **20** 169–79

[8] Bishop S L, Richler J, Cain A C and Lord C 2007 Predictors of perceived negative impact in mothers of children with autism spectrum disorder *Am. J. Mental Retard.* **112** 450–61

[9] Hastings R P 2003 Child behaviour problems and partner mental health as correlates of stress in mothers and fathers of children with autism *J. Intell. Disabil. Res.* **47** 231–7

[10] Brobst J B, Clopton J R and Hendrick S S 2009 Parenting children with autism spectrum disorders: the couple's relationship *Focus Autism Other Dev. Disab.* **24** 38–49

[11] Mowery B D 2011 Post-traumatic stress disorder (PTSD) in parents: Is this a significant problem? *Pediatric Nursing* **37** 89–92

[12] Smith T 2014 Field report: Promoting evidence-based interventions: The Association for Science in Autism Treatment *Behav. Anal. Pract.* **7** 147–81

[13] APA Presidential Task Force on Evidence-Based Practice 2006 Evidence-based practice in psychology *Am. Psychol.* **61** 271–85

[14] Horner R H and Kratochwill T R 2012 Synthesizing single-case research to identify evidence-based practices: Some brief reflections *J. Behav. Educ.* **21** 266–72

[15] Mackintosh V, Myers B and Goin-Kochel R 2005 Sources of information and support used by parents of children with autism spectrum disorders *J. Dev. Disabil.* **12** 41–51

[16] Bowker A, D'Angelo N, Hicks R and Wells K 2011 Treatments for autism: parental choices and perceptions of change *J. Autism Dev. Disord.* **41** 1373–82

[17] Green V, Pituch K, Itchon J, Choi A, O'Reily M and Sigafoos J 2006 Internet survey of treatments used by parents of children with autism *Res. Dev. Disabil.* **27** 70–84

[18] Schreck K A and Mazur A 2008 Behavior analyst use of and beliefs in treatments for people with autism *Behav. Interv.* **23** 201–12

[19] Matson J L, Adams H L, Williams L W and Rieske R D 2013 Why are there so many unsubstantiated treatments in autism? *Res. Autism Spectr. Disord.* **7** 466–74

[20] Zane T, Davis C and Rosswurm M 2008 The cost of fad treatments in autism *J. Early Intensive Behav. Interv.* **5** 44–51

[21] Vyse S 2015 Where do fads come from? *Controversial Therapies for Autism and Intellectual Disabilities: Fad, Fashion, and Science in Professional Practice* ed R M Foxx and J A Mulick (New York: Routledge) pp 23–36

[22] Lovaas O I 1987 Behavioral treatment and normal educational and intellectual functioning in young autistic children *J. Consult. Clin. Psychol.* **55** 3–9

[23] Shyu Y, Tsai J and Tsai W 2010 Explaining and selecting treatments for autism: parental explanatory models in Taiwan *J. Autism Dev. Disord.* **40** 1323–31

[24] Lack C W and Rousseau J 2016 *Critical Thinking, Science, and Pseudoscience* (New York: Springer)

[25] Jacobson J W, Foxx R M and Mulick J A 2015 Facilitated communication: the ultimate fad treatment *Controversial Therapies for Autism and Intellectual Disabilities: Fad, Fashion, and Science in Professional Practice* ed R M Foxx and J A Mulick (New York: Routledge) pp 283–302

[26] Reichow B 2012 Overview of meta-analyses on early intensive behavioral intervention for young children with autism spectrum disorders *J. Autism Dev. Disord.* **42** 512–20

[27] Allen K D and Warzak W J 2000 The problem of parental nonadherence in clinical behavior analysis: Effective treatment is not enough *J. Appl. Behav. Anal.* **33** 373–91

[28] Gena A, Galanis P, Tsirempolou E, Michalopoulou E and Sarafidou K 2016 Parent training for families with a child with ASD: a naturalistic systemic behavior analytic model *Eur. J. Counsel. Psychol.* **4** 4–31

[29] National Research Council 2001 Educating children with autism *Committee on Education and Intervention for Children with Autism, Division of Behavioral and Social Sciences and Education* (Washington, DC: National Academy Press)

[30] Ingersoll B and Dvortcsak A 2006 Including parent training in the early childhood special education curriculum for children with spectrum autism disorders *J. Posit. Behav. Interv.* **8** 79–87

[31] Schreibman L 2000 Intensive behavioral/psychoeducational treatments for autism: research needs and future directions *J. Autism Dev. Disord.* **30** 373–8

[32] Stocco C S and Thompson R H 2015 Contingency analysis of caregiver behavior: Implications for parent training and future directions *J. Appl. Behav. Anal.* **48** 417–35

[33] Johnson B M, Miltenberger R G, Egemo-Helm K, Jostad C M, Flessner C and Gatheridge B 2005 Evaluation of behavioral skills training for teaching abduction-prevention skills to young children *J. Appl. Behav. Anal.* **38** 67–78

[34] Mueller M M *et al* 2003 Training parents to implement pediatric feeding protocols *J. Appl. Behav. Anal.* **36** 545–62

[35] Miles N I and Wilder D A 2009 The effects of behavioral skills training on caregiver implementation of guided compliance *J. Appl. Behav. Anal.* **42** 405–10

[36] Lafasakis M and Sturmey P 2007 Training parent implementation of discrete trial teaching: Effects on generalization of parent teaching and child correct responding *J. Appl. Behav. Anal.* **40** 685–9

[37] Dogan R K, King M L, Fischetti A T, Lake C M, Mathews T L and Warzak W J 2017 Parent-implemented behavioral skills training of social skills *J. Appl. Behav. Anal.* **50** 805–18

[38] Gerow S, Hagan-Burke S, Rispoli M, Gregori E, Mason R and Ninci J 2018 A systematic review of parent-implemented functional communication training for children with ASD *Behav. Modif.* **42** 335–63

[39] Nefdt N, Koegel R, Singer G and Gerber M 2010 The use of a self-directed learning program to provide introductory training in pivotal response treatment to parents of children with autism *J. Posit. Behav. Interv.* **12** 23–32

[40] Wainer A L and Ingersoll B R 2013 Disseminating ASD interventions: a pilot study of a distance learning program for parents and professionals *J. Autism Dev. Disord.* **43** 11–24

[41] Heitzman-Powell L S, Buzhardt J, Rusinko L C and Miller T M 2014 Formative evaluation of an ABA outreach training program for parents of children with autism in remote areas *Focus Autism Other Dev. Disabil.* **29** 23–38

[42] Vismara L A, Young G S, Stahmer A C, Griffith E M and Rogers S J 2009 Dissemination of evidence-based practice: can we train therapists from a distance? *J. Autism Dev. Disord.* **39** 1636

[43] Addison L and Lerman D C 2009 Descriptive analysis of teachers' responses to problem behavior following training *J. Appl. Behav. Anal.* **42** 485–90

[44] Carr E G, Taylor J C and Robinson S 1991 The effects of severe behavior problems in children on the teaching behavior of adults *J. Appl. Behav. Anal.* **24** 523–35

[45] Kestner K and Peterson S M 2017 A review of resurgence literature with human participants *Behav. Anal.: Res. Pract.* **17** 1–17

[46] Mace F C, McComas J J, Mauro B C, Progar P R, Taylor B, Ervin R and Zangrillo A N 2010 Differential reinforcement of alternative behavior increases resistance to extinction: clinical demonstration, animal modeling and clinical test of one solution *J. Exp. Anal. Behav.* **93** 349–67

IOP Publishing

Neurological Disorders and Imaging Physics, Volume 3
Application to autism spectrum disorders and Alzheimer's
Ayman El-Baz and Jasjit S Suri

Chapter 9

Visual saliency for medical imaging and computer-aided diagnosis

Olfa Ben-Ahmed, Christine Fernandez-Maloigne, Adrien Julian and Marc Paccalin

Computational modeling of visual attention is an active research topic in the field of computer vision. It has been successfully explored in many applications such as image and video analysis, object detection, human–machine interaction, and robotics. However, most of the visual attention models were dedicated to natural scenes and their investigation in the context of medical images for computer-aided diagnosis is still limited. In clinical practice, radiologists visually analyze a large number of clinical images in a limited time. Such a fastidious task can lead to mistakes, as the radiologist may overlook subtle abnormalities that they might not wish to omit. Hence, proposing new techniques that deploy the radiologist's visual attention is necessary to develop better training programs and to create new tools to assist clinical decision making. In this chapter, we present the relevance of computational saliency models in medical imaging in the context of abnormality detection and computer-aided diagnosis. Then, we present a new saliency-based method to automatically modulate the clinician's visual inspection of regions of interest (ROIs) for Alzheimer's disease (AD) detection. The proposed method follows the visual analysis made by radiologists when analyzing magnetic resonance images (MRIs), allowing in addition a quantitative determination of the brain's ROIs which are different between subjects. The proposed saliency model combines bottom-up and top-down attention models using domain knowledge in brain MRI analysis and Alzheimer's disease diagnosis. The proposed approach detects and qualifies the state-of-the-art ROIs for AD diagnosis and provides discriminating brain patterns for an efficient AD/mild cognitive impairment (MCI) subject classification.

9.1 Introduction

Alzheimer's disease (AD) is the most common form of dementia. Structural MRI (sMRI) is an integral part of the clinical assessment providing a way for clinicians to detect brain abnormalities for an early AD diagnosis. In this context, several methods of visual brain MRI analysis have been proposed in the literature. The main goal of such methods is to examine and identify anatomical brain differences that can be associated with the presence or the absence of pathology. Voxel based morphometry (VBM) [1] is a computational neuroimaging analysis that compares regional patterns of the brain between groups of subjects by performing statistical tests across all voxels in the MRI scans. VBM has been widely applied to study the gray matter (GM) density variation [2–8]. Traditional voxel-wise methods are the gold standard methods for brain MRI analysis but they are still far from emulating the clinician diagnosis process for a diagnosis at the individual level. The short-coming of such methods is that every voxel in the image is analyzed individually. Usually, the clinician identifies neurodegenerative diseases in MRI scans by looking for a disease specific pattern of neurodegeneration in the brain. This suggests that the decision relevant information comprises patterns and not only in single voxels. In addition, some pathologies may affect not only a single anatomical structure or interconnected regions, but specific structures localized far away from each other. These kinds of patterns are difficult to find and to analyze with the standard morphometric techniques. Actually, most of the methods cited above were proposed for group analysis and cannot be used to classify individual patients. To cope with these issues, computer vision tools and visual features-based approaches have been proposed to model the individual brain atrophy patterns. The development and analysis of visual features in medical imaging have been extensively studied for different MRI modalities [9–18].

In clinical practice, visual assessment and analysis of MRI data is the most widely used method for brain atrophy evaluation [19]. In fact, the radiologist focuses attention on brain regions susceptible to change in specific dementia and then achieves structured reporting of these findings [20]. The development of an automated approach for objective visual interpretation could potentially augment the visual skills of radiologists by extracting image features that may be relevant to the diagnosis. This also makes diagnosis easier for clinicians without expertise to extract diagnostically useful information. The visual assessment strategies used by radiologists for AD diagnosis consist usually in restricting eye movements to some regions of interest in the image while scrolling through slices of a given MRI projection [21]. This process includes two distinct processes: the image perception process to recognize image patterns and the process of reasoning to make decisions based on the perceived patterns. The first search process involves the so-called 'visual attention' which is the cognitive process of selectively attending to a region of interest (ROI) while filtering out irrelevant information [22]. In a second step, the clinician identifies relationships between the perceived patterns and possible diagnosis.

The visual attention system is inspired by human behavior and the brain neural architecture to selectively process only the salient visual stimuli [23]. The visual attention models have been primarily influenced by feature integration theory (FIT) [24]. The latter suggests that attention must be directed serially to elementary visual features of the stimulus such as colors, contrast, and orientations. The first visual attention experiments were conducted by Yarbus, one of the founders of modern eye movement research, using the human gaze. He proved that eyes fixate on those scene elements which carry the most important information. From this figure, one can see that fixations (saccades) are on the most important parts of the face such as the eyes, nose, and mouth. The visual attention models output a static two-dimensional saliency map through the use of the obtained saccadic model of eye movements. The obtained saliency maps can indicate informative regions and filter out irrelevant regions and thus predict where observers look.

Computational visual attention models are based on the concept of saliency map to predict fixation patterns and visual search behavior. There are two main categories of saliency models: the bottom-up saliency model which is image-driven based on only low-level features such as contact, intensity, color, and orientation, and the top-down model which is a knowledge-driven model [25]. Visual computational models have been used to detect and characterize salient regions in natural images and videos scenes [25–29], but their investigation in medical images for computer-aided diagnosis has remained very limited. In the context of medical images, the bottom-up influences correspond to image features whereas top-down influences correspond to the knowledge and expertise of radiologists [30]. Modeling the clinician's visual attention when reading medical images can automate pathology detection and hence the diagnosis process [31].

In this chapter, we introduce the visual attention paradigm and we review the most relevant works on the use of visual saliency in medical imaging for computer-aided diagnosis. Then, we present our recent visual saliency-based method for Alzheimer's disease detection and classification. The proposed method tends to emulate the radiologist's first examination step where she/he defines highly informative diagnostic regions [21]. Indeed, it spots and describes the affected ROIs in structural MRI images which could assist the clinician in AD diagnosis making. The proposed saliency model combines a bottom-up and a top-down saliency map using domain knowledge in AD diagnosis. The first takes advantage of low-level MRI characterization (texture and edge) and the second is based on an embedded learning process to identify and localize the subset of gray matter regions that provide optimal discrimination between normal and affected groups. Obtained saliency maps are then used for AD/MCI subject classification in a multiple kernel learning framework.

9.2 Visual saliency for medical image analysis

Modeling expert knowledge and perceptual expertise in medical imaging helps to improve anatomical abnormality detection and then to assist clinicians in diagnostics [32, 33]. Visual saliency has been explored to model the radiologist's visual attention for computer-aided diagnosis by making the area of interest stand out in the foreground from the rest of the background medical image [34]. The saliency method, which detects lesions and tissue abnormalities, has attracted more and more attention and has achieved promising results, in particular in tumor detection in various human organs and on different medical imaging types. Mehmood *et al* [35] proposed a prioritization based approach to help clinicians to quickly determine and access the required level of visual information of a particular brain tumor case from brain MRI. They build a brain MRI visual attention model based on multi-scale contrast, motion, and contour features. Chung *et al* [36] proposed a novel saliency-based method for identifying suspicious regions in multi-parametric MR prostate images based on statistical texture distinctiveness. In [37], the authors proposed a novel saliency-based detection method for the delineation of whole tumor regions from multi-channel brain MR images (FLAIR, T2, and T1C). They used a pseudo-coloring strategy for multi-channel brain MRI for salient region selection. A binary proto-object is then generated for ROI segmentation. Figure 9.1 illustrates the visual results of saliency maps computed on T1-weighted MRI for two high-grade glioma cases. The proposed saliency-based method enables ROI extraction and quantification in terms of tumor size, shape, and position.

Moreover, the study of visual saliency was extended in the context of retinal image analysis. Deepak *et al* [38] proposed a visual saliency-based framework for detecting potential locations of abnormalities in retinal images. They successfully detected lesions based on their saliency values and local binary pattern features using an unsupervised classifier. More recently, Zou *et al* [39] proposed a learning-based visual saliency model method for detecting diagnostic diabetic macular edema (DME) regions of interest in retinal images. Other saliency methods have shown promising results for ROI segmentation and lesion detection for various diseases. In [40], a novel framework is developed to automatically detect masses from mammograms even in the presence of regions with pectoral muscles. The proposed framework used saliency-based segmentation and feature extraction for mass description.

Saliency methods have also been investigated for region of interest segmentation and localization. For instance, Mehmood *et al* [41] proposed an automatic segmentation method of ROIs in MRI using saliency information and active contours while Fouquier *et al* [42] integrated visual saliency in a graph knowledge-based framework for internal brain structures recognition and segmentation from 3D MRI. Mahapatra *et al* [43] proposed an active learning method for prostate segmentation from structural MRI. Shao *et al* [44] modulated the radiologist's visual attention from breast ultrasound (BUS) images to automatically locating the suspicious lesions. Automatic extraction of focus tissues from CT liver images using a visual attention model are proposed in [45]. The latter first extracted texture

MR image	Saliency map	Segmentation result	Ground truth

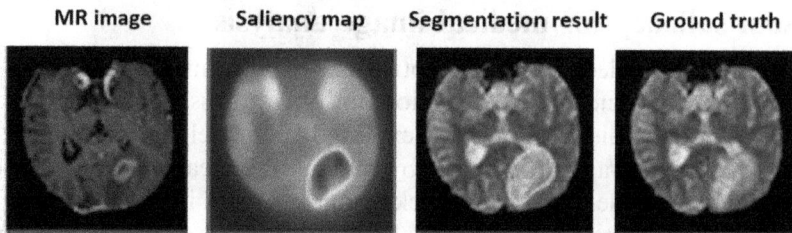

Figure 9.1. Examples of saliency maps for MRIs of two high-grade glioma cases. From left to right: MRI slice, saliency map, proto-object, and corresponding ground truth [37].

features from liver regions from a Gaussian pyramid of feature-component maps then a saliency map was generated by combinations of several sub-saliency maps. Finally, ROI candidates were located by labeling the saliency map. Jampani *et al* [46] investigated the relevance of computational saliency models in medical images in the context of lesion detection in chest x-ray images. In [47], the authors proposed a visual saliency-based computer-aided detection system to detect ulcers from WCE images for human digestive tract diagnosis. They applied the achieved saliency map to better encode the image features for ulcer image classification task. The proposed saliency method is based on multi-level superpixel color and texture representation.

In addition to predicting visual attention via computational models, eye-tracking research has also been conducted for decades to understand how clinicians process visual information from medical images [48] for decision making [31, 48–51]. Some works in the literature have investigated the radiologist's fixations in medical images using the eye-tracking technique for salient region prediction [52]. For instance, Khosravan *et al* [50] combined a visual attention map computed using radiologist gaze information with a computer-derived saliency map to perform gray-scale CT image segmentation. They proved that gaze information is useful to address the problem of image segmentation. Figure 9.2 presents an example of qualitative evaluation of their proposed segmentation method. Drew *et al* [53] used clinicians' gazes for lung nodule detection in chest CT scans. In [54], the authors proposed a model for tumor detection in chest x-ray images. They collected the eye-tracking data of radiologists visually analyzing images in the presence of tumors. The extracted data were used to develop a model for predicting the sequence of events from the time of viewing the x-ray image up to the diagnostic decision making. Chung *et al* [55] presented a new technique for deriving information on visual saliency with experimental eye-tracking data. An eye-tracking system was employed to determine which features in endoscopy video images are considered to be salient to a group of human observers. Bernal *et al* [56] used a saliency model to generate WM-DOVA maps for accurate polyp highlighting in a colonoscopy video. The proposed method has been validated versus saliency maps derived from the physician's gaze. Moreover, Bicacro *et al* [57] proposed a feature selection approach based on an expert physician's eye-tracking procedure. Visual attention approaches do not only help in diagnosis decision making but can also provide training for resident radiologists and optimize teaching for less experienced physicians [33]. Tourassi *et al* [58] proposed a machine learning framework that combines

| CT image | Attention map | Saliency map | Segmentation result |

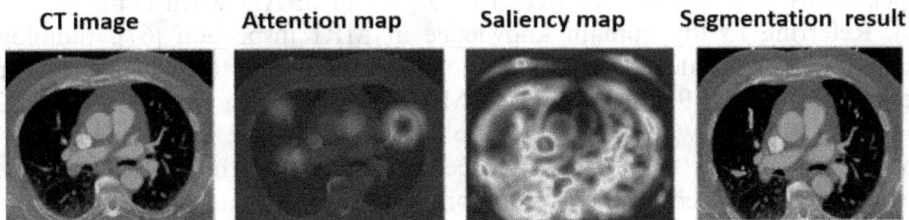

Figure 9.2. Qualitative evaluation of the medical image segmentation proposed in [50].

radiologists' gaze behavior and textural characteristics of the image to predict radiologist errors during the diagnosis of mammographic lesions.

Recently, saliency models have been proposed to classify AD subjects [59–61]. Rueda *et al* [61] extracted relevant information from brain MRI using a regional saliency method. They performed classification of brain MRIs, based on finding pathology-related patterns through the identification of regional structural changes. Later, in [62], they proposed an automatic image analysis method based on saliency maps for group diagnosis. The GBVS algorithm [63] is used to generate saliency maps that highlight particular regions. However, the previous cited works are based on the direct application of the same state-of-the-art saliency model as the GBVS. They did not include any domain knowledge regarding the pathology which makes the model weak for the specific AD diagnosis task. In addition, all those works are proposed for group analysis and thus are not applicable for a diagnosis at the individual level. In addition, they did not include the classification of the most challenging group (i.e. mild cognitive impairment (MCI)).

9.3 Saliency model for Alzheimer's disease detection from structural MRI

In this section, we explain first the visual atrophy patterns of Alzheimer's disease then we present our proposed saliency-based approach for AD detection [64]. Finally, we use the generated AD saliency maps as signatures in a classification framework to distinguish AD subjects from MCI and normal control (NC) ones.

9.3.1 Visual assessment of brain atrophy for AD diagnosis

From a practical point of view, a clinician with some experience may be able to detect brain abnormalities and then identify the more affected brain areas. They do so by looking for structural brain variation in MRI data. The major contributors to brain atrophy, for AD, are neuronal losses, in particular the gray matter, which is commonly regarded as a relevant contribution to this aspect of AD [65].

According to [66], the pattern of cell neurodegeneration seen with structural MRI in several brain areas may be a sensitive bio-marker for AD. In visual assessment-based MRI analysis, the brain shrinkage (i.e. cell degeneration) could be seen as a variation of tissue (gray or/and white matter) properties (i.e. density). For example, a brain region with decreased density reflects a reduced volume in this structure (increasing medial temporal atrophy, the MRI cortical thickness shrivels up, loss of hippocampus volume, and ventricular enlargement in AD when compared to an NC). Referring to the domain knowledge in MRI inspection [67], radiologists usually look for the degeneration and volume loss of such cells to identify brain atrophy [66]. Figure 9.3 shows typical MRI scans of a cognitively normal (CN) subject and a patient with AD. In the case of AD, the volume losses are traduced by a loss of gray matter (GM) [8, 68]. Hence, in this chapter, the proposed top-down saliency model is based on the assumption that brain shrinkage could be seen as a local brain tissue distribution variation meaning that, for example, locally shrunk brain structures will display a different proportion of gray matter and cerebrospinal

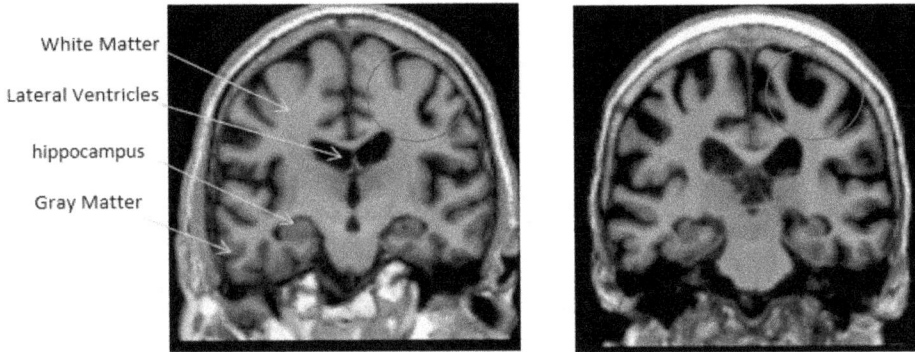

White Matter

Lateral Ventricles

hippocampus

Gray Matter

Figure 9.3. Normalized axial slices of T1-weighted MRI of normal and AD subjects from the Alzheimer's Disease Neuroimaging Initiative (ADNI) dataset.

fluid (CSF) compared to when they are unaffected. The bottom-up saliency model is based on low-level features extracted from MRI slices.

9.3.2 AD saliency map generation

In this section, we present the proposed method to build the visual saliency map from anatomical MRI for computer-aided diagnosis of Alzheimer's disease. The saliency map is a fusion of a bottom-up and a top-down saliency map [64]. The first one relies on low-level MRI characterization (visual features) while the second is based on an embedded learning process to identify and localize the subset of gray matter (GM) regions that provide optimal discrimination between subjects.

1. *Image preprocessing*: In order to reduce the visual domain search so as to consider only the most relevant information for atrophy detection, a set of processing steps were applied to the MRI data. For each subject, the preprocessing included corrections for eddy currents and head motion, and skull stripping with the Brain Extraction Tool (BET) Software Library FSL[1]. Then, all MR images are co-registered to the MNI standard space using the MNI 512 brain template [6], using the freely available VBM8 toolbox[2] with the Statistical Parametric Mapping (SPM)[3] software running in Matlab. After processing, all MR images have a size of $121 \times 145 \times 121$ voxels with a $(1.5\,\text{mm} \times 1.5\,\text{mm} \times 1.5\,\text{mm})$ of voxel size.

2. *Bottom-up saliency*: The bottom-up influences correspond to MRI low-level visual features. Many bottom-up saliency models have been proposed in the literature to simulate human intelligence in visual attention [63, 69]. For example, figure 9.4 illustrates the process of saliency map generation using the Itti saliency model [69]. This model is inspired by the human visual system and was designed originally for natural images and videos [32]. It uses a pyramid-based representation for low-level feature extraction. Moreover, Itti's model uses three different feature

---

[1] http://www.fmrib.ox.ac.uk/fsl.

[2] http://dbm.neuro.uni-jena.de/vbm/.

[3] http://www.fil.ion.ucl.ac.uk/spm/software/.

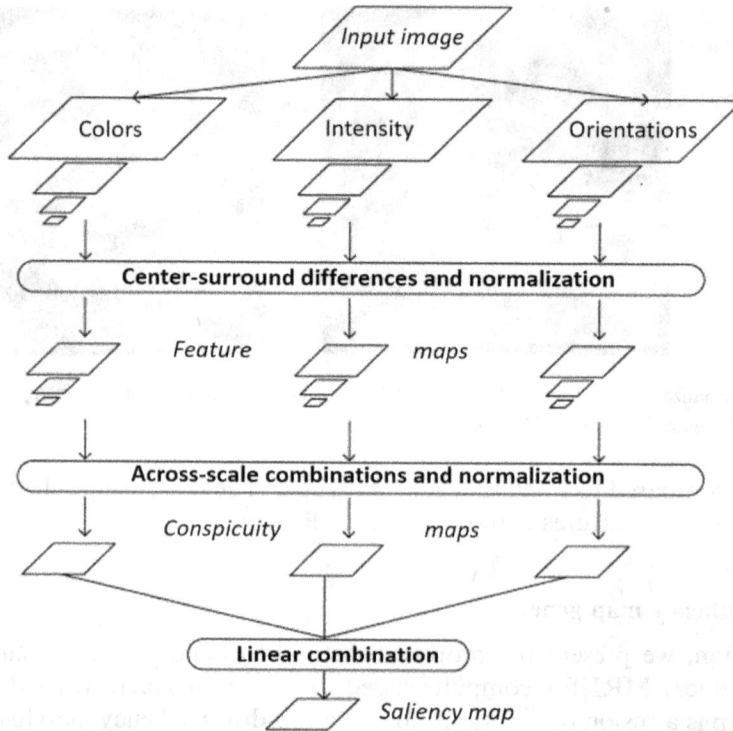

Figure 9.4. Itti saliency map model. Features are extracted and combined across different scales in a center-surround manner. The obtained feature maps are linearly combined to build the saliency map.

channels, opposition of colors, intensity, and orientation (a Gabor filter with four different orientations), to construct center-surround feature maps. Then the obtained maps are linearly combined and normalized to yield the final saliency map.

In this work, we take a cue from the original Itti's model and we propose an adopted version of the saliency model for MRI data. Since there is no color in MRI, we removed the color feature channel while the intensity feature channel is the same as in the original model.

Texture is an important characteristic for human visual perception [70]. A different arrangement of image pixels for different textures would provide us with saliency information. Hence, we add a new texture feature channel. To describe texture in the MR images we resort to the local binary pattern (LBP) descriptor which has proved to be the most used and efficient descriptor to analyze MRI texture [71]. In addition, edge detection is one of the most essential steps for extracting structural features for human perception [72]. Edges are efficient for finding discontinuities in gray level images in particular MRI images. Hence, we add a new edge feature channel. We propose to build MRI edge feature maps using a Canny detector. We limited our Gaussian pyramid levels to five scales instead of the eight levels used in the original Itti's model. All features are extracted on multiple scales of the MRI slice and stored in separate feature maps. A unique saliency map is

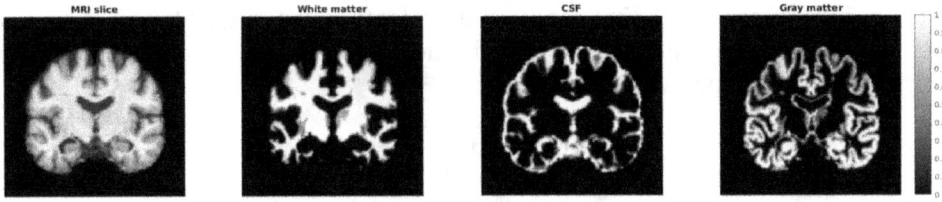

Figure 9.5. Spatial maps of GM, WM, and CSF distribution.

generated through the combination of center-surround feature maps (conspicuity maps). Finally, a weighted mean of conspicuity maps produces the bottom-up saliency maps S_{BU}:

$$S_{BU} = \frac{1}{2}(\text{Map}(\text{Texture}) + \text{Map}(\text{Edge})). \tag{9.1}$$

3. *Top-down saliency map*: In addition to the image-driven features, human visual attention is also influenced by goal-driven top-down features. In the context of medical images, top-down features correspond to the domain knowledge and the expertise of the radiologist. Referring to the domain knowledge in Alzheimer's disease, we propose to investigate whether GM tissue from AD patients could be differentiated from that of NCs to build the top-down saliency map.

First, GM, white matter (WM), and CSF maps are extracted using the SPM 8 software (Welcome Trust Centre for Neuroimaging, Institute of Neurology, UCL, London, UK[4]). The original MRI and the obtained GM map are both normalized to the MNI space using an affine transformation. The obtained probability maps contain values in the range of zero to one, representing the prior probability of a voxel being either GM, WM, or CSF after an image has been normalized to the same space. Figure 9.5 presents the obtained probability maps. As the intensity value reported in each voxel of the GM segment image is proportional to the amount of GM in that specific location (the modulation option has been selected in the SPM segmentation), a higher/lower value in the discrimination map indicates that patients have a higher/lower GM volume in that specific location.

Based on the obtained GM maps, we propose to modulate the clinician's preference location through a spatial ranking map. This is the first stage to build the top-down saliency map. To do so, we learn whether brain GM tissue from AD patients could be differentiated from that of NCs in a standard spatial space. Therefore, we develop a recursive feature elimination (RFE) approach [73] to recursively learn the relevant regions of GM. Our goal is different from what is usually done to eliminate nonrelevant features, it consists in ranking MRI GM voxels according to their contribution to separating AD and NC subjects with a support vector machine (SVM) classifier. The ranking criterion used in SVM-RFE is derived from the SVM model [74].

The larger the absolute magnitude of a weight vector is, the more strongly it affects the final discrimination. At each step, the coefficients $c_i = (w_i)^2$ of the weight

---

[4] http://www.fil.ion.ucl.ac.uk/spm/.

vector w of a linear SVM [75] are used as a feature ranking criterion. For more general proposes, we use a ten-cross validation approach to compute the ranking criterion as an average over the ten runs:

$$w_i = \frac{\sum_{k=1}^{10} w_i^k}{10}. \tag{9.2}$$

The higher the ranking score is, the greater a contribution the feature makes in classification training. The features are eliminated according to their contribution and the SVM is re-trained at each step. Indeed, according to experiments, the best classification performances are obtained with a percentage of retained features between 20% and 35%. We finally obtain a ranked list R of relevant features and only the top 30% most relevant features provided by the ranking method are retained.

It is important to note that the described saliency models (top-down and bottom-up) operate only in 2D, given that they are originally proposed for natural images. To work with 3D medical image volumes, the model needs to be applied slice-by-slice following each volume direction (sagittal, coronal, and axial). Hence, saliency maps are computed for every slice for all **MRI** projections.

For each projection p, the obtained feature ranked list R_p is an $H*W$ length vector. The first element of this list is the most relevant voxel in the slice S_p^s with W and H, respectively, the weight and the height of the slice s and $p \in$ (axial, sagittal, coronal). Algorithm 1 (below) presents the steps of building the top-down saliency map.

The ranking map $Rmap^{MNI}$ is obtained by projecting the obtained stored ranked features R_p into the MNI coordinate space in which each pixel is represented by its final rank idx in the features list (the more relevant the pixel is the higher is its intensity). Hence, the vector R_p is transformed to a matrix $Rmap_p^{MNI}$ according to the following transformation:

$$
\begin{aligned}
T&: \text{vector} \rightarrow \text{matrix} \\
(x, y)_p &\leftarrow T(R_p(idx, W, H)) \\
Rmap_p^{MNI}(x, y) &\leftarrow idx
\end{aligned}
\tag{9.3}
$$

The pixel relevance illustrates the clinician's preference order in brain areas inspection. The first ranked areas could be the most important for the clinician and consequently the first target to examine. Figure 9.6 shows samples of ranking maps for the three MRI projections. Colors bars represent the ranks (the most important regions are highlighted; red: rank = 1).

The top-down saliency map S_{TD} is generated (step 3 of algorithm 1, below) for each subject using the obtained standard $Rmap^{MNI}$. Therefore, we compute the top-down saliency map by conserving on the $Rmap^{MNI}$ only the relevant features that represent regions with an important amount of GM (visible voxels) where Prob(GM) $>= 0.5$ (GM density lower than 0.5 in some regions means that there is neurodegeneration in those regions) [1].

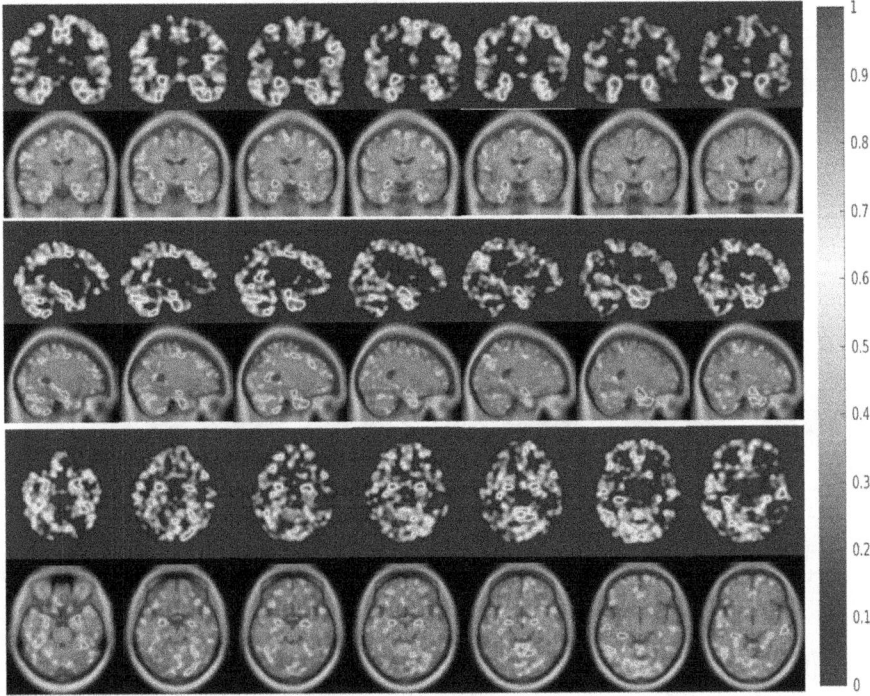

Figure 9.6. Ranking maps of MRI regions for AD and NC discrimination. Red indicates higher ranked voxels.

Algorithm 1. S_{TD} map estimation.

Input: Voxels ranked list for a projection p: R_p

1. Build ranking map:

   $Rmap_p^{MNI}(x, y) \leftarrow T(R(idx))$ for all idx.

2. $Rmap_p^{MNI}$ normalization into a fixed range of $[0\dots1]$.

3. **if** $Prob(GM) >= 0.5$ **then**

4. $S_{TD} \leftarrow Rmap_p^{MNI}$

5. **else**

6. $S_{TD} \leftarrow 0$

7. **end if**

8. $S_{TD} \leftarrow S_{TD}*GF$

9. **return** S_{TD}

4. *Data used to learn the top-down saliency map*: The data used to build the ranking map come from the ADNI dataset[5]. We use 366 structural T1-weighted MRIs of 200 AD subjects and 166 aged NCs. The AD subjects were aged between 65

---

[5] http://adni.loni.usc.edu/.

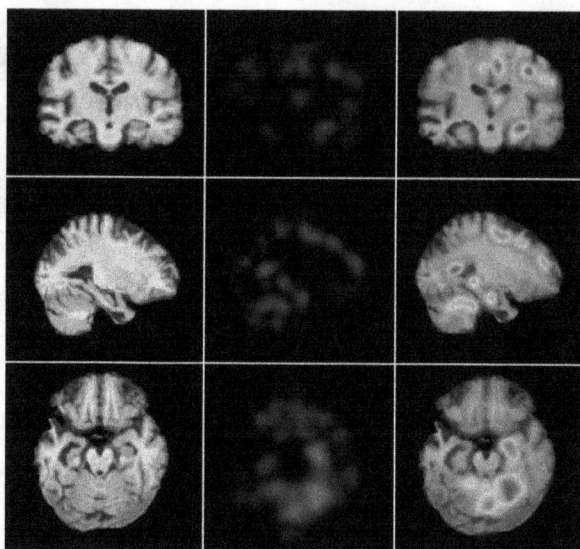

Figure 9.7. Saliency maps of an AD subject. Reprinted with permission from [64].

and 90 with mini mental state examination (MMSE) scores between 18 and 27, while the NC group is aged between 60 and 91 with MMSE scores ranging from 25 to 30. The groups' statistical difference test was performed using the t-test and the p-value between groups for both age and MMSE was < 0.001.

5. *AD-related saliency map*: The final saliency map is obtained by fusing of the aforementioned top-down and bottom-up saliency maps. The used fusion is a geometric mean between both maps:

$$S_{AD} \leftarrow \sqrt{S_{TD} \cdot S_{BU}}. \tag{9.4}$$

9.4 Visual interpretation of visual saliency

9.4.1 Region of interest detection and quantification

Figures 9.7 and 9.8 present, respectively, samples of saliency maps on structural MRI slices for both AD and NC subjects from the ADNI dataset [64]. Saliency maps are color-coded according to the relevance of brain regions. For example, red spots represent the most visually salient (relevant for diagnosis) areas of the MRI. The obtained saliency maps effectively detect and quantify regions of interest that are known to be altered in the degenerative disease and could be more prominent for the clinician's attention. When the brain regions are not yet shrunk (this is the case for NC MRI), clinicians tend to pay more attention to details within the salient regions (the hippocampus regions are more salient in the NC MRI compared to the AD ones). From these saliency maps, we can see that the detected regions of interest differ between AD and NC patients depending on the degree of brain atrophy.

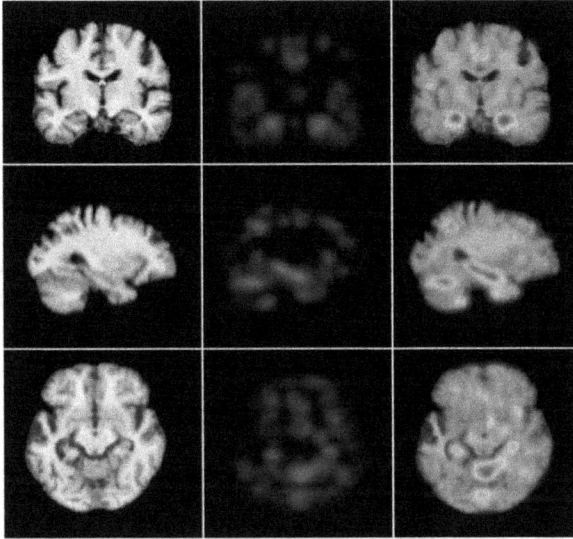

Figure 9.8. Saliency maps of an NC subject. Reprinted with permission from [64].

Table 9.1. ROIs identified by mapping the saliency maps with AAL.

Index in AAL	Brain ROI
37, 38	Left and right hippocampus
39, 40	Parahippocampus
13, 14	Inferior frontal gyrus
44, 45	Precuneus cortex
41, 42	Left and right amygdale

Hence, the saliency maps provide different patterns of brain structure that could be helpful to discriminate AD subjects from NC.

By mapping the obtained saliency maps to the Automated Anatomical Labeling Atlas (AAL) [75], we can identify regions of interest in which saliency maps vary between subjects. Table 9.1 gives some detected regions. The ROIs captured by the saliency maps correspond intimately to regions indicated by the literature as more affected by AD: the hippocampus, entorhinal cortex, and parahippocampal areas, among other regions such as the amygdale.

In addition, the saliency map shows the ROI preference order (hierarchy) for the diagnosis, hence, the lateral ventricle and the cortex are detected later (less salient). In addition, the saliency maps help to detect the real volume of the hippocampus ROI which constitutes a key step for its effective segmentation. From these figures, one can see that the patterns detected and spotted by saliency maps are different between subjects and thus could be useful for AD subject classification.

9.4.2 Comparison with state-of-the-art saliency models

We compare the performance of our proposed AD saliency model with state-of-the-art saliency models for such a specific problem. Figure 9.9 presents examples of MRI saliency maps of sagittal, axial, and coronal projections generated by our proposed method, and by two famous saliency models, namely the original Itti model and the GBVS model. Those models were proposed for natural scene interpretation. From the illustrated results, it is obvious that both the Itti and GBVS algorithms failed to detect relevant regions for Alzheimer's disease compared to our proposed approach. This confirms the role of domain knowledge in improving automatic MRI content analysis and interpretation.

9.5 AD classification using saliency maps

9.5.1 Proposed method

In this section, we propose using the obtained saliency maps as features to classify AD/MCI subjects. Figure 9.10 presents the pipeline of the proposed saliency-based classification framework. The system is composed of three main sections: the image preprocessing, saliency map generation, and finally subject classification. The saliency maps are computed on all slices for each MRI projection (axial, sagittal, and coronal) separately. This process results in a multi-view saliency map (or 3D saliency map) per subject. The obtained representations are then fed into a classifier to build a model for subject classification into three classes (AD, NC, and MCI).

The obtained saliency map provides a very useful source of information on locations of abnormalities in the MRI and thus it should also help in discriminating between AD/MCI subjects. To take advantage of information coming from different

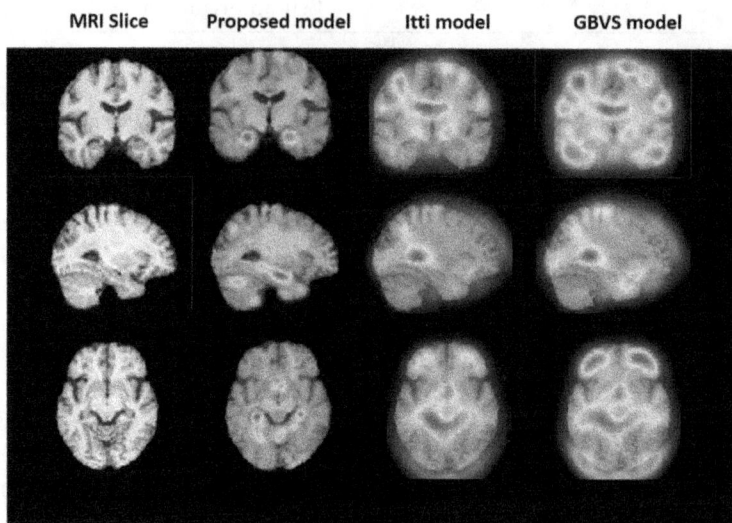

Figure 9.9. Comparison of the output of our proposed model with regards to two widely used saliency models: Itti [70] and GBVS [63]. Reprinted with permission from [64].

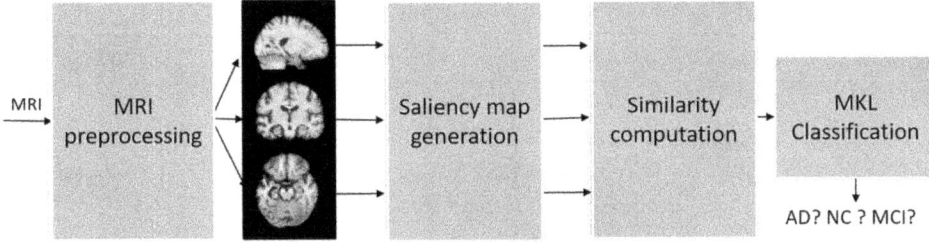

Figure 9.10. Steps of saliency-based classification method.

projections, we opt for a multiple kernel learning (MKL) [77] based classification solution. The MKL uses kernels as inputs. Kernels are matrices of size N by N (N being the number of training samples), representing the pair-wise similarity between samples. Here, we use a histogram intersection kernel (HIK) to compute the saliency maps' similarity. Therefore, the kernel between two saliency maps S and S' would be calculated as

$$k(S', S) = \sum_{i=1}^{n} \min(S_i, S_i').$$
(9.5)

Accordingly, we construct a single kernel for each projection, and then we aggregate all these kernels through a weighted average:

$$K(S, S') = \sum_{p=1}^{3} \beta_p k_p(S, S')$$
(9.6)

$$\text{with } \beta_p \geqslant 0, \; \sum_{p=1}^{3} \beta_p = 1.$$

Kernel k_p are the HIK kernels of projection p. β_p defines the weight for each projection p. The decision function is defined as

$$f(u) = \sum_{i=1}^{l} \alpha_i^* K(u, u_i) + b^*,$$
(9.7)

where α_i^* and b^* are coefficients to be learned from data. MKL aims to simultaneously optimize the α_i and the β_p subject to $\beta_p \geqslant 0$, $\sum_{p=1}^{3} \beta_p = 1$. In the current work, we use the SimpleMKL algorithm proposed by [77] to solve the classification problem.

9.5.2 MRI data

The MRI data used in this work contain 137 AD patients, 162 NCs, and 210 subjects with MCI. More detailed information about MRI acquisition procedures is available on the ADNI website[6]. Table 9.2 presents a summary of the demographic

---

[6] http://adni.loni.ucla.edu/.

Table 9.2. Demographic description of the ADNI subset.

Diagnosis	Number	Age	Gender (M/F)	MMSE
AD	137	range = [55 91]	67/70	range = [18 27]
NC	162	range = [60 90]	76/86	range = [25 30]
MCI	210	range = [55 88]	127/83	range = [23 30]

characteristics of the selected subjects (including the number, age, gender, and MMSE (mini mental state examination) of subjects).

9.5.3 Classification results

Saliency-based classification was applied separately in each of two groups (NC versus AD, NC versus MCI, and MCI versus AD). We used ten-fold cross validation to evaluate classification performance. Therefore, the original sample set was randomly divided into ten parts. Then, one part was used as the test and the remaining nine parts were used as training data. To evaluate the performance of our classification method we compute:

- Accuracy = (TP + TN)/(TP + TN + FN + FP)
- Sensitivity = TP/(TP + FN)
- Specificity = TN/(TN + FP)
- Balanced accuracy = 0.5 * (Sensitivity + Specificity)

True positives are AD patients correctly identified as AD, TN are controls correctly classified as controls, FN are AD patients incorrectly identified as controls, and FP are controls incorrectly identified as AD. Similar definitions hold for the other binary classification problems NC versus MCI and AD versus MCI. Figure 9.11 presents the obtained results.

For classifying AD from NC, our method achieves a classification accuracy of 88.98%, a sensitivity of 83.46%, and a specificity of 94.4%. On the other hand, for classifying MCI from NC our method achieves a classification accuracy of 81.31%, a sensitivity of 74.21%, and a specificity of 84.22%. For the most challenging task (AD versus MCI), we achieved accuracy of 79.85%, a sensitivity of 64.02%, and a specificity of 79.93%.

The obtained results show that our saliency-based atrophy detection approach allows us to consistently distinguish AD/MCI subjects from normal controls.

9.6 Conclusion

In this chapter, we presented the importance of computational saliency models in medical imaging for abnormality detection and computer-aided diagnosis. We reviewed the most relevant work on this topic. Then, we presented our saliency-based approach for AD/MCI subject classification. The proposed framework is based on the fusion of bottom-up and top-down saliency maps using domain knowledge in AD diagnosis. In the bottom-up approach, information comes from low-level MRI characterization (texture and edge) and the top-down approach

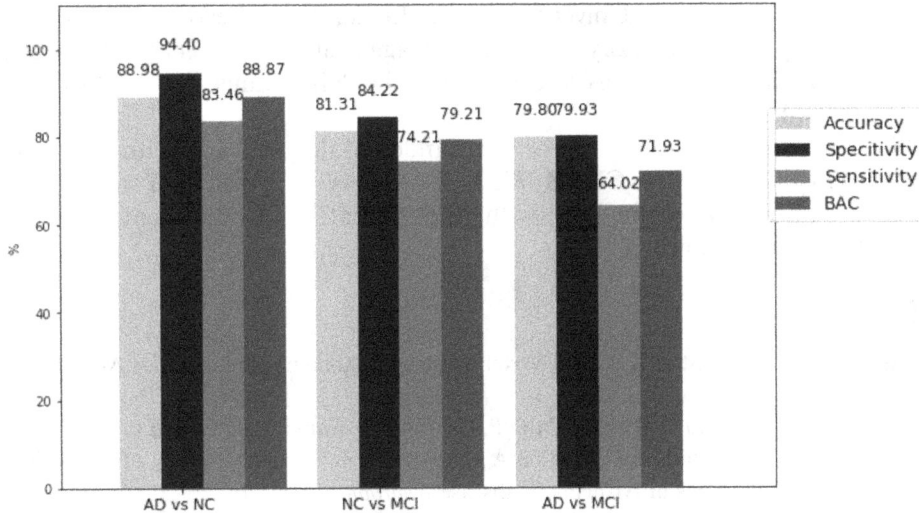

Figure 9.11. Classification results using MKL on all projections.

includes a learning process to localize GM regions that provide optimal separation between groups. Saliency maps were used as signatures for AD/MCI subject classification. According to the obtained classification results, the proposed method could help clinicians to evaluate their diagnosis findings. This also makes the diagnosis easier for clinicians with limited expertise to extract diagnostically useful and objective information. Future works will consist in improving the current model with clinician's gaze tracking.

Acknowledgement

Data collection and sharing for this work was funded by the Alzheimer's Disease Neuroimaging Initiative (ADNI) (National Institutes of Health Grant U01 AG024904). ADNI is funded by the National Institute on Aging, the National Institute of Biomedical Imaging and Bioengineering, and through generous contributions from the following: Abbott; Alzheimer's Association; Alzheimer's Drug Discovery Foundation; Amorfix Life Sciences Ltd.; AstraZeneca; Bayer HealthCare; BioClinica, Inc.; Biogen Idec Inc.; Bristol-Myers Squibb Company; Eisai Inc.; Elan Pharmaceuticals Inc.; Eli Lilly and Company; F Hoffmann-La Roche Ltd and its affiliated company Genentech, Inc.; GE Healthcare; Innogenetics, NV; IXICO Ltd.; Janssen Alzheimer Immunotherapy Research and Development, LLC.; Johnson and Johnson Pharmaceutical Research and Development LLC.; Medpace, Inc.; Merck and Co., Inc.; Meso Scale Diagnostics, LLC.; Novartis Pharmaceuticals Corporation; Pfizer Inc.; Servier; Synarc Inc.; and Takeda Pharmaceutical Company. The Canadian Institutes of Health Research is providing funds to support ADNI clinical sites in Canada. Private sector contributions are facilitated by the Foundation for the National Institutes of Health http://www.fnih.org. The grantee organization is the Northern California Institute for Research and Education, and the study is coordinated by the Alzheimer's Disease

Cooperative Study at the University of California, San Diego. ADNI data are disseminated by the Laboratory for Neuro Imaging at the University of California, Los Angeles. This research was also supported by NIH grants P30 AG010129 and K01 AG030514.

Olfa Ben-Ahmed and Christine Fernandez-Maloigne, are with the XLIM Research Institute, UMR CNRS 7252, University of Poitiers, France, Adrien Julian and Marc Paccalin are with the Geriatrics Department, University Hospital, Poitiers, France.

References

[1] Ashburner J and Friston K J 2000 Voxel based morphometry: the methods *Neuroimage* **11** 805–21

[2] Busatto G F, Garrido G E, Almeida O P, Castro C C, Camargo C H, Cid C G, Buchpiguel C A, Furuie S and Bottino C M 2003 A voxel-based morphometry study of temporal lobe gray matter reductions in Alzheimer's disease *Neurobiol. Aging* **24** 221–31

[3] Shiino A, Watanabe T, Maeda K, Kotani E, Akiguchi I and Matsuda M 2006 Four subgroups of Alzheimer's disease based on patterns of atrophy using VBM and a unique pattern for early onset disease *NeuroImage* **33** 17–26

[4] Mechelli A, Price C J, Friston K J and Ashburner J 2005 Voxel-based morphometry of the human brain: methods and applications *Curr. Med. Imaging Rev.* **1** 105–13

[5] Vasconcelos L D G, Jackowski A P, Oliveira M O D, Flor Y M R, Bueno O F A and Brucki S M D 2011 Voxel-based morphometry findings in Alzheimer's disease: neuropsychiatric symptoms and disability correlations—preliminary results *Clinics* **66** 1045–50

[6] Frisoni G B, Testa C, Sabattoli F, Beltramello A, Soininen H and Laakso M P 2005 Structural correlates of early and late onset Alzheimer's disease: voxel based morphometric study *J. Neurol. Neurosurg. Psychiatry* **76** 112–4

[7] Wee C, Yap P and Shen D 2013 Prediction of Alzheimer's disease and mild cognitive impairment using cortical morphological patterns *Hum. Brain Mapp.* **34** 3411–25

[8] Karas G, Burton E, Rombouts S, Van Schijndel R, O'Brien J, Scheltens P, McKeith I, Williams D, Ballard C and Barkhof F 2003 A comprehensive study of gray matter loss in patients with Alzheimer's disease using optimized voxel-based morphometry *Neuroimage* **18** 895–907

[9] Ben Ahmed O, Benois-Pineau J, Allard M, Catheline G and Ben Amar C 2017 Recognition of Alzheimer's disease and mild cognitive impairment with multimodal image-derived biomarkers and multiple kernel learning *Neurocomputing* **220** 98–110

[10] Ben Ahmed O *et al* 2015 Alzheimer's disease diagnosis on structural MR images using circular harmonic functions descriptors on hippocampus and posterior cingulate cortex *Comput. Med. Imaging Graph.* **44** 13–25

[11] Qin Y-Y *et al* 2013 Gross feature recognition of Anatomical Images based on Atlas grid (GAIA): incorporating the local discrepancy between an atlas and a target image to capture the features of anatomic brain MRI *NeuroImage* **3** 202–11

[12] Chen Y, Storrs J, Tan L, Mazlack L J, Lee J-H and Lu L J 2014 Detecting brain structural changes as biomarker from magnetic resonance images using a local feature based SVM approach *J. Neurosci. Methods* **221** 22–31

[13] Toews M, Wells W, Collins D L and Arbel T 2010 Feature-based morphometry: discovering group-related anatomical patterns *NeuroImage* **2** 2318–27

[14] Zhang J, Xia Y, Xie Y, Fulham M and Feng D D 2018 Classification of medical images in the biomedical literature by jointly using deep and handcrafted visual features *IEEE J. Biomed. Health Inform.* **22** 1521–30

[15] Unay D 2010 Augmenting clinical observations with visual features from longitudinal MRI data for improved dementia diagnosis *Proc. of the Int. Conf. on Multimedia Information Retrieval, ser. MIR '10* (New York: ACM) pp 193–200

[16] Agarwal M and Mostafa J 2010 Image retrieval for Alzheimer disease detection *Proc. of the First MICCAI Int. Conf. on Medical Content-Based Retrieval for Clinical Decision Support, ser. MCBR-CDS'09* (Berlin: Springer) pp 49–60

[17] Mizotin M, Benois-Pineau J, Allard M and Catheline G 2012 Feature-based brain MRI retrieval for Alzheimer disease diagnosis *19th IEEE Int. Conf. on Image Processing (ICIP)* pp 1241–4

[18] Ben Ahmed O, Benois-Pineau J, Allard M, Catheline G and Amar C B 2010 Diffusion tensor imaging retrieval for Alzheimer's disease diagnosis *2014 12th Int. Workshop on Content-Based Multimedia Indexing (CBMI)* (Piscataway, NJ: IEEE) pp 1–6

[19] Harper L *et al* 2016 MRI visual rating scales in the diagnosis of dementia: evaluation in 184 post-mortem confirmed cases *Brain* **139** 1211–25

[20] Bresciani L, Rossi R, Testa C, Geroldi C, Galluzzi S, Laakso M P, Beltramello A, Soininen H and Frisoni G B 2005 Visual assessment of medial temporal atrophy on MR films in Alzheimer's disease: comparison with volumetry *Aging Clin. Exp. Res.* **17** 8–13

[21] Pena G P and Andrade-Filho J S 2009 How does a pathologist make a diagnosis? *Arch. Pathol. Lab. Med.* **133** 124–32

[22] Carrasco M 2011 Visual attention: the past 25 years *Vis. Res.* **51** 1484–525

[23] Behrmann M and Haimson C 1999 The cognitive neuroscience of visual attention *Curr. Opin. Neurobiol.* **9** 158–63

[24] Treisman A M and Gelade G 1980 A feature-integration theory of attention *Cogn. Psychol.* **12** 97–136

[25] Borji A and Itti L 2013 State-of-the-art in visual attention modeling *IEEE Trans. Pattern Anal. Mach. Intell.* **35** 185–207

[26] Borji A, Cheng M, Jiang H and Li J 2015 Salient object detection: a benchmark *IEEE Trans. Image Process.* **24** 5706–22

[27] Obeso A M, Benois-Pineau J, Vázquez M S G and Acosta A A R 2018 Introduction of explicit visual saliency in training of deep CNNs: application to architectural styles classification *2018 Int. Conf. on Content-Based Multimedia Indexing (CBMI)* pp 1–5

[28] Obeso A M, Benois-Pineau J, Guissous K, Gouet-Brunet V, Vázquez M S G and Acosta A A R 2018 Comparative study of visual saliency maps in the problem of classification of architectural images with deep CNNs *2018 Eighth Int. Conf. on Image Processing Theory, Tools and Applications (IPTA)* pp 1–6

[29] Gbèhounou S, Lecellier F, Fernandez-Maloigne C and Courboulay V 2013 Can salient interest regions resume emotional impact of an image? *Int. Conf. on Computer Analysis of Images and Patterns* (Berlin: Springer) pp 515–22

[30] Alzubaidi M, Balasubramanian V N, Patel A, Panchanathan S and Black J A 2010 What catches a radiologist's eye? a comprehensive comparison of feature types for saliency prediction *Medical Imaging: Computer-Aided Diagnosis* vol 7624 (Bellingham, WA: International Society for Optics and Photonics) p 76240W

[31] Le Callet P and Niebur E 2013 Visual attention and applications in multimedia technologies *Proc. IEEE* **101** 2058–67

[32] Li R, Shi P and Haake A R 2013 Image understanding from experts' eyes by modeling perceptual skill of diagnostic reasoning processes *2013 IEEE Conf. on Computer Vision and Pattern Recognition (CVPR)* pp 2187–94

[33] Lala D and Nakazawa A 2016 Heat map visualization of multi-slice medical images through correspondence matching of video frames *Proc. of the Ninth Biennial ACM Symp. on Eye Tracking Research and Applications, ser. ETRA '16* (New York: ACM) pp 119–22

[34] Wen G, Aizenman A, Drew T, Wolfe J M, Haygood T M and Markey M K 2016 Computational assessment of visual search strategies in volumetric medical images *J. Med. Imaging* **3** 015501

[35] Mehmood I, Ejaz N, Sajjad M and Baik S W 2013 Prioritization of brain MRI volumes using medical image perception model and tumor region segmentation *Comput. Biol. Med.* **43** 1471–83

[36] Chung A G, Scharfenberger C, Khalvati F, Wong A and Haider M A 2015 *Proc. of 12th Int. Conf. on Image Analysis and Recognition, ICIAR 2015 (Niagara Falls, ON, Canada, July 22–24, 2015)* (Cham: Springer) pp 368–76

[37] Banerjee S, Mitra S, Shankar B U and Hayashi Y 2016 A novel GBM saliency detection model using multi-channel MRI *PLoS One* **11** e0146388

[38] Deepak K S *et al* 2013 Visual saliency based bright lesion detection and discrimination in retinal images *2013 IEEE 10th Int. Symp. on Biomedical Imaging (ISBI)* (Piscataway, NJ: IEEE) pp 1436–9

[39] Zou X, Zhao X, Yang Y and Li N 2016 Learning-based visual saliency model for detecting diabetic macular edema in retinal image *Comput. Intell. Neurosci.* **2016** 7496735

[40] Agrawal P, Vatsa M and Singh R 2014 Saliency based mass detection from screening mammograms *Signal Process.* **99** 29–47

[41] Mehmood I, Baik R and Baik S W 2013 *Automatic Segmentation of Region of Interests in MR Images Using Saliency Information and Active Contours* (Dordrecht: Springer) pp 537–44

[42] Fouquier G, Atif J and Bloch I 2012 Sequential model-based segmentation and recognition of image structures driven by visual features and spatial relations *Comput. Vis. Image Underst.* **116** 146–65

[43] Mahapatra D and Buhmann J M 2015 Visual saliency based active learning for prostate MRI segmentation *Proc. 6th Int. Workshop on Machine Learning in Medical Imaging: MLMI 2015, Held in Conjunction with MICCAI 2015 (Munich, Germany, October 5, 2015)* (Cham: Springer) pp 9–16

[44] Shao H, Zhang Y, Xian M, Cheng H D, Xu F and Ding J 2015 A saliency model for automated tumor detection in breast ultrasound images *2015 IEEE Int. Conf. on Image Processing (ICIP)* pp 1424–8

[45] Ma L, Wang W, Zou S and Zhang J 2009 Liver focus detections based on visual attention model *2009 3rd Int. Conf. on Bioinformatics and Biomedical Engineering* pp 1–5

[46] Jampani V *et al* 2012 Assessment of computational visual attention models on medical images *Proc. of the Eighth Indian Conf. on Computer Vision, Graphics and Image Processing* (New York: ACM) p 80

[47] Yuan Y, Wang J, Li B and Meng M Q H 2015 Saliency based ulcer detection for wireless capsule endoscopy diagnosis *IEEE Trans. Med. Imaging* **34** 2046–57

[48] Lévêque L, Bosmans H, Cockmartin L and Liu H 2018 State of the art: eye-tracking studies in medical imaging *IEEE Access* **6** 37023–34

[49] Brunyé T T, Mercan E, Weaver D L and Elmore J G 2017 Accuracy is in the eyes of the pathologist: the visual interpretive process and diagnostic accuracy with digital whole slide images *J. Biomed. Inform.* **66** 171–9

[50] Khosravan N *et al* 2017 Gaze2segment: a pilot study for integrating eye-tracking technology into medical image segmentation *Medical Computer Vision and Bayesian and Graphical Models for Biomedical Imaging* ed H Müller *et al* (Cham: Springer) pp 94–104

[51] McLaughlin L, Bond R, Hughes C, McConnell J and McFadden S 2017 Computing eye gaze metrics for the automatic assessment of radiographer performance during x-ray image interpretation *Int. J. Med. Inform.* **105** 11–21

[52] van der Gijp A, Ravesloot C J, Jarodzka H, van der Schaaf M F, van der Schaaf I C, van Schaik J P J and ten Cate T J 2017 How visual search relates to visual diagnostic performance: a narrative systematic review of eye-tracking research in radiology *Adv. Health Sci. Educ.* **22** 765–87

[53] Drew T, Vo M L-H, Olwal A, Jacobson F, Seltzer S E and Wolfe J M 2013 Scanners and drillers: characterizing expert visual search through volumetric images *J. Vis.* **13** 3

[54] Nodine C F and Kundel H L 1987 Using eye movements to study visual search and to improve tumor detection *Radiographics* **7** 1241–50

[55] Chung A J, Deligianni F, Hu X-P and Yang G-Z 2005 Extraction of visual features with eye tracking for saliency driven 2D/3D registration *Image Vis. Comput.* **23** 999–1008

[56] Bernal J, Sanchez F J, Fernandez-Esparrach G, Gil D, Rodriguez C and Vilarino F 2015 WM-DOVA maps for accurate polyp highlighting in colonoscopy: validation versus saliency maps from physicians *Comput. Med. Imaging Graph.* **43** 99–111

[57] Bicacro E, Silveira M, Marques J S and Costa D C 2012 3D brain image-based diagnosis of Alzheimer's disease: bringing medical vision into feature selection *2012 9th IEEE Int. Symp. on Biomedical Imaging (ISBI)* pp 134–7

[58] Voisin S, Pinto F, Morin-Ducote G, Hudson K B and Tourassi G D 2013 Predicting diagnostic error in radiology via eye-tracking and image analytics: preliminary investigation in mammography *Med. Phys.* **40** 101906

[59] Pulido A, Rueda A and Romero E 2013 Classification of Alzheimer's disease using regional saliency maps from brain MR volumes *SPIE Medical Imaging* (Bellingham, WA: International Society for Optics and Photonics) pp 86700R

[60] Rueda A, Arevalo J E, Cruz-Roa A, Romero E and González F A 2012 Bag of features for automatic classification of Alzheimer's disease in magnetic resonance images *CIARP* pp 559–66

[61] Daza J C and Rueda A 2016 Classification of Alzheimer's disease in MRI using visual saliency information *2016 IEEE 11th Colombian Computing Conference (CCC)* (Piscataway, NJ: IEEE) pp 1–7

[62] Rueda A *et al* 2014 Extracting salient brain patterns for imaging-based classification of neurodegenerative diseases *IEEE Trans. Med. Imaging* **33** 1262–74

[63] Harel J, Koch C and Perona P 2007 Graph-based visual saliency *Advances in Neural Information Processing Systems* pp 545–52

[64] Ben-Ahmed O, Lecellier F, Paccalin M and Fernandez-Maloigne C 2017 Multi-view visual saliency-based MRI classification for Alzheimer's disease diagnosis *2017 Seventh Int. Conf. on Image Processing Theory, Tools and Applications (IPTA)* pp 1–6

[65] Karas G, Scheltens P, Rombouts S, Visser P, Van Schijndel R, Fox N and Barkhof F 2004 Global and local gray matter loss in mild cognitive impairment and Alzheimer's disease *Neuroimage* **23** 708–16

[66] Braak H and Braak E 1998 Evolution of neuronal changes in the course of Alzheimer's disease *Neurology* **53** 127–40

[67] Blennow K, Hampel H, Weiner M and Zetterberg H 2010 Cerebrospinal fluid and plasma biomarkers in Alzheimer disease *Nat. Rev. Neurol.* **6** 131–44

[68] Eckerström C, Andreasson U, Olsson E, Rolstad S, Blennow K, Zetterberg H, Malmgren H, Edman Å and Wallin A 2010 Combination of hippocampal volume and cerebrospinal fluid biomarkers improves predictive value in mild cognitive impairment *Dement. Geriatr. Cogn. Disord.* **29** 294–300

[69] Itti L *et al* 1998 A model of saliency-based visual attention for rapid scene analysis *IEEE Trans. Pattern Anal. Mach. Intell.* **20** 1254–9

[70] Parkhurst D J and Niebur E 2004 Texture contrast attracts overt visual attention in natural scenes *Eur. J. Neurosci.* **19** 783–9

[71] Unay D, Ekin A, Cetin M, Jasinschi R and Ercil A 2007 Robustness of local binary patterns in brain MR image analysis *2007 29th Annual Int. Conf. of the IEEE Engineering in Medicine and Biology Society* pp 2098–101

[72] Park S-J, Shin J-K and Lee M 2002 Biologically inspired saliency map model for bottom-up visual attention *Int. Workshop on Biologically Motivated Computer Vision* (Berlin: Springer) pp 418–26

[73] Zacharaki E, Wang S, Chawla S, Yoo D S, Wolf R, Melhem E and Davatzikos C 2009 MRI-based classification of brain tumor type and grade using SVM-RFE *ISBI '09. IEEE Int. Symp. on Biomedical Imaging: From Nano to Macro, 2009* pp 1035–8

[74] Hearst M A, Dumais S T, Osman E, Platt J and Scholkopf B 1998 Support vector machines *Intell. Syst. Appl. IEEE* **13** 18–28

[75] Tzourio-Mazoyer N *et al* 2002 Automated anatomical labeling of activations in SPM using a macroscopic anatomical parcellation of the MNI MRI single-subject brain *NeuroImage* **15** 273–89

[76] Scholkopf B and Smola A J 2001 *Learning with Kernels: Support Vector Machines, Regularization, Optimization, and Beyond* (Cambridge, MA: MIT Press)

[77] Rakotomamonjy A, Bach F, Canu S and Grandvalet Y 2008 SimpleMKL *J. Mach. Learn. Res.* **9** 2491–521

IOP Publishing

Neurological Disorders and Imaging Physics, Volume 3
Application to autism spectrum disorders and Alzheimer's
Ayman El-Baz and Jasjit S Suri

Chapter 10

The early diagnosis of Alzheimer's disease using advanced biomedical engineering technology

C S Sandeep and A Sukesh Kumar

Geriatrics deals with the numerous clinical issues that are common in the elderly population and plenty of these follow an orthodox pattern in clinical observations. Patients characteristically have poor insight and frequently attribute their early symptoms of memory loss to normal aging. Alzheimer's disease (AD) is a common kind of senile dementia. There are many causes of this disease. Although our understanding of the key steps underlying neurodegeneration in AD is incomplete, it is clear that it begins long before symptoms are noticed by the patient. Any disease-modifying treatments that are developed will presumably achieve success if initiated early enough in the process, and this requires developing reliable, valid, and economical ways to diagnose Alzheimer's-type pathology. However, despite comprehensive searches, no single test has shown adequate sensitivity and specificity, and it is likely that a combination is required. There are a lot of tests and neuroimaging modalities to be performed for an efficient diagnosis of the disease. Standard clinical decision-making systems are more manual in nature and an ultimate conclusion in terms of actual diagnosis is difficult. In this case, the utilization of advanced biomedical engineering technology will certainly be useful for making a diagnosis. Profiling of human body parameters using computers can be used for the early diagnosis of AD. There are several neuroimaging techniques employed in clinical practice for the diagnosis of Alzheimer's-type pathology. The most prominent of them are magnetic resonance imaging (MRI), positron emission tomography (PET), single photon emission computed tomography (SPECT), and optical coherence tomography (OCT). In addition to the imaging modalities, there are various laboratory tests and neuropsychological tests that can be used for diagnosing the disease. In this scenario, a combination of the above tests with the help of biomedical engineering technology can be utilized for developing an expert system for the early diagnosis of AD.

10.1 Introduction

Alzheimer's disease (AD) is an irreversible age connected neurodegenerative disorder of the human brain that results in memory loss and impairs the power to perform routine functions. The cognitive impairment of patients might cause disturbances for family members as well as caretakers. The key factor driving research toward the early diagnosis of AD is how to prevent the disease from its progression to a later stage. The progression of AD is classified into four different stages: mild cognitive impairment (MCI), and mild, moderate, and severe AD. In MCI there is only a little cognitive impairment which becomes much worse by the severe stage of AD, leading to complete memory impairment. When the disease progresses from MCI to severe AD, the condition of the patient becomes more critical. Alzheimer's disease was first discovered in 1906 by a German neurologist and psychiatrist called Alois Alzheimer [1]. According to one report, at present, nearly thirty-six million individuals (35.6m) are believed to be living with Alzheimer's disease, and those suffering from different types of dementia might increase to about sixty-six million (65.7m) by 2030, and over one hundred and fifteen million (115.4m) by 2050 [2]. In another study, the number of individuals with dementia might double by 2030, and more than triple by 2050 [3]. AD is the sixth leading reason for death and accounts for 70% of the prevailing cases of dementia in the world [4]. Therefore, a new method or technique needs to be developed to effectively diagnose AD at the earliest possible stage to slow down disease progression and to decrease the mortality rate.

10.2 Literature review

Different methods are used for the diagnosis of AD using biomedical engineering technology. A work by Lebed, Jacova, Wang, and Faisal Beg on 'Novel surface-smoothing based local gyrification index' for the early identification of AD investigates the importance of the quantification of cortical surface folding in distinguishing and classifying several neurodegenerative diseases, such as AD. In this paper, mapping of the native gyrification index on the cortical surface of subjects with mild AD, very mild dementia, and age-matched healthy subjects was performed. They speculate that in Alzheimer's disease, the folding of the whole cortical mantle undergoes dynamic changes as regional atrophy begins and expands, with both decreases and increases in gyrification. A detailed longitudinal study of gyrification computed for healthy and AD subjects is needed to analyze this hypothesis [5].

Fletcher *et al*, in 'Combining boundary-based ways with tensor-based morphometry in the measurement of longitudinal brain change', explored a tensor-based morpho-metric technique for automatically computing longitudinal modification in brain structure. A penalty term and inverse consistency are required to manage the over-reporting of non-biological modification. This might force a trade-off between the intrinsic sensitivity and specificity, potentially resulting in an under-reporting of authentic biological modification with time. Therefore, it is necessary to incorporate previous information regarding tissue boundaries that aims to keep the lustiness and specificity contributed by the penalty term and inverse consistency while maintaining localization and sensitivity. The boundary shift approach needs careful user-delineated brain masks to properly find edges and computes brain volume change only within the edge masks [6].

Escudero et al, in 'Machine learning-based method for personalized and cost-effective detection of Alzheimer's disease', describe a machine learning approach for customized and efficient identification of AD. It uses regionally weighted learning to tailor a classifier model to each patient and computes the sequence of biomarkers that are most informative or cost-effective to diagnose patients. The approach is performed similarly to considering all the information right away, while considerably reducing the number (and cost) of the biomarkers required to attain an assured identification for each patient. The limitation of this work is that alternative classifiers will be tested as base learners. 'Modified cost' will be developed to account for extra factors in the choice of biomarkers and an independent validation set ought to be used to optimize the approach by considering that values are clinically acceptable [7].

Abuhassan, Coyle, and Maguire, in 'Investigating the neural correlates of pathological cortical networks in Alzheimer's disease using heterogeneous neuronal models', investigated the pathophysiological causes of abnormal cortical oscillations in AD using two heterogeneous neural network models. The impact of stimulative circuit disruption on the beta band power (13–30 Hz) employing a conductance-based network model of two hundred neurons is assessed. Then, the neural correlates of abnormal cortical oscillations in several frequency bands supporting a larger network model of a thousand neurons, consisting of various varieties of cortical neurons, are also analyzed. In this work, despite the nonuniformity of the network models, the beta band power suffers considerably from stimulative neural and synaptic loss. Second, the results of modeling a functional impairment in the stimulative circuit show that beta band power exhibits the most decrease compared to alternative bands [8].

Padilla et al, in 'NMF-SVM based CAD tool applied to functional brain images for the diagnosis of Alzheimer's disease' conferred a unique computer-aided diagnosing (CAD) technique for the early diagnosis of AD based on nonnegative matrix factorization (NMF) and support vector machines (SVMs) with bounds of confidence. The CAD tool is intended for the study and classification of functional brain pictures. These databases are analyzed by applying the Fisher discriminant ratio (FDR) and nonnegative matrix factorization (NMF) for feature choice and extraction of the foremost relevant features. The resulting NMF-transformed sets of information, that contain a reduced range of features, are classified by means of an SVM-based classifier with bounds of confidence for resolution [9].

Morra et al, in the paper 'Comparison of AdaBoost and support vector machines for detecting Alzheimer's disease through automated hippocampal segmentation', manually divided gold standard hippocampal tracings which were available for all subjects (training and testing). Then they regenerated the segmentations into constant quantity surfaces to map the disease effects on anatomy. After surface reconstruction, they computed the significance maps and overall corrected-values for the 3D profile of shape variations between AD and normal subjects. Additive-value plots, in conjunction with the false discovery rate methodology, were used to examine the ability of every methodology to find correlations with diagnosis and cognitive scores [10].

Tahaei, Jalili, and Knyazeva, in 'Synchronizability of EEG-based functional networks in early Alzheimer's disease', proposed electroencephalography (EEG) with its high temporal resolution for the analysis of functional interdependencies between totally different brain regions. The cross-correlation of artifact-free EEGs was used to construct the brain's useful networks. The extracted networks were then tested for their synchronization properties by calculating the eigenratio of the Laplacian matrix of the association graph, i.e. the largest eigenvalue divided by the second smallest one [11].

Duchesne *et al*, in 'MRI-based automated computer classification of probable AD versus normal controls', used an automated computer classification (ACC) technique in the context of cross-sectional analysis of magnetic resonance images (MRIs) in neurodegenerative diseases, in particular AD. In this paper, the accuracy of the methodology is assessed compared to reality and imperfect information, i.e. cohorts of MRI with variable acquisition parameters and imaging quality. The comparative methodology uses the Jacobian determinants derived from dense deformation fields and scaled gray-level intensity from a particular volume of interest focused on the medial lobe. The ensuing accuracy is 92, employing an SVM classifier based on the method of least squares optimization [12].

Simpson *et al*, in 'Ensemble learning incorporating uncertain registration', proposed an approach for improving the accuracy of statistical prediction strategies in spatially normalized analysis for discriminating subjects with AD from age-matched healthy controls. A probabilistic registration methodology is employed to estimate a distribution of probable mappings between the subject and atlas house. From this distribution, samples are drawn to be used as training examples. The creation of multiple predictors, that are subsequently combined using an ensemble learning approach, has been investigated [13].

Faro *et al*, published a paper on 'Basal ganglia activity measurement by automatic 3D striatum segmentation in SPECT images' for the diagnosis of AD. This paper investigated an automatic system for quantitative measurement of basal ganglia activity by 3D corpus striatum activity reconstruction from 123I FP-CIT SPECT images on a nigrostriatal dopaminergic system. In this method-ology, image improvement is administered by means of active contour models, 2D corpus striatum segmentation in the HSV color space, and surface rendering by the extended marching cubes algorithmic rule, wherever the isovalue of the striatum's isosurface and therefore the final 3D volume is automatically com-puted. The low resolution of the SPECT image does not continually allow the clear identification of the two structures composing the striatum: the putamen and caudate [14].

The above-discussed methods are currently used for the diagnosis of AD in the biomedical engineering field.

10.3 Causes and effects of AD

Dementia is the common term for impaired brain function and encompasses symptoms such as memory pathology, confusion of places and things, the inablity to perform routine tasks, a loss of intellectual functions, and impaired judgment. However, this condition could be a symptom of many underlying neurological disorders apart from Alzheimer's disease [15]. The different types of dementia in addition to Alzheimer's disease are Parkinson's disease, dementia with Lewy bodies, Huntington's disease, Creutzfeldt–Jakob disease, frontotemporal dementia, Down syndrome dementia, and Korsakoff syndrome. Figure 10.1 shows a pictorial representation of the different types of dementia brain diseases. Among the different types of dementia diseases, AD is the most prevalent, and involves a gradual loss of brain functions as well as changes in behavior and personality [16]. The two most vital neuropathological proteins that are associated with AD are tangles and plaques. Tangles are filament-like bundles that accumulate inside the nerve cell and plaques are present outside the nerve cell. When these two proteins are deposited on the nerve cells, the brain cannot communicate or send signals to other parts of the body and vice versa [17]. The following increase the risk of AD: genetic mutations, advanced age, small head size, any previous record of trauma, cardiovascular problems, type 2 diabetes, high blood pressure, high cholesterol, lower education, alcohol consumption, smoking, depression, a lack of physical activity and exercise, and the female gender [18]. The symptoms associated with AD might result in problems with language, impairment in recognizing an object, impaired motor

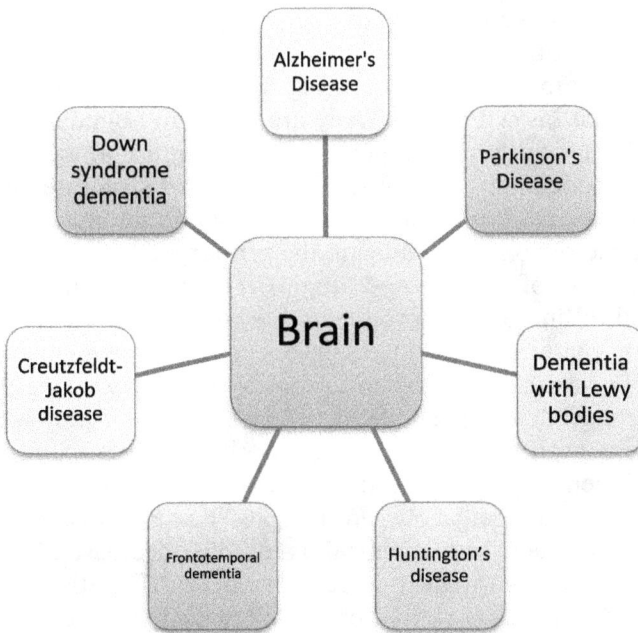

Figure 10.1. Dementia type diseases.

function, and most vital of all, memory impairment. With disease progression, patients suffer disability and also immobility. The brain structure of such patients shows gross atrophy in the brain with a compensative enlargement of ventricles [19].

10.4 Hallmarks of AD

A positive analysis of Alzheimer disease can be made only by autopsy of a patient's brain. This neuropathological assessment discloses gross cortical atrophy, indicating damage of neurons. The two important hallmarks of AD are the two types of proteins commonly called plaques and tangles. A microscopic assessment can reveal the presence of large numbers of plaques and tangles in the nerve cells. Plaques and tangles are seen mostly in the temporal and frontal lobes, including the hippocampus [20]. When the disease progresses, these proteins extend to other parts of the brain including the occipital and parietal lobes. Plaques are undissolvable deposits of an amino acid peptide called β-amyloid (Aβ). Plaques can be classified as diffuse or classical. Diffuse plaques are collections of Aβ which are typically not associated with the degeneration of neurons. Classical plaques, on the other hand, are deposits that are related to the degeneration of brain cells. In early-onset AD, more plaques are formed than in late-onset AD [21].

The other hallmark of AD is tangles, which are deposited inside the brain cells. Tangles can cause problems in the communication between the brain and other body parts, and also cause damage to transentorhinal regions, the hippocampus, amygdala, and neocortical areas. Tangles are usually proteins of microtubule-associated tau (τ). Tau is normally found in great abundance in neurons, and is essential in cellular transport and axonal growth. Normally, plaques and tangles occur in the case of normal subjects, but are more abundant and more commonly distributed in the brains of AD patients [22]. Even though the role of plaques and tangles in AD is not known exactly, they are found in abundance in the areas of the temporal and frontal lobes, hippocampus, and occipital and parietal lobes of the brain. The hippocampi are tiny sea-horse-shaped structures in the temporal lobes, which play a significant role in establishing and maintaining memory. The hippocampi are the regions that indicate the primary changes in AD and also the highest concentration of plaques and tangles. Therefore, examining plaques and tangles is helpful to find out the progression of memory loss in AD patients. The accumulation of plaques and tangles in the cortical areas can explain other symptoms of AD, such as irregular visuospatial orientation, trouble with trained tasks, and language deformities. Another problem associated with AD is the reduction in the concentrations of neurotransmitters due to neuronal losses [23]. This creates problems with neuronal interconnections, known as synapses, the chemical signals that are sent between neurons. For example, acetylcholine is a neurotransmitter that acts as a chemical agent. The decrease of such a chemical agent causes the decline of intellectual behavior in an AD patient. Accordingly, in the treatment of AD drugs are commonly used to increase the level of acetylcholine in the brain [24].

10.5 The retina and AD

We know that the dementia of AD is a brain connected disease, however, several reports show that there is proof of visual issues associated with the early stages of AD. These include issues in reading, the inability to search out objects or see deep objects, and issues finding moving objects and recognizing color features [25]. Studies that have been performed previously shown that this happens because of pathology in the visual cortex. However, the investigations made on the retina show that there is evidence that the anterior visual pathways are also connected to AD. This leads to the loss of nerves in the retinal layers [26]. The main reason for the visual disabilities is the deposits of Aβ protein and acetylcholine that have been found in the retina. The property by which Aβ causes the death of retinal neurons is not understood completely, however, the same kind of neuronic pathology occurring in the retina and brain of AD patients has been revealed [27]. The beginning of AD pathology might occur within the visual cortical region. In early AD, there are problems with vision, in addition to cognitive losses. Like neuronal losses in glaucoma and age-related macular degeneration, AD show similar neuronal losses in the retina. In normal cases, without memory impairment, there was an absence of such pathology in the hippocampus area where the symptoms of AD begin, the area that processes memory. The loss of visual function other than acuteness could also be an initial vital indication of AD [28]. Age-related degeneration degrades all frequencies of contrast sensitivity, demonstrates color disorders across all wavelengths, and decreases foveal detection of motion. Researchers are stating that rather than beginning within the brain regions that process memory, such as the hippocampus, AD will begin earlier in the brain region that integrates visual function. In individuals with AD, contrast sensitivity degrades in the lower spatial frequencies, motion perception, i.e. the flexibility to observe movement, is reduced, there are field of vision defects, and color discrimination of blue, short wavelength hues is found to be reduced [29]. There have been findings of Aβ and neurofibrillary tangles in the retina, the presence of which is related to Alzheimer's disease. Therefore, by examining the retina of AD subjects, it can be possible to detect AD at its early stages [30].

A biomarker is an indicator to measure the severity or presence of some disease state. More generally a biomarker is something which will be used as an indicator of a particular disease state or some other physiological condition of an organism. The retina of the eye will be considered as a biomarker as there are changes within the layer of the retina of AD patients, in particular in the retinal nerve fiber loss (RNFL) layer [31]. The different neurological diseases that affect the deterioration of the retina are diabetic retinopathy (DR), multiple sclerosis (MS), Parkinson's disease (PD), and Alzheimer's disease (AD). In all cases, there are neuronal losses, but in AD the loss is prominent at the retinal nerve fiber layer (RNFL). Retina related neurological disorders are shown in figure 10.2.

Diabetic retinopathy presents with vascular and neuronal alterations in the retina, as a result of chronically high glucose concentrations that form advanced glycosylation end products [32]. These concentrations will harm the basement membranes

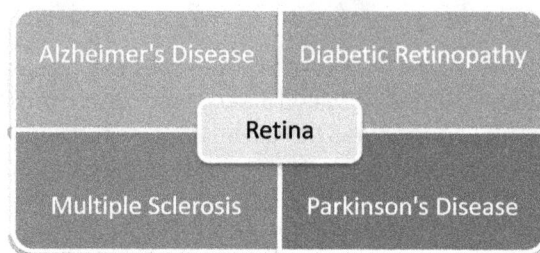

Figure 10.2. Retina related neurological diseases.

and can become proliferative, in which new abnormal vessels form and can bleed easily. Multiple sclerosis, on the other hand, is a neurodegenerative disorder affecting more than 1.3 million people worldwide. It occurs due to an aberrant immune process, causing antibodies and inflammatory factors to form on myelin-related proteins, resulting in the demyelination of neurons [33]. Parkinson's disease is the second most typical neurodegenerative disorder after Alzheimer's disease. Parkinson's is primarily a progressive motor disorder related to the degeneration of dopaminergic neurons within the neural structure. Research has shown Parkinson's to be a multisystem disorder with additional nonmotor impairments and pathology occurring outside the basal ganglia. Dopamine plays a significant role in motor function, but it is also an important neurotransmitter/neuromodulator in the retina. So neurological changes in the brain conjointly affect attention and also the cortical region. The disease progression of the above diseases is examined using completely different imaging modalities, such as magnetic resonance imaging, scanning laser ophthalmoscopy, fundus imaging, and optical coherence tomography. Based on the above circumstances, the retina can also provide information and sufficient results regarding the early diagnosis of AD [34].

Neuropsychological tests and scans will facilitate indicating the memory recall of a patient and also the potential areas where the patient has deficiency. Using these tests will be useful to seek out the types of treatment plans which can be given, but neuropsychological tests alone are not useful in detecting early AD. Therefore trials must be conducted by combining neuropsychological tests with clinical tests and various imaging modalities. For efficient and early diagnosis of AD, a population-based study is required which provides a concept regarding the various tests involved in diagnosing AD [35].

10.6 Tests for diagnosing AD

There are different types of tests for the diagnosis of AD. They include clinical tests, neuropsychological tests, and imaging tests. The clinical tests include routine blood tests and testing of the cerebrospinal fluid (CSF) of the brain [36]. CSF aggregates on the brain and spinal cord. In AD patients, the CSF level increases sharply. Neuropsychological tests are mainly used for determining the specific type and level of cognitive impairment of the AD patient. Some of the tests include the Rey auditory verbal learning test, trial making test parts A and B, category fluency, digit span forward and backward, digit symbol substitution test, the clock drawing task, and mini-mental state examination [37]. These tests are noninvasive and can be used in the screening process for determining AD. Another type of test for determining AD is imaging tests. These include computed tomography (CT), magnetic resonance imaging (MRI), positron emission tomography (PET), single photon emission computed tomography (SPECT), and optical coherence tomography (OCT). CT scans are used for determining the atrophy of the brain as well as for finding the enlarged ventricles in the case of AD [38]. Patients with dementia might not have cerebral atrophy, at least in the earlier stages of the disease. Therefore it cannot be used for early stage diagnosis of the disease. Also, it is difficult to differentiate a healthy patient and a patient with dementia in the early stage. Another imaging method that is widely used is structural MRI. This performs scans of the brain tissue region to find the progression of AD from MCI to the severe stage. Structural MRI also scans the medial temporal lobe, ventricular volumes, and whole brain volumes. Therefore an MRI scan can be used for the early diagnosis of AD. Also with MRI, it is easier to differentiate between healthy brain atrophy and a patient with AD. In addition, positron emission tomography (PET) is an advanced imaging method that uses a biochemical means of acquiring pictures instead of structural information [39]. PET technology involves the detection of photons by a camera-like device that records the amount of emission originating from given points in space and time. Positron emitting radioisotopes are used to generate radioactivity. A PET scan will measure different compounds in the brain, in the case of AD PET is used to measure fluorodeoxyglucose (FDG). With dementia patients, the neurons' intake of glucose and FDG becomes impaired. By highlighting the regions of decreased FDG uptake, PET can theoretically help in the diagnosis of dementia, even in the absence of gross structural damage. PET has been used widely to study AD, and it is evolving into an effective tool for early diagnosis [40]. PET has been used to find individuals in danger of AD even before onset of symptoms. PET can be a very expensive scan to perform, although it has been one of the most helpful in providing visual pictures in the detection of AD. There are certain advantages to PET, but the scan is still an invasive technique. In this scan, various radioisotopes are injected into the body to determine the glucose level, and are very dangerous and may produce various side effects in the future.

We have already discussed the use of the retina as a biomarker for the early diagnosis of AD in section 10.5. From that point of view, optical coherence tomography (OCT) scans can be very useful for taking images of the retina [41]. OCT is a promising and noninvasive imaging technique that provides cross-sectional images of the eye retina with high resolution images that can be used for the diagnosis of AD. During the OCT process, six linear scans centered on the optical nerve head (ONH) are obtained, and the OCT software derives the ONH parameters in an automatic manner for the measurements of the retinal nerve fiber layer (RNFL), which is affected in AD patients [42, 43]. Importantly, some new techniques are still being developed with the help of advanced biomedical engineering technology that can not only detect AD but possibly explain the symptoms and how the disease works.

Based on the above observations, if it is possible to combine the features of different tests with the help of biomedical engineering technology, we can predict the disease in its earlier stages [44].

10.7 Early diagnosis of AD

In AD research, a lot of has been accomplished in the last twenty years, but a good deal remains to be done to improve its diagnosis and treatment within the early stages. There is increasing proof that early diagnosis of AD is going to be the key issue to maximize treatment advantages [45, 46]. However, patients are often diagnosed in the later stages of the disease, when disabling symptoms and neuropathological changes have become well established. AD affects a considerable and increasing a part of the population. Despite the shortage of disease-modifying treatments, discovering sensitive and specific markers of early AD would be a significant breakthrough because it would permit us to cut down or maybe even arrest the degenerative process before dementia develops [47]. Moreover, current symptomatic treatments, such as acetylcholine esterase inhibitors, could also be more economical and effective when administered within the early stages of AD [48]. The diagnosis of clinically probable AD will currently be made in living subjects only once the stage of dementia has been reached.

For the early diagnosis of AD using numerous imaging modalities, a screening test is needed. In this case, the researchers will use neuropsychological tests as a screening method. Because it is a noninvasive technique, it will be used with minimum effort. Also, the neuropsychological tests will completely differentiate different types of dementia from Alzheimer's. Also, if it is possible to develop a computer code or tool based on neuropsychological tests, physicians will use this tool as an associate aid for screening patients. The development of the tool is done in such a way that it can be employed in any operating system with minimum memory space [49]. Also, it is necessary to provide a choice for saving the records of the patient. The early, correct diagnosis of AD is particularly vital to patients and their families. It helps them organize for the long term while the patient can still participate in making decisions [50].

10.8 Medical imaging techniques

Medical imaging is the technique of making visual representations of the inside of a body for clinical analysis and medical intervention, and as a visual illustration of the function of some organs or tissues (physiology) [51]. Medical imaging seeks to reveal the internal structures hidden by the skin and bones, as well as diagnose and treat disease. Medical imaging additionally establishes information on normal anatomy and physiology to make it possible to spot abnormalities. Although imaging of removed organs and tissues may be performed for medical reasons, such procedures are typically considered part of pathology rather than medical imaging. After the screening process is completed using the preliminary screening process, the next process is to find a better solution to predict disease progression. Also, it is necessary to state whether a patient has AD or not. In this scenario, there are invasive and

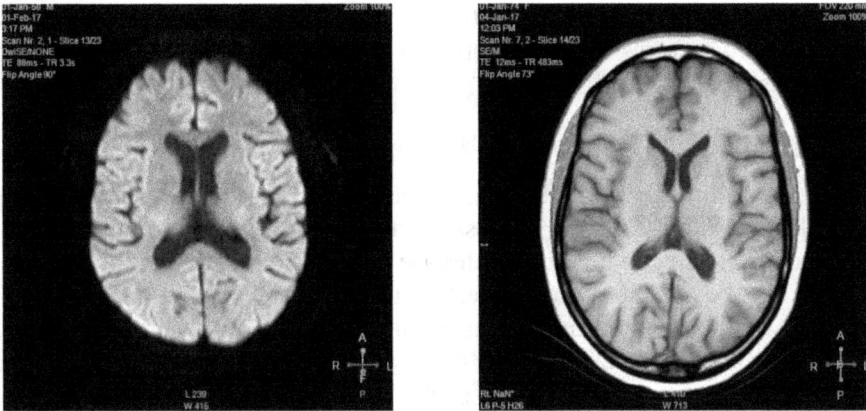

Figure 10.3. MRI of an AD brain (left) and a normal brain (right). (Image source: Sree Gokulam Medical College and Research Foundation.)

noninvasive imaging modalities to determine disease progression. Based on the permission of the patient, the researchers can use both techniques. CT, MRI, and OCT are noninvasive while PET and SPECT are invasive. If the patient goes with the invasive techniques, there is a chance of experiencing side effects in the future due to the injection of radiotracers into the body. Therefore it is safest to adopt a noninvasive imaging technique such as MRI or OCT.

MRI is a medical imaging practice employed in radiology to create images of the anatomy and therefore the physiological processes of the body in health and sickness. MRI scanners use magnetic fields, force field gradients, and radio waves to obtain pictures of the organs in the body [52]. MRI does not involve x-rays or the utilization of radiation, which distinguishes it from CT scans and PET scans. While the hazards of x-rays are currently well-controlled in most medical contexts, an MRI scan should still be seen as an improved alternative to a CT scan. MRI is widely employed in hospitals and clinics for diagnosis and staging of disease and follow-up without exposing the body to radiation. However, MRI might yield completely different diagnostic information than CT [53]. Compared to CT scans, MRI scans generally take longer and are louder, and they sometimes require the subject to enter a narrow, confining tube. The MRI scans of an AD and a normal subject are shown in figure 10.3.

OCT is an imaging technique that uses low-coherence light to capture micrometer-resolution, two- and three-dimensional pictures from optical scattering media (e.g. biological tissue). It is now widely used for medical imaging, in particular the internal layers of the retina [54]. OCT relies on low-coherence interferometry, usually using near-infrared light. The utilization of comparatively long wavelength light permits it to penetrate into the scattering medium. Depending on the properties of the light source OCT uses sub-micrometer resolution. OCT is one among a class of optical tomographic procedures. Also, OCT allows a better signal-to-noise ratio and provides quicker signal acquisition. Commercially obtainable optical coherence imaging systems are used in various applications, including art conservation and diagnostic medication, notably in ophthalmology and optometry where it is used to obtain elaborated pictures inside the retina. The OCT of an AD and a normal retina is shown in figure 10.4.

10.9 Analysis of MRI and OCT images

After selecting the image modality, such as MRI or OCT, the next step is to send the patients selected in the screening process to the scanning section to take the required images. The physician can predict the disease progression from the obtained images. However, to make an automated expert system for the early diagnosis of AD, we have to analyze the image using advanced biomedical engineering technology. It is possible to analyze the images using an image database that is available on the internet with the necessary permission or we have to create a database from images taken from hospitals. A database of MRI images can be easily obtained from the Alzheimer's Disease Neuroimaging Initiative (ADNI) and Open Access Series of Imaging Studies (OASIS), but a database of OCT images on the internet is not easily available.

For the analysis of the obtained images, it is necessary to follow different steps, such as image acquisition, image pre-processing, image segmentation, image post-processing, feature extraction, feature selection, and finally the classification of images. The block diagram used for the process of image analysis is shown in figure 10.5.

10.9.1 Image acquisition

Image acquisition is the first step in the process of image analysis. It is the process of acquiring the images. It is done using an MRI scanner or OCT scanner. MRI is a technique which is used to find deteriorated tissues in the AD brain and OCT is an imaging technique used to find the loss of retinal layers in AD victims. The physician can use these techniques for the diagnosis of AD with more efficiency. The images obtained through MRI or OCT can be directly stored on the computer which can later be used for analyzing the images.

10.9.2 Pre-processing

Pre-processing is the second step in image analysis. The obtained images are in digital format with different file size and image formats. First of all, we have to make

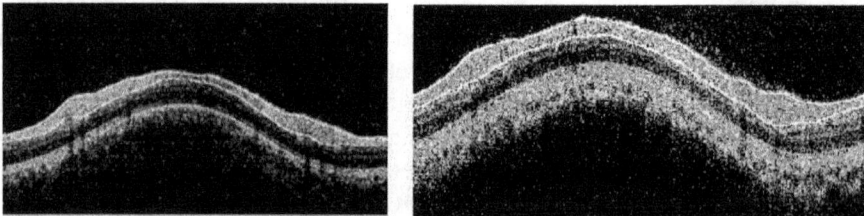

Figure 10.4. OCT of AD (left) and normal (right) retinas. (Source: Sree Gokulam Medical College and Research Foundation.)

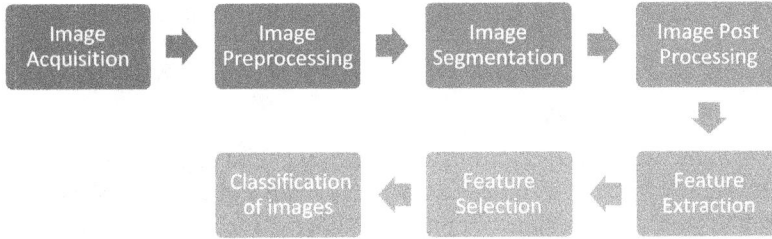

Figure 10.5. Block diagram for the analysis of MRI or OCT images.

the digital images into a standard format and standard size for further processing. The standard image size can be taken as 360×360 pixels. Usually, the obtained image contains noise in the form of hairs, bubbles, etc. This noise causes inaccuracies in classification. In order to avoid this, images are subjected to pre-processing techniques. Image pre-processing is the removal of noise from the image. For example, hair pixels present in MRI or OCT images occlude some of the information of the necessary tissue or layers, such as the boundary and texture. Hence, the removal of hair is an important pre-processing step in such systems. The first method is digitally removing hairs and then smoothing the final result using different available filters. Such noise reduction is a typical pre-processing step to improve the results of later processing (for example, edge detection on an image). Filtering is very widely used in digital image processing because, under certain conditions, it preserves edges while removing noise.

10.9.3 Image segmentation

Image segmentation is the third step of image analysis. It is the most important part of image analysis. Segmentation is the partitioning of an image into meaningful information. It is the procedure of separating a digital image into multiple segments (sets of pixels, also known as super-pixels). The goal of segmentation is to simplify and/or modify the illustration of an image into meaningful information so that it is easier to analyze. Image segmentation is normally used to trace objects and boundaries (lines, curves, etc) in images. More exactly, image segmentation is the practice of conveying a label to every single pixel in an image such that pixels with a similar label share definite characteristics. The output of image segmentation is a set of fragments that together cover the whole image or a group of contours extracted from the image [55]. Every single pixel in a region is similar with respect to some distinctive character or computed property. This includes color, intensity, and texture. When segmentation is applied to a group of images, as in medical imaging, the resulting contours that are created after image segmentation can be used to create 3D reconstructions with the help of interpolation algorithms. The segmentation techniques are broadly classified into threshold based segmentation, edge-based segmentation, region-based segmentation, segmentation based on clustering techniques, and matching. As there are different segmentation methods, it is necessary to select an appropriate method by comparing all the methods or by performing a literature search to find which method is suitable for the early diagnosis of AD.

10.9.4 Image post-processing

Post-processing is the fourth step in image analysis. The segmented images are post-processed so that they can be further used for the subsequent stages. Since extracting the features is the most essential part of diagnosing AD, extracting the exact boundary of the tissue or layer is a vital task. For this, after segmentation, with the space between adjacent shapes filled, extra parts are eliminated, and the noise is removed. Then, the exact boundary of the tissue or layer is extracted. This is done using acceptable morphological processes, as well as erosion, dilation, closing and opening, and region filling. Therefore, this stage makes the image suitable for input to the feature extraction stage so that all artifacts are cleared.

10.9.5 Feature extraction

Feature extraction is the fifth step in image analysis. There are some distinctive features that distinguish an AD image from a normal image. Feature extraction extracts the eminent and necessary features of image information from the segmented image. By extracting features, the image information is narrowed right down to a collection of features. The various features which will be extracted in the case of MRI or OCT images for the early identification of AD are space, perimeter, ellipticity, centroid, and texture features, etc.

10.9.6 Feature selection

Feature selection is the sixth step in image analysis. Different features can be extracted from MRI or OCT images, but we do not need all the features available in the image. We should select the necessary and appropriate features, otherwise the analysis becomes overly complex.

10.9.7 Classification

The final step in image analysis is the classification of images. The classification process is considered the final significant task in building and developing reliable automated systems for the early diagnosis of AD. The classifier is used to classifying AD subjects from normal subjects. There are different approaches to classifying the images. The most prominent of these are neural networks, fuzzy logic, support vector machines, etc. It is necessary to choose the classification methods according to the MRI or OCT images obtained [56].

10.10 Discussion

Although there are numerous different neuropathological tests, imaging modalities, biomarkers, and drug therapies, etc, for the identification of AD, they are inadequate for a precise diagnosis. However, if we are able to combine the features of all of the above using soft computing techniques such as fuzzy logic, neural computing, biological process computation, and probabilistic reasoning, it is going to be possible to achieve early diagnosis of the disease by very convenient means [57–59]. Soft computing differs from conventional (hard) computing in that, unlike hard computing, it is tolerant of impreciseness, uncertainty, partial truth, and approximation. The guideline of soft computing is: 'Exploit the tolerance for impreciseness, uncertainty, partial truth, and approximation to attain flexibility, lustiness, and low solution costs'. The clinical information could consist of missing, incorrect, and typically incomplete value sets, therefore using soft computing is the better alternative to handle such information. The principal constituent methodologies in soft computing are complementary instead of competitive. Fuzzy logic handles impreciseness, neural computing deals with learning, biological process computation is for improvement, and probabilistic reasoning handles uncertainty.

In the literature review, we have discussed some of the early works in the field of early AD diagnosis, which still present some limitations. Therefore a new method has to be developed by combining the different tests with advanced biomedical engineering technology. If we can develop an expert system with new approaches, it can be an aid to the physician in diagnosis. Therefore patients can be diagnosed using an expert system and they do need not be sent for scanning.

10.11 Conclusion

There are plenty of clinical tests, drug therapies and diagnostic tools such as biomarkers and neuroimaging techniques available for the diagnosis of Alzheimer's disease. However, these techniques are inadequate for the definite identification of AD at the earlier stages. Thus a new reliable and economical technique should be developed so as to diagnose the disease using advanced biomedical engineering technology combining various clinical tests, neuroimaging techniques such as SPECT, MRI, PET, OCT, etc, databases such as ADNI, OASIS, etc, and soft computing tools. With the assistance of the above strategies, profiling of human body parameters for diagnosis of AD may be created. The screening tests for identifying AD patients early may be conducted with minimum effort. A clinical follow-up for carrying out diagnosis can be set using the above approaches. As AD can be classified as mild, moderate, or severe, identification of these stages is a difficult task. The early identification of these stages can be made with the above described methods in a reliable and effective way. As we know that the prevalence of this disease is increasing worldwide with no appropriate diagnosis, a good approach towards this will be achieved with an approach that can diagnose AD with the minimum effort, cost, and time.

Acknowledgments

We are grateful to Dr K Mahadevan, Professor, Department of Ophthalmology, and Dr P Manoj, Professor, Department of Neurology of Sree Gokulam Medical College and Research Foundation, Trivandrum, Kerala, India for their support in writing this chapter.

References

[1] Alzheimer's Association 1906 Über einen eigenartigen schweren Erkrankungsprozeß der Hirnrinde *Neur. Centralbl.* **23** 1129–36

[2] Alzheimer's Association 2010 2010 Alzheimer's disease facts and figures *Alzheimer's Dement.* **6** 158–94

[3] Alzheimer's Association 2012 *Alzheimer's Facts and Figures* (Chicago, IL: Alzheimer's Association)

[4] Alzheimer's Disease International 2010 ADI press release for 'Alzheimer's Disease International World Alzheimer Report 2010: The Global Economic Impact of Dementia' 21 September http://alz.co.uk/media/nr100921.html

[5] Lebed E, Jacova C, Wang L and Beg M F 2013 Novel surface-smoothing based local gyrification index *IEEE Trans. Med. Imaging* **32** 660–9

[6] Fletcher E *et al* 2013 Combining boundary-based methods with tensor-based morphometry in the measurement of longitudinal brain change *IEEE Trans. Med. Imaging* **32** 223–36

[7] Escudero J, Ifeachor E, Zajicek J P, Green C, Shearer J and Pearson SAlzheimer's Disease Neuroimaging Initiative 2013 Machine learning-based method for personalized and cost-effective detection of Alzheimer's disease *IEEE Trans. Biomed. Eng.* **60** 164–8

[8] Abuhassan K, Coyle D and Maguire L P 2012 Investigating the neural correlates of pathological cortical networks in Alzheimer's disease using heterogeneous neuronal models *IEEE Trans. Biomed. Eng.* **59** 890–6

[9] Padilla P, Lopez M, Gorriz J M, Ramirez J, Salas-Gonzalez D and Alvarez I 2012 NMF-SVM based CAD tool applied to functional brain images for the diagnosis of Alzheimer's disease *IEEE Trans. Med. Imaging* **31** 207–16

[10] Morra J H, Tu Z, Apostolova L G, Green A E, Toga A W and Thompson P M 2010 Comparison of AdaBoost and support vector machines for detecting Alzheimer's disease through automated hippocampal segmentation *IEEE Trans. Med. Imaging* **29** 30–43

[11] Tahaei M S, Jalili M and Knyazeva M G 2012 Synchronizability of EEG-based functional networks in early Alzheimer's disease *IEEE Trans. Neural Syst. Rehabil. Eng.* **20** 636–41

[12] Duchesne S, Caroli A, Geroldi C, Barillot C, Frisoni G B and Collins D L 2008 MRI-based automated computer classification of probable AD versus normal controls *IEEE Trans. Med. Imaging* **27** 509–20

[13] Simpson I, Woolrich M W, Andersson J L R, Groves A R and Schnabel J 2012 Ensemble learning incorporating uncertain registration *IEEE Trans. Med. Imaging* **32** 748–56

[14] Faro A, Giordano D, Spampinato C, Ullo S and Di Stefano A 2011 Basal ganglia activity measurement by automatic 3-D striatum segmentation in SPECT images *IEEE Trans. Instrum. Meas.* **60** 3269–80

[15] Shimokawa *et al* 2001 Influence of deteriorating ability of emotional comprehension on interpersonal behavior in Alzheimer-type dementia *Brain Cogn.* **47** 423–33

[16] Frosch M P, Anthony D C and Girolami U D 2010 The central nervous system *Robbins and Cotran Pathologic Basis of Disease* ed S L Robbins, V Kumar, A K Abbas, R S Cotran and N Fausto (Philadelphia, PA: Elsevier) pp 1313–7

[17] Harvey R A, Champe P C and Fisher B D 2006 *Lippincott's Illustrated Reviews: Microbiology* 2nd edn (Philadelphia, PA: Lippincott Williams and Wilkins) p 432

[18] Cummings J L, Vinters H V, Cole G M and Khachaturian Z S 1998 Alzheimer's disease: etiologies, pathophysiology, cognitive reserve and treatment opportunities *Neurology* **51** 2–17

[19] Yaari R and Corey-Bloom J 2007 Alzheimer's disease: pathology and pathophysiology *Semin. Neurol.* **27** 32–41

[20] Larson E B *et al* 2006 Exercise is associated with reduced risk for incident dementia among persons 65 years of age and older *Ann. Intern. Med.* **144** 73–81

[21] Harvey R J, Skelton-Robinson M and Rossor M N 2003 The prevalence and causes of dementia in people under the age of 65 years *J. Neurol. Neurosurg. Psychiatry* **74** 1206–9

[22] Chiu H F *et al* 1998 Prevalence of dementia in Chinese elderly in Hong Kong *Neurology* **50** 1002–9

[23] Chu L W *et al* 2010 Bioavailable testosterone predicts a lower risk of Alzheimer's disease in older men *J. Alzheimers Dis.* **21** 1335–45

[24] McKhann G, Drachman D, Folstein M, Katzman R, Price D and Stadlan E M 1984 Clinical diagnosis of Alzheimer's disease. Report of the NINCDS-ADRDA Work Group under the auspices of Department of Health and Human Services Task Force on Alzheimer's disease *Neurology* **34** 939–44

[25] Oliveira L T, Louzada P R, Mello F G and Ferreira S T 2011 Amyloid-β decreases nitric oxide production in cultured retinal neurons: a possible mechanism for synaptic dysfunction in Alzheimer's disease? *Neurochem. Res.* **36** 163–9

[26] Berisha F, Feke G T, Trempe C L, McMeel J W and Schepens C L 2007 Retinal abnormalities in early Alzheimer's disease *Invest. Ophthalmol. Vis. Sci.* **48** 2285–9

[27] Kesler A, Vakhapova V, Korczyn A D, Naftalive E and Neudorfer M 2011 Retinal thickness in patients with mild cognitive impairment and Alzheimer's disease *Clin. Neurol. Neurosurg.* **113** 523–6

[28] Guo L, Duggan J and Corderio M F 2010 Alzheimer's disease and retinal neurodegeneration *Curr. Alzheimer Res.* **7** 3–14

[29] McKee A *et al* 2006 Visual association pathology in preclinical Alzheimer disease *J. Neuropathol. Exp. Neurol.* **65** 621–30

[30] Frost S, Martins R N and Kanagasingam Y 2010 Ocular biomarker for early detection of Alzheimer's disease *J. Alzheimers Dis.* **22** 1–16

[31] Ohno Matsu K 2011 Parallel findings in age-related macular degeneration and Alzheimer's disease *Prog. Retin. Eye Res.* **30** 217–38

[32] Valenti D A 2004 Anterior visual system and circadian function with reference to Alzheimer's disease *Vision in Alzheimer's disease. Interdisciplinary Topics in Gerontology* **vol 34** pp 1–29

[33] Mei M and Leat S 2007 Suprathreshold contrast matching in maculopathy *Invest. Ophthalmol. Vis. Sci.* **48** 3419–24

[34] Feigl B, Brown B and Lovie-Kitchin J 2005 Monitoring retinal function in early age-related maculopathy: visual performance after one year *Eye* **199** 1169–77

[35] Sandeep C S and Sukesh Kumar A 2015 A review on the early diagnosis of Alzheimer's disease (AD) through different tests, techniques and databases *AMSE J.* **76** 1–22

[36] American Psychiatric Association 1994 *Diagnostic and Statistical Manual of Mental Disorders* 4th edn (Washington, DC: American Psychiatric Association)

[37] Mak H K, Zhang Z, Yau K K, Zhang L, Chan Q and Chu L W 2011 Efficacy of voxel-based morphometry with DARTEL and standard registration as imaging biomarkers in Alzheimer's disease patients and cognitively normal older adults at 3.0 Tesla MR imaging *J. Alzheimers Dis.* **23** 655–64

[38] Scheltens P *et al* 1992 Atrophy of medial temporal lobes on MRI in 'probable' Alzheimer's disease and normal ageing: diagnostic value and neuropsychological correlates *J. Neurol. Neurosurg. Psychiatry* **55** 967–72

[39] Tartaglia M C, Rosen H J and Miller B L 2011 Neuroimaging in dementia *Neurotherapeutics* **8** 82–92

[40] Mazziotta J C and Phelps M E 1986 Positron emission tomography studies of the brain *Positron Emission Tomography and Autoradiography: Principles and Applications for the Brain and Heart* ed M E Phelps, J C Mazziotta and H Schelbert (New York: Raven) pp 493–579

[41] Cronin-Golomb A, Rizzo J and Corkin S 1991 Visual function in Alzheimer's disease and normal aging *Ann. NY Acad. Sci.* **640** 28–35

[42] Gilmore G C, Wend H E and Naylor L 1994 Motion perception and Alzheimer's disease *J. Gerontol.* **49** 52–7

[43] Bradshaw J R, Thomson J L G and Campbell M J 1983 Computed tomography in the investigation of dementia *Br. Med. J.* **286** 277–80

[44] Sandeep C S, Sukesh Kumar A, Mahadevan K and Manoj P 2018 Analysis of retinal OCT images for the early diagnosis of Alzheimer's disease *Springer-Advances in Intelligent Systems and Computing book series (AISC)* **vol 749** 509–20

[45] Villemagne V L *et al* 2008 Aβ deposits in older non-demented individuals with cognitive decline are indicative of preclinical Alzheimer's disease *Neuropsychologia* **46** 1688–97

[46] Morris J C *et al* 2009 Pittsburgh compound B imaging and prediction of progression from cognitive normality to symptomatic Alzheimer disease *Arch. Neurol.* **66** 1469–75

[47] Sperling R A *et al* 2011 Toward defining the preclinical stages of Alzheimer's disease: recommendations from the National Institute on Aging-Alzheimer's Association workgroups on diagnostic guidelines for Alzheimer's disease *Alzheimers Dement.* **7** 280–92

[48] Fjell A M *et al* 2010 CSF biomarkers in prediction of cerebral and clinical change in mild cognitive impairment and Alzheimer's disease *J. Neurosci.* **30** 2088–101

[49] Sandeep C S and Sukesh Kumar A 2017 A psychometric assessment method for the early diagnosis of Alzheimer's disease *Int. J. Sci. Eng. Res.* **8** 901–5

[50] de Souza L C *et al* 2011 CSF tau markers are correlated with hippocampal volume in Alzheimer's disease *Neurobiol. Aging* **33** 1253–7

[51] Mattsson N *et al* 2009 CSF biomarkers and incipient Alzheimer disease in patients with mild cognitive impairment *J. Am. Med. Assoc.* **302** 385–93

[52] de Toledo-Morrell L *et al* 1997 Alzheimer's disease: *in vivo* detection of differential vulnerability of brain regions *Neurobiol. Aging* **18** 463–8

[53] Jack C R Jr. *et al* 1992 MR-based hippocampal volumetry in the diagnosis of Alzheimer's disease *Neurology* **42** 183–8

[54] Sandeep C S, Sukesh Kumar A, Mahadevan K and Manoj P 2017 Dimensionality reduction of optical coherence tomography images for the early diagnosis of Alzheimer's disease *Am. J. Electr. Electron. Eng.* **5** 58–63

[55] Monteiro M L, Cunha L P, Costa-Cunha L V, Maia O O Jr and Oyamada M K 2009 Relationship between optical coherence tomography, pattern electroretinogram and automated perimetry in eyes with temporal hemianopia from chiasmal compression *Invest. Ophthalmol. Vis. Sci.* **50** 3535–41

[56] Sandeep C S, Sukesh Kumar A, Mahadevan K and Manoj P 2018 Analysis of MRI and OCT images for the early diagnosis of Alzheimer's disease using wavelet networks *AMSE Journal on Lectures on Modelling and Simulation* pp 31–40

[57] Baron R and Girau B 1998 Parameterized normalization: application to wavelet networks *Proc. IEEE Int. Conf. Neural Netw.* **2** 1433–7

[58] Zhang Q H 1997 Using wavelet network in nonparametric estimation *IEEE Trans. Neural Netw.* **8** 227–36

[59] Davanipoor M, Zekri M and Sheikholeslam F 2012 Fuzzy wavelet neural network with an accelerated hybrid learning algorithm *IEEE Trans. Fuzzy Syst.* **20** 463–70

IOP Publishing

Neurological Disorders and Imaging Physics, Volume 3
Application to autism spectrum disorders and Alzheimer's
Ayman El-Baz and Jasjit S Suri

Chapter 11

A local/regional computer aided system for the diagnosis of mild cognitive impairment

Fatma El-Zahraa A El-Gamal, Mohammed M Elmogy, Hassan Hajjdiab, Ashraf Khalil, Mohammed Ghazal, Ali Mahmoud, Hassan Soliman, Ahmed Atwan, Jasjit S Suri, Gregory N Barnes and El-Baz for the Alzheimer's Disease Neuroimaging Initiative[1]Ayman

Alzheimer's disease (AD) is an irreversible disorder that targets the central nervous system (CNS) in addition to being one of the leading causes of dementia. Based on the clinical and pathological features that characterize the disease, AD is considered to go through three main stages: early (mild), intermediate (moderate), and late (severe). Among these stages, the early diagnosis of the disease is considered as one of the main obstacles facing researchers in this field for a number of reasons, including the variability of the disease among AD sufferers as well as the possible emergence of pathological features 10–15 years before being diagnosed clinically. This chapter aims to present a computer aided diagnosis (CAD) system with the main goal of addressing personalized diagnosis through providing local/regional based diagnosis for early diagnosis of AD, in the mild cognitive impairment (MCI) stage. In this system, ^{11}C Pittsburgh compound B positron emission tomography (PiB-PET) scans are used to build the system that in turn goes into five stages: the preprocessing stage (through data standardization and de-noising processes), brain labeling (using automatic anatomical labeling (AAL) to parcellate the brain scans into 116 different anatomical regions), feature extraction (using a scale-invariant Laplacian of Gaussian (LoG) to detect the significantly greater retention of PiB in the scans in the form of a blob detection), statistical analysis (using a two-sample

---

[1] Data used in preparation of this article were obtained from the Alzheimer's Disease Neuroimaging Initiative (ADNI) database(adni.loni.usc._edu). As such, the investigators within the ADNI contributed to the design and implementation of ADNI and/or provided data but did not participate in analysis or writing of this report. A complete listing of ADNI investigators can be found at: http://adni.loni.usc.edu/wp-content/uploads/howtoapply/ADNIAcknowledgementList.pdf.

t-test to determine the number of regions that are found to be significantly influenced by the disease), and finally the diagnosis stage that in turn goes into two diagnosis levels to provide local/regional diagnosis followed by global diagnosis of the subjects (using a probabilistic support vector machine (pSVM) followed by a standard SVM). Evaluating the system's performance was performed on 19 normal control (NC) and 65 MCI scans where the results show superior results of the system in general for accuracy, specificity, and sensitivity as compared to other related work.

11.1 Introduction

Alzheimer's disease (AD) is a chronic neurodegenerative disorder marked by cognitive and behavioral impairments [1, 2]. Statistically, 42% of AD sufferers are people over 85 years of age with the percentage decreasing to only 6% for people 70–74 years old. Although the probability is small, younger individuals may also be affected by AD [3].

AD is characterized by clinical symptoms and pathological features, both of which vary among patients [4]. In the clinical presentation, the patient faces progressive deficits in cognition as well as disturbances in thought, perception, and behavior. Neuropathological abnormalities include the formation of neuro-fibrillary tangles and neuritic plaques as well as neuronal loss and granulovacuolar degeneration. Indeed, the quantity and location of neurofibrillary tangles and neuritic plaques represent neurodegenerative features that distinguish AD from other types of dementia. Efforts to establish an early diagnosis of AD have been thwarted by the fact that pathological features of the disease occur 10–15 years before the emergence of clinical symptoms.

Mild cognitive impairment (MCI) can lead to AD. MCI can be defined as an impairment of cognition that is more severe than expected from normal aging and a person's education with objective evidence of impairment in one or more cognitive domains including memory, executive function, attention, language, or visuospatial skills. MCI does not interfere with independence and daily activities including social or occupational functioning [5]. In this regard, MCI represents an intermediate stage between the cognitive decline observed with normal aging and the severe impairment observed in dementia [5, 6]. It is important to note that not all MCI cases proceed to AD, although studies have found that it increases the risk of later developing AD [6, 7]. MCI due to AD is a progressive decline in cognition over months to years. MCI due to AD has a lack of significant vascular factors, vascular imaging findings, parkinsonism, visual hallucinations, and prominent behavioral or language disorders [8].

There are a number of tests that have to be considered when trying to establish a diagnosis of AD. These tests include neuropsychological screening (to measure related cognitive impairments), the patient's medical history, and mental/physical examination. In addition, blood tests, as well as brain imaging, are usually evaluated to rule out other neurological or physiological disorders [9, 10]. In general, these tests help to classify the subjects along the disease cascade into MCI, when the appearance of some cognitive decline does not fulfill the dementia criteria and other stages of AD [10]. A recent meta-analysis of neuropsychological measures suggested that verbal memory measures and other language tests yield high predictive accuracy for those MCI subjects who will progress to AD. Other domains including executive function and visual memory showed better specificity than sensitivity [11]. These data show that there is a clinical need to identify biomarkers of neural circuits involved in MCI, which ultimately lead to the development of AD.

Brain biomarkers have been postulated to help in the diagnosis of AD throughout the disease's natural history. For example, [12] presented a study in which positron emission tomography (PET) amyloid imaging and cerebrospinal fluid (CSF) amyloid beta 42 (Aβ42) revealed Aβ abnormalities in the brain. These abnormalities are the earliest pathological features observed in the AD-related disease cascade. Additionally, the study reported that both the increase in CSF-tau and cerebral atrophy serve as biomarkers of neuronal injury and neurodegeneration. Finally, the decrease in 2-[^{18}F]fluoro-2-deoxy-D-glucose PET (FDG-PET), as demonstrated by the study, helps to reveal the synaptic dysfunction that accompanies the neuro-degeneration. Therefore, according to this study, sMRI measures abnormalities of the brain's structure, and FDG-PET/CSF-tau identifies tau-mediated neuronal injury and dysfunction. The authors concluded that PET amyloid imaging could be considered an early identifier of AD-related abnormalities.

PET is a main scanning application of the emission computed tomography (ECT) methodology. Despite the role of the PET amyloid imaging in the early diagnosis of AD, arguments for and against the implementation of this scan modality in clinics should be carefully considered. False positive diagnoses of AD may occur since normal elderly subjects can have elevated Aβ levels. Fortunately, the introduction of carbon-11 labeled Pittsburgh compound B (^{11}C PiB), a neutral analog of the thioflavin T, caused a noteworthy conversion in studies related to AD [13]. The compound assists in visualizing the pathological hallmarks related to AD and consequently helps in quantifying the neuropathological burden during subsequent AD stages [14].

Numerous research efforts have been proposed to help in the differentiation between normal control (NC), MCI, and AD utilizing PET scans. These efforts include studies focused on testing computer aided diagnosis (CAD) systems, such as the automatic classification system proposed by [15]. For this purpose, principal component analysis (PCA) and support vector machines (SVMs) were utilized. To evaluate the system, the PiB and FDG related scans were used to compare their results regarding early diagnosis. Although PiB and FDG showed similar accu-racies, the PiB had a higher power of discrimination in the very early cases. Again, [16] utilized PET scans to construct a CAD system that relied on the eigenimage framework and was composed of feature extraction, dimensionality reduction, and classification stages. In this system, PCA and independent component analysis (ICA) was utilized for image projection (feature reduction), eigenimage based decomposition for feature reduction, and SVM for classification. Through these stages, the system achieved an accuracy of 88.24%. Another study by [17] demonstrated a CAD system that improved the classification accuracy of AD using PCA, ICA, and SVM. PCA was used for dimension reduction, ICA for feature extraction, and SVM using linear and radial basis function (RBF) kernels for classification. The results of their study showed either better or equivalent performance as compared to the competitive CAD methods with higher accuracy than the traditional visual assessment methods.

In the same context, [18] proposed a CAD system aimed at improving the accuracy of three-way classification between NC, MCI, and AD. The study used PCA and linear discriminant analysis/Fisher discriminant ratio for the feature selection process followed by an artificial neural network/SVM for the classification purpose. Testing this methodology with FDG-PET scans led to a classification accuracy of 89.52%. In [19], the researchers presented a CAD system composed of three stages: the Mann–Whitney–Wilcoxon U-test for voxel selection in order to exclude outliers, factor analysis for feature extraction, and linear-SVM for classification. Testing this system on PET scans achieved an accuracy of 92.9%. Also, [20] exploited association rule mining in a CAD system used in the early diagnosis of AD. Testing the system using two datasets, including PET scans, showed better results compared to other related works. In [21] the researchers introduced a CAD system to serve the early diagnosis of AD by combining nonnegative matrix factorization (NMF) for feature selection and reduction, and SVM with confidence bounds for classification. Application of the system led to an accuracy of 86%.

In addition to the aforementioned studies, various researchers have used voxel analysis for the early diagnosis of AD. For example, [22] performed voxel-wise interregional correlations through statistical parametric mapping to extract relevant information. The study's results illustrated the association between the pathophysiological process of AD and alterations of the functional brain networks. According to the study, the default mode network (DMN) and memory function-related networks are the leading causes of such alterations. The authors of [23] compared AD versus NC subjects to find the brain regions that showed significant increases in the uptake of ^{11}C PiB by applying voxel-based analysis. A statistical parametric mapping (SPM) analysis was performed using an automated region of interest (ROI). The study found that the voxel-based analysis showed widespread distribution regarding the increased ^{11}C PiB uptake. In [24], the researchers statistically evaluated the amyloid imaging agents' (i.e. PiB) retention differences throughout the brain. In addition, they compared the PiB results with the FDG-based scans of glucose metabolism. The results revealed that the statistical significance of the PiB analysis was both greater than others and had a more considerable spatial extent. The results also showed that the PiB significance was retained after corrections of family-wise error and false discovery rate.

Forsberg et al [25] studied amyloid deposition in patients with MCI. They used ^{11}C PiB and FDG-based PET scans with AD, which were compared with NC scans. The analysis showed intermediate retention of the mean cortical PiB in the MCI as compared to NC and AD. Also, the study found significantly higher PiB retention in the MCI conversion to an AD group, comparable to that of AD patients ($p > 0.01$) and much less in MCI subjects who did not convert to AD. The authors of [26] presented a voxel-based analysis relying on FDG and PiB along with another tracer known as 2-(1-6-[(2-(^{18}F) fluoroethyl) (methyl) amino]-2-naphthylethylidene) malononitrile (FDDNP). These tracers were utilized to address the pathological hallmarks of AD, beta-amyloid plaques, potential neurofibrillary tangles, and glucose metabolism related impairments. The experimental results demonstrated the available capacity to develop and test disease-modifying drugs targeting both tau and amyloid pathology, and/or energy metabolism when using the same subject based PET imaging with these three tracers.

Despite the achievements of the investigations above, the studies only supported a global diagnosis that indicates whether or not the subject belongs to a specific studied group. The main objectives of this chapter are summarized in the following points. First, it provides a personalized diagnosis to help individualize diagnostic options as well as monitoring disease progression. Second, it improves the final global diagnosis results compared to previously published studies. To achieve these goals, our system utilized PiB-PET scans due to the superior role of this brain imaging modality, as compared to other scanning modalities, when applied to the early diagnosis of the disease. Therefore, the chapter is organized as follows. Section 11.2 starts by describing the materials used for the preparation of the study and defines the methods used in the proposed CAD system. In section 11.3, different experimental results are presented to evaluate the performance as well as the efficiency of our system. Finally, the discussion of the applied tests and future work are highlighted in section 11.4.

11.2 Material and methods

11.2.1 Materials

A set of ^{11}C PiB-PET scans were used to validate the proposed framework. These scans were collected from the Alzheimer's Disease Neuroimaging Initiative (ADNI) database (www.loni.ucla.edu/ADNI). The database of ADNI was initially launched in 2003 as a public–private partnership, led by principal investigator Michael Weiner, MD. The goal of demonstrating ADNI was to test whether serial MRI, PET, or other markers, in addition to clinical and neuropsychological assessment, can measure the progression of MCI and AD by combining them. For further information, see www.adni-info.org. The used dataset was obtained from ADNI 1 where it contains a total number of 84 scans obtained from 19 NC and 65 MCI subjects. NC comprises those subjects who do not show any signs of depression, cognitive impairment, or dementia. The MCI group, in general, comprises those subjects with subjective memory concerns, whether self-reported or through an informant or a clinician. These subjects display neither significant impairment levels in other cognitive domains nor signs of dementia. The logical memory II subtest of the Wechsler memory scale (WMS) was performed on the participants to document the normality/abnormality of their memory function with respect to their level of education. The demography of the used dataset is presented in table 11.1. To

11.2.2 Methods

The main aim of this study was to present local (i.e. region) based diagnosis of MCI regarding AD, to assist clinicians in the personalized treatment of the disease. To

Table 11.1. The demographic data of the NC and MCI groups of the ^{11}C PiB-PET scans.

(N = 84)	Average age ± SD	Gender N (%)		WMS logical memory II results (based on years of education)		
		Male	Female	≥16 years	8–15 years	0–7 years
NC (19)	78.3 ± 5.01	11 (57.89)	8 (42.1)	≥ 9	≥ 5	≥ 3
MCI (65)	75.78 ± 7.67	44 (67.69)	21 (32.30)	≥ 8	≥ 4	≥ 2

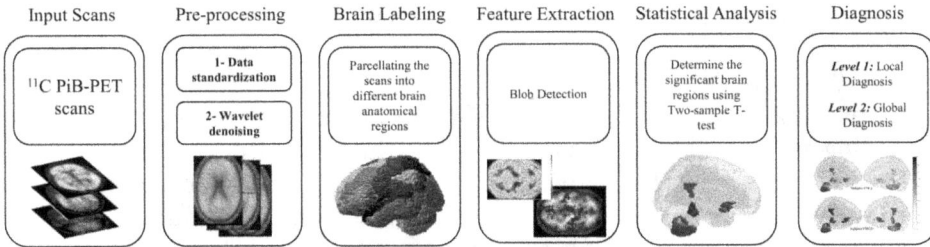

Figure 11.1. The framework of the proposed early diagnosis system for AD-based on local and global region analysis using ^{11}C PiB-PET scans.

achieve this goal, five main steps were performed, as illustrated in figure 11.1. First, the scans were preprocessed through data standardization and de-noising. Data standardization aimed to prepare the scans for the labeling step, while the de-noising process aimed to improve the scan's quality and consequently the system's accuracy. Second, the atlas of Automated Anatomical Labeling (AAL) was used for brain parcellation to serve the local diagnosis goal. After labeling the brain, we used a Laplacian of Gaussian (LoG) with the automatic scale to extract the discriminant features from the scans. Then, a statistical analysis was performed to determine the significant brain regions to analyze rather than using all the labeled regions in the decision-making process. Finally, these regions were used to construct two decision-making levels using a probabilistic version of SVM (pSVM) and standard SVM to provide local followed by global diagnosis. The details of the proposed system are presented in the following subsections.

11.2.2.1 Preprocessing

Scans underwent some preprocessing to orient the data and reduce noise. For orienting the data in a standard coordinate system, the SPM MATLAB toolbox [27] was used to perform re-orientation, co-registration, spatial normalization, and re-slicing. Noise reduction was accomplished through wavelet shrinkage. The details are as follows: PET scans associated sMRI data were re-oriented so that the anterior–posterior axis coincided with the AC–PC line. This differs from the ADNI pipeline, which only ensures the axis is parallel to the AC–PC line. The associated PiB-PET scan was re-oriented to the resulting sMRI scan to produce a re-orientation matrix that was then used to re-orient the remaining PiB-PET scans. Precise co-registration between the PET scans and the previously used sMRI scans was performed using rigid body transformations (translations and rotations) to maximize the mutual information. Then, the spatial normalization and re-slicing were applied to the sMRI and PET scans to align the scans to the MNI-152 standard space. In this step, general affine transformation (translations, rotations, non-uniform scaling, and shears) was used, followed by nonlinear deformations. After data standardization, wavelet-based de-noising was applied using the symlet8 mother wavelet with Stein's unbiased risk estimate as a threshold selection rule and soft thresholding [28]. The aim was to retain image detail while removing artifacts of image acquisition and/or transmission [29]. At this point, the scans were ready for voxel-wise comparison and labeling steps.

11.2.2.2 Brain labeling

Due to the local diagnosis based goal of the proposed system, the brain labeling/partitioning needs to be performed. Through this aim, detailed diagnosis of the subjects could be achieved. For this purpose, any of the detailed based brain atlases can be used, such as the AAL, Talairach Daemon, and Brodmann areas atlases [30–32]. In this chapter, the AAL atlas was used to label each of the preprocessed scans' voxel's positions to the matched anatomical regions. The AAL atlas provides a total of 116 brain regions: 45 per cerebral hemisphere, 9 per cerebellar hemisphere, and 8 in the vermis of the cerebellum. AAL provides a detailed parcellation of the brain and is recommended for use with PET scans. To accomplish the labeling procedure, the xjView MATLAB toolbox [33] was utilized.

11.2.2.3 Blob detection based feature extraction

Each of the labeled regions is individually fed to the scale-invariant blob detector, which employs LoG with automatic scale selection for feature extraction. Blob detection aims to separate structures (i.e. blobs) from the image background. Each blob is itself a radially symmetric distribution of image intensity about a local minimum or maximum [34]. Blobs corresponding to local maxima could reveal the targeted abnormalities, given that significantly greater retention of PiB in a brain region is linked with greater incidence of $A\beta$ plaques within that region [26].

11.2.2.4 Statistical analysis

AAL regions where mean PiB uptake differs significantly in MCI with respect to NC was determined using two-sample t-tests. Each atlas region was tested independently. The Bonferroni method was applied to identify a region as significant when the p-value was less than 0.000 43 (i.e. 0.05/116). Significant regions were subsequently used in building a classifier.

11.2.2.5 Diagnosis

A two-level diagnosis was performed to make local (region-specific abnormalities) and global (level of cognitive impairment) diagnoses. For this purpose, SVM and one of its variants (pSVM) were utilized. Standard SVM is an abstract machine learning technique where the training data are used for the learning followed by a generalization attempt for a correct prediction on other novel data [35]. In SVM, a hyperplane (known as maximal margin hyperplane) is used for the binary separation of the labeled training data. The goal is to build a decision function $f: R^S \rightarrow \pm 1$, according to S-dimensional training patterns pi and t_i, capable of performing classification for the new example (p, t): $(p_1, t_1), (p_2, t_2), \ldots, (p_s, t_s) \in R^S \pm 1$. The decision hyperplanes in multidimensional feature space can be defined through using either a linear separation of the training data, using linear discriminant functions, or combining SVM with kernel techniques that produce a nonlinear decision boundary (hyperplane) in the input space [36]. In addition to its classification power, a variation of SVM, which produces a posterior probability output of the classifier (pSVM), is useful to allow further post processes.

In the proposed system, a separate pSVM model was constructed, in the first diagnosis level, for each significant region. Each pSVM produces a probabilistic result for the incidence of MCI given the features from its brain region independently of all others. In the second level, the scores obtained from the first level were fused concerning each subject and used to train and test a single SVM model to produce the global diagnosis. Classifier performance, i.e. accuracy, sensitivity, and specificity, were estimated using both leave-one-subject-out (LOSO) and K-fold cross-validation, with two- and four-fold validation.

11.3 Results

According to the Bonferroni corrected two-sample t-tests, the AAL brain regions of significance, nine regions, were Cerebelum_3_L (left alar central lobule), Cerebelum_8_R (right biventer lobule), Cingulum_Post_L (left posterior cingulate gyrus), Olfactory_L and Olfactory_R (bilateral olfactory cortex), plus Vermis_1_2, Vermis_3, Vermis_8, and Vermis_9 (lobules I–II, III, VIII, and IX of the vermis) as visualized in figure 11.2. The performance of the resulting classifier was evaluated under a number of different kernels selected for the SVM, with the best results being obtained when the linear kernel was used at both level 1 and level 2 (table 11.2). This classifier was used to compare three cases, using data from all brain regions, using all regions except the significant ones, and using pre-selected, specific regions. The third case was found to outperform the other two cases (table 11.3).

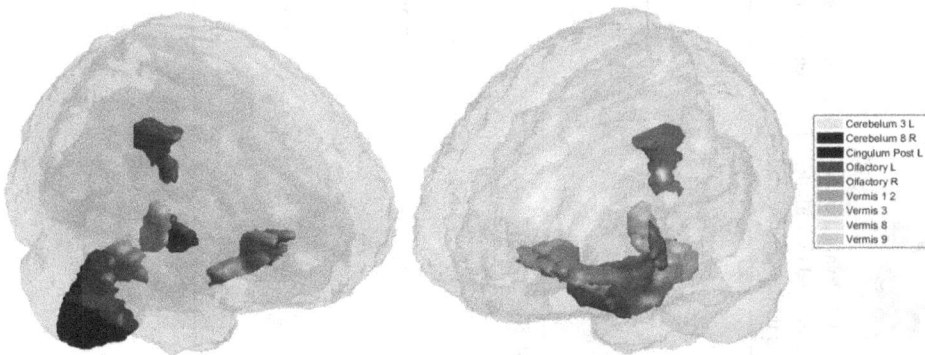

Figure 11.2. The significant regions as obtained through the two-sample t-test.

Table 11.2. Evaluation of SVM classifier performance when different kernels are used. Classifier accuracy (ACC), sensitivity (Sens.), and specificity (Spec.), in (%), were estimated over all the labeled regions using LOSO and *K*-fold cross-validation. The bolded results (linear-pSVM, linear-SVM) indicates the best combination of kernels used in the two levels.

			LOSO			Level 2					
						K-fold					
						K = 2			*K* = 4		
Level 1			RBF	Linear	Polynomial	RBF	Linear	Polynomial	RBF	Linear	Polynomial
	RBF	ACC	72.61	50	22.61	59.52	22.61	77.38	77.38	22.61	77.38
		Spec.	26.31	57.89	100	57.89	100	0	0	100	0
		Sens.	86.15	47.69	0	60	0	100	100	0	100
	Linear	ACC	77.38	**88.09**	95.23	77.38	**79.76**	77.38	77.38	**89.28**	77.38
		Spec.	0	**47.36**	78.94	0	**10.52**	0	0	**52.63**	0
		Sens.	100	**100**	100	100	**100**	100	100	**100**	100
	Poly	ACC	77.38	88.09	77.38	77.38	80.95	76.19	77.38	88.09	77.38
		Spec.	0	47.36	0	0	26.31	0	0	47.36	0
		Sens.	100	100	100	100	96.92	98.46	100	100	100

Table 11.3. Using LOSO and *K*-fold cross-validation methods to evaluate the classification results, in (%), (using linear-pSVM linear-SVMs) using the features of different cases of input regions: all the labeled regions, all the regions except the significant ones, and the significant regions only.

		LOSO	K-fold	
			$K = 2$	$K = 4$
All labeled regions	ACC	88.09	79.76	89.28
	Spec.	47.36	10.52	52.63
	Sens.	100	100	100
The resulting significant regions	ACC	100	97.61	98.80
	Spec.	100	94.73	94.73
	Sens.	100	98.46	100
Excluding the significant regions	ACC	82.14	77.38	83.33
	Spec.	21.05	0	26.31
	Sens.	100	100	100

Table 11.4. A comparison of the proposed system's performance, in (%), against other related studies using the LOSO cross-validation method.

	ACC	Spec.	Sens.
[20]	90.48	100	87.69
[17]	89.17	—	—
The proposed system	100	100	100

The efficiency of the proposed system was compared to two other published methods [17, 20]. Our methodology was found to distinguish MCI from NC as well or better than any previous techniques (table 11.4). Note that the performance of the compared classifiers is taken directly from the respective publications since each of them used the same database (i.e. ^{11}C PiB-PET scans from ADNI), and LOSO cross-validation. Finally, figure 11.3 provides examples of the local diagnostic results (i.e. the results of the first diagnosis level) for two NC and two MCI cases. The color bar in both figures represents the degree of abnormality starting from 0 (unaffected) to 1 (indicative of MCI). These examples indicate the varying abnormality effects in each significant region for each case independently.

11.4 Discussion

This chapter discusses a personalized MCI diagnosis system while improving diagnostic performance as compared to other available methods. The personalized diagnosis is achieved through regional/local measurements, using the AAL atlas, which reflects how the disease affects different brain regions. To enhance this procedure, a statistical analysis was initially performed to determine the salient brain

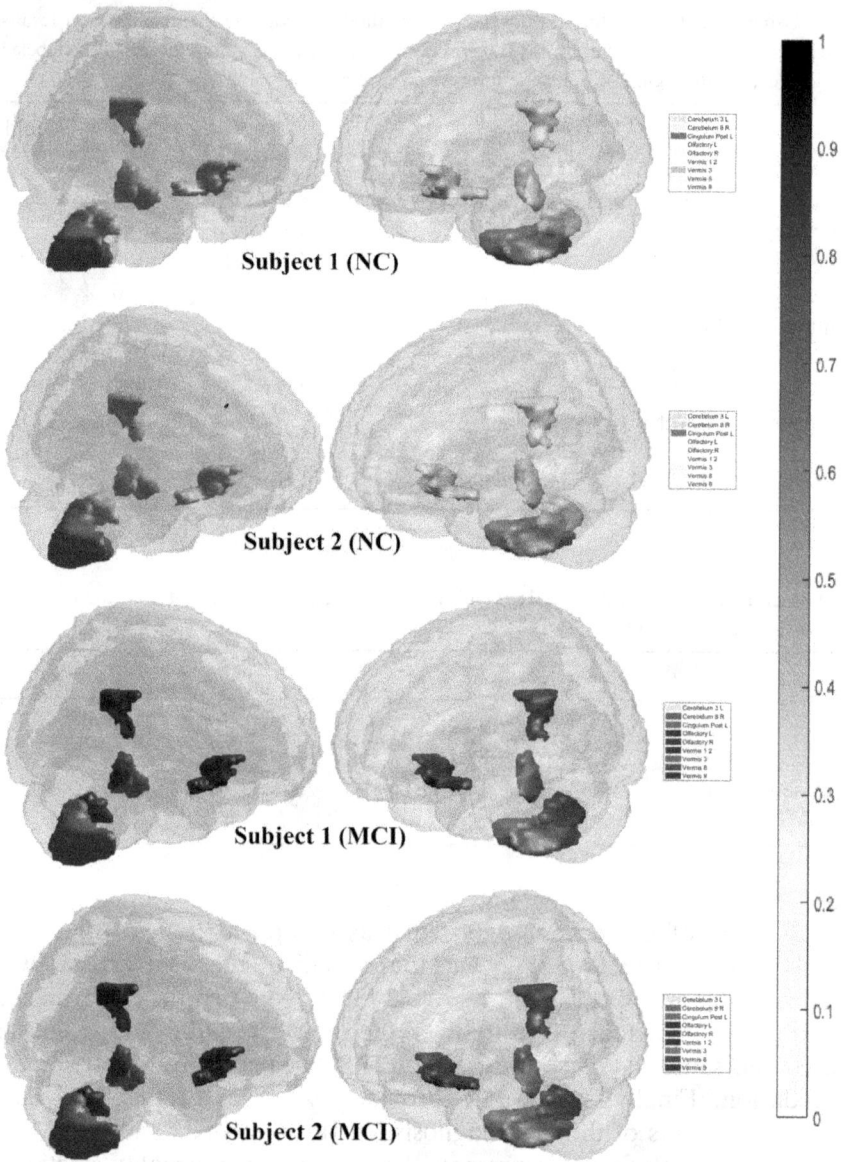

Figure 11.3. Examples of local diagnosis results for NC versus MCI classification problem.

regions and consequently analyze the influence of the disease on them through the first diagnosis level, as shown in figure 11.3. Apart from being the first stage of the MCI diagnosis procedure, the reported local diagnoses are helpful in the personalized management of the disease. This has been represented through the color bar that shows the degree of abnormality from 0 (unaffected) to 1 (indicative of MCI) in each of the nine significant regions separately. Finally, the system fuses the regional based probabilistic results obtained from the first diagnosis level to produce the final global diagnosis of each subject.

Regarding the cerebellar regions identified as significant, there have been several studies that focused on cerebellar abnormalities in dementia. Some of these studies were neuropathological-based studies that found that the neuronal shrinkage and loss represent well-known changes that accompany AD [37]. Morphological studies have targeted the prominent neuropathological AD-related hallmarks, such as amyloid plaques [38, 39]. According to [38], most AD-based pathological features that are found in the cerebellum include diffuse $A\beta$ deposits. Considering the following finding and since the regions with increased $A\beta$ plaques are represented as the high retention in the PiB, we could justify this finding in the early stage of AD. In addition to these findings and according to [40], structural based changes, mainly involved in the vermis, were judged to represent the progression of the disease.

Dysfunction in the olfactory cortex was found, through some studies, to probably be one of the earliest symptoms that can clinically obtained regarding AD [41, 42]. According to [43], the combination of the olfactory function tests along with the conventional diagnostic methods provides the ability to improve the sensitivity as well as the specificity of diagnosing AD. This consequently facilitates both the early recognition and diagnosis of AD. More details about the research and the future directions of the olfactory dysfunction in AD can be found in [44].

Finally, as regards the posterior cingulate gyrus, some neuroimaging studies, which targeted several cortical regions, have identified that the posterior cingulate regions of the medial parietal cortex are among the earliest regions affected in AD [45]. These studies include one that supports the involvement of the posterior cingulate in the very early progression of AD [46].

In general, as shown in table 11.2, the linear kernel shows the best results, either using it on both levels of the classifier or along with the polynomial kernel, while the RBF kernel performed poorly even when used conjointly with another kernel. The ability of the extracted features to differentiate the two groups (NC and MCI) and consequently make them linearly separate could justify the highest results of the linear kernel. For the polynomial as well as the RBF kernels, nonlinear kernels, the obtained separable features in addition to the small size of the dataset caused the outperformance of the polynomial kernel compared to the RBF kernel, which can show better results with the large size of the dataset. According to these results and since linear–linear-SVM-based classifiers show in general the best results, they were used to build the CAD system.

As expected, the significant region-based diagnostic performance outperforms that of the other two trial classifiers (table 11.3) with a maximum performance of 100% of accuracy, specificity, and sensitivity. This finding could be due to the discrimination power of the features extractor in addition to the demonstration of alpha that helped in obtaining these salient regions. Regarding the two other tests, using all the labels shows better results than excluding the significant regions but worse results than using the significant regions. This finding could be due to the presence of the labeled regions that are not significant enough to differentiate between the groups and consequently lead to misclassification results that could finally affect the overall performance of the system. In addition, the same case occurs when excluding the significant regions, but in addition to the presence of these not significant enough regions, the significant regions are excluded causing this drop of performance compared to the other two tested cases.

The fact that our system improved the performance compared to other techniques while using the same dataset is of crucial significance. The superior results of the proposed system over prior works (table 11.4) could be explained through the capabilities of the combined components of the system that could extract the most discriminant features, identify the significant brain regions, and then perform the classification using the SVM along with the linear kernel.

Since the primary goal of the chapter was to provide local diagnosis, figure 11.3 illustrates various examples that demonstrate the variability of the abnormality in each of the salient regions among the subjects. Figure 11.3 shows sample cases of the NC and MCI groups, proving this point, where the color bar indicates the degree of the disease's effect, starting with the dark blue for no effect, to dark red for total effect. This figure demonstrates the power of the proposed system in revealing the local diagnosis of the disease in the significant brain regions. The analysis results that could be derived from the figures show the ability of the proposed analysis to reflect the current degree of dysfunctionality that each region reaches through the disease. This visualization consequently can be considered an effective assistant for the individualized/personalized diagnosis process.

According to the above results, the proposed system may be of assistance in the diagnosis of MCI. In other words, the system offers a personalized diagnosis of the subjects in a short computation time using a subset of the brain's regions rather than using all regions. Additionally, the system provides high global diagnosis results compared to other related studies. Due to the promising results of the proposed system and pilot nature of our data, we plan to examine two goals in future work. First, the system's performance will be evaluated on larger PET scan datasets. Second, the results will be incorporated with other AD-related scanning modalities, to provide more analysis based assistance for the early diagnosis of AD.

This work could also be applied to various other applications in medical imaging, such as the kidney, heart, prostate, lung, and retina, as well as several non-medical applications [47–50].

One application is renal transplant functional assessment, in particular developing noninvasive CAD systems for renal transplant function assessment, utilizing different image modalities (e.g. ultrasound, computed tomography (CT), MRI, etc). Accurate assessment of renal transplant function is critically important for graft survival. Although transplantation can improve a patient's wellbeing, there is a potential post-transplantation risk of kidney dysfunction that, if not treated in a timely manner, can lead to the loss of the entire graft, and even patient death. In particular, dynamic and diffusion MRI-based systems have been clinically used to assess transplanted kidneys with the advantage of providing information on each kidney separately. For more details about renal transplant functional assessment, see [51–78].

The heart is also an important application to this work. The clinical assessment of myocardial perfusion plays a major role in the diagnosis, management, and prognosis of ischemic heart disease patients. Thus, there have been ongoing efforts to develop automated systems for accurate analysis of myocardial perfusion using first-pass images [79–95].

Another application for this work could be the detection of retinal abnormalities. The majority of ophthalmologists depend on visual interpretation for the identification of diseases types. However, inaccurate diagnosis will affect the treatment procedure which may lead to fatal results. Hence, there is a crucial need for computer automated diagnosis systems that yield highly accurate results. Optical coherence tomography (OCT) has become a powerful modality for the noninvasive diagnosis of various retinal abnormalities such as glaucoma, diabetic macular edema, and macular degeneration. The problem with diabetic retinopathy (DR) is that the patient is not aware of the disease until the changes in the retina have progressed to a level that treatment tends to be less effective. Therefore, automated early detection could limit the severity of the disease and assist ophthalmologists in investigating and treating it more efficiently [96–98].

Abnormalities of the lung could also be another promising area of research and a related application to this work. Radiation-induced lung injury is the main side effect of radiation therapy for lung cancer patients. Although higher radiation doses increase the radiation therapy effectiveness for tumor control, this can lead to lung injury as a greater quantity of normal lung tissue is included in the treated area. Almost 1/3 of patients who undergo radiation therapy develop lung injury following radiation treatment. The severity of radiation-induced lung injury ranges from ground-glass opacities and consolidation at the early phase to fibrosis and traction bronchiectasis in the late phase. Early detection of lung injury will thus help to improve the management of treatment [99–141].

This work can also be applied to other brain abnormalities, such as dyslexia in addition to autism. Dyslexia is one of the most complicated developmental brain disorders that affect children's learning abilities. Dyslexia leads to the failure to develop age-appropriate reading skills in spite of the normal intelligence level and adequate reading instructions. Neuropathological studies have revealed an abnormal anatomy of some structures, such as the corpus callosum in dyslexic brains. There has been a lot of work in the literature that aims at developing CAD systems for diagnosing such a disorder, along with other brain disorders [142–164].

For the vascular system [165], this work could also be applied for the extraction of blood vessels, e.g. from phase contrast (PC) magnetic resonance angiography (MRA). Accurate cerebrovascular segmentation using noninvasive MRA is crucial for the early diagnosis and timely treatment of intracranial vascular diseases [147, 148, 166–171].

Acknowledgments

The funding of the collected and shared data of this article was provided by the Alzheimer's Disease Neuroimaging Initiative (ADNI) (National Institutes of Health Grant U01 AG024904) and DOD ADNI (Department of Defense award number W81XWH-12-2-0012). ADNI is funded by the National Institute on Aging, the National Institute of Biomedical Imaging and Bioengineering, and through generous contributions from the following: AbbVie, Alzheimer's Association; Alzheimer's Drug Discovery Foundation; Araclon Biotech; BioClinica, Inc.; Biogen; Bristol-Myers Squibb Company; CereSpir, Inc.; Cogstate; Eisai Inc.; Elan Pharmaceuticals, Inc.; Eli Lilly and Company; EuroImmun; F. Hoffmann-La Roche Ltd and its affiliated company Genentech, Inc.; Fujirebio; GE Healthcare; IXICO Ltd.; Janssen Alzheimer Immunotherapy Research & Development, LLC.; Johnson & Johnson Pharmaceutical Research & Development LLC.; Lumosity; Lundbeck; Merck & Co., Inc.; Meso Scale Diagnostics, LLC.; NeuroRx Research; Neurotrack Technologies; Novartis Pharmaceuticals Corporation; Pfizer Inc.; Piramal Imaging; Servier; Takeda Pharmaceutical Company; and Transition Therapeutics. The Canadian Institutes of Health Research is providing funds to support ADNI clinical sites in Canada. Private sector contributions are facilitated by the Foundation for the National Institutes of Health (www.fnih.org). The grantee organization is the Northern California Institute for Research and Education, and the study is coordinated by the Alzheimer's Therapeutic Research Institute at the University of Southern California. ADNI data are disseminated by the Laboratory for Neuro Imaging at the University of Southern California.

References

[1] WHO 2017 *World Health Organization* http://who.int/ (Accessed: 6 March 2017)

[2] Hodler J, Schulthess G and Zollikofer C 2012 *Diseases of the Brain, Head and Neck, Spine 2012-2015: Diagnostic Imaging and Interventional Techniques* (Milan: Springer)

[3] Brown D 2013 *Brain Diseases and Metalloproteins* (Singapore: Pan Stanford)

[4] Lu L and Bludau J 2011 *Alzheimer's Disease* 1st edn (Westport, CT: ABC-CLIO)

[5] Yaffe K 2013 *Chronic Medical Disease and Cognitive Aging: Toward a Healthy Body and Brain* (New York: Oxford University Press)

[6] Anderson N, Murphy K and Troyer A 2012 *Living with Mild Cognitive Impairment: A Guide to Maximizing Brain Health and Reducing Risk of Dementia* (New York: OOxford University Press)

[7] NIH 2017 *National Institute on Aging* http://nia.nih.gov/ (Accessed: 6 November 2017)

[8] Langa K M and Levine D A 2014 The diagnosis and management of mild cognitive impairment: a clinical review *JAMA* **312** 2551–61

[9] Turkington C and Harris J 2001 *The Encyclopedia of the Brain and Brain Disorders* (New York: Infobase)

[10] Wegrzyn R and Rudolph A 2012 *Alzheimer's Disease: Targets for New Clinical Diagnostic and Therapeutic StrategiesFrontiers in Neuroscience* (Boca Raton, FL: CRC Press)

[11] Belleville S *et al* 2017 Neuropsychological measures that predict progression from mild cognitive impairment to Alzheimer's type dementia in older adults: a systematic review and meta-analysis *Neuropsychol. Rev.* **27** 328–53

[12] Jack C, Knopman D, Jagust W, Shaw L, Aisen P, Weiner M, Petersen R and Trojanowski J 2010 Hypothetical model of dynamic biomarkers of the Alzheimer's pathological cascade *Lancet Neurol.* **9** 119–28

[13] Johnson K, Minoshima S, Bohnen N, Donohoe K, Foster N, Herscovitch P, Karlawish J, Rowe C, Carrillo M and Hartley D E A 2013 Appropriate use criteria for amyloid PET: a

report of the amyloid imaging task force, the society of nuclear medicine and molecular imaging, and the Alzheimer's association *Alzheimer's Dement.* **9** E1–16

[14] Varghese T, Sheelakumari R, James J S and Mathuranath P 2013 A review of neuro-imaging biomarkers of Alzheimer's disease *Neurol Asia* **18** 239–48

[15] Illán I, Gorriz J, Ramirez J, Chaves R, Segovia F, López M, Salas-Gonzalez D, Padilla P and Puntonet C 2010 Machine learning for very early Alzheimer's disease diagnosis; a ^{18}F-FDG and PiB PET comparison *2010 IEEE Nuclear Science Symp. Conf. Record (NSS/MIC)* (Piscataway, NJ: IEEE) pp 2334–7

[16] Illán I, Górriz J, Ramírez J, Salas-Gonzalez D, López M, Segovia F, Chaves R, Gómez-Rio M and Puntonet C 2011 ^{18}F-FDG PET imaging analysis for computer aided Alzheimer's diagnosis *Inf. Sci.* **181** 903–16

[17] Jiang J, Shu X, Liu X and Huang Z 2015 A computed aided diagnosis tool for Alzheimer's disease based on ^{11}C-PiB pet imaging technique *IEEE Int. Conf. on Information and Automation*

[18] López M, Ramrez J, Górriz J M, Álvarez I, Salas-Gonzalez D, Segovia F, Chaves R, Padilla P and Gómez-Ro M 2011 Principal component analysis-based techniques and supervised classification schemes for the early detection of Alzheimer's disease *Neurocomputing* **74** 1260–71

[19] Martnez-Murcia F, Górriz J, Ramrez J, Puntonet C and Salas-González D 2012 Computer aided diagnosis tool for Alzheimer's disease based on Mann–Whitney–Wilcoxon U-test *Expert Syst. Appl.* **39** 9676–85

[20] Chaves R, Ramirez J, Górriz J, Illán I and Salas-Gonzalez D 2012 FDG and PiB biomarker PET analysis for the Alzheimer's disease detection using association rules *Nuclear Science Symp. and Medical Imaging Conf. (NSS/MIC), 2012 IEEE* pp 2576–9

[21] Padilla P, Lopez M, Gorriz J, Ramirez J, Salas-Gonzalez D and Alvarez I 2012 NMF-SVM based cad tool applied to functional brain images for the diagnosis of Alzheimer's disease *IEEE Trans. Med. Imaging* **31** 207–16

[22] Morbelli S *et al* 2012 Resting metabolic connectivity in prodromal Alzheimer's disease. a European Alzheimer disease consortium (EADC) project *Neurobiol. Aging* **33** 2533–50

[23] Kemppainen N *et al* 2006 Voxel-based analysis of PET amyloid ligand [^{11}C] PiB uptake in Alzheimer disease *Neurology* **67** 1575–80

[24] Ziolko S K, Weissfeld L A, Klunk W E, Mathis C A, Hoge J A, Lopresti B J, DeKosky S T and Price J C 2006 Evaluation of voxel-based methods for the statistical analysis of PiB PET amyloid imaging studies in Alzheimer's disease *Neuroimage* **33** 94–102

[25] Forsberg A, Engler H, Almkvist O, Blomquist G, Hagman G, Wall A, Ringheim A, Långström B and Nordberg A 2008 PET imaging of amyloid deposition in patients with mild cognitive impairment *Neurobiol. Aging* **29** 1456–65

[26] Shin J, Lee S-Y, Kim S J, Kim S-H, Cho S-J and Kim Y-B 2010 Voxel-based analysis of Alzheimer's disease PET imaging using a triplet of radiotracers: PiB, FDDNP, and FDG *Neuroimage* **52** 488–96

[27] Wellcome Centre for Human NeuroImaging 2017 Spm12—statistical parametric mapping http://fil.ion.ucl.ac.uk/spm/software/spm12// (Accessed: 7 November 2017)

[28] Bagci U and Mollura D J 2013 Denoising PET images using singular value thresholding and Steins unbiased risk estimate *Int. Conf. on Medical Image Computing and Computer-Assisted Intervention* (Berlin: Springer) pp 115–22

[29] Agrawal S and Bahendwar Y 2011 Denoising of MRI images using thresholding techniques through wavelet transform *Int. J. Comput. Appl. Eng. Sci.* **1** 361–4

[30] Su S-S, Chen K-W and Huang Q 2014 Discriminant analysis in the study of Alzheimer's disease using feature extractions and support vector machines in positron emission tomography with ^{18}F-FDG *J. Shanghai Jiaotong Univ.* **19** 555–60

[31] Zhang Y, Dong Z, Phillips P, Wang S, Ji G, Yang J and Yuan T-F 2015 Detection of subjects and brain regions related to Alzheimer's disease using 3D MRI scans based on eigenbrain and machine learning *Front. Comput. Neurosci.* **9** 66

[32] Salas-Gonzalez D, Segovia F, Martnez-Murcia F J, Lang E W, Gorriz J M and Ramrez J 2016 An optimal approach for selecting discriminant regions for the diagnosis of Alzheimer's disease *Curr. Alzheimer Res.* **13** 838–44

[33] Alivelearn.net 2017 XJview—a viewing program for SPM http://alivelearn.net/xjview/ (Accessed: 23 January 2017)

[34] Toennies K D 2012 *Guide to Medical Image Analysis: Methods and Algorithms* (Berlin: Springer)

[35] Campbell C and Ying Y 2011 *Learning with Support Vector MachinesSynthesis Lectures on Artificial Intelligence and Machine Learning* (San Rafael, CA: Morgan & Claypool)

[36] Illán I, Górriz J, López M, Ramrez J, Salas-Gonzalez D, Segovia F, Chaves R and Puntonet C G 2011 Computer aided diagnosis of Alzheimers disease using component based SVM *Appl. Soft Comput.* **11** 2376–82

[37] BaldacSara L, Borgio J G F, Moraes W A d S, Lacerda A L T, Montano M B M M, Tufik S, Bressan R A, Ramos L R and Jackowski A P 2011 Cerebellar volume in patients with dementia *Rev. Bras. Psiquiatr.* **33** 122–9

[38] Wang H-Y, D'Andrea M R and Nagele R G 2002 Cerebellar diffuse amyloid plaques are derived from dendritic Aβ42 accumulations in Purkinje cells *Neurobiol. Aging* **23** 213–23

[39] Cole G, Neal J W, Singhrao S K, Jasani B and Newman G R 1993 The distribution of amyloid plaques in the cerebellum and brain stem in Down's syndrome and Alzheimer's disease: a light microscopical analysis *Acta Neuropathol.* **85** 542–52

[40] Sjöbeck M and Englund E 2001 Alzheimer's disease and the cerebellum: a morphologic study on neuronal and glial changes *Dement. Geriatr. Cogn. Disord.* **12** 211–8

[41] Serby M, Larson P and Kalkstein D 1991 The nature and course of olfactory deficits in Alzheimer's disease *Am. J. Psychiatry* **148** 357

[42] Devanand D, Michaels-Marston K S, Liu X, Pelton G H, Padilla M, Marder K, Bell K, Stern Y and Mayeux R 2000 Olfactory deficits in patients with mild cognitive impairment predict Alzheimers disease at follow-up *Am. J. Psychiatry* **157** 1399–405

[43] Velayudhan L 2015 Smell identification function and Alzheimer's disease: a selective review *Curr. Opin. Psychiatry* **28** 173–9

[44] Zou Y-M, Lu D, Liu L-P, Zhang H-H and Zhou Y-Y 2016 Olfactory dysfunction in Alzheimer's disease *Neuropsychiatr. Dis. Treat.* **12** 869–75

[45] Scheff S W, Price D A, Ansari M A, Roberts K N, Schmitt F A, Ikonomovic M D and Mufson E J 2015 Synaptic change in the posterior cingulate gyrus in the progression of Alzheimer's disease *J. Alzheimer's Dis.* **43** 1073–90

[46] Rami L *et al* 2012 Distinct functional activity of the precuneus and posterior cingulate cortex during encoding in the preclinical stage of Alzheimer's disease *J. Alzheimer's Dis.* **31** 517–26

[47] Mahmoud A H 2014 Utilizing radiation for smart robotic applications using visible, thermal, and polarization images *PhD Dissertation* University of Louisville, KY

[48] Mahmoud A, El-Barkouky A, Graham J and Farag A 2014 Pedestrian detection using mixed partial derivative based histogram of oriented gradients *2014 IEEE Int. Conf. on Image Processing (ICIP)* (Piscataway, NJ: IEEE) pp 2334–7

[49] El-Barkouky A, Mahmoud A, Graham J and Farag A 2013 An interactive educational drawing system using a humanoid robot and light polarization *2013 IEEE Int. Conf. on Image Processing* (Piscataway, NJ: IEEE) pp 3407–11

[50] Mahmoud A H, El-Melegy M T and Farag A A 2012 Direct method for shape recovery from polarization and shading *2012 19th IEEE Int. Conf. on Image Processing* (Piscataway, NJ: IEEE) pp 1769–72

[51] Ali A M, Farag A A and El-Baz A 2007 Graph cuts framework for kidney segmentation with prior shape constraints *Proc. of Int. Conf. on Medical Image Computing and Computer-Assisted Intervention, (MICCAI'07) (Brisbane, Australia, 29 October–2 November 2007)* **vol 1** pp 384–92

[52] Chowdhury A S, Roy R, Bose S, Elnakib F K A and El-Baz A 2012 Non-rigid biomedical image registration using graph cuts with a novel data term *Proc. of IEEE Int. Symp. Biomedical Imaging: From Nano to Macro, (ISBI'12) (Barcelona, Spain, 2–5 May 2012)* pp 446–9

[53] El-Baz A, Farag A A, Yuksel S E, El-Ghar M E, Eldiasty T A and Ghoneim M A 2007 Application of deformable models for the detection of acute renal rejection *Deformable Models* (New York: Springer) pp 293–333

[54] El-Baz A, Farag A, Fahmi R, Yuksel S, El-Ghar M A and Eldiasty T 2006 Image analysis of renal DCE MRI for the detection of acute renal rejection *Proc. of IAPR Int. Conf. on Pattern Recognition (ICPR'06) (Hong Kong, 20–24 August 2006)* pp 822–5

[55] El-Baz A, Farag A, Fahmi R, Yuksel S, Miller W, El-Ghar M A, El-Diasty T and Ghoneim M 2006 A new CAD system for the evaluation of kidney diseases using DCE-MRI *Proc. of Int. Conf. on Medical Image Computing and Computer-Assisted Intervention, (MICCAI'08) (Copenhagen, Denmark, 1–6 October 2006)* pp 446–53

[56] El-Baz A, Gimel'farb G and El-Ghar M A 2008 A novel image analysis approach for accurate identification of acute renal rejection *Proc. of IEEE Int. Conf. on Image Processing, (ICIP'08) (San Diego, California, USA, 12–15 October 2008)* pp 1812–5

[57] El-Baz A, Gimel'farb G and El-Ghar M A 2008 Image analysis approach for identification of renal transplant rejection *Proc. of IAPR Int. Conf. on Pattern Recognition, (ICPR'08) (Tampa, Florida, USA, 8–11 December 2008)* pp 1–4

[58] El-Baz A, Gimel'farb G and El-Ghar M A 2007 New motion correction models for automatic identification of renal transplant rejection *Proc. of Int. Conf. on Medical Image Computing and Computer-Assisted Intervention, (MICCAI'07) (Brisbane, Australia, 29 October–2 November 2007)* pp 235–43

[59] Farag A, El-Baz A, Yuksel S, El-Ghar M A and Eldiasty T 2006 A framework for the detection of acute rejection with dynamic contrast enhanced magnetic resonance imaging *Proc. of IEEE Int. Symp. on Biomedical Imaging: From Nano to Macro, (ISBI'06) (Arlington, Virginia, USA, 6–9 April 2006)* pp 418–21

[60] Khalifa F, Beache G M, El-Ghar M A, El-Diasty T, Gimel'farb G, Kong M and El-Baz A 2013 Dynamic contrast-enhanced MRI-based early detection of acute renal transplant rejection *IEEE Trans. Med. Imaging* **32** 1910–27

[61] Khalifa F, El-Baz A, Gimel'farb G and El-Ghar M A 2010 Non-invasive image-based approach for early detection of acute renal rejection *Proc. of Int. Conf. Medical Image Computing and Computer-Assisted Intervention, (MICCAI'10) (Beijing, China, 20–24 September 2010)* pp 10–8

[62] Khalifa F, El-Baz A, Gimel'farb G, Ouseph R and El-Ghar M A 2010 Shape-appearance guided level-set deformable model for image segmentation *Proc. of IAPR Int. Conf. on Pattern Recognition, (ICPR'10) (Istanbul, Turkey, 23–26 August 2010)* pp 4581–4

[63] Khalifa F, El-Ghar M A, Abdollahi B, Frieboes H, El-Diasty T and El-Baz A 2013 A comprehensive non-invasive framework for automated evaluation of acute renal transplant rejection using DCE-MRI *NMR Biomed.* **26** 1460–70

[64] Khalifa F, El-Ghar M A, Abdollahi B, Frieboes H B, El-Diasty T and El-Baz A 2014 Dynamic contrast-enhanced MRI-based early detection of acute renal transplant rejection *2014 Annu. Scientific Meeting and Educational Course Brochure of the Society of Abdominal Radiology, (SAR'14) (Boca Raton, Florida, 23–28 March 2014)* CID: 1855 912

[65] Khalifa F, Elnakib A, Beache G M, Gimel'farb G, El-Ghar M A, Sokhadze G, Manning S, McClure P and El-Baz A 2011 3D kidney segmentation from CT images using a level set approach guided by a novel stochastic speed function *Proc. of Int. Conf. Medical Image Computing and Computer-Assisted Intervention, (MICCAI'11) (Toronto, Canada, 18–22 September 2011)* pp 587–94

[66] Khalifa F, Gimel'farb G, El-Ghar M A, Sokhadze G, Manning S, McClure P, Ouseph R and El-Baz A 2011 A new deformable model-based segmentation approach for accurate extraction of the kidney from abdominal CT images *Proc. of IEEE Int. Conf. on Image Processing, (ICIP'11) (Brussels, Belgium, 11–14 September 2011)* pp 3393–6

[67] Mostapha M, Khalifa F, Alansary A, Soliman A, Suri J and El-Baz A 2014 Computer-aided diagnosis systems for acute renal transplant rejection: challenges and methodologies *Abdomen and Thoracic Imaging* ed A El-Baz, L Saba and J Suri (Berlin: Springer) pp 1–35

[68] Shehata M, Khalifa F, Hollis E, Soliman A, Hosseini-Asl E, El-Ghar M A, El-Baz M, Dwyer A C, El-Baz A and Keynton R 2016 A new non-invasive approach for early classification of renal rejection types using diffusion-weighted MRI *IEEE Int. Conf. on Image Processing (ICIP), 2016* (Piscataway, NJ: IEEE) pp 136–40

[69] Khalifa F, Soliman A, Takieldeen A, Shehata M, Mostapha M, Shaffie A, Ouseph R, Elmaghraby A and El-Baz A 2016 Kidney segmentation from CT images using a 3D NMF-guided active contour model *IEEE 13th Int. Symp. on Biomedical Imaging (ISBI), 2016* (Piscataway, NJ: IEEE) pp 432–5

[70] Shehata M, Khalifa F, Soliman A, Takieldeen A, El-Ghar M A, Shaffie A, Dwyer A C, Ouseph R, El-Baz A and Keynton R 2016 3D diffusion MRI-based CAD system for early diagnosis of acute renal rejection *2016 IEEE 13th Int. Symp. on Biomedical Imaging (ISBI)* (Piscataway, NJ: IEEE) pp 1177–80

[71] Shehata M, Khalifa F, Soliman A, Alrefai R, El-Ghar M A, Dwyer A C, Ouseph R and El-Baz A 2015 A level set-based framework for 3D kidney segmentation from diffusion MR images *IEEE Int. Conf. on Image Processing (ICIP), 2015* (Piscataway, NJ: IEEE) pp 4441–5

[72] Shehata M, Khalifa F, Soliman A, El-Ghar M A, Dwyer A C, Gimel'farb G, Keynton R and El-Baz A 2016 A promising non-invasive CAD system for kidney function assessment *Int. Conf. on Medical Image Computing and Computer-Assisted Intervention* (Berlin: Springer) pp 613–21

[73] Khalifa F, Soliman A, Elmaghraby A, Gimel'farb G and El-Baz A 2017 3D kidney segmentation from abdominal images using spatial-appearance models *Computat. Math. Methods Med.* **2017** 1–10

[74] Hollis E, Shehata M, Khalifa F, El-Ghar M A, El-Diasty T and El-Baz A 2016 Towards non-invasive diagnostic techniques for early detection of acute renal transplant rejection: a review *Egypt. J. Radiol. Nucl. Med.* **48** 257–69

[75] Shehata M, Khalifa F, Soliman A, El-Ghar M A, Dwyer A C and El-Baz A 2017 Assessment of renal transplant using image and clinical-based biomarkers *Proc. of 13th Annu. Scientific Meeting of American Society for Diagnostics and Interventional Nephrology (ASDIN'17) (New Orleans, LA, USA, 10–12 February 2017)*

[76] Shehata M, Khalifa F, Soliman A, El-Ghar M A, Dwyer A C and El-Baz A 2017 Early assessment of acute renal rejection *Proc. of 12th Annu. Scientific Meeting of American Society for Diagnostics and Interventional Nephrology (ASDIN'16) (Pheonix, AZ, USA, 19–21 February 2016)*

[77] Eltanboly A, Ghazal M, Hajjdiab H, Shalaby A, Switala A, Mahmoud A, Sahoo P, El-Azab M and El-Baz A 2019 Level sets-based image segmentation approach using statistical shape priors *Appl. Math. Comput.* **340** 164–79

[78] Shehata M, Mahmoud A, Soliman A, Khalifa F, Ghazal M, El-Ghar M A, El-Melegy M and El-Baz A 2018 3D kidney segmentation from abdominal diffusion MRI using an appearance-guided deformable boundary *PLoS One* **13** e0200082

[79] Khalifa F, Beache G, El-Baz A and Gimel'farb G 2010 Deformable model guided by stochastic speed with application in cine images segmentation *Proc. of IEEE Int. Conf. on Image Processing, (ICIP'10) (Hong Kong, 26–29 September 2010)* pp 1725–8

[80] Khalifa F, Beache G M, Elnakib A, Sliman H, Gimel'farb G, Welch K C and El-Baz A 2013 A new shape-based framework for the left ventricle wall segmentation from cardiac first-pass perfusion MRI *Proc. of IEEE Int. Symp. on Biomedical Imaging: From Nano to Macro, (ISBI'13) (San Francisco, CA, 7–11 April 2013)* pp 41–4

[81] Khalifa F, Beache G M, Elnakib A, Sliman H, Gimel'farb G, Welch K C and El-Baz A 2012 A new nonrigid registration framework for improved visualization of transmural perfusion gradients on cardiac first–pass perfusion MRI *Proc. of IEEE Int. Symp. on Biomedical Imaging: From Nano to Macro, (ISBI'12) (Barcelona, Spain, 2–5 May 2012)* pp 828–31

[82] Khalifa F, Beache G M, Firjani A, Welch K C, Gimel'farb G and El-Baz A 2012 A new nonrigid registration approach for motion correction of cardiac first-pass perfusion MRI *Proc. of IEEE Int. Conf. on Image Processing, (ICIP'12) (Lake Buena Vista, Florida, 30 September–3 October 2012)* pp 1665–8

[83] Khalifa F, Beache G M, Gimel'farb G and El-Baz A 2012 A novel CAD system for analyzing cardiac first-pass MR images *Proc. of IAPR Int. Conf. on Pattern Recognition (ICPR'12) (Tsukuba Science City, Japan, 11–15 November 2012)* pp 77–80

[84] Khalifa F, Beache G M, Gimel'farb G and El-Baz A 2011 A novel approach for accurate estimation of left ventricle global indexes from short-axis cine MRI *Proc. of IEEE Int. Conf. on Image Processing, (ICIP'11) (Brussels, Belgium, 11–14 September 2011)* pp 2645–9

[85] Khalifa F, Beache G M, Gimel'farb G, Giridharan G A and El-Baz A 2011 A new image-based framework for analyzing cine images *Handbook of Multi Modality State-of-the-Art Medical Image Segmentation and Registration Methodologies* vol 2 ed A El-Baz, U R Acharya, M Mirmedhdi and J S Suri (New York: Springer) ch 3, pp 69–98

[86] Khalifa F, Beache G M, Gimel'farb G, Giridharan G A and El-Baz A 2012 Accurate automatic analysis of cardiac cine images *IEEE Trans. Biomed. Eng.* **59** 445–55

[87] Khalifa F, Beache G M, Nitzken M, Gimel'farb G, Giridharan G A and El-Baz A 2011 Automatic analysis of left ventricle wall thickness using short-axis cine CMR images *Proc. of IEEE Int. Symp. on Biomedical Imaging: From Nano to Macro, (ISBI'11) (Chicago, Illinois, 30 March–2 April 2011)* pp 1306–9

[88] Nitzken M, Beache G, Elnakib A, Khalifa F, Gimel'farb G and El-Baz A 2012 Accurate modeling of tagged CMR 3D image appearance characteristics to improve cardiac cycle strain estimation *2012 19th IEEE Int. Conf. on Image Processing (ICIP)* (Piscataway, NJ: IEEE) pp 521–4

[89] Nitzken M, Beache G, Elnakib A, Khalifa F, Gimel'farb G and El-Baz A 2012 Improving full-cardiac cycle strain estimation from tagged CMR by accurate modeling of 3D image appearance characteristics *2012 9th IEEE Int. Symp. on Biomedical Imaging (ISBI)* (Piscataway, NJ: IEEE) pp 462–5

[90] Nitzken M J, El-Baz A S and Beache G M 2012 Markov–Gibbs random field model for improved full-cardiac cycle strain estimation from tagged CMR *J. Cardiovasc. Magn. Reson.* **14** 1–2

[91] Sliman H, Elnakib A, Beache G, Elmaghraby A and El-Baz A 2014 Assessment of myocardial function from cine cardiac MRI using a novel 4D tracking approach *J. Comput. Sci. Syst. Biol.* **7** 169–73

[92] Sliman H, Elnakib A, Beache G M, Soliman A, Khalifa F, Gimel'farb G, Elmaghraby A and El-Baz A 2014 A novel 4D PDE-based approach for accurate assessment of myocardium function using cine cardiac magnetic resonance images *Proc. of IEEE Int. Conf. on Image Processing (ICIP'14) (Paris, France, 27–30 October 2014)* pp 3537–41

[93] Sliman H, Khalifa F, Elnakib A, Beache G M, Elmaghraby A and El-Baz A 2013 A new segmentation-based tracking framework for extracting the left ventricle cavity from cine cardiac MRI *Proc. of IEEE Int. Conf. on Image Processing, (ICIP'13) (Melbourne, Australia, 15–18 September 2013)* pp 685–9

[94] Sliman H, Khalifa F, Elnakib A, Soliman A, Beache G M, Elmaghraby A, Gimel'farb G and El-Baz A 2013 Myocardial borders segmentation from cine MR images using bi-directional coupled parametric deformable models *Med. Phys.* **40** 1–13 .

[95] Sliman H, Khalifa F, Elnakib A, Soliman A, Beache G M, Gimel'farb G, Emam A, Elmaghraby A and El-Baz A 2013 Accurate segmentation framework for the left ventricle wall from cardiac cine MRI *Proc. Int. Symp. on Computational Models for Life Science, (CMLS'13) (Sydney, Australia, 27–29 November 2013)* vol **1559** pp 287–96

[96] Eladawi N, Elmogy M, Ghazal M, Helmy O, Aboelfetouh A, Riad A, Schaal S and El-Baz A 2018 Classification of retinal diseases based on OCT images *Front. Biosci.* **23** 247–64

[97] ElTanboly A, Ismail M, Shalaby A, Switala A, El-Baz A, Schaal S, Gimelfarb G and El-Azab M 2017 A computer-aided diagnostic system for detecting diabetic retinopathy in optical coherence tomography images *Med. Phys.* **44** 914–23

[98] Sandhu H S, El-Baz A and Seddon J M 2018 Progress in automated deep learning for macular degeneration *JAMA Ophthalmol.* **136** 1366–7

[99] Abdollahi B, Civelek A C, Li X-F, Suri J and El-Baz A 2014 PET/CT nodule segmentation and diagnosis: a survey *Multi Detector CT Imaging* ed L Saba and J S Suri (London: Taylor and Francis) ch 30, pp 639–51

[100] Abdollahi B, El-Baz A and Amini A A 2011 A multi-scale non-linear vessel enhancement technique *2011 Annu. Int. Conf. of the IEEE Engineering in Medicine and Biology Society, EMBC* (Piscataway, NJ: IEEE) pp 3925–9

[101] Abdollahi B, Soliman A, Civelek A, Li X-F, Gimel'farb G and El-Baz A 2012 A novel Gaussian scale space-based joint MGRF framework for precise lung segmentation *Proc. of IEEE Int. Conf. on Image Processing, (ICIP'12)* (Piscataway, NJ: IEEE) pp 2029–32

[102] Abdollahi B, Soliman A, Civelek A, Li X-F, Gimel'farb G and El-Baz A 2012 A novel 3D joint MGRF framework for precise lung segmentation *Machine Learning in Medical Imaging* (Berlin: Springer) pp 86–93

[103] Ali A M, El-Baz A S and Farag A A 2007 A novel framework for accurate lung segmentation using graph cuts *Proc. of IEEE Int. Symp. on Biomedical Imaging: From Nano to Macro, (ISBI'07)* (Piscataway, NJ: IEEE) pp 908–11

[104] El-Baz A, Beache G M, Gimel'farb G, Suzuki K and Okada K 2013 Lung imaging data analysis *Int. J. Biomed. Imaging* **2013** 1–2

[105] El-Baz A, Beache G M, Gimel'farb G, Suzuki K, Okada K, Elnakib A, Soliman A and Abdollahi B 2013 Computer-aided diagnosis systems for lung cancer: challenges and methodologies *Int. J. Biomed. Imaging* **2013** 1–46

[106] El-Baz A, Elnakib A, Abou El-Ghar M, Gimel'farb G, Falk R and Farag A 2013 Automatic detection of 2D and 3D lung nodules in chest spiral CT scans *Int. J. Biomed. Imaging* **2013** 1–11

[107] El-Baz A, Farag A A, Falk R and La Rocca R 2003 A unified approach for detection, visualization, and identification of lung abnormalities in chest spiral CT scans *International Congress Series* vol 1256 (Amsterdam: Elsevier) pp 998–1004

[108] El-Baz A, Farag A A, Falk R and La Rocca R 2002 Detection, visualization and identification of lung abnormalities in chest spiral CT scan: phase-I *Proc. of Int. Conf. on Biomedical Engineering (Cairo, Egypt)* **vol 12**

[109] El-Baz A, Farag A, Gimel'farb G, Falk R, El-Ghar M A and Eldiasty T 2006 A framework for automatic segmentation of lung nodules from low dose chest CT scans *Proc. of Int. Conf. on Pattern Recognition, (ICPR'06)* vol 3 (Piscataway, NJ: IEEE) pp 611–4

[110] El-Baz A, Farag A, Gimel'farb G, Falk R and El-Ghar M A 2011 A novel level set-based computer-aided detection system for automatic detection of lung nodules in low dose chest computed tomography scans *Lung Imaging and Computer Aided Diagnosis* vol 10 pp 221–38

[111] El-Baz A, Gimel'farb G, Abou El-Ghar M and Falk R 2012 Appearance-based diagnostic system for early assessment of malignant lung nodules *Proc. of IEEE Int. Conf. Image Processing, (ICIP'12)* (Piscataway, NJ: IEEE) pp 533–6

[112] El-Baz A, Gimel'farb G and Falk R 2011 A novel 3D framework for automatic lung segmentation from low dose CT images *Lung Imaging and Computer Aided Diagnosis* ed A El-Baz and J S Suri (London: Taylor and Francis) ch 1, pp 1–16

[113] El-Baz A, Gimel'farb G, Falk R and El-Ghar M 2010 Appearance analysis for diagnosing malignant lung nodules *Proc. of IEEE Int. Symp. on Biomedical Imaging: From Nano to Macro (ISBI'10)* (Piscataway, NJ: IEEE) pp 193–6

[114] El-Baz A, Gimel'farb G, Falk R and El-Ghar M A 2011 A novel level set-based CAD system for automatic detection of lung nodules in low dose chest CT scans *Lung Imaging and Computer Aided Diagnosis* vol 1 ed A El-Baz and J S Suri (London: Taylor and Francis) ch 10, pp 221–38

[115] El-Baz A, Gimel'farb G, Falk R and El-Ghar M A 2008 A new approach for automatic analysis of 3D low dose CT images for accurate monitoring the detected lung nodules *Proc. of Int. Conf. on Pattern Recognition, (ICPR'08)* (Piscataway, NJ: IEEE) pp 1–4

[116] El-Baz A, Gimel'farb G, Falk R and El-Ghar M A 2007 A novel approach for automatic follow-up of detected lung nodules *Proc. of IEEE Int. Conf. on Image Processing, (ICIP'07)* **vol 5** (Piscataway, NJ: IEEE) pp V–501

[117] El-Baz A, Gimel'farb G, Falk R and El-Ghar M A 2007 A new CAD system for early diagnosis of detected lung nodules *ICIP 2007. IEEE Int. Conf. on Image Processing, 2007* **vol 2** (Piscataway, NJ: IEEE) pp II–461

[118] El-Baz A, Gimel'farb G, Falk R, El-Ghar M A and Refaie H 2008 Promising results for early diagnosis of lung cancer *Proc. of IEEE Int. Symp. on Biomedical Imaging: From Nano to Macro, (ISBI'08)* (Piscataway, NJ: IEEE) pp 1151–4

[119] El-Baz A, Gimel'farb G L, Falk R, Abou El-Ghar M, Holland T and Shaffer T 2008 A new stochastic framework for accurate lung segmentation *Proc. of Medical Image Computing and Computer-Assisted Intervention, (MICCAI'08)* pp 322–30

[120] El-Baz A, Gimel'farb G L, Falk R, Heredis D and Abou El-Ghar M 2008 A novel approach for accurate estimation of the growth rate of the detected lung nodules *Proc. of Int. Workshop on Pulmonary Image Analysis* pp 33–42

[121] El-Baz A, Gimel'farb G L, Falk R, Holland T and Shaffer T 2008 A framework for unsupervised segmentation of lung tissues from low dose computed tomography images *Proc. of British Machine Vision, (BMVC'08)* pp 1–10

[122] El-Baz A, Gimel'farb G, Falk R and El-Ghar M A 2011 3D MGRF-based appearance modeling for robust segmentation of pulmonary nodules in 3D LDCT chest images *Lung Imaging and Computer Aided Diagnosis* ch 3, pp 51–63

[123] El-Baz A, Gimel'farb G, Falk R and El-Ghar M A 2009 Automatic analysis of 3D low dose CT images for early diagnosis of lung cancer *Pattern Recogn.* **42** 1041–51

[124] El-Baz A, GGimel'farb A, Falk R, El-Ghar M A, Rainey S, Heredia D and Shaffer T 2009 Toward early diagnosis of lung cancer *Proc. of Medical Image Computing and Computer-Assisted Intervention, (MICCAI'09)* (Berlin: Springer) pp 682–9

[125] El-Baz A, Gimel'farb G, Falk R, El-Ghar M A and Suri J 2011 Appearance analysis for the early assessment of detected lung nodules *Lung Imaging and Computer Aided Diagnosis* ch 17, pp 395–404

[126] El-Baz A, Khalifa F, Elnakib A, Nitkzen M, Soliman A, McClure P, Gimel'farb G and El-Ghar M A 2012 A novel approach for global lung registration using 3D Markov–Gibbs appearance model *Proc. of Int. Conf. Medical Image Computing and Computer-Assisted Intervention, (MICCAI'12) (Nice, France, 1–5 October 2012)* pp 114–21

[127] El-Baz A, Nitzken M, Elnakib A, Khalifa F, Gimel'farb G, Falk R and El-Ghar M A 2011 3D shape analysis for early diagnosis of malignant lung nodules *Proc. Int. Conf. Medical Image Computing and Computer-Assisted Intervention, (MICCAI'11) (Toronto, Canada, 18–22 September 2011)* pp 175–82

[128] El-Baz A, Nitzken M, Gimel'farb G, Van Bogaert E, Falk R, El-Ghar M A and Suri J 2011 Three-dimensional shape analysis using spherical harmonics for early assessment of detected lung nodules *Lung Imaging and Computer Aided Diagnosis* ch 19, pp 421–38

[129] El-Baz A, Nitzken M, Khalifa F, Elnakib A, Gimel'farb G, Falk R and El-Ghar M A 2011 3D shape analysis for early diagnosis of malignant lung nodules *Proc. of Int. Conf. on*

Information Processing in Medical Imaging, (IPMI'11) (Monastery Irsee, Germany (Bavaria), 3–8 July 2011) pp 772–83

[130] El-Baz A, Nitzken M, Vanbogaert E, Gimel'Farb G, Falk R and Abo El-Ghar M 2011 A novel shape-based diagnostic approach for early diagnosis of lung nodules *2011 IEEE Int. Symp. on Biomedical Imaging: From Nano to Macro* (Piscataway, NJ: IEEE) pp 137–40

[131] El-Baz A, Sethu P, Gimel'farb G, Khalifa F, Elnakib A, Falk R and El-Ghar M A 2011 Elastic phantoms generated by microfluidics technology: validation of an imaged-based approach for accurate measurement of the growth rate of lung nodules *Biotechnol. J.* **6** 195–203

[132] El-Baz A, Sethu P, Gimel'farb G, Khalifa F, Elnakib A, Falk R and El-Ghar M A 2010 A new validation approach for the growth rate measurement using elastic phantoms generated by state-of-the-art microfluidics technology *Proc. of IEEE Int. Conf. on Image Processing, (ICIP'10) (Hong Kong, 26–29 September 2010)* pp 4381–3

[133] El-Baz A, Sethu P, Gimel'farb G, Khalifa F, Elnakib A, Falk R and Suri M A E-G J 2011 Validation of a new imaged-based approach for the accurate estimating of the growth rate of detected lung nodules using real CT images and elastic phantoms generated by state-of-the-art microfluidics technology *Handbook of Lung Imaging and Computer Aided Diagnosis* vol 1 ed A El-Baz and J S Suri (New York: Taylor and Francis) ch 18, pp 405–20

[134] El-Baz A, Soliman A, McClure P, Gimel'farb G, El-Ghar M A and Falk R 2012 Early assessment of malignant lung nodules based on the spatial analysis of detected lung nodules *Proc. of IEEE Int. Symp. on Biomedical Imaging: From Nano to Macro, (ISBI'12)* (Piscataway, NJ: IEEE) pp 1463–6

[135] El-Baz A, Yuksel S E, Elshazly S and Farag A A 2005 Non-rigid registration techniques for automatic follow-up of lung nodules *Proc. of Comput. Assisted Radiology and Surgery, (CARS'05)* **vol 1281** (Amsterdam: Elsevier) pp 1115–20

[136] El-Baz A S and Suri J S 2011 *Lung Imaging and Computer Aided Diagnosis* (Boca Raton, FL: CRC Press)

[137] Soliman A, Khalifa F, Dunlap N, Wang B, El-Ghar M and El-Baz A 2016 An iso-surfaces based local deformation handling framework of lung tissues *2016 IEEE 13th Int. Symp. on Biomedical Imaging (ISBI)* (Piscataway, NJ: IEEE) pp 1253–9

[138] Soliman A, Khalifa F, Shaffie A, Dunlap N, Wang B, Elmaghraby A and El-Baz A 2016 Detection of lung injury using 4D-CT chest images *2016 IEEE 13th Int. Symp. on Biomedical Imaging (ISBI)* (Piscataway, NJ: IEEE) pp 1274–7

[139] Soliman A, Khalifa F, Shaffie A, Dunlap N, Wang B, Elmaghraby A, Gimel'farb G, Ghazal M and El-Baz A 2017 A comprehensive framework for early assessment of lung injury *2017 IEEE Int. Conf. on Image Processing (ICIP)* (Piscataway, NJ: IEEE) pp 3275–9

[140] Shaffie A, Soliman A, Ghazal M, Taher F, Dunlap N, Wang B, Elmaghraby A, Gimel'farb G and El-Baz A 2017 A new framework for incorporating appearance and shape features of lung nodules for precise diagnosis of lung cancer *2017 IEEE Int. Conf. on Image Processing (ICIP)* (Piscataway, NJ: IEEE) pp 1372–6

[141] Soliman A, Khalifa F, Shaffie A, Liu N, Dunlap N, Wang B, Elmaghraby A, Gimel'farb G and El-Baz A 2016 Image-based CAD system for accurate identification of lung injury *2016 IEEE Int. Conf. on Image Processing (ICIP)* (Piscataway, NJ: IEEE) pp 121–5

[142] Dombroski B, Nitzken M, Elnakib A, Khalifa F, El-Baz A and Casanova M F 2014 Cortical surface complexity in a population-based normative sample *Transl. Neurosci.* **5** 17–24

[143] El-Baz A, Casanova M, Gimel'farb G, Mott M and Switala A 2008 An MRI-based diagnostic framework for early diagnosis of dyslexia *Int. J. Comput. Assist. Radiol. Surg.* **3** 181–9

[144] El-Baz A, Casanova M, Gimel'farb G, Mott M, Switala A, Vanbogaert E and McCracken R 2008 A new CAD system for early diagnosis of dyslexic brains *Proc. Int. Conf. on Image Processing (ICIP'2008)* (Piscataway, NJ: IEEE) pp 1820–3

[145] El-Baz A, Casanova M F, Gimel'farb G, Mott M and Switwala A E 2007 A new image analysis approach for automatic classification of autistic brains *Proc. IEEE Int. Symp. on Biomedical Imaging: From Nano to Macro (ISBI'2007)* (Piscataway, NJ: IEEE) pp 352–5

[146] El-Baz A, Elnakib A, Khalifa F, El-Ghar M A, McClure P, Soliman A and Gimel'farb G 2012 Precise segmentation of 3-D magnetic resonance angiography *IEEE Trans. Biomed. Eng.* **59** pp 2019–29

[147] El-Baz A, Farag A, Gimel'farb G, El-Ghar M A and Eldiasty T 2006 Probabilistic modeling of blood vessels for segmenting MRA images *18th Int. Conf. on Pattern Recognition (ICPR'06)* **vol 3** (Piscataway, NJ: IEEE) pp 917–20

[148] El-Baz A, Farag A A, Gimel'farb G, El-Ghar M A and Eldiasty T 2006 A new adaptive probabilistic model of blood vessels for segmenting MRA images *Medical Image Computing and Computer-Assisted Intervention–MICCAI 2006* **vol 4191** (Berlin: Springer) pp 799–806

[149] El-Baz A, Farag A A, Gimel'farb G and Hushek S G 2005 Automatic cerebrovascular segmentation by accurate probabilistic modeling of TOF-MRA images *Medical Image Computing and Computer-Assisted Intervention–MICCAI 2005* (Berlin: Springer) pp 34–42

[150] El-Baz A, Farag A, Elnakib A, Casanova M F, Gimel'farb G, Switala A E, Jordan D and Rainey S 2011 Accurate automated detection of autism related corpus callosum abnormalities *J. Med. Syst.* **35** 929–39

[151] El-Baz A, Farag A and Gimelfarb G 2005 Cerebrovascular segmentation by accurate probabilistic modeling of TOF-MRA images *Image Analysis* **vol 3540** (Berlin: Springer) pp 1128–37

[152] El-Baz A, Gimel'farb G, Falk R, El-Ghar M A, Kumar V and Heredia D 2009 A novel 3D joint Markov–Gibbs model for extracting blood vessels from PC–MRA images *Medical Image Computing and Computer-Assisted Intervention–MICCAI 2009* **vol 5762** (Berlin: Springer) pp 943–50

[153] Elnakib A, El-Baz A, Casanova M F, Gimel'farb G and Switala A E 2010 Image-based detection of corpus callosum variability for more accurate discrimination between dyslexic and normal brains *Proc. IEEE Int. Symp. on Biomedical Imaging: From Nano to Macro (ISBI'2010)* (Piscataway, NJ: IEEE) pp 109–12

[154] Elnakib A, Casanova M F, Gimel'farb G, Switala A E and El-Baz A 2011 Autism diagnostics by centerline-based shape analysis of the corpus callosum *Proc. IEEE Int. Symp. on Biomedical Imaging: From Nano to Macro (ISBI'2011)* (Piscataway, NJ: IEEE) pp 1843–6

[155] Elnakib A, Nitzken M, Casanova M, Park H, Gimel'farb G and El-Baz A 2012 Quantification of age-related brain cortex change using 3D shape analysis *2012 21st Int. Conf. on Pattern Recognition (ICPR)* (Piscataway, NJ: IEEE) pp 41–4

[156] Mostapha M, Soliman A, Khalifa F, Elnakib A, Alansary A, Nitzken M, Casanova M F and El-Baz A 2014 A statistical framework for the classification of infant DT images *2014 IEEE Int. Conf. on Image Processing (ICIP)* (Piscataway, NJ: IEEE) pp 2222–6

[157] Nitzken M, Casanova M, Gimel'farb G, Elnakib A, Khalifa F, Switala A and El-Baz A 2011 3D shape analysis of the brain cortex with application to dyslexia *2011 18th IEEE Int. Conf. on Image Processing (ICIP)* (Piscataway, NJ: IEEE) pp 2657–60

[158] El-Gamal F E-Z A, Elmogy M M, Ghazal M, Atwan A, Barnes G N, Casanova M F, Keynton R and El-Baz A S 2017 A novel CAD system for local and global early diagnosis of Alzheimer's disease based on PiB-PET scans *2017 IEEE Int. Conf. on Image Processing (ICIP)* (Piscataway, NJ: IEEE) pp 3270–4

[159] Ismail M, Soliman A, Ghazal M, Switala A E, Gimelfarb G, Barnes G N, Khalil A and El-Baz A 2017 A fast stochastic framework for automatic MR brain images segmentation *PloS one* **12** e0187391

[160] Ismail M M, Keynton R S, Mostapha M M, ElTanboly A H, Casanova M F, Gimel'farb G L and El-Baz A 2016 Studying autism spectrum disorder with structural and diffusion magnetic resonance imaging: a survey *Front. Hum. Neurosci.* **10** 211

[161] Alansary A *et al* 2016 Infant brain extraction in T_1-weighted MR mages using BET and refinement using LCDG and MGRF models *IEEE J. Biomed. Health Inform.* **20** 925–35

[162] Ismail M, Soliman A, ElTanboly A, Switala A, Mahmoud M, Khalifa F, Gimel'farb G, Casanova M F, Keynton R and El-Baz A 2016 Detection of white matter abnormalities in MR brain images for diagnosis of autism in children *2016 IEEE 13th Int. Symp. on Biomedical Imaging (ISBI)* pp 6–9

[163] Ismail M, Mostapha M, Soliman A, Nitzken M, Khalifa F, Elnakib A, Gimel'farb G, Casanova M and El-Baz A 2015 Segmentation of infant brain MR images based on adaptive shape prior and higher-order MGRF *2015 IEEE Int. Conf. on Image Processing (ICIP)* pp 4327–31

[164] Asl E H, Ghazal M, Mahmoud A, Aslantas A, Shalaby A, Casanova M, Barnes G, Gimelfarb G, Keynton R and El-Baz A 2018 Alzheimers disease diagnostics by a 3D deeply supervised adaptable convolutional network *Front. Biosci.* **23** 584–96

[165] Mahmoud A, El-Barkouky A, Farag H, Graham J and Farag A 2013 A non-invasive method for measuring blood flow rate in superficial veins from a single thermal image *Proc. of the IEEE Conf. on Computer Vision and Pattern Recognition Workshops* pp 354–9

[166] El-baz A, Shalaby A, Taher F, El-Baz M, Ghazal M, El-Ghar M A, Takieldeen A and Suri J 2017 Probabilistic modeling of blood vessels for segmenting magnetic resonance angiography images *Med. Res. Arch.* **5**

[167] Chowdhury A S, Rudra A K, Sen M, Elnakib A and El-Baz A 2010 Cerebral white matter segmentation from MRI using probabilistic graph cuts and geometric shape priors *ICIP* pp 3649–52

[168] Gebru Y, Giridharan G, Ghazal M, Mahmoud A, Shalaby A and El-Baz A 2018 Detection of cerebrovascular changes using magnetic resonance angiography *Cardiovascular Imaging and Image Analysis* (Boca Raton, FL: CRC Press) pp 1–22

[169] Mahmoud A, Shalaby A, Taher F, El-Baz M, Suri J S and El-Baz A 2018 Vascular tree segmentation from different image modalities *Cardiovascular Imaging and Image Analysis* (Boca Raton, FL: CRC Press) pp 43

[170] Taher F, Mahmoud A, Shalaby A and El-Baz A 2018 A review on the cerebrovascular segmentation methods *2018 IEEE Int. Symp. on Signal Processing and Information Technology (ISSPIT)* (Piscataway, NJ: IEEE) pp 359–64

[171] Kandil H, Soliman A, Fraiwan L, Shalaby A, Mahmoud A, ElTanboly A, Elmaghraby A, Giridharan G and El-Baz A 2018 A novel MRA framework based on integrated global and local analysis for accurate segmentation of the cerebral vascular system *2018 IEEE 15th Int. Symp. on Biomedical Imaging (ISBI 2018)* (Piscataway, NJ: IEEE) pp 1365–8

IOP Publishing

Neurological Disorders and Imaging Physics, Volume 3
Application to autism spectrum disorders and Alzheimer's
Ayman El-Baz and Jasjit S Suri

Chapter 12

Identifying Alzheimer's disease using feature reduction of GLCM and supervised classification techniques

Yasmeen Farouk and Sherine Rady

Alzheimer's disease causes progressive disorder in brain cells. Long before the symptoms of the disease appear, the structure of these cells starts to degenerate. Studying biomarkers using magnetic resonance imaging (MRI) can help in detecting early cell changes in people at high risk of developing Alzheimer's disease, and thus slowing its progression and controlling the symptoms. This chapter presents a supervised-learning method for identifying Alzheimer's disease which combines textural features extracted from a gray level co-occurrence matrix with voxel-based morphometry neuroimaging analysis. The analysis is locally applied on structural MRI, targeting certain anatomical brain regions that have proved to have more discriminative power than analyzing whole brain MRI. Principal component analysis is introduced to reduce the high dimensionality of the acquired features. This chapter also compares the classification ability of three different classifiers, support vector machine, k-nearest neighbor, and decision tree, by tuning the underlying parameters of each classifier. The proposed method managed to successfully differentiate between Alzheimer's disease patients and normal controls with an accuracy of 93%.

12.1 Introduction

Dementia is a slow decline in brain functions. Every three seconds a new dementia case develops worldwide. The estimated number of people living with dementia in 2030 will be 131 million, most of them in low and middle income countries [43]. Alzheimer's disease (AD) is the most common type of dementia and accounts for an estimated 60%–80% of dementia cases [10]. One third of elderly people die with AD or other related dementia with only quarter of AD patients being correctly diagnosed [2]. Another form of dementia that appears at an earlier state than AD is mild cognitive impairment (MCI). It is an early form of dementia that also affects the brain cells. MCI is thought of as an intermediate period between normal cognition and AD development. About one third of patients diagnosed with MCI develop AD every year [41]. Recent studies have proved that diagnosing MCI at an early stage will help in slowing down the conversion from MCI to AD.

AD causes progressive disorder in neurons that act as the basic working cells of the brain. In a healthy elderly aging people, the brain typically shrinks and loses neurons. However, brains affected by AD shrink dramatically due to neuron death and tissue loss. Neurodegeneration in AD is proven to begin in the medial temporal lobe [36]. The loss of structure and function of neurons progresses to the entorhinal cortex, which functions as the hub for memory and navigation. It moves then to the hippocampus, which is responsible for autobiographical events and spatial memory, and its volume reduces abnormally in AD patients. Such deterioration leads to poor memory functions including short term, long term, and even spatial memory, which is responsible for navigation. In the advanced stage, the damage reaches the cerebral cortex area responsible for language, reasoning, and social interaction, leaving the patient unable to communicate or even perform daily tasks independently. The average life expectancy of AD patients is three to nine years [48]. Thus, understanding the degeneration of brain cells in patient with Alzheimer's can offer treatments that slow the disease progression and control its symptoms, promising a better quality of life for patients and their families. Pre-specifying a set of anatomical regions of interest (ROIs) for analysis can lead to better classification performance results by focusing the analysis on the relevant anatomical brain regions which have proved to have more discriminative power than whole brain analysis methods [20].

In order to evaluate a patient's cognitive status, many physicians use some brief bedside tests. Mini-mental state examination (MMSE) is the most widely used, which has proved to have reasonable reliability. The MMSE total score can be used to distinguish between NC, MCI, and AD [13].

Studying the biomarkers found in structural MRI (sMRI) can detect early changes in the ROIs in the brains of people at high risk of developing AD. Various neuroimaging techniques have been used in AD studies, which fall into two broad categories: (i) structural imaging that provides information about the shape, position, and volume of brain tissue such as computed tomography (CT) and magnetic resonance imaging (MRI); and (ii) functional imaging that measures an aspect of brain function, observing the relationship between activity in certain brain areas and specific mental functions such as functional magnetic resonance imaging (fMRI), single-photon emission computed tomography (SPECT), and positron emission tomography (PET). sMRI describes various tissue types; gray matter (GM) and white matter (WM) are among the most important types as they play an important role in AD diagnosis. GM contains more cell bodies than WM. The latter functions as bundles connecting various GM areas together and has a relatively light appearance in sMRI compared to GM. Morphometric techniques measure the volume, shape, and thickness of GM and WM structures.

The assessment of these structures in AD patients can be compared with normal (healthy) controls (NCs) in an automated fashion using the voxel-based morphometry

(VBM) technique [38]. It is widely used to examine MRI of the human brain for Alzheimer's [11, 15, 31], autism [4, 40], the effects of aging [22, 37], post-traumatic stress disorder [52], schizophrenia [33], and depression [50, 53]. VBM performs a voxel-based statistical analysis on GM and WM tissue density. It allows investigation of the local concentration of GM and WM in brain anatomy by comparing different brain images on a voxel-by-voxel basis after spatially normalizing the images. The output is a statistical parametric map (SPM) showing regions where the GM and WM concentrations differ significantly between AD and NCs.

The spatial relationship of pixels in the gray level sMRI is also examined using a gray level co-occurrence matrix (GLCM) as an additional feature extraction analysis. This technique is employed to asses the texture of an image by measuring how often pairs of pixels with specific values and in a specified spatial order occur in an image. This analysis results in creating a matrix upon which statistical measures are extracted. The GLCM approach is used in the medical domain to detect brain tumors [56] and other brain disorders.

The feature vector generated by GLCM suffers from the high dimensionality problem. Hence, it is preferable to use a feature reduction mechanism in order to remove redundant predictor variables and experimental noise and thus achieve shorter training times and enhance the classifier's performance. Feature reduction avoids over-fitting as well and improves the model's prediction generalization ability. Principal component analysis (PCA) is a well-known method for dimensionality reduction, which can find patterns in datasets summarizing their main characteristics and detecting relations between them. A new compressed set of significant features is created from the high-dimensional feature space which can be used in the classification model construction. Previous neuroimaging machine learning studies state that feature reduction is a fundamental step before applying a predictive model to any neuroimaging data [39].

Machine learning techniques have been widely used to classify AD; here, the support vector machine (SVM), k-nearest neighbors (K-NN), and decision tree are investigated. SVM is a supervised statistical classification method. This binary classifier is used in several works to classify AD and mild cognitive impairment (MCI) [36] or classify AD and NC [28]. K-NN is another supervised instance-based learning technique [5] where hypotheses are constructed directly from the training set. Decision trees are successful predictive modeling approaches used in machine learning.

This work presents a supervised-learning method for identifying AD by combining textural features extracted from GLCM and VBM neuroimaging analysis. The work in [21] investigated these combinations with an entropy-based feature selection process to identify an optimal set of features from the high dimension VBM feature vector. The accuracy achieved using such an approach has been recorded as 88%, which is higher than using GLCM features alone or VBM features alone. The former work used an SVM classifier. The work in [20] studied two more supervised classification techniques: K-NN and decision tree. The K-NN classifier obtained the best accuracy of 92% and the lowest running time. The SVM classifier comes next and finally decision tree. The work concluded as well that certain anatomical ROIs have more powerful discriminative power than the whole brain analysis. Certain

ROIs recorded an increase in the accuracy and a decrease in the feature vector size that led to a high speed-up factor in the running time. This work extends both studies by introducing a feature reduction technique using PCA to the GLCM feature vector. It explores as well the effect of tuning the parameters of the supervised classification techniques.

The rest of the chapter is organized as follows. Section 2 gives an insight of the related work presented in AD classification. Section 3 gives a description of the scientific approach and the methods used. Section 4 explains the data used and presents the experimental study and the performance results obtained. Finally, section 5 concludes the work presented in the chapter and discusses future work.

12.2 Related work

Several attempts have been made in the literature to identify patients suffering from AD at the prodromal stage of MCI. Subjects diagnosed with MCI have cognitive impairments beyond those expected for their age and education. However, these impairments do not meet the neuropathological criteria for AD. Several approaches are proposed to automatically classify patients with AD and/or MCI from anatomical MRI and other neuroimaging techniques. The work presented by [54] combines three neuroimaging modalities: MRI, FDG-PET, and CSF biomarkers. It used a kernel combination method to classify AD, MCI, and NC.

Other studies seek to automatically classify sMRI as AD or NC based on the analysis of certain brain ROIs. Brain atrophy in AD and prodromal AD is spatially distributed over many brain regions [28]. The brain regions that AD can affect include the entorhinal cortex, the hippocampus, lateral and inferior temporal structures, and the anterior and posterior cingulate. Depending on the type of features extracted from MRI, the classification approaches can be categorized into ROI-based, vertex-based, or voxel-based approaches. The ROI-based approach applies nonlinear registration to register MRI to a predefined brain region template. One example of the studies based on volumetric measurements of ROI is [28], where computing the ROIs from MRI is performed by combining image segmentation and a wrapped classification algorithm. Considerable effort has been made to study the atrophy rate in volumetric measurements of ROIs in AD [45] and in MCI [47].

Many researchers have tried to investigate whether the atrophy in the early stages of AD is confined to the hippocampus and the entorhinal cortex or not [36]. Our previous work in [20] proved that ROI analysis has more discriminative power than the whole brain methods for identifying AD. The best set of discriminators was the hippocampus, cerebellum left, cerebellum right, and calcarine. This recorded a 4% increase in accuracy compared to the whole brain method with only 7% of the whole brain feature vector and a 70× speed-up of the running time. In [35] a whole brain hierarchical network is employed to represent different subjects in the brain. Vertex-based approaches [18, 42, 51] obtain information regarding disease progression by using cortical thickness, sulcal depth, or cortical surface area as features. These approaches measure the vertex atrophy on the cortical surface. The cortical thickness represents a crucial dementia biomarker. Jason *et al* [34] showed a significant cortical thickness decline in AD in various parts of the brain: the allocortica, temporal, orbitofrontal, and parietal regions. Their voxel-based technique was able to detect changes in cortical thickness in AD without prior anatomical definitions. Olivier *et al* in [44] measured the cortical thickness in MRI volume and the resultant map was parcellated into 22 regions. The study was successfully able to differentiate between stable MCI from progressive MCI by computing a normalized thickness index using the subset of the parcellated regions. The study also included a relationship investigation between the normalized thickness index, the education level, and the timeline of conversion to AD, acquiring 85% accuracy differentiating between AD and NC. The voxel-based approach statistically compares voxel distributions of major brain tissues, such as GM, WM, and cerebrospinal fluid (CSF). The voxel-based morphometry (VBM) method proposed by Ashburner *et al* [9] allows investigation of focal differences in GM and WM in brain anatomy by comparing different brain images on a voxel-by-voxel basis after spatially normal-izing the images. The output is a statistical parametric map (SPM) showing regions where GM and WM concentration differs significantly between AD and NC. The voxel-based approach has been widely used in AD diagnosis using different classification techniques [16, 17, 29]. Texture analysis is combined with other feature extraction techniques for AD classification. Farouk *et al* in [21] presented an investigation of combining GLCM image texture feature extraction and VBM neuroimaging analysis technique by the means of SVM classifier. Toro *et al* in [49] proposed the use of histons as a textural characteristic as well to classify AD from NC using T1-weighted MRI. The study performed over-segmentation using GM, WM, and CSF for the histon-calculation process and acquired 88% accuracy.

Several machine learning techniques are capable of dealing with the high-dimensional data encountered in MRI studies, such as SVM [19], independent component analysis (ICA) [32], linear discriminant analysis (LDA) [6], wavelet transform [24], and decision trees [29]. Artificial neural networks have been successfully used to diagnose AD [25, 46]. In [55] NN is used to classify MRI as normal or abnormal by first employing discrete wavelet transform to extract features from images, and then the technique of principal component analysis (PCA) is applied to reduce the dimensions of the features. K-NN has proved its robustness in many biomedical fields. In the study [7], pools of mass spectrometer saliva data were analyzed using an extended K-NN algorithm for AD diagnosis and a modification of Euclidean distance formula was presented as well. The research conducted by Breslow and Aha in [14] summarizes the approaches used for tree simplification in terms of size and complexity without compromising accuracy.

Feature reduction techniques have been widely used before applying the classi-fication model to avoid the curse of dimensionality and over-fitting penalties. PCA has been successfully used to extract relevant features in neuroimaging classification problems, not only in AD but also in schizophrenia, psychosis, and ADHD. As for MCI conversion prediction and AD classification, in [12] Beheshti *et al* used PCA to evaluate a feature selection method for high-dimensional data. The presented method combines feature ranking with a genetic algorithm to reduce the dimension-ality and select optimal features for the classifier model. In [55], PCA is used to reduce the dimensions of features and NN is used to classify MRI as normal or abnormal by first employing discrete wavelet transform to extract features from images. The work in [49] applied PCA on calculated histons and the resulting vector was used to train an SVM classifier model.

12.3 The proposed supervised-learning approach for AD identification

MRI dataset is analyzed in two pipelined tracks. One track analyzes images using the VBM analysis technique which generates one part of the feature vector describing the morphometric features. The other pipelined track uses GLCM textural analysis to generate the other part of the feature vector describing the textural features. The GLCM feature vector alone undergoes a feature reduction process to obtain a reduced features set.

For exploring the significance of different anatomical ROIs, an ROI masking process is applied on the combined feature vector.

As fed into the classifier, images are finally classified into two classes: AD patient or NC. The framework of the explained scenario is illustrated in figure 12.1.

12.3.1 Voxel-based morphometric feature extraction

VBM allows investigation of focal differences in brain structure. It can be applied on any of the three anatomy measures GM, WM, and CSF density. VBM of MRI data involves four steps: normalization, segmentation, smoothing, and finally statistical analysis [9, 38].

1. *Spatial normalization*: All the images are spatially normalized to the same stereotactic space. A nonlinear registration of each of the images is applied to the same template image. The resultant normalized images are followed by a later smoothing step to compensate for the inexact nature of the spatial normalization.

2. *Segmentation*: The spatially normalized images undergo a segmentation process. The output is three classes representing the three brain tissue types: GM, WM, and CSF. GM images are selected for further processing.

3. *Smoothing*: The smoothed image voxel contains the average concentration of GM from around the voxel. This is known as GM density. Smoothing has the effect of rendering the data more normally distributed, increasing the parametric statistical tests' validity. Every voxel of the image after spatial smoothing contains the mean concentration of GM from voxel statistics, which is called 'gray matter' density.

4. *Statistical analysis*: A statistical parametric map is obtained by performing voxel-wise parametric statistical tests based on the general linear model. The feature values after segmentation represent the probability of belonging to the tissue (GM or WM) for every voxel of the image. These features generate one part of the feature vector.

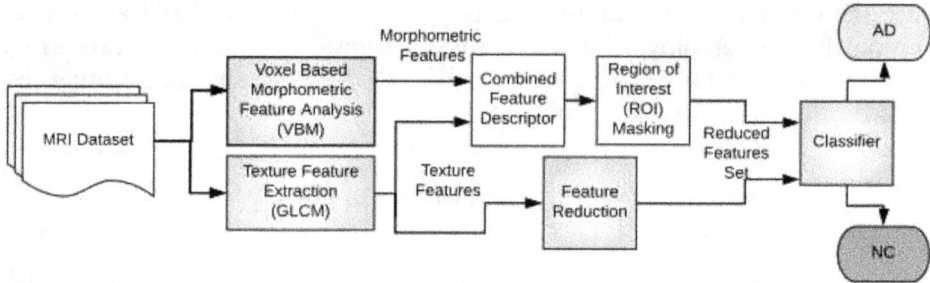

Figure 12.1. Block diagram showing the processing for AD classification.

The output from this analysis is a statistical parametric map showing regions where GM concentration differs significantly between AD and NC.

12.3.2 Texture feature extraction

Texture features reflect the regular changes of gray values in images. These changes in the values are correlated statistically and spatially. A textural feature vector is constructed from the gray level co-occurrence matrix (GLCM). The GLCM estimates image properties related to second-order statistics. It accounts for the spatial inter-dependency of two pixels at specific neighboring positions. It is created from high resolution gray scale MRI. It is then normalized and used to calculate five different texture features [27]: contrast (12.1), correlation (12.2), energy (12.3), homogeneity (12.4), and entropy (12.5). Each (i, j)th entry in the matrix represents the probability of transferring from one pixel with a gray level of i to another with a gray level of j under a given distance and angle. N is the number of gray levels in the image.

1. *Contrast*: A measure of the intensity of contrast between a pixel and its neighbor reflecting the quantity of local changes in an image. It shows the sensitivity of the texture in relation to changes in the intensity:

$$\sum_{i}^{N-1} \sum_{j}^{N-1} |i - j|^2 p(i, j).$$ (12.1)

2. *Correlation*: A measure of how correlated a pixel is to its neighborhood. Correlation is 1 or −1 for a perfectly positively or negatively correlated image. It is often used to measure deformation and displacement such as the motion of an optical mouse:

$$\sum_{ij}^{N-1} \frac{(i - \mu i)(j - \mu j)p(i, j)}{\sigma_i \sigma_j}.$$ (12.2)

3. *Energy*: Also known as uniformity, uniformity of energy, or angular second moment (ASM). It measures the image homogeneity, the more homogeneous

$$\sum_{ij}^{N-1} P(i, j)^2.$$ (12.3)

4. *Homogeneity*: A measure of the similarity of pixels; the closeness of the distribution of elements in the GLCM to the GLCM diagonal. A diagonal GLCM gives homogeneity of 1:

$$\sum_{ij}^{N-1} \frac{p(i, j)}{1 + |i - j|}.$$ (12.4)

5. *Entropy*: A measure of the randomness of the intensity image. It measures the image information. An inhomogeneous image has low first order entropy, while a homogeneous image has high entropy:

$$-\sum_{i}^{N-1} \sum_{j}^{N-1} p(i, j) \log(p(i, j)).$$ (12.5)

12.3.3 Feature reduction

Principal component analysis (PCA) is used for feature reduction. It constructs a new set of relevant features by linearly transforming the correlated variables in the observed data into a smaller number of uncorrelated variables. The new space of features has a lower dimensionality where the variance of the data in this new representation is maximized [30].

In order to obtain the principal components from the high-dimensional data, the original features have to be normalized by subtracting the mean and dividing by the standard deviation. Eigenvalues and eigenvectors of the covariance matrix of the normalized data are then calculated. The decreasing variance in the data is represented by sorting the eigenvalues in a decreasing order. Principal components are finally obtained by multiplying the originally normalized data with the eigenvectors. The top eigenvectors describe most of the data variance. Thus, the decreased set of uncorrelated principal components represents the original high-dimensional correlated dataset.

12.3.4 Classification

The classifier block represents one of the three different classification techniques under investigation: SVM, K-NN, and decision tree [26].

12.3.4.1 SVM
SVM is a supervised-learning model that can perform binary linear or nonlinear classification. SVM supports both regression and classification tasks. It is based on

the concept of decision planes that define decision boundaries. It maps the training set into a new maximal margin hyperplane where a clear gap between the two different classes can be created. SVM creates a model to classify new images. Linear SVM parameters define decision hyperplanes in the multidimensional feature space by the equation

$$g(x) = w^T x + b, \tag{12.6}$$

where x denotes the feature vector, w is the weight vector, and b is the threshold. The hyperplane position is determined by vectors w and b. The vector is orthogonal to the decision plane and b determines its distance to the origin. To construct an optimal hyperplane, SVM employs an iterative training algorithm, which is used to minimize an error function. According to the form of the error function, SVM classification models have two types: C-SVM and nu-SVM. C and nu are regularization parameters which help implement a penalty on the mis-classifications that are performed while separating the classes. This helps in improving the accuracy of the output.

For nonlinearly separable data, the approach described for linear SVMs can be extended. The nonlinear SVM kernels, polynomial, and Gaussian are used for finding nonlinear decision hyperplanes in the input space. The polynomial kernel function for degree-h polynomials is defined in

$$K(X_i, X_j) = K(X_i \cdot X_j + c)^h, \tag{12.7}$$

where X_i and X_j are vectors in the input space and $c \geqslant 0$ is a free parameter trading off the influence of higher-order versus lower-order terms in the polynomial. When $c = 0$, the kernel is called homogeneous. The Gaussian kernel function is defined in

$$K(X_i, X_j) = e^{-\|X_i - X_j\|^2 / 2\sigma^2}, \tag{12.8}$$

where X_i and X_j are vectors in the input space. It may be recognized as the squared Euclidean distance between the two feature vectors whereas σ is a free parameter.

12.3.4.2 K-NN

K-NN is a supervised instance-based learning model where hypotheses are constructed directly from the training set. The output of K-NN is a class membership determined by the dominant class label of its neighbors. The exhaustive comparison of K-NN can be time consuming. Moreover, storing the feature vector sets in the training phase consumes memory too. The training phase of the algorithm stores the feature vectors and class labels of the training set. In the classification phase, the unlabeled test sample is assigned to the most frequent class label in the k training samples nearest to that point. If $k = 1$, then the test sample is assigned to the class of that single nearest neighbor.

The distance metrics upon which the nearest neighbors are evaluated highly affect the accuracy of the K-NN classifier. Eight distance functions are explored with their formulas: Euclidean (12.9), Hamming (12.10), Correlation (12.11), Cosine (12.12),

Cityblock (12.13), Chebychev (12.14), Jaccard (12.15), and Spearman (12.16). $d(x, y)$ is the distance between x and y. N is the total number of variables.

1. *Euclidean*: Measures the straight-line distance in Euclidean space:

$$d(x, y) = \sqrt{\sum_{i=1}^{k} (y_i - x_i)^2}. \tag{12.9}$$

2. *Hamming*: Measures the minimum number of substitutions required to change one object into the other, i.e. the number of mismatches among their pairs of variables:

$$d(x, y) = \sum_{k=0}^{N-1} [i_{x,k} \neq i_{y,k}]. \tag{12.10}$$

3. *Correlation*: Measures the statistical dependence between two vectors. It is obtained by dividing their distance covariance by the product of their distance standard deviations. The correlation distance is one minus this obtained value:

$$d(x, y) = 1 - \frac{d\text{Cov}(x, y)}{\sqrt{d\text{Var}(x)d\text{Var}(y)}}. \tag{12.11}$$

4. *Cosine*: Measures the similarity between two vectors by obtaining the cosine of the angle between them. The cosine distance is one minus this value:

$$d(x, y) = 1 - \cos\theta = 1 - \frac{\mathbf{x} \cdot \mathbf{y}}{\|\mathbf{x}\| \, \|\mathbf{y}\|}. \tag{12.12}$$

5. *Cityblock*: Also known as the Manhattan distance. It is obtained by measuring the sum of the absolute difference of Cartesian coordinates between two points. Unlike Euclidean distance, the shortest distance between any two points, the city block distance is the distance in x plus the distance in y.

$$d(x, y) = 1 - \sum_{k=0}^{N-1} |x_k - y_k|. \tag{12.13}$$

6. *Chebychev*: Also known as Chessboard distance. It is obtained by measuring the maximum of differences along any coordinate dimension:

$$d(x, y) = \max_i(|x_i - y_i|). \tag{12.14}$$

7. *Jaccard*: Measures the dissimilarity between datasets. The Jaccard coefficient is the percentage of nonzero coordinates that differ. The Jaccard distance is one minus this value:

$$d(x, y) = 1 - \frac{|x \cup y| - |x \cap y|}{|x \cup y|}. \tag{12.15}$$

8. *Spearman*: Describes the relationship between two variables using a monotonic function. It is obtained by subtracting Spearman's rank r_s correlation coefficient from one. $\mathrm{cov}(rg_x, rg_y)$ is the covariance of the rank variables and σ_{rg_x}, σ_{rg_y} are their standard deviations:

$$d(x, y) = 1 - r_s = 1 - \frac{\mathrm{cov}(rg_x, rg_y)}{\sigma_{rg_x}\sigma_{rg_y}}. \tag{12.16}$$

12.3.4.3 Decision tree

A decision tree represents the data in a tree-like structure, where each internal node denotes a test on a feature, each branch represents an outcome of the test, and each leaf node holds the class label. Given a test sample, X, with an unknown class label, its feature values are tested against the decision tree. The root node is tested first and continues down until reaching a leaf node holding the predicted class label. The decision tree can be converted to classification rules and can handle multidimensional data. The complexity of a decision tree is defined by its number of splits. Simpler trees are better as they are less likely to overfit the data.

Two splitting predictors are explored: standard CART and curvature. The training speed of both predictors is similar.

1. *Standard CART*: It is the foundation for other decision trees such as the bagged decision trees, random forest, and boosted decision trees. It selects the split predictor that maximizes the information gain over all possible splits of other predictors. Information gain is the most popular feature selection measure. It is based on Shannon's information theory. The feature with the highest information gain is chosen as the splitting feature for the root node. This feature minimizes the number of tests needed to classify a given sample. The expected information needed to classify a sample in D is given by equation (12.17), where pi is the probability that a sample in D belongs to class C_i:

$$\mathrm{Info}(D) = -\sum_{i=1}^{n} p_i \log_2(p_i). \tag{12.17}$$

2. *Curvature*: It selects the split predictor that minimizes the p-value of chi-square tests for curvature. A chi-square test assesses independence by comparing two variables to measure their relationship. A small value indicates that there is a relationship. The data are first classified into mutually exclusive classes. Assuming a null hypothesis, a probability is given that any data fall into the corresponding class. The chi-square formula is given by equation (12.18), where c is the degree of freedom, O is the observed value, and E is the expected one:

$$X_c^2 = \sum_{i=1}^{k} \frac{(O_i - E_i)^2}{E_i}. \tag{12.18}$$

12.4 Experimental results and discussion

12.4.1 Dataset

In MRI, tissue relaxation properties contribute to image contrast. There are two different relaxation times; T1 and T2. These times measure the time taken for spinning protons to lose phase coherence among the nuclei spinning perpendicular to the main field. The most common MRI sequences are T1-weighted and T2-weighted scans. In T1-weighted images GM appears darker than WM. In this study, a total of 275 T1-weighted MR images are considered. The dataset used in this work is obtained from the Alzheimer's Disease Neuroimaging Initiative (ADNI) database [1]. Table 12.1 shows the demographics of the dataset.

12.4.2 Experimental work

Two performance tests are carried out. One to explore the effect of tuning the classifiers' parameters and the other to evaluate the effect of the feature reduction technique on the GLCM features.

The k-fold cross-validation strategy is adopted for evaluating the classification performance. Cross-validation evaluates predictive models by iteratively partitioning the original sample into $k - 1$ training sets to train the model and a single test set to evaluate it. Hence, all observations are used for both training and validation. In this work, k is set to 10.

Classification accuracy, sensitivity, and specificity are then calculated using the following formulas [26]:

$$\text{Accuracy} = (TP + TN)/(P + N) \tag{12.19}$$

$$\text{Sensitivity}, = TP/P \tag{12.20}$$

$$\text{Specificity} = TN/N, \tag{12.21}$$

where TP, TN, P, and N refer to the number of samples representing true positive, true negative, positive, and negative, respectively.

Table 12.1. Demographics of the dataset.

	Normal $N = 113$	AD $N = 162$
Male/female	72/41	71/91
Age: $\mu(\sigma)$	77.49(5.88)	73.82(7.63)
MMSE: $\mu(\sigma)$	25.74(7.74)	21.54(3.92)

12.4.2.1 Performance tests of tuning classifier parameters

AD patients and NC samples undergo a statistical *t*-test at the final step in the VBM block in the proposed framework, shown in figure 12.1. This statistical analysis shows the significant regions of GM loss between AD and NC in the whole brain. ROI binary masks are used on these regions to extract ROIs. Eight ROIs are defined for masking: hippocampus, cerebellum left, cerebellum right, medulla, calcarine, pons, occipital lobe, and frontal lobe. The masked ROIs are grouped into five groups for testing:

- ROI1 refers to the hippocampus region.
- ROI2 refers to the hippocampus, cerebellum left, and cerebellum right regions.
- ROI3 refers to the hippocampus, cerebellum left, cerebellum right, and calcarine regions.
- ROI4 refers to the hippocampus, cerebellum left, cerebellum right, calcarine, and frontal lobe regions.
- ROI5 refers to the hippocampus, cerebellum left, cerebellum right, calcarine, frontal lobe, pons, occipital lobe, and medulla regions.

The performance tests to measure the effect of tuning the parameters of the three classifiers are carried out on these five groups of ROIs.

Tables 12.2–12.4 show the details of K-NN performance using different distance functions. Correlation, Euclidean, and Cosine distance functions scored the best accuracy with almost the same average accuracy among all ROIs, 91%. Cityblock and Spearman rank next with average accuracies of 88% and 86%, respectively. Jaccard scored the worst average accuracy of 75%. Correlation and Cosine functions scored the best sensitivity values with averages of 92.4% and 91%, respectively, among all ROIs. Euclidean scored the best specificity values with an average of 90% among all ROIs.

ROI3, including the hippocampus, cerebellum left, cerebellum right, and calcarine, continued to prove its high discriminative power, concluded in [20]. The Cosine distance function obtained the best accuracy with ROI3 of 93%. The performance of various K-NN distance functions on ROI3 is illustrated in figure 12.2. Correlation and Euclidean scored similar running times as their score in accuracy, with an average value of 1.2 s. Despite having the same average accuracy as Correlation and Euclidean, Cosine recorded an increase of about 40% in running time than its peers. Spearman recorded the slowest running time with an average of 5.6 s. Its running time also increases rapidly as feature vector size increases in different ROIs. The running time of other classifiers recorded a slight increase. The running time of different K-NN distance functions across all ROIs is illustrated in figure 12.3.

Tables 12.5–12.7 show the details of the two different tree splitting functions. The standard CART function recorded better accuracy in the middle ROIs. The curvature function scored better in the first and last ROI. The average accuracy for both functions is almost the same, 77.6% for CART and 76.6% for curvature.

Table 12.2. Accuracy measurements for K-NN distance functions on ROIs.

	Correlation	Euclidean	Hamming	Cosine	Cityblock	Chebychev	Jaccard	Spearman
ROI1	**88.00%**	86.55%	77.45%	86.18%	85.45%	83.64%	79.64%	81.82%
ROI2	90.91%	**91.27%**	81.09%	89.82%	89.82%	81.82%	76.36%	90.55%
ROI3	92.73%	92.73%	78.91%	**93.09%**	89.09%	84.36%	74.55%	86.18%
ROI4	91.64%	**92.00%**	82.55%	91.64%	88.73%	82.55%	71.64%	86.55%
ROI5	90.55%	90.55%	80.73%	**92.00%**	86.91%	80.73%	70.55%	85.45%

Table 12.3. Sensitivity measurements for K-NN distance functions on ROIs.

	Correlation	Euclidean	Hamming	Cosine	Cityblock	Chebychev	Jaccard	Spearman
ROI1	**90.74%**	90.12%	77.45%	87.65%	90.12%	84.57%	79.01%	83.33%
ROI2	**92.59%**	91.36%	79.01%	89.51%	90.12%	83.33%	68.52%	90.74%
ROI3	**94.44%**	93.21%	75.93%	**94.44%**	87.65%	84.57%	64.81%	85.8%
ROI4	92.59%	91.98%	80.25%	**93.83%**	88.27%	83.33%	61.11%	88.89%
ROI5	**91.98%**	88.89%	77.78%	**91.98%**	86.42%	80.86%	59.88%	87.04%

Table 12.4. Specificity measurements for K-NN distance functions on ROIs.

	Correlation	Euclidean	Hamming	Cosine	Cityblock	Chebychev	Jaccard	Spearman
ROI1	**84.07%**	81.42%	80.53%	**84.07%**	78.76%	82.30%	80.53%	79.65%
ROI2	88.5%	**91.15%**	84.07%	90.27%	89.38%	79.65%	87.61%	90.27%
ROI3	90.27%	**92.04%**	83.19%	91.15%	91.15%	84.07%	88.5%	86.73%
ROI4	90.27%	**92.04%**	85.84%	88.5%	89.38%	81.42%	86.73%	83.19%
ROI5	88.5%	**92.92%**	84.96%	**92.04%**	87.61%	80.53%	85.84%	83.19%

Both functions achieved the same average sensitivity across ROIs, 80%. Standard CART achieved better specificity than curvature by a 4% increase in the average value. Despite their similar average accuracy, the average running time of curvature is less than that of CART by more than 33%. CART recorded 9 s while Curvature recorded only 6 s. The performance of various tree splitting functions on ROI3 is illustrated in figure 12.4 and their running time across all ROIs is illustrated in figure 12.5.

The SVM linear kernel function showed very high accuracy compared to Gaussian and polynomial kernels. It achieved 90% average accuracy across all ROIs while the Gaussian and polynomial kernels achieved 64% and 57%, respectively. However, the Gaussian kernel achieved similar accuracy to the linear kernel in the small sized ROI1. The accuracy measurements of the three kernels on the five

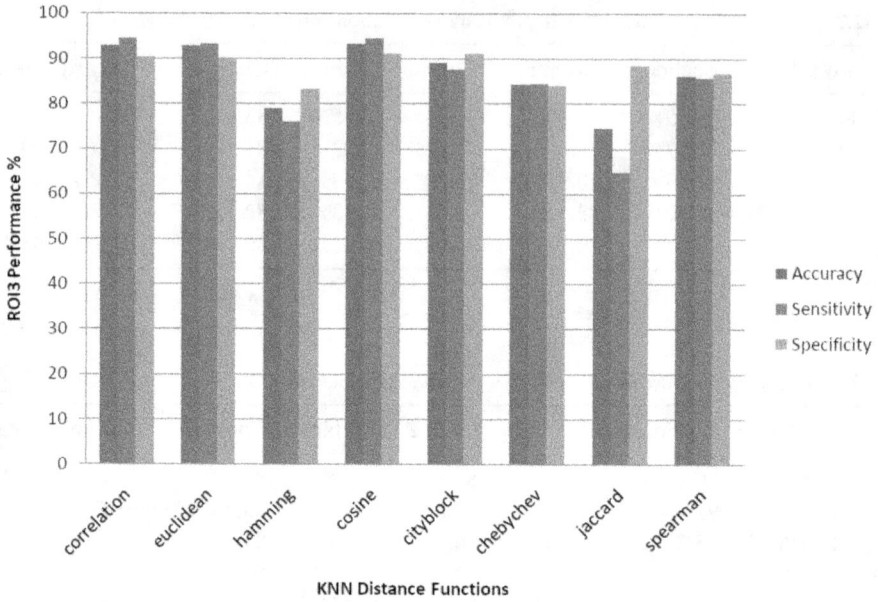

Figure 12.2. Performance measurements of K-NN distance functions on ROI3.

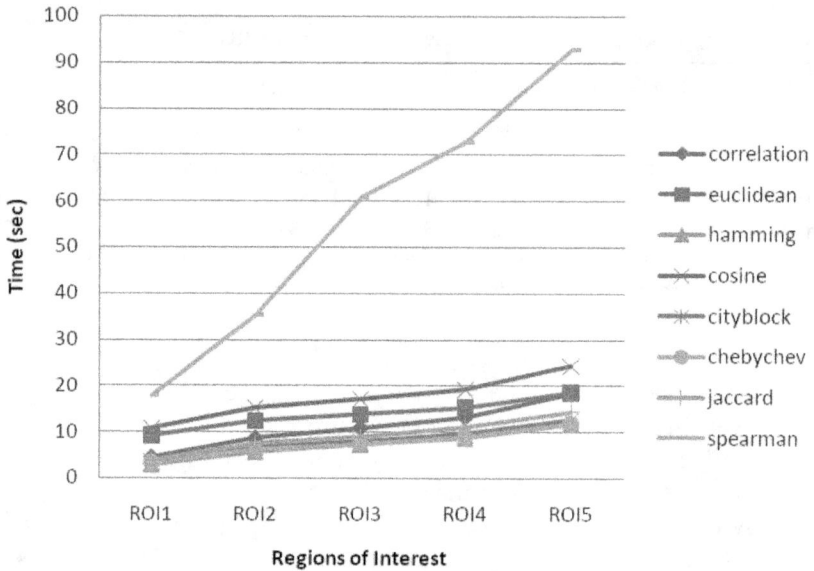

Figure 12.3. Running time of different K-NN distance functions.

ROIs are illustrated in figure 12.6. The linear kernel also consumes less time than the Gaussian and polynomial kernels. However, the increased rate of running time for different sized ROIs is much slower in the polynomial compared to the other functions. Running time is illustrated in figure 12.7.

Table 12.5. Accuracy measurements for tree splitting functions on ROIs.

	CART	Curvature
ROI1	74.18%	**77.45%**
ROI2	**79.27%**	75.64%
ROI3	**78.91%**	76.00%
ROI4	**77.82%**	74.91%
ROI5	77.82%	**79.27%**

Table 12.6. Sensitivity measurements for tree splitting functions on ROIs.

	CART	Curvature
ROI1	74.69%	**77.78%**
ROI2	79.63%	**80.25%**
ROI3	**83.33%**	81.48%
ROI4	**83.95%**	82.72%
ROI5	78.4%	**83.33%**

Table 12.7. Specificity measurements for tree splitting functions on ROIs.

	CART	Curvature
ROI1	73.45%	**76.99%**
ROI2	**78.76%**	69.03%
ROI3	**72.57%**	68.14%
ROI4	**69.03%**	63.72%
ROI5	**76.99%**	73.45%

12.4.2.2 Performance tests of PCA on GLCM features

In this experiment, the features obtained from GLCM texture feature analysis undergo feature reduction by means of PCA before applying classification, as illustrated in the proposed framework shown in figure 12.1. In PCA, the first principal component accounts for most of the data variance and each consecutive principal component accounts for as much of the remaining variance as possible. PCA assessment examined four sets of different numbers of principal components (PC), k. The first set of PC contains 4352 PCs, $k = 4352$. The three remaining sets are $k = 3352$, $k = 2352$, and $k = 1352$.

Performance measurements for the three different classifiers are taken to examine PCA effectiveness. The linear SVM kernel function is used in the SVM classifier and

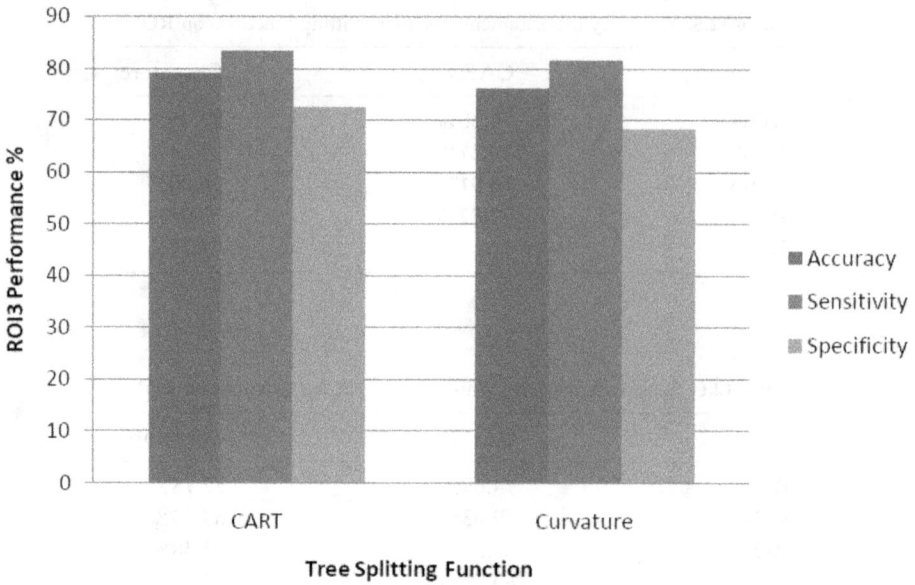

Figure 12.4. Performance measurements of tree splitting functions on ROI3.

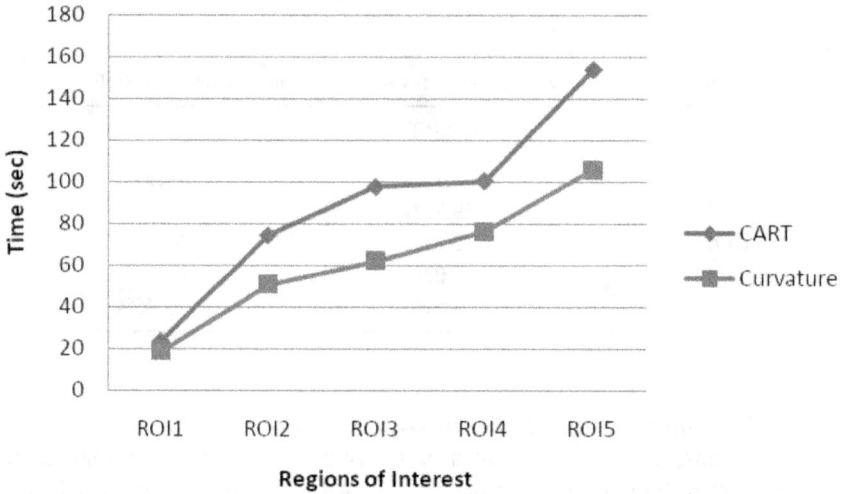

Figure 12.5. Running time of different tree splitting functions.

the standard CART splitting function is used in the tree classifier. Two distance functions are used in the K-NN classifier: Correlation and Euclidean.

Figure 12.8 illustrates the accuracy of the classifiers on different PC sets. The second set of PC ($k = 3352$) achieved better results in the K-NN classifier while reducing 23% of the feature vector size. The best accuracy obtained is 90.55% with the second set of PCs ($k = 3352$) by the K-NN classifier with the Euclidean distance function. The third PC set ($k = 2352$) achieved very comparable accuracy results of

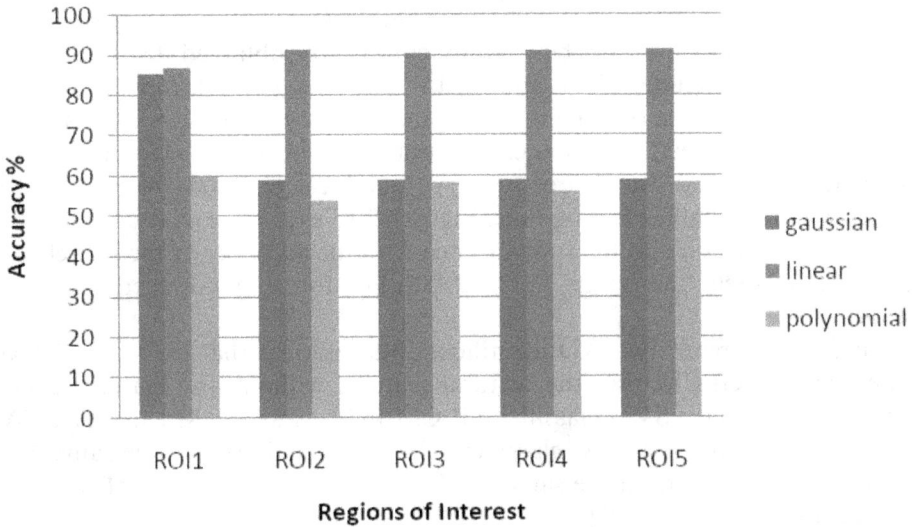

Figure 12.6. Accuracy measurements of SVM kernel functions on ROIs.

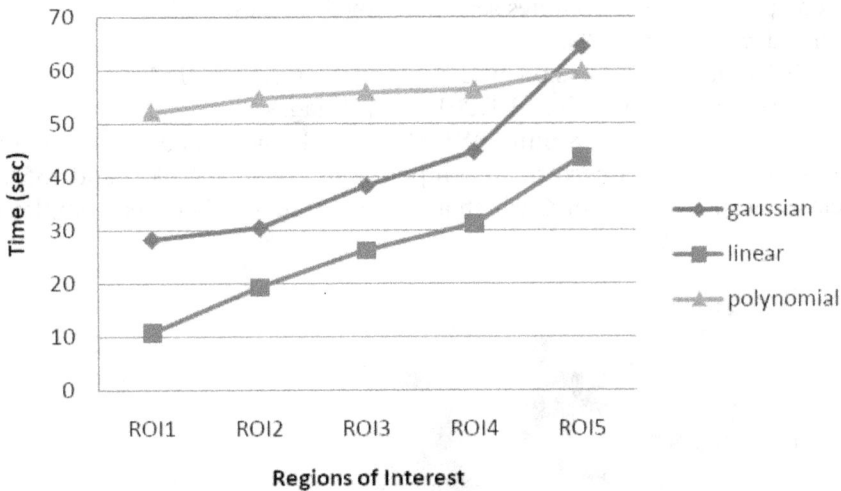

Figure 12.7. Running time of different SVM kernel functions.

89.45% with the same classifier parameters and 46% reduced feature vector size. The K-NN classifier with Euclidean distance function achieved the highest accuracy in all PC sets.

Figure 12.9 illustrates the running time of such classifiers. The SVM classifier achieved the fastest running time while the tree classifier achieved the slowest running time. Figure 12.10 shows the relationship between PCA feature reduction and the running time with the accuracy of the SVM classifier. More PCs included in the classification requires more running time but does not guarantee better accuracy. The figure suggests that the best accuracy can be obtained by the range of [3200–3400] PCs. Figures 12.11 and 12.12 show the relationship between PCA feature reduction and the running time with the accuracy of the K-NN classifier. The former uses the correlation distance function and the latter uses the Euclidean distance function. The two figures suggest the PC range [3200–3400] for the best accuracy with reasonable running time. Figure 12.13 shows the relationship between PCA feature reduction and the running time with the accuracy of the tree classifier. The figure suggests the range [2200–2500] PC for a good accuracy–running time ratio.

Table 12.8 compares two AD identification techniques that use GLCM texture feature analysis. GLCM uses the textural features without any feature reduction techniques and uses the SVM classifier. GLCM-PCA uses GLCM features, PCA for feature reduction, and the SVM classifier. The two techniques use the same dataset and the tests are applied on the same machine specification. Details of the GLCM technique are discussed in [21].

The combined GLCM-PCA technique provides a higher accuracy of 90.55%. The achieved accuracy increased by more than 7% from using GLCM features only. The combined approach also promotes sensitivity by 3% and specificity by 13%. Feature vector size decreased by 23%.

All VBM data pre-processing was performed using the DARTEL toolbox in SPM12 [8] running on MATLAB R2018a [3]. Images were registered to the MNI space and smoothed by an 8 mm FWHM Gauss kernel function. Selected spatial directions for the GLCM performed computations are 0°, 45°, 90°, and 135°. In the SVM classifier, the coefficient C which affects the trade-off between complexity and

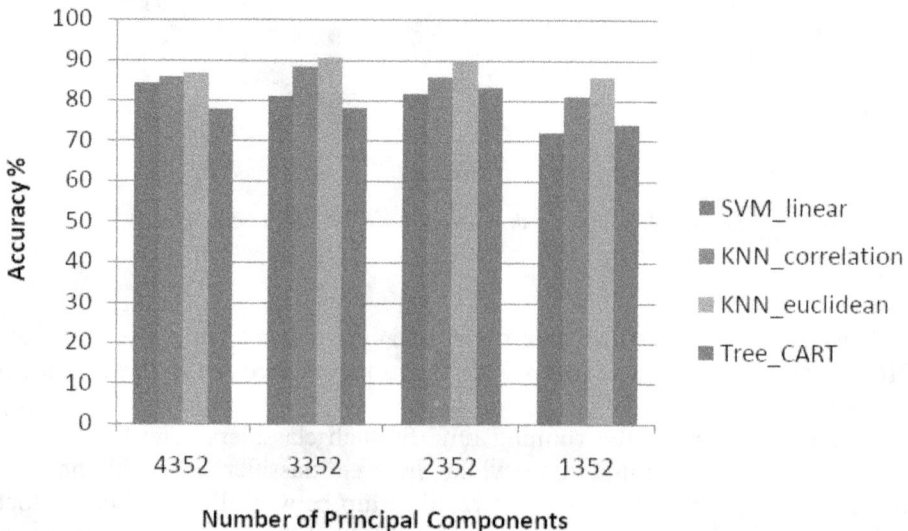

Figure 12.8. Accuracy measurements of classifiers on different numbers of principal components.

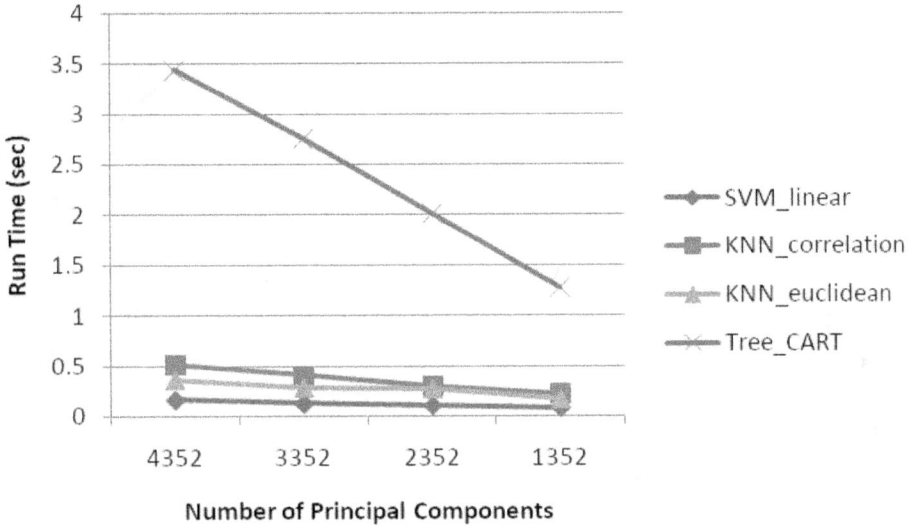

Figure 12.9. Running time of classifiers with different numbers of principal components.

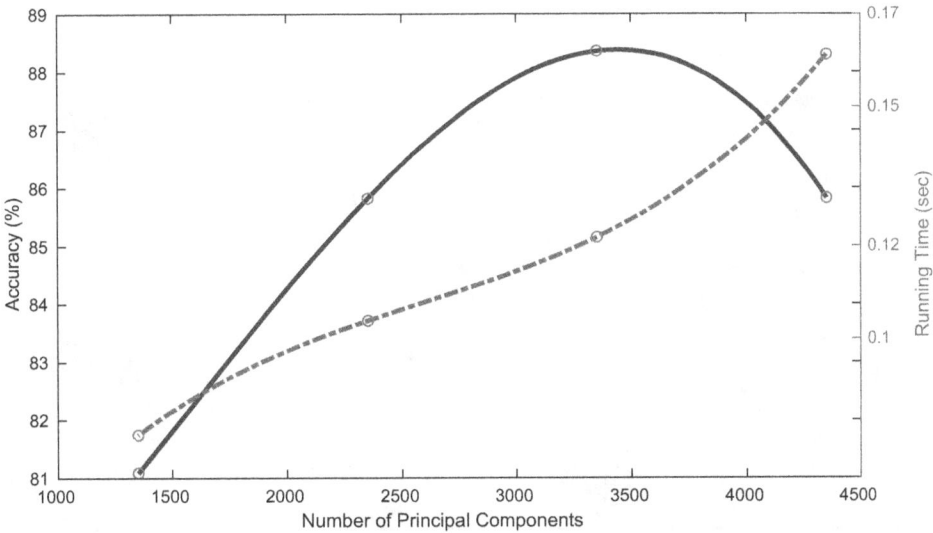

Figure 12.10. Relationship between the number of principal components and running time with the accuracy of the SVM classifier.

proportion of non-separable samples is used as fixed with a value of 1 in all experiments.

The presented experiments are all executed on a 64 bit Intel Core i5-4200M CPU @ 2.50GHz x4 machine, with 11.5 GiB memory and the Ubuntu 18.04 operating system.

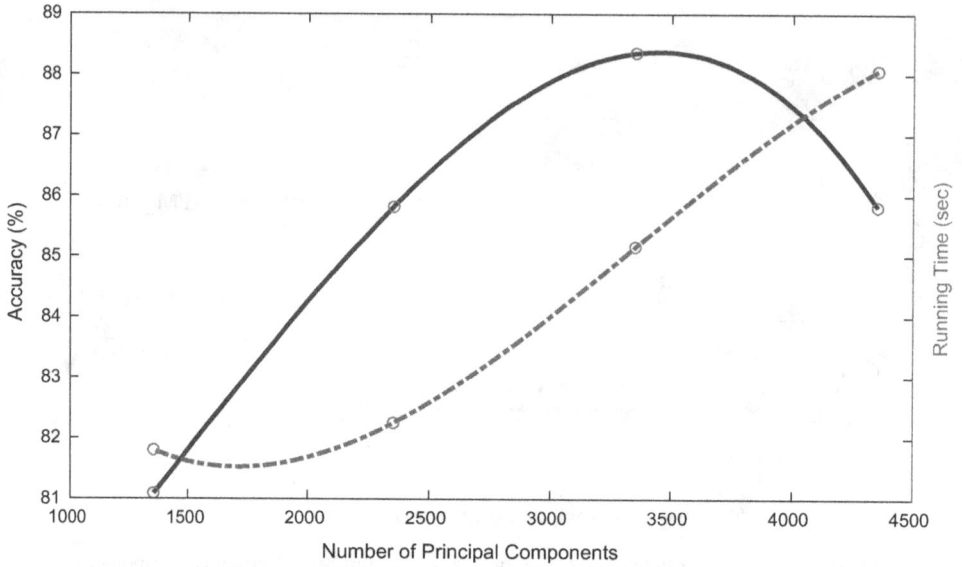

Figure 12.11. Relationship between the number of principal components and running time with the accuracy of the K-NN–correlation classifier.

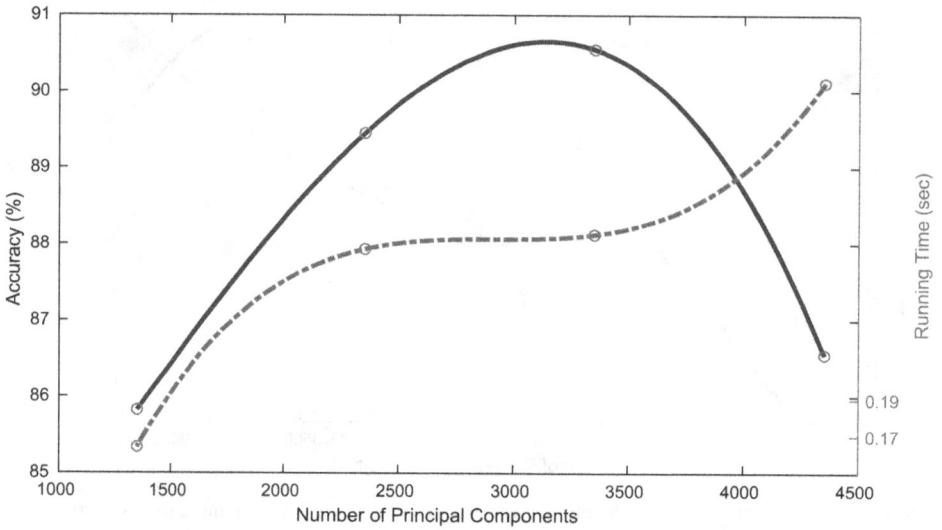

Figure 12.12. Relationship between the number of principal components and running time with the accuracy of the K-NN–Euclidean classifier.

12.5 Conclusion and future work

This chapter proposed a framework for identifying Alzheimer's disease that combines textural features extracted from a gray level co-occurrence matrix and voxel-based morphometry neuroimaging analysis. Principal component analysis is

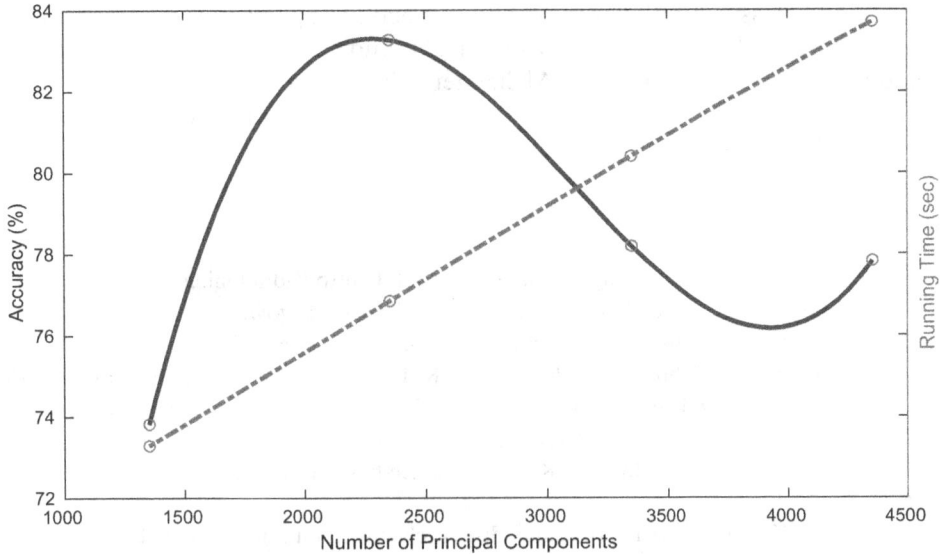

Figure 12.13. Relationship between the number of principal components and running time with the accuracy of the tree classifier.

Table 12.8. Comparison of the results of PCA feature reduction.

	GLCM	GLCM-PCA
Accuracy	83.27%	**90.55%**
Sensitivity	87.65%	**90.74%**
Specificity	76.99%	**90.27%**
Feature vector size	4352	**3352**

integrated to reduce the high dimensionality of the acquired features and enhance both accuracy and running time. The combined GLCM-PCA feature reduction technique provided higher accuracy than using GLCM features alone. The achieved accuracy increased by 7% with an additional 23% reduction in the feature space. Using the combined approach, the sensitivity and specificity increased as well.

The classification power of three different classifiers, support vector machine, k-nearest neighbor, and decision tree, is explored and their underlying parameters are tuned. Eight distance functions in the K-NN classifier are explored. Correlation, Euclidean, and Cosine acquired the best accuracy. Two splitting functions are examined in the decision tree classifier: the standard CART and curvature function. Their performance measurements showed dependency on feature vector size. Three kernel functions are explored in the SVM classifier. The linear kernel function showed very high accuracy compared to the other two.

The K-NN classifier with Cosine distance function applied on the ROIs, including hippocampus, cerebellum left, cerebellum right, and calcarine, proved to be the best technique to differentiate between Alzheimer's disease patients and normal controls with an accuracy of 93%. Introducing PCA to such a technique promises to increase the obtained accuracy.

References

[1] Alzheimers Disease Neuroimaging Initiative (ADNI) http://adni.loni.usc.edu/
[2] Centers for Disease Control and Prevention http://www.cdc.gov/.
[3] Matlab software http://www.mathworks.com/products/matlab/.
[4] Abell F, Krams M, Ashburner J, Passingham R, Friston K, Frackowiak R, Happé F, Frith C and Frith U 1999 The neuroanatomy of autism: a voxel-based whole brain analysis of structural scans *Neuroreport* **10** 1647–51
[5] Aha D W, Kibler D and Albert M K 1991 Instance-based learning algorithms *Mach. Learn.* **6** 37–66
[6] Alam S and Kwon G-R 2017 Alzheimer disease classification using KPCA, LDA, and multi-kernel learning SVM *Int. J. Imaging Syst. Technol.* **27** 133–43
[7] Anyaiwe D E O, Wilson G D, Geddes T J and Singh G B 2018 Harnessing mass spectra data using KNN principle: diagnosing Alzheimer's disease *ACM SIGBioinform. Record* **7** 2–9
[8] Ashburner J *et al* 2014 *Spm12 manual* https://doi.org/10.1016/B978-0-12-372560-8.50052-8.
[9] Ashburner J and Friston K J 2000 Voxel-based morphometry: the methods *Neuroimage* **11** 805–21
[10] Alzheimer's Association *et al* 2017 2017 Alzheimer's disease facts and figures *Alzheimer's Dement.* **13** 325–73
[11] Baron J C, Chetelat G, Desgranges B, Perchey G, Landeau B, De La Sayette V and Eustache F 2001 *In vivo* mapping of gray matter loss with voxel-based morphometry in mild Alzheimer's disease *Neuroimage* **14** 298–309
[12] Beheshti I *et al* 2016 Feature-ranking-based Alzheimer's disease classification from structural MRI *Magn. Reson. Imaging* **34** 252–63
[13] Benson A D, Slavin M J, Tran T-T, Petrella J R and Doraiswamy P M 2005 Screening for early Alzheimer's disease: is there still a role for the mini-mental state examination? *Prim. Care Companion J. Clin. Psychiatry* **7** 62
[14] Breslow L A and Aha D W 1997 Simplifying decision trees: a survey *Knowl. Eng. Rev.* **12** 1–40
[15] Busatto G F, Diniz B S and Zanetti M V 2008 Voxel-based morphometry in Alzheimer's disease *Expert Rev. Neurother.* **8** 1691–702
[16] Busatto G F, Garrido G E J, Almeida O P, Castro C C, Camargo C H P, Cid C G, Buchpiguel C A, Furuie S and Bottino C M 2003 A voxel-based morphometry study of temporal lobe gray matter reductions in Alzheimer's disease *Neurobiol. Aging* **24** 221–31
[17] Chetelat G, Landeau B, Eustache F, Mezenge F, Viader F, De La Sayette V, Desgranges B and Baron J-C 2005 Using voxel-based morphometry to map the structural changes associated with rapid conversion in MCI: a longitudinal MRI study *Neuroimage* **27** 934–46

[18] Cho Y *et al* 2012 Individual subject classification for Alzheimer's disease based on incremental learning using a spatial frequency representation of cortical thickness data *Neuroimage* **59** 2217–30

[19] Cuingnet R *et al* 2011 Automatic classification of patients with Alzheimer's disease from structural MRI: a comparison of ten methods using the ADNI database *Neuroimage* **56** 766–81

[20] Farouk Y and Rady S 2018 Supervised classification techniques for identifying Alzheimer's disease *Int. Conf. on Advanced Intelligent Systems and Informatics* (Berlin: Springer) pp 189–97

[21] Farouk Y, Rady S and Faheem H 2018 Statistical features and voxel-based morphometry for Alzheimer's disease classification *2018 9th Int. Conf. on Information and Communication Systems (ICICS)* (Piscataway, NJ: IEEE) pp 133–8

[22] Good C D, Johnsrude I S, Ashburner J, Henson R N A, Fristen K J and Frackowiak R S J 2002 A voxel-based morphometric study of ageing in 465 normal adult human brains *5th IEEE EMBS Int. Summer School on Biomedical Imaging, 2002* (Piscataway, NJ: IEEE) p 16

[23] Gray K R *et al* 2013 Random forest-based similarity measures for multi-modal classification of Alzheimer's disease *NeuroImage* **65** 167–75

[24] Hackmack K *et al* 2012 Multi-scale classification of disease using structural MRI and wavelet transform *Neuroimage* **62** 48–58

[25] Hamilton D, O'mahony D, Coffey J, Murphy J, O'hare N, Freyne P, Walsh B and Coakley D 1997 Classification of mild Alzheimer's disease by artificial neural network analysis of SPET data *Nucl. Med. Commun.* **18** 805–10

[26] Han J, Pei J and Kamber M 2011 *Data Mining: Concepts and Techniques* (Amsterdam: Elsevier) doi: 10.1016/B978-0-12-381479-1.00020-4

[27] Haralick R M, Shanmugam K and Dinstein I H 1973 Textural features for image classification *IEEE Trans. Syst. Man Cybern.* **6** 610–21

[28] Hidalgo-Muñoz A R, Ramírez J, Górriz J M and Padilla P 2014 Regions of interest computed by SVM wrapped method for Alzheimers disease examination from segmented MRI *Front. Aging Neurosci.* **6**

[29] Hirata Y, Matsuda H, Nemoto K, Ohnishi T, Hirao K, Yamashita F, Asada T, Iwabuchi S and Samejima H 2005 Voxel-based morphometry to discriminate early Alzheimer's disease from controls *Neurosci. Lett.* **382** 269–74

[30] Jolliffe I 2011 *Principal Component Analysis* (Berlin: Springer) doi: 10.1007/978-3-642-04898-2_455

[31] Karas G B, Burton E J, Rombouts S A R B, Van Schijndel R A, O'Brien J T, Scheltens P H, McKeith I G, Williams D, Ballard C and Barkhof F 2003 A comprehensive study of gray matter loss in patients with Alzheimer's disease using optimized voxel-based morphometry *Neuroimage* **18** 895–907

[32] McKeown M J, Hansen L K and Sejnowsk T J 2003 Independent component analysis of functional MRI: what is signal and what is noise? *Curr. Op. Neurobiol.* **13** 620–9

[33] Koelkebeck K *et al* 2013 Impact of gray matter reductions on theory of mind abilities in patients with schizophrenia *Soc. Neurosci.* **8** 631–9

[34] Lerch J P, Pruessner J C, Zijdenbos A, Hampel H, Teipel S J and Evans A C 2004 Focal decline of cortical thickness in Alzheimer's disease identified by computational neuro-anatomy *Cereb. Cortex* **15** 995–1001

[35] Liu J, Li M, Lan W, Wu F-X, Pan Y and Wang J 2016 Classification of Alzheimer's disease using whole brain hierarchical network *IEEE/ACM Trans. Comput. Biol. Bioinform.* **15** 624–32

[36] Magnin B, Mesrob L, Kinkingnéhun S, Pélégrini-Issac M, Colliot O, Sarazin M, Dubois B, Lehéricy S and Benali H 2009 Support vector machine-based classification of Alzheimer's disease from whole-brain anatomical MRI *Neuroradiology* **51** 73–83

[37] Matsuda H 2013 Voxel-based morphometry of brain MRI in normal aging and Alzheimer's disease *Aging Dis.* **4** 29–37 ISSN: 2152-5250 http://europepmc.org/articles/PMC3570139

[38] Mechelli A, Price C J, Friston K J and Ashburner J 2005 Voxel-based morphometry of the human brain: methods and applications *Curr. Med. Imaging Rev.* **1** 105–13

[39] Mwangi B, Tian T S and Soares J C 2014 A review of feature reduction techniques in neuroimaging *Neuroinformatics* **12** 229–44

[40] Nickl-Jockschat T, Habel U, Michel T M, Manning J, Laird A R, Fox P T, Schneider F and Eickhoff S B 2012 Brain structure anomalies in autism spectrum disorder: a meta-analysis of VBM studies using anatomic likelihood estimation *Hum. Brain Mapp.* **33** 1470–89

[41] Palmer K, Musicco M and Caltagirone C 2010 Are guidelines needed for the diagnosis and management of incipient Alzheimer's disease and mild cognitive impairment? *Int. J. Alzheimers Dis* **2010**

[42] Park H, Yang J-J, Seo J and Lee J-M 2012 Dimensionality reduced cortical features and their use in the classification of Alzheimer's disease and mild cognitive impairment *Neurosci. Lett.* **529** 123–7

[43] Prince M, Comas-Herrera A, Knapp M, Guerchet M and Karagiannidou M 2016 *World Alzheimer Report 2016: Improving Healthcare for People Living with Dementia: Coverage, Quality and Costs Now and in The Future* doi: 10.13140/RG.2.2.22580.04483

[44] Querbes O *et al* 2009 Early diagnosis of Alzheimer's disease using cortical thickness: impact of cognitive reserve *Brain* **132** 2036–47

[45] Rusinek H, Endo Y, De Santi S, Frid D, Tsui W-H, Segal S, Convit A and de Leon M J 2004 Atrophy rate in medial temporal lobe during progression of Alzheimer disease *Neurology* **63** 2354–9

[46] Savio A, García-Sebastián M, Hernández C, Graña M and Villanúa J 2009 Classification results of artificial neural networks for Alzheimer's disease detection *IDEAL* vol 5788 (Berlin: Springer) pp 641–8

[47] Tapiola T *et al* 2008 MRI of hippocampus and entorhinal cortex in mild cognitive impairment: a follow-up study *Neurobiol. Aging* **29** 31–8

[48] Todd S, Barr S, Roberts M and Passmore A P 2013 Survival in dementia and predictors of mortality: a review *Int. J. Geriatr. Psychiatry* **28** 1109–24

[49] Toro C, Gonzalo-Martín C, García-Pedrero A and Ruiz E M 2018 Supervoxels-based histon as a new Alzheimer's disease imaging biomarker *Sensors* **18** 1752

[50] Wang L, Wang T, Liu S, Liang Z, Meng Y, Xiong X, Yang Y, Lui S and Ji Y 2014 Cerebral anatomical changes in female asthma patients with and without depression compared to healthy controls and patients with depression *J. Asthma* **51** 927–33

[51] Westman E, Muehlboeck J-S and Simmons A 2012 Combining MRI and CSF measures for classification of Alzheimer's disease and prediction of mild cognitive impairment conversion *Neuroimage* **62** 229–38

[52] Yamasue H *et al* 2003 Voxel-based analysis of MRI reveals anterior cingulate gray-matter volume reduction in posttraumatic stress disorder due to terrorism *Proc. Natl Acad. Sci.* **100** 9039–43

[53] Yang J, Yin P, Wei D, Wang K, Li Y and Qiu J 2016 Effects of parental emotional warmth on the relationship between regional gray matter volume and depression-related personality traits *Soc. Neurosci.* **12** 337–48

[54] Zhang D *et al* 2011 Multimodal classification of Alzheimer's disease and mild cognitive impairment *Neuroimage* **55** 856–67

[55] Zhang Y, Dong Z, Wu L and Wang S 2011 A hybrid method for MRI brain image classification *Expert Syst. Appl.* **38** 10049–53

[56] Preethi G and Sornagopal V 2014 MRI image classification using GLCM texture features *2014 Int. Conf. on Green Computing Communication and Electrical Engineering (ICGCCEE)* 1–6

IOP Publishing

Neurological Disorders and Imaging Physics, Volume 3
Application to autism spectrum disorders and Alzheimer's
Ayman El-Baz and Jasjit S Suri

Chapter 13

Current trends and considerations of Alzheimer's disease

Fatma El-Zahraa A El-Gamal, Mohammed M Elmogy, Ashraf Khalil, Hassan Hajjdiab, Mohammed Ghazal, Ali Mahmoud, Hassan Soliman, Ahmed Atwan, Gregory N Barnes and Ayman El-Baz

Currently, Alzheimer's disease (AD) is considered one of the most well known neurodegenerative diseases; its victims are primarily elderly people. Over time, various studies have been carried out to demonstrate the formation, causes, and medical treatments of such a fatal disorder. Despite the existence of such studies, the accurate understanding and early diagnosis of AD are still areas open for further study. Until now, one of the factors that added to the ambiguity surrounding AD is that the pathological changes of AD begin to occur in the patient's brain nearly ten to fifteen years before clinical diagnosis. This factor, along with other known and unknown factors, challenge our ability to diagnose AD early. This chapter aims to review the literature that deals with AD from different perspectives in addition to demonstrating the obstacles that are still facing the researchers who are interested in such a research area.

13.1 Introduction

Alzheimer's disease (AD) is one of the most well known neurodegenerative diseases affecting the central nervous system (CNS) and lies under the umbrella of dementia. Generally, dementia is considered a significant problem for public health across socioeconomic and ethnic lines for both genders. The term dementia refers to the loss of previous intellectual capacity in broad cognitive domains including memory, language, etc [1]. Dementia itself is not a single disease, but rather it is a symptom of various conditions that in turn lead to disruption of brain functionality [2].

Among these dementia-related conditions and disorders is AD, which is a progressive disorder characterized by both clinical and pathological features. Neuropsychological deficits, non-cognitive neuropsychiatric disorders, and a steady progression rate are examples of the clinical features, while the formation of neurofibrillary tangles and neuritic plaques, neuronal loss, and amyloid angiopathy are examples of the pathological features that differ from one patient to another [3].

Like any other disease, the early diagnosis of AD can protect the patient's life and help them to continue their lives safely. Unfortunately, the early diagnosis of AD is considered a difficult task. Such difficulty arises from a number of factors including, but not limited to, the diversity of clinical as well as pathological features that changes from one patient to another [3]. Additionally, the appearance of pathological symptoms nearly ten to fifteen years before clinical ones may also hinder early diagnosis, due to the serious effects that occur in the patient brain before being detected clinically. Also, the ambiguity regarding how exactly the disease forms and its accurate causes, are also obstacles for early diagnosis of AD.

Over time, significant scientific efforts have been made with the ultimate goal of clarifying such a complex disease. These efforts include advances in scanning technologies, medical studies, as well as biomedical engineering efforts. Despite all of these advances in the context of understanding the nature of AD in addition to the treatment, this research area is still open since no accurate description of the disease has yet been achieved.

This chapter aims to describe the recent research efforts targeting AD that contribute advances that serve to clarify our understanding of AD. To achieve this, the chapter starts by presenting a brief background on nervous system diseases. Then, it focuses on studies concerning the central nervous system and the neurodegenerative diseases that encompass dementia. Next, due to the important role of scanning technologies in diagnosis in general and in AD diagnosis in particular, the second part of this chapter presents brain related imaging technologies along with their role in the study of AD. The third part of this chapter aims to present some of the common databases that are used in the field of AD research. The fourth part of this chapter will then introduce the scientific achievements in the field of AD, while the fifth and final part discusses the findings and concludes the chapter.

13.2 Anatomical background

This section presents brief explanation of the anatomical background of neurological diseases as an introduction to the subsequent detailed discussion of Alzheimer's disease and its related subcategories. The section is divided into three subsections. An introduction on neurological diseases is presented first to explore the primary pathological category that encapsulates AD. Then, a subsection of the neurological diseases, namely neurodegenerative diseases and disorders, is studied. The purpose of such an examination is to strictly focus on the subcategory of neurological diseases that demonstrates the main properties that characterize AD and other related disorders. Finally, a detailed presentation of AD along with its symptoms, disease levels, and the risks that faces Alzheimer's patients, their families, and their countries, is presented.

13.2.1 Neurological diseases

The National Institute of Health (NIH) described neurological diseases, also called nervous system diseases, as problems that affect the brain, spinal cord, or nerves that in turn make up the nervous system [4]. In general, the nervous system components can be divided into two main parts: the central nervous system and peripheral nervous system. The central nervous system is the part of the nervous system that is composed of the brain, cranial nerves, and spinal cord. This part is responsible for controlling the entire functioning of the body [5]. The other part, the peripheral nervous system, contains the nerves of the whole body outside the brain and the spinal cord. This part has the function of sending impulses to the peripheral structures including body organs and muscles [5]. Due to the functions of the parts of the nervous system, an infection or any abnormality in the nervous system has a direct impact on different bodily functions, including learning, thinking, speaking, moving, etc.

Generally speaking, the abnormalities that affect the nervous system can be divided according to different criteria. One criterion classifies the abnormalities according to the part of the nervous system that is affected by these abnormalities. The other criterion uses the nature of the abnormalities as a basis for such classification. This chapter classifies neurological diseases and disorders according to the affected part of the nervous system. In other words, the diseases and disorders are mainly classified into central nervous system diseases and peripheral nervous system diseases, as presented in figure 13.1. Such a classification was performed by taking into account the ninth International Classification of Diseases (ICD) report that was presented by the World Health Organization (WHO) and aimed to classify diseases globally [6].

Due to the objective of the current chapter, the remainder of this subsection focuses on the first category, central nervous system diseases. Focusing on such diseases aims to differentiate between the diseases under this category and to introduce the degenerative diseases that involve Alzheimer's disease.

13.2.1.1 Central nervous system diseases

As said previously, the diseases under this category affect the central part of the nervous system, i.e. the brain and the spinal cord. These diseases include:

1. *Inflammatory diseases*: Inflammation is the reaction of the body as a response to an infection or injury that affects the body organs [7]. The inflammation starts when a certain type of pathogen, such as tissue injury, or bacterial or viral products, trigger the blood components, vascular system, and tissue cells to respond to them. Such a response is continued as long as the stimulus still exists and while the pathogen is neither exhausted nor discharged from the triggered tissue [8]. In general, there are two types into which inflammation can be classified: acute inflammation and chronic inflammation. Such a classification is carried out based on the duration of the inflammation. In acute inflammation, the body provides an immediate response to the injury/ infection to protect against pathogens. This target is achieved through the

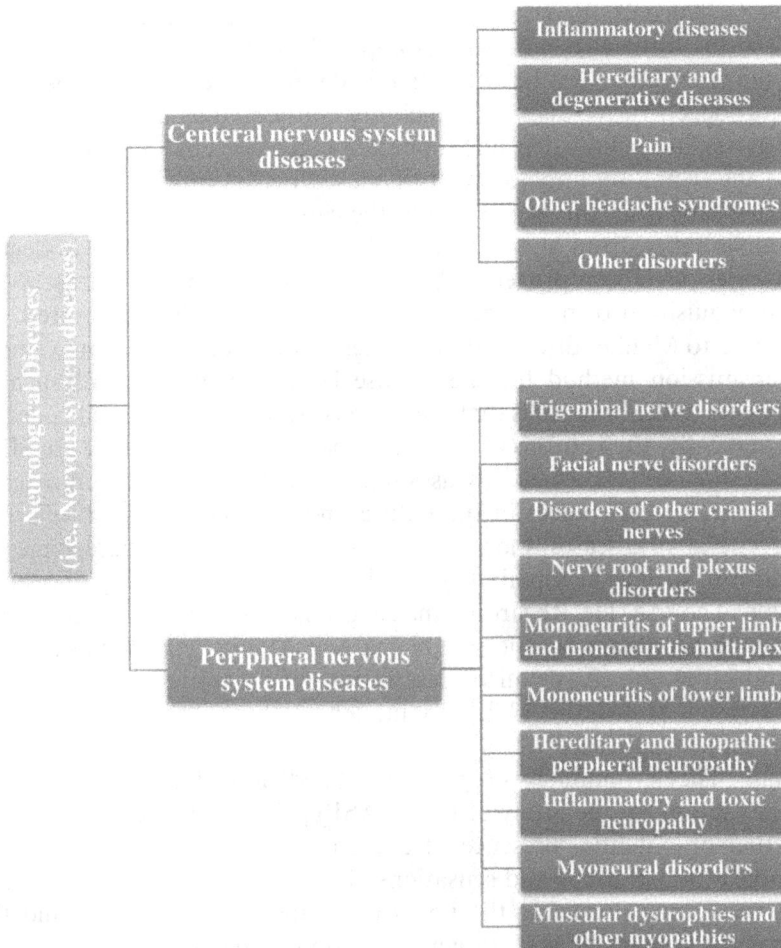

Figure 13.1. The classification of nervous system diseases (neurological diseases).

plasma and leukocytes, which access the infected/injured site in the tissue leaving their location in the blood [8]. On the other hand, in chronic inflammation the body takes a long time to respond. In this case, the cells inside the damaged site will be affected, and their type will be changed [8]. In the central nervous system, there are different types of inflammatory diseases, such as meningitis, encephalitis, myelitis, and encephalomyelitis [6].

2. *Hereditary and degenerative diseases*: Hereditary diseases, also known as genetic diseases, are the diseases that occur when inherited genes or chromosomes happen to be abnormal, whether or not such abnormality appears at birth [9]. There are three main subgroups of this category of neurological diseases. The first subgroup concerns abnormality in a single gene (e.g. sickle cell anemia). In the second subgroup, several genes are diagnosed with abnormality (e.g. hypertension). The final subgroup relates to

the existence of abnormality in the chromosomes. Such an abnormality can be in the form of the absence or copy of an entire chromosome (e.g. Down syndrome) [10]. On the other hand, degenerative diseases, also known as neurodegenerative diseases, which are considered as the main class of brain diseases, are mainly dependent on the progressive changes related to age [11]. As the definition implies, most of these diseases affect people in their later years. However, this is not always the case as there are cases where younger people suffer from such diseases. Also, some forms of neurodegenerative diseases have been proven to be inherited, such as prion disease. Moreover, the acquisition of neurodegenerative disease was also discovered, but only related to Menkes disease. In this latter context, blood was an example of the transmission method for the disease [12]. The influence of such diseases appears in the brain cells that are concerned with thinking and consciousness. Specifically, the neurons are the main part of brain cells that are affected by such diseases [12]. Many diseases fall under the neurodegenerative category. Despite the differences in the nature and the diagnosing criteria of neuro-degenerative diseases, the diseases under this category share some common features, such as histopathology, molecular mechanisms of pathogenesis, and clinical course [11]. Regarding the central nervous system, hereditary diseases include hereditary spastic paraplegia and Ataxia–telangiectasia (Louis–Bar syndrome) [6]. A detailed description of neurodegenerative diseases is presented in section 13.2.2 to introduce the subsequent explanation of Alzheimer's disease.

3. *Pain*: The terminology of pain is defined according to the International Association for the Study of Pain (IASP) [13] as the association between the actual or potential presence of a damaging tissue and the corresponding unpleasant emotions and sensations. The physiological role of pain appears in the form of warning the body about the damaged tissue and therefore protects human life [14]. In general, pain can be classified into nociceptive, neuropathic, idiopathic, or continuous pain. Nociceptive pain is the class where the sensory receptors (i.e. primary sensory neurons) are activated and cause pain. The pain under this category can be in the form of somatic pain (where the skin or musculoskeletal system is considered as the source of pain). Alternatively, it can arise through the visceral system and therefore be called visceral pain. Also, the continuous pain associated with damage in the sensory system lies under the umbrella of nociceptive pain [14]. Changes to the peripheral nervous system are considered as the originators of neuro-pathic pain (e.g. the formulation of a neuroma). It is worth noting here that this type of pain is not necessarily proportional to the extent of tissue damage and can continue even after the end of the noxious stimulus. Such a property prevents this type of pain from providing any form of protection to human health [15]. In contrast to the previously mentioned pain types, in some conditions there can be a pain without any noticeable damage to a tissue or nerve. This type of pain is known as idiopathic pain [14]. Finally, when different pains combine, the resulting pain is known as mixed pain [14].

4. *Other headache syndromes*: This category of the central nervous system involves, as the name implies, some syndromes that cause different types of headache. These syndromes include cluster headache, tension-type headache, post-traumatic headache, complicated headache syndromes, hypnic headache, primary cough headache, primary exertional headache, and primary stabbing headache [14].

5. *Other disorders*: In addition to the previously described diseases, there is another group of disorders related to the parts of the central nervous system. These disorders include multiple sclerosis, hemiplegia, and epilepsy. Multiple sclerosis, for example, is a disease where scar tissue is formed due to a breaking down of the myelin, which causes serious morbidity. The occurrence of such breaking down prevents or at least reduces the signals flowing through the central nervous system and consequently affects the vision, strength, or coordination functions [16]. Hemiplegia is a type of palsy that is considered as the most frequently occurring type and that mostly affects the left side rather than the right [17]. Finally, epilepsy is known to be one of the most common neurological disorders, where the patient suffers from two or more unprovoked repeated epileptic seizures within two years [17].

It is important to note that there are various categories and consequently subcategories of all the central nervous system disorders.

13.2.1.2 *Peripheral nervous system diseases*

Rather than affecting the central nervous system, peripheral nervous system diseases affect the nerves of the whole body outside both the brain and the spinal cord. Due to the scope of this chapter, the following paragraphs present only a brief summary of the disorders that affect the different components of the peripheral nervous system. As illustrated in figure 13.1, various disorders fall under this category. The first three disorders that are illustrated in the figure can be merged under a much broader category, called cranial nerve root disorders.

Generally speaking, the cranial nerve root is the root of one of the twelve pairs or nerves that leave either from the brainstem or the highest spinal cord level [18]. These twelve pairs of nerves are olfactory, optic, oculomotor, trochlear, trigeminal, abducens, facial vestibulocochlear, glossopharyngeal, vagus, accessory, and hypoglossal, respectively [19]. Different types of disorders can affect the nerves of the cranial nerve root. For example, the trigeminal nerve, the fifth nerve, can experience trigeminal neuralgia. The facial nerve, the seventh nerve, can experience the Bell's palsy disorder. Also, the other nerve roots can suffer from various disorders such as glossopharyngeal neuralgia (on the glossopharyngeal nerve, the ninth nerve), disorders of the pneumogastric nerve (the tenth nerve), disorders of accessory nerve (the 11th nerve) and disorders of the hypoglossal nerve (the 12th nerve).

In addition to the cranial nerve root disorders, there are some disorders that affect both the nerve root and the plexus. The nerve root is the initial nerve segment that leaves the central nervous system, and the cranial nerve root is a subtype of it. The plexus is a branching network of vessels, blood, or lymphatic vessels, or the axons of nerves outside the central nervous system [18]. The examples of disorders that can emerge in such peripheral nervous system regions include brachial plexus lesions, lumbosacral plexus lesions, neuralgic amyotrophy, and phantom limb syndrome.

In addition to these disorders, the upper and lower limbs of the human body can be affected by different types of mononeuritis. For the upper limb, such mono-neuritis include carpal tunnel syndrome, lesions of the median, ulnar, and radial nerves, upper limb causalgia, and mononeuritis multiplex. On the other hand, the lower limbs can be affected by lesions of the sciatic, femoral, lateral popliteal, medial popliteal, and plantar nerves, meralgia paresthetica, tarsal tunnel syndrome, and Morton's neuroma [20].

As in central nervous system disorders, heredity also plays a role in peripheral nervous system disorders. Examples include hereditary peripheral and sensory neuro-pathies, Refsum's disease, and peroneal muscular atrophy. On the other hand, some disorders are considered idiopathic, including notalgia paresthetica, idiopathic chronic neuropathy, Strachan's syndrome, and supranuclear paralysis. The inflammatory and toxic neuropathy category involves some disorders such as Guillain–Barré syndrome, alcoholic polyneuropathy, and inflammatory polyneuropathy.

In terms of myoneural disorders, myasthenia gravis, either with or without (acute) exacerbation, is the main disorder in this category. Other disorders include toxic myoneural disorders and other disorders with an unspecified cause, such as hypoventilation during sleep due to neuromuscular disorder. Finally, for the category of muscular dystrophies and other myopathies, heredity appears to again play a role in some disorders, such as hereditary progressive muscular dystrophy. For the other myopathies, heredity plays a role in some disorders including benign congenital myopathy, central core disease, and centronuclear myopathy [20].

13.2.2 Neurodegenerative diseases

Currently, due to the increase in mean age, studying neurodegenerative diseases as well as the early diagnosis of such diseases are become more essential. Such an early diagnosis helps in early therapeutic intervention to address the degeneration that can appear before the discovery of clinical symptoms. To achieve early diagnosis, imaging technologies provide great assistance, facilitating the acquisition of precise details and even subtle changes of the brain. Before going through the details of imaging technologies presented section 13.3, this section focuses on understanding the nature of neurodegenerative diseases as well as discriminating between related disorders. Of these disorders, the upcoming subsection will focus on Alzheimer's disease, due to the complicated structure and the rapid spreading of this disease [21]. Initially, it is important to introduce normal aging since there can be misunder-standing between its features and neurodegenerative diseases.

13.2.2.1 Normal aging

Aging is considered as a common disorder that may affect elderly persons. In other words, not all elderly persons suffer from the abnormalities that relate to their progress in age. Some older adults can have normal brains to the extent that it is impossible to differentiate between the brains of such people and normal young people. Older adults with such brains are said to have successful aging that includes having a minimal loss of brain physiology compared to young people's brains [21].

In contrast to successful aging, normal aging is associated with specific abnormalities that assist radiologists in their diagnosis of such disorders. These abnormalities arise from the fact that the brain volume completes its physiological growth in adolescence and remains reasonably stable until reaching the early forties. From this time onward and as a result of normal aging, the brain's volume begins to decrease slowly [11].

For neurodegenerative diseases, the fundamental risk factor is normal aging. Such a risk creates the challenge of differentiating between normal declines and the disease atrophy. Additionally, it highlights the difficulty of comparing the normal variation in structural integrity (i.e. the variations in iron content, brain volume, and the white matter amount) against subtle morphological changes [11, 21]. Figure 13.2 illustrates the brain areas that are affected by these changes and that can represent an indication of neurodegenerative disease.

Gradually, with the progress of human age, cerebral iron deposition starts to increase in an unknown manner. Based on this fact, any person up to the age of seventy years who is diagnosed with such symptoms must undergo extra

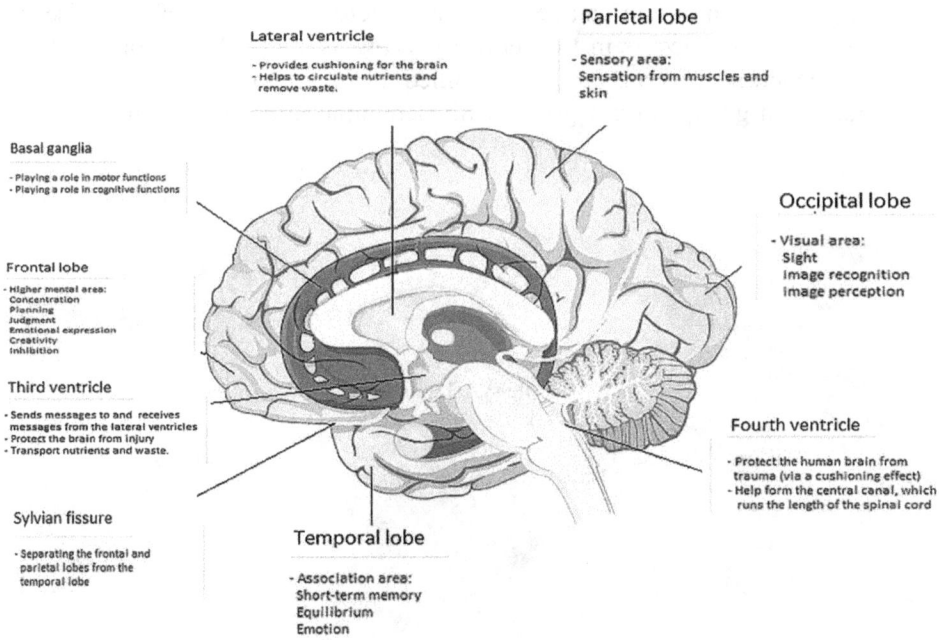

Lateral ventricle

- Provides cushioning for the brain
- Helps to circulate nutrients and remove waste.

Parietal lobe

- Sensory area:
Sensation from muscles and skin

Basal ganglia

- Playing a role in motor functions
- Playing a role in cognitive functions

Occipital lobe

- Visual area:
Sight
Image recognition
Image perception

Frontal lobe

- Higher mental area:
Concentration
Planning
Judgment
Emotional expression
Creativity
Inhibition

Third ventricle

- Sends messages to and receives messages from the lateral ventricles
- Protect the brain from injury
- Transport nutrients and waste.

Fourth ventricle

- Protect the human brain from trauma (via a cushioning effect)
- Help form the central canal, which runs the length of the spinal cord

Sylvian fissure

- Separating the frontal and parietal lobes from the temporal lobe

Temporal lobe

- Association area:
Short-term memory
Equilibrium
Emotion

Figure 13.2. A sagittal view of the normal brain sections that are influenced by aging changes.

examinations. These examinations, in particular testing basal ganglia, aim to evaluate any abnormal reduction and thus the possibility of suffering from neuro-degenerative disease. Additionally, changing brain volume is related to age. In other words, we can consider the third decade as a bridge where the brain achieves its maximum weight and starts to gradually decrease after that. Such a change in volume takes place between the ages of 65–70 and is represented through a widening in both the Sylvian fissure and the basal cisterns, followed by widening of the interhemispheric fissures and associated ventricles. In addition, the frontal lobes and then the parietal lobes are exposed to normal atrophy. The changes also affect the lateral ventricles, which are continuously enlarged and the temporal horns, which are continuously spared. The abnormality in these types of changes appears in the temporal horns, where a change in this part of the brain is an identifier of neurodegenerative disease [21].

Finally, white matter changes are one of the changes related to age and at the same time are a common feature of dementia. Such abnormality is observed with a focal or even diffusion of hyperintensity of the white matter. The diagnosis of such an abnormality is impossible since it is a typical change in both elderly people as well as dementia patients [21]. Generally speaking, different factors influence the occurrence and the evolution of these changes, such as lifestyle (e.g. exercise, smoking, etc), blood pressure, and stress. These factors lead to minimizing, delaying, or driving age related changes [22].

13.2.2.2 Neurodegenerative diseases

As previously mentioned, neurodegenerative diseases are the category of disease where the progression of human age is the main factor. The cause of such diseases is the degeneration that occurs in the central nervous system [11]. According to [23], the neurodegenerative diseases can be classified into three main groups, as shown in figure 13.3. Such groups are: a group with dementia as the common symptom, a

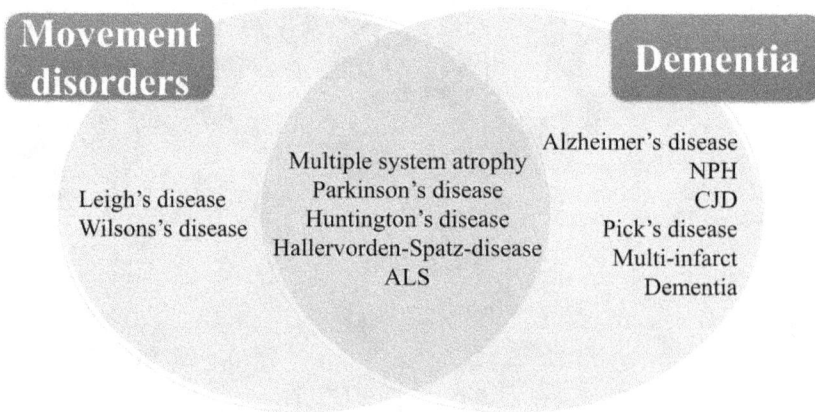

Figure 13.3. The classes of neurodegenerative diseases [23].

group with movement disorders as the common symptom, and a final group that is characterized by both dementia and movement disorders.

1. *Group A (dementia)*: Dementia mainly refers to the changes in the human brain that change the functionality of memory, personality, and behavior. Consequently, it leads to a negative impact on social interaction and living a normal life. There are 70–80 types of dementia which can result due to neurodegenerative diseases, heart attacks, frequent blows to the head, or damage resulting from the long-term abuse of alcohol [5]. Concentrating on neurodegenerative diseases, the diseases that cause dementia include Alzheimer's disease, normal pressure hydrocephalus (NPH), Creutzfeldt–Jakob disease (CJD), and Pick's disease (FTD).

   (a) *Alzheimer's disease* is one of the two most common neurodegenerative diseases, the other being Parkinson's disease. It is also considered the most common neurodegenerative disease whose sufferers are elderly people. Alzheimer's disease is found to be the cause of 70% of all cases of dementia. The possibility of suffering from such a disease reaches its maximum frequency of 42% among elderly people who exceed 85 years of age. Before this period, the probability decreases 6% among elderly people who are between 70–74 years old. Although most cases affect elderly people, younger people can also suffer from Alzheimer's disease [12].

   (b) *Normal pressure hydrocephalus (NPH)*, also known as symptomatic hydrocephalus, is a disease that features walking difficulties, urinary incontinence, and impairments related to cognition [11]. It is important to note that the actual cause of many NPH cases is still unknown and the diagnosis of such disease is considered troublesome since its main features are not unique to NPH and are common in many elderly people. NPH can lead to dementia in only 6% of patients. Since this is not a high percentage, NPH is infrequently used in the differential diagnosis of dementia [24].

   (c) *Creutzfeldt–Jakob disease (CJD)* is a rare disease that is caused when prion proteins of the brain are accumulated. The transmissible spongiform encephalopathies (TSEs) or prion diseases are the other names of the pathologic category that contains such disease [25]. To date, prion diseases have been found to be the only class of diseases that can be transmitted (e.g. through blood) between people [12]. The characteristics of CJD are dementia and difficulties in walking [5].

   (d) *Frontotemporal dementia (FTD)* is a neurodegenerative disease that is also known as Pick's disease because it was discovered by the physician Arnold Pick. FTD is caused by damage in the nerve cells that consequently causes loss of frontotemporal brain function. Such injuries variably impair behavior and personality, disturb language abilities, and change muscle or motor functions [26].

2. *Group B (movement disorders)*: As the name implies, movement disorders are the diseases or syndromes that affect the production and control of

human movement. For example, Leigh's syndrome is a disorder that influences the mitochondria, which in turn is a small organelle that resides in the cell body and produced the required energy for the cells and tissues of the body. Therefore, the impairment of this organelle affecting the brain leads to mental and developmental impairments, and affects the regulation of motor performance [5]. Wilson's disease [27] is a movement disorder, and its affects appear in various organs including the brain, eyes, liver, kidney, and skin. Many disorders are associated with Wilson's disease, which in turn helps in the diagnosis of the disease. Among these disorders are the movement disorders that include exhibiting tremors, chorea, and athetosis. It is important to know that all of these disorders and others mainly occur due to abnormalities in the copper levels in the body, that in turn are caused by problems in copper metabolism.

3. *Group C (dementia and movement disorders)*: The patients suffering from any of the disorders that belong to the third group have both dementia as well as movement disorders.

  (a) *Parkinson's disease* causes gait changes, resting tremors, late postural instability, and bradykinesia (a reduction in the speed and spontaneity of voluntary movement). The reduction of transmitted dopaminergic levels from the structural or functional nigrostriatal pathways is the responsible factor for all of these disorders [28].

  (b) *Huntington's disease*: The disorders associated with this disease are caused by CAG repeats (polyglutamine). The expansion of such repeats within the gene coding region on chromosome 4 (144) is considered the source of the abnormalities associated with Huntington's disease [24]. Dementia followed by choreoathetosis and rigidity are the main consequences of such abnormalities in Huntington's patients [21].

  (c) *Hallervorden–Spatz syndrome (HSS)* is a rare disease of children. A mutation in molecules affects the homeostasis of iron which causes subsequent disorders of rigidity, dystonia, and choreoathetosis [29].

  (d) *Others*: Multiple system atrophy (a collection of diseases that affects various body systems) and ALS disease are other examples in this category. In ALS disease, progressive weakness and wasting of the muscles are the main characteristics [29]. In multiple system atrophy, balance, the automatic functions of the body, and movement abilities are the main aspects affected in the sufferers of these diseases [23].

13.2.3 Alzheimer's disease

Alzheimer's disease (AD) is considered the most common neurodegenerative disease that affects elderly people, although small numbers of younger people are affected. Over time, various studies have been carried out with the aim of clarifying the nature, causes, and types of AD. The problem of clearly identifying the disease in its early stages still exists, which in turn prevents the early medical intervention to save the patient's life. An overview of AD is presented here to describe the nature and consequences of the disease.

The seriousness of AD is highlighted by its ultimately fatal effect on the brain and the progressive dementia leading up to this, as mentioned earlier. This dementia means that the patient will experience memory loss in addition to the loss of daily mental abilities [30].

13.2.3.1 Alzheimer's disease formation

Pathologically, two lesions are used to diagnose AD: extracellular neuritic plaques and neurofibrillary tangles [11]. The brain consists of a vast network of neurons where the connections between these neurons are made through the synapses. The task of such synapses is to transmit information between the neurons and the lesions affect this communication. Ten to fifteen years before the clinical appearance of the disease, these extracellular neuritic plaques consisting of beta-amyloid protein and neurofibrillary tangles consisting of tau protein are formed on the neurons.

Initially, the plaques are formed through the surface of the neuron and specifically speaking through a large protein called Amyloid Precursor Protein (APP) that resides there. Normally, there are two enzymes in the neuron surface called Alpha-secretase and Gamma-secretase that release soluble fragment of APP, APP alpha, away from the surface of the neuron. In AD, the process is performed differently. First, the release of APP starts most often through Beta-secretase enzyme rather than Alpha-secretase. Such release produces a short insoluble fragment of APP called beta-amyloid protein that is then regulated with each other forming toxic extracellular neuritic plaques (senile plaques) that in turn interferes the functionality of the neurons [31, 32, 254].

Abnormalities inside the neuron cause neurofibrillary tangles. In the standard procedure, a signal known as soma is sent within the neuron to the synapse in order to transfer information between the neurons. In order for the soma signal to arrive at the synapse, it has to pass through the microtubules that in turn are found in the skeleton of the neuron and are stabilized by the tau protein. In AD, the tau protein becomes defective and consequently is no longer attached to the microtubules, leading to a breaking of the microtubules. A gradual aggregation of these defective detached tau proteins occurs, producing filaments in the neuron. The abnormal accumulation of these filaments forms neurofibrillary tangles within the neuron that in turn suppress the transport system and finally causing cell death. This leads to loss of memory and as the process continues the brain will shrink in size and gradually lose its function [31, 32]. Figure 13.4 summarizes the steps of extracellular neuritic plaque and neurofibrillary tangle formation and the consequent effects on the brain.

13.2.3.2 The stages of Alzheimer's disease

As mentioned previously, the formation of the primary AD lesions, extracellular neuritic plaques and neurofibrillary tangles, begins ten to fifteen years before the clinical diagnosis of AD. Based on this, there are three classifications of AD: three, six, and seven stage classifications. Among these classifications, the three and seven stage classifications are the most common [33]. The main characteristics of each stage are

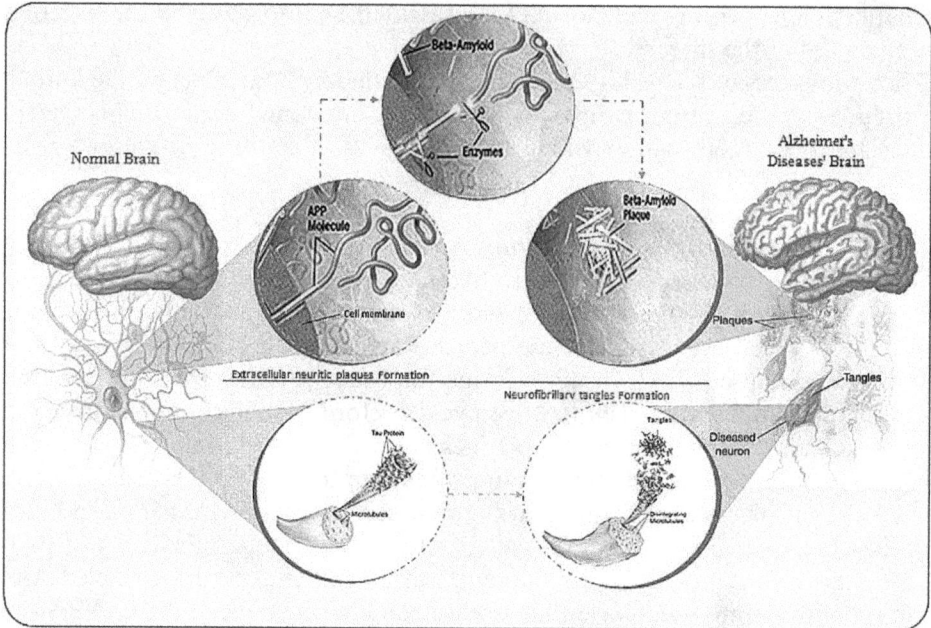

Figure 13.4. Alzheimer's disease formation.

presented later, and the description is going to use the three-stage classification strategy [30] as the base while describing the seven based classification strategy [26] within it.

13.3 Medical imaging modalities for AD

The concept of medical imaging refers to the class of images whose target is the human body. In such images, the internal details of the body that are not explicitly obvious are captured. These captured details play an essential role in further diagnosis, treatment, and monitoring of human medical conditions. Over time, numerous technologies have emerged. Despite the differences of each technology, there are common similarities that help in grouping these technologies into some main categories. In general, these similarities represent the nature of the information that can be extracted through the technologies under each category. From the extracted information, the experts can decide whether or not this imaging technology can clarify specific details about the patient's condition.

For the brain, the related imaging modalities are categorized as illustrated in figure 13.5. A detailed explanation of these categories along with their usefulness in the study brain diseases and specifically in the context of AD is the goal of the present section.

13.3.1 Radiology

Medically speaking, radiology is the branch of medicine where imaging technology represents the backbone of the diagnosis and treatment process. Within this

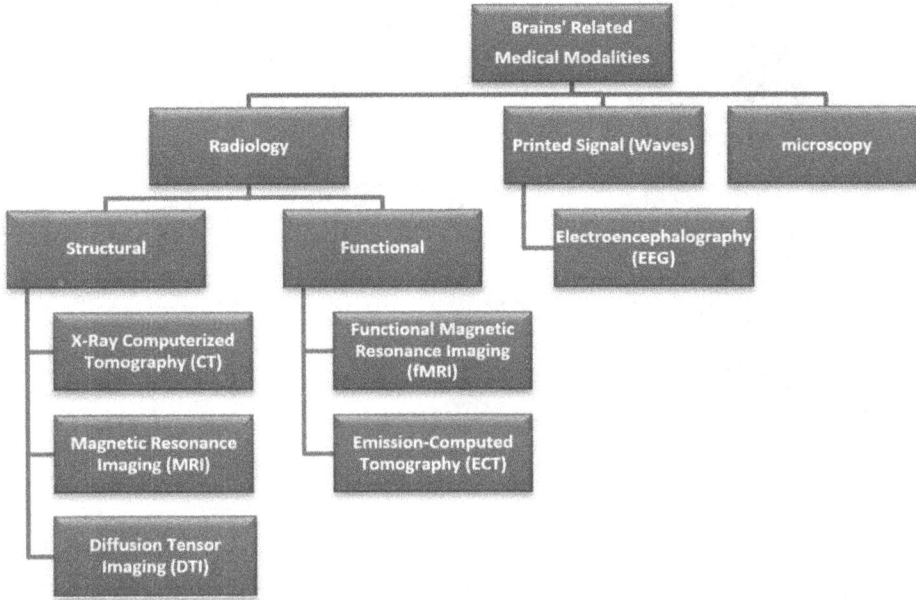

Figure 13.5. The common brain related medical imaging modalities.

category lie many modalities that can mainly be partitioned into structural and functional modalities, according to the common characteristics of the obtained results. Structural imaging modalities, as the name implies, primarily target anatomical or morphological data and construct a spatial resolution based on such data. In functional modalities, data or biomarkers represented in physiological, metabolical, or biological term are the basis for the derived spatial resolution of these modalities [34].

In addition to such partitioning of the radiology-related modalities, they can also be divided into two categories based on the usage of the obtained images: diagnostic and interventional radiology. In the diagnostic category, the images of the internal structure of the body are utilized in different diagnosis processes, such as interpreting symptoms and monitoring the illness and treatment stages. In the other category, interventional radiology, the obtained images are used in upcoming medicine procedures, such as inserting catheters [4].

It is important to know that despite the existence of different partitioning criteria, the most well known criterion is the one that relies on the characteristics of the obtained images (i.e. the division into structural and functional categories). There are various modalities under each of these categories. For example, the structural category includes the x-ray, x-ray computed tomography (CT), magnetic resonance imaging (MRI), and ultrasound (US) modalities. The functional category includes functional MRI (fMRI), emission computed tomography (ECT), and diffusion tensor imaging (DTI). Figure 13.5 illustrates the most utilized modalities in brain imaging in each of the categories, while the following sections present a brief presentation of the main characteristics of each.

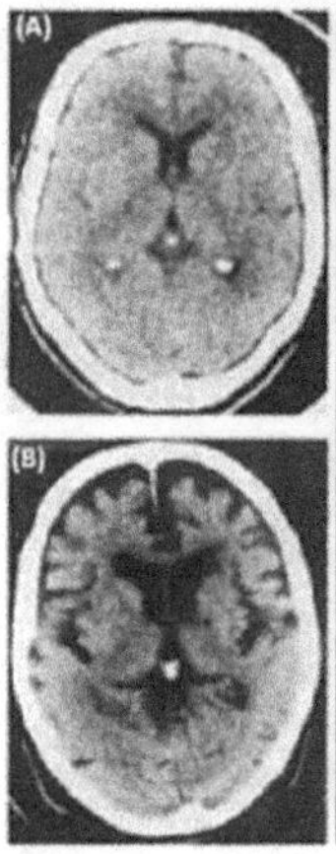

	Main characteristics	Producing cross-sections of the body through scanning using x-rays.
	Advantages	- Wide field of view. - Detecting even subtle differences between the tissues of the body. - High spatial resolution. - High penetration depth. - Clinical translation.
	Disadvantages	- Radiation. - High examination dose. - Limited sensitivity. - Poor soft tissue contrast.

Figure 13.6. A general description of a CT scan [37] (right) and an example (left) [38]: (A) a scan for a normal subject and (B) a scan for an AD patient with frontal atrophy.

13.3.1.1 Brain related structural modalities

1. *X-ray computerized tomography (CT).*

CT or computed axial tomography (CAT) is one of the well known structural modalities that aims to output cross-sectional images to reflect the anatomy of the body's internal structure. CT is a subcategory of x-ray imaging modalities. CT follows the same approach as ordinary x-rays except that the x-rays in CT are emitted in all possible directions [4]. Finally, a 3D scan that represents an accumulation of a series of obtained 2D axial slices is produced [35]. Regarding dementia, CT scans have the primary advantage of helping to differentially diagnose dementia through, for example, ruling out paramedian tumors or normal pressure hydrocephalus. Additionally, it mainly assists in excluding potential surgically treatable causes in the context of dementia (e.g. tumors, subdural hemorrhage, etc). In AD, the CT scan is not considered as a standard technique for the early diagnosis of the disease since it does not provide any role to assist in such a diagnosis. However, this does not cancel its usefulness regarding the late changes of AD since the analysis of CT scans can help in revealing the diffusion of cerebral atrophy, the enlargements associated with the cortical sulci, and the increase of the size of the ventricles [36]. Figure 13.6 shows a general description of a CT scan [37] in addition to an example that illustrates the difference between normal and AD patients' scans [38].

2. *Magnetic resonance imaging (MRI).*

MRI is one of the most recently developed medical imaging modalities. It has become a powerful assistant technology in diagnosing different body pathologies. This role is due to the obtained details that capture a tissue spatial map of the hydrogen nuclei in addition to the ability to address the viscosity, stiffness, and protein content of the tissues [39]. The characteristic of capturing proton density or magnetization properties, such as spin–spin (T2) or spin–lattice (T1) relaxation times, allows MRI to discriminate the tissues. This characteristic adds value to MRI since it allows the modality to show subtle changes of tissues even with a high degree of tissue similarity [40].

The mechanism behind MRI technology involves orientating protons within a strong magnetic field. In such a field and through the assistance of resonant radio-frequency waves, the orientated protons can be manipulated. After this, the realignment of the manipulated protons to their equilibrium state will be measured. It is important to note here that the constants of relaxation time are highly dependent on the tissue and therefore MRI offers excellent contrast for the soft tissue [41]. In addition to showing a detailed description of the human body, the resulting images provide unparalleled soft tissues images on a non invasive basis. MRI is also of great value for studying AD, and has been used as the main modality for a wide range of scientific studies. MRI is a non invasive medical imaging technique that assists in the structural analysis of AD [36].

Investigating the decline from a normal control to mild cognitive impairment (MCI) and AD was achieved using MRI. Most studies that utilize MRI technology demonstrated the common occurrence of atrophy in the medial structures of the temporal lobe (i.e. the hippocampus and entorhinal cortex) in AD patients. The analysis of MRI demonstrated the association between such atrophy and an increased risk of developing AD, which in turn helps in predicting the future decline in healthy adults' memories. In addition, investigations such as the volumetric analysis of the structural MRI technique may assist in detecting significant changes in the size of brain regions, which represents a promising indicator for the diagnosis process during the progression of AD [36]. Figure 13.7 illustrates a description as well as an example of an MRI scan [1, 37]. The example shows the volumetric difference as well as general atrophy between a normal subject and AD patient scans.

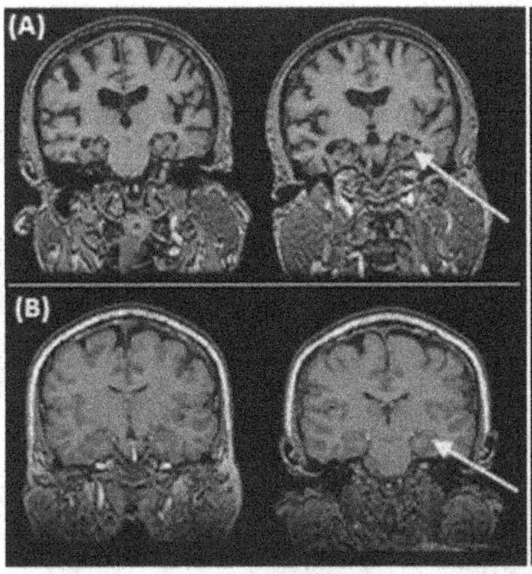

	Main characteristics	Producing "slices" of the human body through magnetic signals.
	Advantages	- Good resolution. - Shows anatomical details. - Non-ionizing radiation. - No observed short term effects. - Clinical translation.
	Disadvantages	- Strong distortion of magnetic field. - Cannot be used on patients with metallic devices like pacemakers. - Low throughput. - Cost.

Figure 13.7. An MRI scan description (right) [37] and an example (left) [1]: (A) an AD patient and (B) a normal subject. The arrows indicate the hippocampal atrophy that appears in the AD patient.

3. *Diffusion tensor imaging (DTI)*:

DTI is a type of MRI that increases our ability to understand the structure of the brain as well as the connectivity of its neurons. DTI deals with representing the microstructure of the brain and mainly targets the organization of brain regions (e.g. tracking areas of white matter). The purpose of DTI is to measure the diffusion tensor of water. To accomplish such a task, DTI measure the diffusion weights of pulse sequences, which in turn are characterized by sensitivity to the motion of the microscopic random water. As a result, diffusion-weighted images (DWIs) are produced that display water diffusion along axes or on encoding directions and the qualification of such diffusion. Such results, in turn, assist in measuring and quantifying the orientation of the tissues in addition to their structure, which consequently assists in the examination process of cerebral white matter, as well as the neural fiber tracts. For more details related to DTI and its working mechanism, see [42].

In general, DTI helps in the assessment process of differential dementia diagnosis (AD as well as vascular dementia) through the tensor maps of DTI. The measurement of fiber tract integrity through DTI helps in directly assessing the fibers of white matter and therefore could potentially be considered as an AD biomarker. In addition, reflecting any disruption in the axons through random movement of water molecules through the tissues helps in characterizing AD since such a disruption causes, in turn, a reduction in the anisotropy (this means that the water molecules' movement along the neural tract is length-wise greater than cross-wise). Finally, the regional analysis of DTI indicates that the changes of the hippocampal microstructure may represent a better predictor of the progression risk of MCI to AD [36]. In figure 13.8, a description of DTI is presented along with an example that shows the brain regions affected by AD [43, 44].

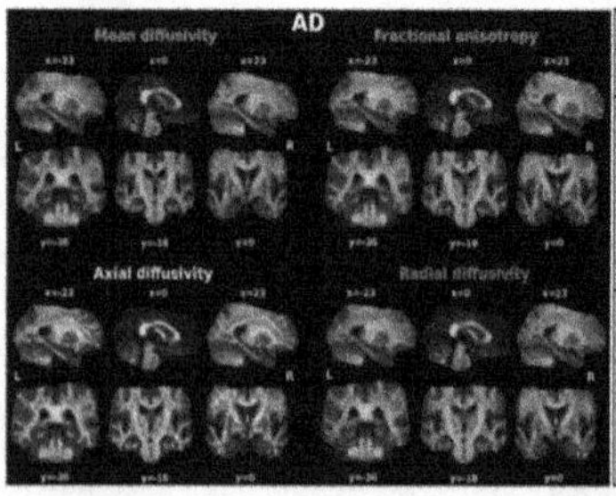

Main characteristics	Representing the microstructure of the brain and mainly targeting the organization of brain regions.	
Advantages	- Shares many advantages of MRI in addition to: 1. Axonal integrity based *in vivo* assessment. 2. Fast at higher strengths of field (~5 min). 3. No need for the same patient compliance as in fMRI.	
Disadvantages	- Shares many disadvantages of MRI in addition to: 1. A lack of knowledge about how certain factors (e.g., edema) influence the local signal. 2. A lack of knowledge about how the lesions of white matter influence the measured signals.	

Figure 13.8. A general description of DTI (right) and an example (left) that shows abnormalities of AD patients compared to normal controls. The spatial statistical tract color maps that illustrate the voxel-wise differences between the two subjects are overlaid on the skeleton of the mean fractional anisotropy (FA). Red indicates the voxels with increased mean diffusivity as well as decreased FA. Yellow indicates voxels of increased axial diffusivity while the blue color indicates voxels of increased radial diffusivity [43, 44].

13.3.1.2 Brain related functional modalities

1. *Functional magnetic resonance imaging (fMRI)*:

fMRI aims to capture the intrinsic changes of the blood signal. It helps in capturing the brain regions that are involved in certain cognitive tasks in addition to capturing the general functions of the brain evolving speech, language, and sensory motion [39]. The basic idea of fMRI relies on intensity changes depending on the oxygenation level of the blood in the brain, known as blood oxygen level dependence (BOLD). fMRI is considered the most powerful technique for studying blood flow as well as perfusion in the human brain [45]. In

dementia, fMRI has a positive role in demonstrating the functional abnormalities associated with dementia subjects, in addition it has a role in monitoring the treatment status of AD patients. Studying both the resting and activation states with fMRI indicates less coordinated activity for AD patients compared to normal subjects in the regions of the hippocampus, inferior parietal lobes, and cingulate cortex. Determining the neural basis for the cognitive behavioral function changes occurring in the early stages of neurodegenerative disorders, as well as correlating them with the neuro-anatomical network, has become possible due to the recent advances related to fMRI [36]. Figure 13.9 presents a general description of fMRI [46] along with an example [1] that contains scans of both a normal subject and an AD patient tested under an activation task of memorizing a series of presented faces. Red represents the areas that are active when performing such task. It is obvious that, in general, the activation is lower in the AD patient than in the normal control subject.

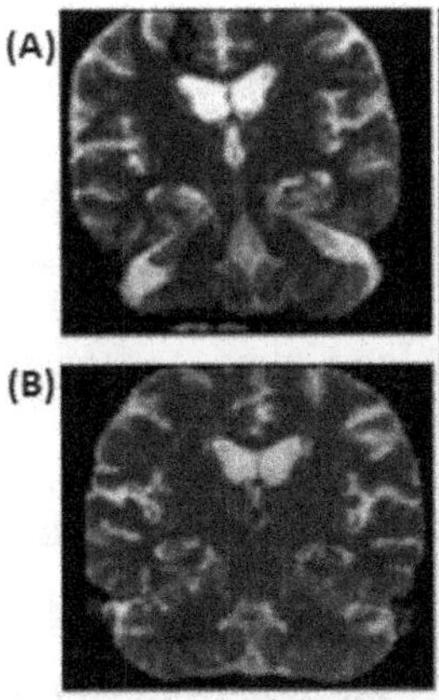

(A) (B)	Main characteristics	Capturing the brain regions that are involved in certain cognitive tasks and the general functions of the brain such as speech, language, etc.
	Advantages	- Wide availability. - Non-invasive nature. - Good spatial resolution. - Powerful for studying blood flow and human brain perfusion.
	Disadvantages	- Cost. - Cannot be used on patients with metallic devices like pacemakers.

Figure 13.9. A general description of fMRI (right) and an example (left) where the two subjects are exposed to an activation task (memorizing faces) with red representing the activation when performing the required task. The scans show that the normal subject (A) shows higher activation than the AD patient (B) [1, 46].

2. *Emission computed tomography (ECT)*:

ECT is a class of medical imaging modalities that deal with the physiology rather than anatomy as in structural modalities. ECT focuses on the distribution of isotopes inside the human body. The main types of ECT are single photon emission computed tomography (SPECT) and positron emission tomography (PET). The difference between SPECT and PET is that in SPECT radioisotopes are administered and consequently decay, emitting a single gamma photon, while in PET isotopes are administered and in each annihilation a couple of photons are produced [35].

Starting with SPECT, such an imaging modality assists in the differential diagnosis of dementia in addition to serving as an accurate diagnosis and measurement tool for the progression of brain changes. On the other hand, PET technology is considered as a powerful tool for demonstrating the changes in the brain functions, whether in healthy controls or patients, through reflecting the brain's condition at the molecular and cellular level. For AD, PET technology is very accurate in diagnosing AD and differentiating between it and the other dementia disorders. Comparing SPECT and PET images, it is useful to mention that although SPECT is cheaper it is less specific than PET, and also that combining SPECT and PET may help in both identifying as well as evaluating the early to late patient condition of the disease [36].

There are three main classes of radiopharmaceuticals that are currently used in functional imaging of the brain in the context of nuclear medicine: regional cerebral blood flow (rCBF), regional cerebral metabolism, and central nervous system receptor binding agents. Generally speaking, the use of such radiopharmaceuticals proceeds as follows: the radiopharmaceutical is injected intravenously, then SPECT or PET is used to measure the regional uptake as well as the distribution of the radiotracers [47].

rCBF relies on lipophilic radiopharmaceutical agents that are diffused from the arterial vascular compartment to the normal compartment of brain tissue. The tracers are then distributed proportionally to the blood flow of the regional tissue with irreversible trapping in the tissue compartment. Measuring regional cerebral metabolism is performed through transporting the applied radiopharmaceuticals through regional cerebral blood flow to the brain tissues. Subsequently, the regional cerebral distribution will reflect the utilization rate of the used tracer in the cerebral metabolic pathway. Finally, using central nervous system receptor binding agents involves measuring the density and binding affinity of the neuronal receptor using suitable radiotracers [47]. Table 13.1 shows the major agents that are used along with the SPECT and PET images for the above mentioned measurement tasks [48] and provides sources for further reading. Figures 13.10 and 13.11 show general descriptions of SPECT and PET [37], respectively, along with examples of AD scans from [49] and [1], respectively.

In addition, Pittsburgh compound B (PiB) is a fluorescent analog of thioflavin T used during PET scans to visualize the pathological hallmarks

Table 13.1. The major SPECT and PET radiotracers for the brain.

	Regional cerebral blood flow	Regional cerebral metabolism	CNS receptor binding agents
SPECT	• Technetium-99m-hexamethylpropylene amine oxime (Tc-99m-HMPAO) • Tc-99m-ethyl cysteinate dimer (Tc-99m-ECD)	• In the normal brain: no tracers to date for measuring normal cerebral metabolism • In brain tumors: thallium-201 Tc-99m-methoxylisobutylnitrile (Tc-99m-MI-BI)	• I-123 q-CIT
PET	• O-15 H2O	• F-18 2-fluoro-2-deoxy-d-glucose (F-18-FDG) • [F-18]-fluoro-3'-deoxy-3'-L-fluorothymidine (FLT)	• Hundreds of PET tracers [47]

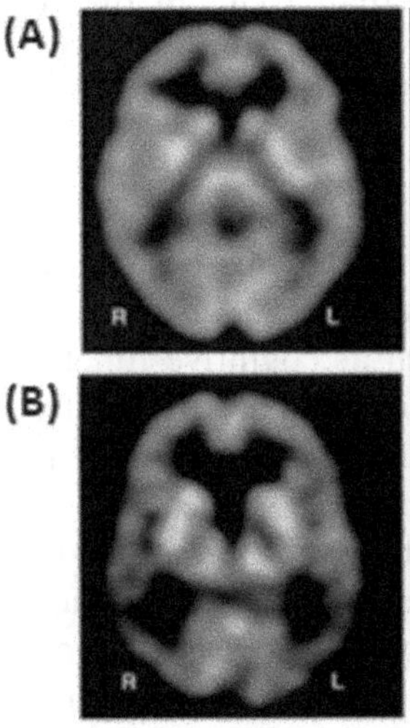

	Main characteristics	Structuring cross-sectional images of a radiotracer within the human body.
(A)	**Advantages**	- The images are free of background. - Higher sensitivity (but lower than PET scans). - Higher penetration depth. - Clinical translation.
(B)	**Disadvantages**	- Produces blurring effects. - Due to multiple scattering of the electrons, attenuation compensation is not possible. - Spatial resolution is limited. - Radiation. - Fails to predict neuropsychological deficits.

Figure 13.10. A general description of SPECT [37] (right) and an example of 99mTc-HMPAO SPECT scans (left) [49]. (A) is a normal subject's scan with in general a uniform distribution of the tracer throughout the cortex along with slight 'hyperperfusion' in the basal ganglia and (B) an AD patient with a general decrease in the perfusion throughout the cortex relative to the cerebellum.

associated with AD. Regarding dementia, applying 18FDG along with PET for studying regional cerebral metabolism through measuring glucose metabolism in different brain regions shows the value of FDG as a metabolic marker in dealing with the preclinical detection as well as the early diagnosis of dementia. Additionally, such a technique can be used in dementia identification, enabling differential diagnosis and the prediction of patients' conversion from MCI to AD. The examination of FDG-PET results shows

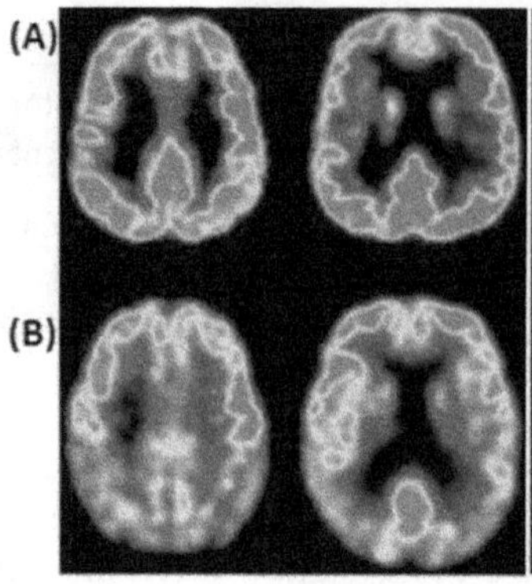

Main characteristics	Producing nuclear imaging of the scanned part of the body.
Advantages	- The ability to image biochemical and physiological phenomena. - High sensitivity. - High penetration depth. - Clinical translation.
Disadvantages	- The most expensive technique. - Limited spatial resolution. - Radiation. - Motion artifacts. - Lower resolution compared to the CT and MRI techniques.

Figure 13.11. A general description of PET [37] (right) and an example of an FDG-PET scan (left) [1]. (A) A normal control scan and (B) an AD patient scan shows prominent hypometabolism in the temporoparietal lobe.

severe reductions of glucose consumption in AD patients' brains compared to normal subjects. Comparing FDG and PIB tracers with PET scans indicates that the latter tracer is more effective than FDG in the context of the early detection of AD progression [36].

13.3.2 Printed signals/waves

Unlike the radiology based modalities that represent the internal tissues of the body in the form of images, the printed signals/waves also known as electrograms produces a set of waves that reflects the physical as well as cognitive functions of the human body. The word electrogram is based on Greek. The first part of the word, 'electro', refers to electricity while the second, 'gram', refers to the writing or recording procedure. Therefore, the whole term can represent the aim of such modalities, which is to record electrical signals regarding the human body [50]. According to the targeted part of the body, various methods have emerged including but not limited to electrocardiograms (ECGs) that target the heart, electromyography (EMG) that targets nervously stimulated muscles, and electroencephalography (EEG) that targets the brain. The following subsection presents a detailed description of the EEG method, as an example of the electrograms that targeted the brain, along with its sensitivity in diagnosing AD.

13.3.2.1 Electroencephalography (EEG)

As mentioned previously, EEG is one of the electrogram methods that targets the brain. The acquired signal records the functional status inside the brain. In other words, it detects the status of the communication between the brain cells that occurs through electrical signals. Such recordings assist in medical diagnoses since any abnormality in the electrical activity within the brain can be detected through such a signal [51]. For AD, different changes affect the EEG signals, which in turn indicate an abnormality. The most common changing feature is what is generally known as 'non-specific diffuse slowing' where normal fast frequencies of the signal (i.e. the

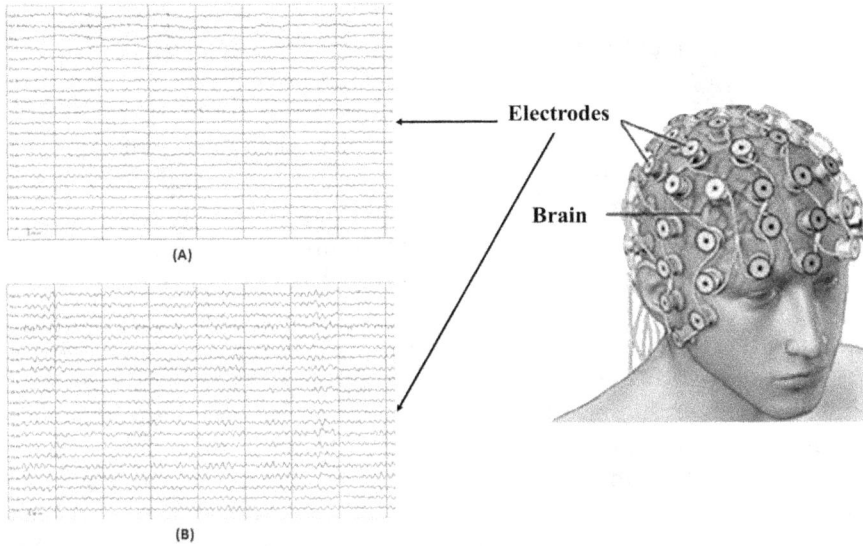

Figure 13.12. A general description of EEG: (A) a signal recorded for a normal elderly person where the slowing out of phase waves in both the F7 and F8 channels occurs because of slow eye movements that characterize the earliest drowsiness phase and (B) the EEG signal of an AD patient that shows diffuse slowing [3, 52].

beta and alpha band) are decreased whiles the normally slow frequencies (i.e. theta and delta) are increased. Figure 13.12 shows a description of EEG as well as a comparison between EEGs that were taken for a normal elderly person who felt drowsiness during the recording and another signal for an AD patient where diffuse slowing appears in the signal [3, 52].

In addition, other features change, such as the absence of synchronization between the recorded signals over different regions of the brain. The source of abnormality here emerges from the fact that the EEG channels' synchronization indicates functional interactions between the brain regions. Therefore, due to the 'disconnection syndrome' that is associated with AD, it is expected to present such an absence of synchronization in the resulting signals. Despite the existence of changing features in the AD based EEG signals, in general, what exactly causes such changes is still unknown in detail, although there are a few principles that are indicated related to EEG. An example of such principles arises from the fact that EEG signals do not directly reflect the neuronal loss or the atrophy that affect the brain, but instead reflect the functionality of the existing neurons even if there is a large number of losses among them. The EEG signal can assist in monitoring the abnormality that is associated with neuron functions.

For the early detection of AD, EEG shows in general low sensitivity as up to 50% of the early AD subjects may present normal EEG signals. On the other hand, EEG signals can assist in differentiating the diagnosis between AD and other sources of cognitive dysfunction (e.g. depression which can be associated with cognitive complaints) despite its absolute inability to differentiate between AD and vascular dementia (VaD) since both disorders may occur together [3].

13.3.2.2 Microscopy

Microscopy is another class of medical modalities where a different procedure is followed for image production process. A microscope is utilized for capturing fine details through enlarging the scanned targets [53]. In addition to utilizing the microscope's characteristic of enlarging small objects, the specimens are also exposed to different types of microscopes' waves that in turn produce different types of microscopic images.

In general, there are two types of microscopes: light (optical) microscopes and electron microscopes. In the light microscopes, optical lenses and light waves are used to produce the resulting images. Depending on how the light system is used, various microscopes have been developed including bright-field, dark-field, fluorescent, and phase contrast microscopes. In electron microscopes, electron beams are utilized rather than light waves. There are in turn different types of electron microscopes including transmission and scanning electron microscopes [54]. The obtained images from electron-based microscopes are more magnified and higher resolution than those of light microscopes [53]. Figure 13.13 shows a comparison between light and electron microscopes in addition to illustrating an example of laser scanning confocal microscopy (a type of light microscope) in relation to AD [55, 56].

13.4 AD literature review

Due to the importance of the early diagnosis of AD disease, various studies have been undertaken. The ultimate goal of these studies is to simplify the experts' task and to assist in the early diagnosis of such a widespread neurodegenerative disease. This section aims to present current studies that apply different methods and techniques to deal with this topic.

For a meaningful characterization of the presented research, the section is divided into two main parts. The first part presents the efforts toward developing methods that provide the main steps of a computer aided diagnosis (CAD) system (i.e. preprocessing, feature representation, and classification steps) as well as the efforts to develop complete CAD systems. The second part presents research with different goals regarding the field of AD (e.g. data analysis, developing a database management system, image retrieval, etc). Finally, it is important to note that the purpose of this division is to collect studies with the same goals and consequently illustrate the achievements of AD related study.

13.4.1 CAD system based studies

Currently, CAD systems represent a powerful tool for diagnosing different diseases. CAD systems are mainly divided into three steps: the preprocessing, feature representation, and classification steps. First, the system prepares the images for further processing steps by applying different image processing and enhancement procedures. Then, the system focuses on obtaining the main features of these images

	Normal	Early	Moderate	Mild

	Light microscope	Electron microscope
Main characteristics	Utilizing a microscope, optical lenses, and lights waves to enlarge small objects and producing resultant images.	Utilizing a microscope and electron beams to enlarge small objects and producing resultant images.
Advantages	- Cheap. - Small and portable. - Unaffected by magnetic fields. - Set-up is quick, simple, and only needs little expertise. - Rare distortion of materials during preparation. - Observing a material's natural color.	- Magnification of scanned objects over 500 000x. - Possibility of investigating greater field depth.
Disadvantages	- Magnification of scanned objects only up to 2000x. - Restricted field depth.	- Expensive. - Very large and therefore requires special operating rooms. - Affected by magnetic fields. - Setting up the materials requires expertise and is time-consuming. - Possibility of material distortion during preparation. - Produces black and white images.

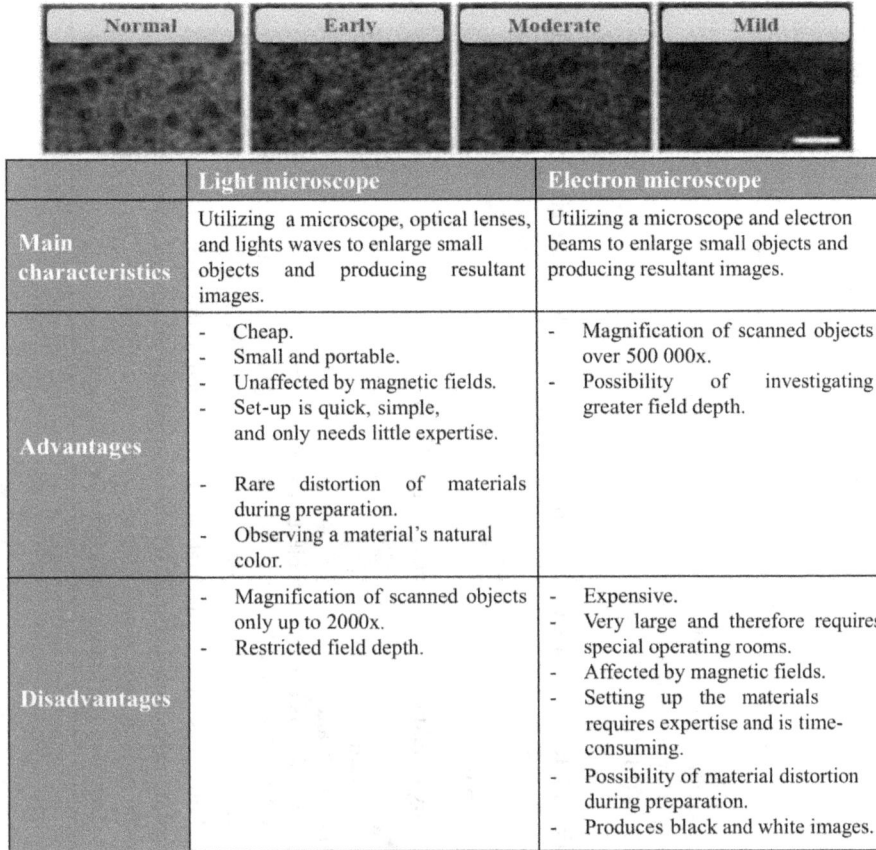

Figure 13.13. A comparison between light and electron microscopy (bottom) and examples of laser scanning confocal microscopy (an example of light microscopes) (top) [55, 56].

through feature extraction, selection, etc. Finally, different classification methods and techniques are used to help in grouping the image features and assist in further prediction processes.

Relying on these steps, various scientific papers were published that either present a complete CAD system or an achievement in one of the CAD systems' steps. Thus, the current section aims to present different achievements regarding these goals, first complete CAD systems and then different achievements regarding the steps of CAD.

13.4.1.1 Complete CAD systems

As previously stated, CAD systems have become a powerful tool for various medical based decision making tasks. Therefore, it is expected that there are numerous papers that address such a topic, and this is the case also regarding such systems for AD. Various CAD systems have been proposed with the aim of improving the distinguishing process between AD and the other subjects. Table 13.2 summarizes various scientific papers that deal with such an objective.

Table 13.2. AD related CAD systems.

Study	Su *et al* [57]
Contribution	Evaluate using statistically based procedures the feature extraction step in a CAD system to help in distinguishing between NC, MCI and AD.

Procedures and methods	1. Preprocessing:
	   • Apply the statistical parametric mapping (SPM) tool to align, smooth and normalize the input images.
	2. Feature extraction:
	   • Region of interest (ROI) selection: statistical analysis (contribution).
	   • Feature extraction: principal component analysis (PCA).
	   • Linear discriminant analysis (LDA).
	3. Feature selection:
	   • Fisher discriminant ratio (FDR): an adaptive threshold for choosing dominant eigenvalues of LDA and PCA.
	4. Classification:
	   • Support vector machine (SVM).

Subjects	119 normal controls (NC), 133 MCI (mild cognitive impairment), 123 AD
Modality	^{18}F-FDG-PET
Availability of data	Alzheimer's Disease Neuroimaging Initiative (ADNI)
Studied medical aspect	Glucose metabolism (function) of the brain
Findings	1. The differentiation between AD subjects and other subjects achieves high accuracy. 2. The differentiation between AD and MCI, and MCI and NC, is improved to some extent, except that the latter case does not achieve high accuracy due to the high similarity between MCI and NC brain metabolisms.
Medical benefit	Discriminating between AD and other cases due to the high accuracy related to this context.
Suggestions and future work	Study the suitability of other feature extraction methods to help in improving the accuracy of differentiating between MCI and NC since the current study found such an achievement difficult due to its limitations in finding distinguishing features between the brain metabolisms of such cases.

(Continued)

Table 13.2. (*Continued*)

Study	Varghese *et al* [58]
Contribution	Evaluate the ability of back propagation neural network BP-NN in identifying the structural changes of the AD subjects related to the gray matter (GM), white matter (WM) and cerebrospinal fluid (CSF).

Procedures and methods	1. Preprocessing:
	    • Skull stripping.
	2. Feature extraction:
	    • Gabor filter.
	3. Classification:
	    • Feed forward BP-NN.

Subjects	20 NC, 25 MCI, 25 AD
Modality	MRI
Availability of data	Not available
Studied medical aspect	Structure of the brain
Findings	The results illustrate that GM tends to lose its volume in an accelerating manner in MCI subjects which in turn assists in differentiation between the three groups of subjects.
Medical benefit	Assists in diagnosing and discriminating between NC, MCI, and AD subjects.
Suggestions and future work	1. Apply the proposed method to other datasets to evaluate its accuracy. 2. Evaluate the method using other modalities to test the possibility for extracting other discriminating features between the three groups.
Study	Martínez *et al* [59]
Contribution	Utilizing two feature selection criteria, two classifiers, factor analysis, and a multivariate normal classifier, to present a CAD system to assist in diagnosing AD.

Procedures and methods	1. Preprocessing and feature extraction:
	    • Dimensionality reduction (reducing the number of voxels) through utilizing the Mann–Whitney–Wilcoxon test and relative entropy for feature ranking.
	    • Feature dimension reduction using factor analysis.
	2. Classification combining the results of the two multivariate quadratic classifiers where:
	    • In classifier A the voxels' ranking (6000 voxels) was performed using Mann–Whitney–Wilcoxon criteria, and factor analysis for extracting eight factor loadings.
	    • In classifier D the voxels' ranking (16 000 voxels) was performed using relative entropy criteria, and factor analysis for extracting six factor loadings.

Subjects	41 NC, 56 AD
Modality	SPECT
Availability of data	Not available
Studied medical aspect	Function of the brain
Findings	The comparison of the proposed method with other presented methods shows better findings in terms of sensibility, specificity, and accuracy.
Medical benefit	Helps in studying the functionality of the brain and consequently assist in diagnosing AD.
Suggestions and future work	1. Evaluate the proposed method with other modalities and other datasets. 2. Utilize other feature reduction methods and compare the results. 3. Combine the obtained features and evaluate the results.

Study	*Jie et al* [60]
Contribution	Propose a new joint based multi-task feature selection method that addresses multi-modality data to improve the classification performance of AD against other disorder stages.

Procedures and methods	1. Preprocessing:
	   • Spatial distortion.
	   • Skull stripping.
	   • Cerebellum removal.
	   • For MR: segment images into GM, WM, and CSF, register them according to atlas images.
	   • For FDG-PET: use a rigid transformation to align them to the MRIs.
	 2. Feature selection:
	   • Supervised and semi-supervised manifold regularized multi-task learning framework.
	   • Group sparsity regularizer.
	   • Laplacian regularization term.
	 3. Classification:
	   SVM.

Subjects	52 NC, 99 MCI (43 MCI converters and 56 MCI non-converters), 51 AD
Modality	MRI-PET
Availability of data	ADNI

(Continued)

Table 13.2. (*Continued*)

Studied medical aspect	Structure and function of the brain.
Findings	The results show the efficiency of the proposed method through utilizing geometric distribution knowledge of the input data.
Medical benefit	Assists in diagnosing AD.
Suggestions and future work	1. Evaluate the method against other datasets to generalize the results. 2. Evaluate fusing the two modalities and evaluate the method against the obtained new image.
Study	Simões *et al* [61]
Contribution	Proposing a 3D texture analysis on local MR images based on local binary patterns (LBP) and combine them in an ensemble classifier without the need for segmenting or nonlinearly aligning the data.
Procedures and methods	1. Preprocessing: • Perform affine registration of the images to a common template. • Produce several 3D cubic patches through subdividing the resulting images, where each patch represents a brain region. 2. Feature extraction: • Local binary patterns–three orthogonal projections (LBP-TOP) and building a feature vector of LBP histogram (LBPH) descriptors for each patch separately. 3. Classification: • Classifier ensemble through utilizing linear SVM and applying weighted majority voting with a fixed parameter.
Subjects	Data divided into two subgroups based on the clinical dementia rate (CDR): 66 NC and 70 AD (CDR = 0.5 and 1), age 60–80 years; 66 NC and 20 AD (CDR = 1), age 60–80 years
Modality	MRI
Availability of data	Open Access Series of Imaging Studies (OASIS)
Studied medical aspect	Structure of the brain

Findings	1. The largest patches produce more informative results when only the mild AD subjects were considered.
	2. When including the very mild group of subjects, small patches with high overlapping can achieve the highest results of classification.
	3. The patches 'hippocampi' and 'ventricles' are characterized by high discrimination which in turn supports their hypothesis of focusing on local patches due to their texture information which is valuable for the discrimination goal.
	4. Locations such as cortical folds, where the anatomical variability is very high even though this may not be associated with early AD, are considered undistinguishing regions.
	5. The ensemble classifier achieves higher accuracy results than the best individual patch.
Medical benefit	Assisting in finding the discriminative regions that help in diagnosing AD at the early development stages.
Suggestions and future work	1. Add the moderate and severe AD data to evaluate the behavior of the proposed method with them and to evaluate the possibility of finding discrimination regions for further assistance in the AD diagnosing processes.
	2. Investigate other strategies for selecting necessary brain patches and/or for combining the results of individual patch classifications.
	3. Investigate other configurations of patches than the simple cube.
	4. Evaluate the method against other databases and other preprocessing steps.
Study	Khedher *et al* [62]
Contribution	Proposing a CAD system through utilizing an SVM classifier on average brain image projection in independent component analysis (ICA).
Procedures and methods	1. Preprocessing:
	• Spatial normalization.
	• Segmentation.
	• Calculating the average images of the NC, MCI, or AD subjects.
	2. Feature extraction:
	• FastICA.
	• Subspace projection.
	3. Classification:
	• SVM.
Subjects	229 NC, 401 MCI (312 stable MCI and 86 progressive MCI), 188 AD

(Continued)

Table 13.2. (*Continued*)

Modality	MRI
Availability of data	ADNI
Studied medical aspect	Structure of the brain
Findings	1. High accuracy results for the ability and the robustness of AD detection.
	2. The results show better findings than, in particular, the voxel-as-features (VAF) approach.
Medical benefit	Assist in the early diagnosis of AD.
Suggestions and future work	Evaluate the method on other modalities.
Study	Ben Ahmed *et al* [63]
Contribution	Exploiting the visual features of the hippocampal area and late fusion for improving the recognition accuracy of AD.
Procedures and methods	1. Preprocessing:
	• Normalization (affine registration).
	2. Feature (hippocampus) extraction:
	• ROI selection: selecting only the voxels that are labeled on the atlas as hippocampus.
	• Feature extraction (signature generation): circular harmonic functions (CHF).
	• CSF volume computation: selecting dark regions using a threshold. Such computation is important since in subjects with AD the hippocampus shrinks and is filled with CSF.
	3. Classification through late fusion using SVM on:
	• SVM (with visual signature).
	• Bayesian classifier (with CSF volume).
Subjects	72 NC, 111 MCI, 35 AD
Modality	MRI
Availability of data	ADNI
Studied medical aspect	Structure of the brain.
Findings	1. The combined features (visual signature and the CSF volume) provide better results, in particular differentiating between AD and MCI, than using each source of features separately.
	2. The proposed method outperforms the accuracy of the volumetric methods.

Medical benefit	Assisting in the early AD diagnosis process.
Suggestions and future work	1. Evaluate extracting multiple ROIs. 2. Investigate the usage of multiple MRI modalities or generally investigate the method against other modalities.
Study	Dyrba et al [64]
Contribution	Evaluating the combination of DTI and MRI with the goal of improving the diagnosis procedure of AD.
Procedures and methods	1. Preprocessing: • Correction of eddy currents and motion of the head. • Skull stripping. • Data fitting to the diffusion tensor model. • Deformation analysis. • Spatial normalization. • Image segmentation (into GM and WM). • Image smoothing. • Image masking. 2. Feature selection: • Plant's approach. • Entropy-based information gain (IG) criterion. 3. Classification: • Multivariate SVM. Note: Combining data: • Scan merging at voxels level. Combining classifiers: • Two layer meta classifier.
Subjects	137 clinically probable AD patients, 143 NC
Modality	DTI-MRI
Availability of data	Not available
Studied medical aspect	Neuronal fiber tract and structure of the brain.
Findings	The combined MTI and MRI modalities of the mild to moderate patients were not found to increase the diagnosis rate of AD.

(*Continued*)

Table 13.2. (*Continued*)

Medical benefit	Avoid relying on working combining results derived from DTI and MRI in AD diagnosing processes.
Suggestions and future work	1. Evaluate the method on the early stages of AD to test the performance of combining the modalities for such patients.
	2. Investigate the accuracy of combining other modalities in the diagnosis process.
Study	Fan [65]
Contribution	Utilizing an ordinal ranking classification method in MRI, FDG-PET, and CSF biomarker data for differentiation between NC, MCI (convertor MCI-C and non-convertor MCI-NC), and AD.
Procedures and methods	1. Preprocessing:
	• Geometric distortion correction of MRI.
	• Co-register the PET and MRI images.
	• MRI segmentation.
	• MRI and PET deformation.
	2. Feature extraction:
	Extract the ROI, calculate the GM volume and the PET value of each ROI. Note: the normalized GM volumes as well as PET value measures of cortical and subcortical ROIs are fed to the ordinal classification as features.
	3. Feature selection:
	Ranking technique.
	4. Classification:
	Ensemble classifier with SVM as the base classifier and voting for output combination.
Subjects	55 NC, 57 MCI-NC, 44 MCI-C, 51 AD
Modality	CSF biomarker, FDG-PET, MRI
Availability of data	ADNI
Studied medical aspect	Structure, function, amyloid-beta and tau of the brain.
Findings	The classification performance can achieve high quality results through utilizing the inherent ordinal severity of the brain damage associated with the progress of AD.

Medical benefit	Assisting in diagnosing AD.
Suggestions and future work	1. Evaluate the correlation between different modalities. 2. Investigate the usage of other feature extraction methods. 3. Evaluate the proposed method against regression problems related to AD.
Study	Varghese *et al* [66]
Contribution	Constructing a CAD system through combining an artificial neural network (ANN) along with a bacterial foraging optimization (BFO) algorithm on MR images to discriminate between AD patients as well as normal controls.

Procedures and methods	1. Preprocessing:
	   • Skull stripping.
	2. Feature extraction:
	   • Gabor filter.
	3. Classification:
	   • ANN tuned through BFO.

Subjects	30 probable MCI, 20 NC
Modality	MRI
Availability of data	Not available
Studied medical aspect	Structure of the brain
Findings	1. The proposed method was able to distinguish between normal and abnormal subjects. 2. The values related to the prediction of MCI within and after 12 months shows significant change.
Medical benefit	The accelerated loss of the GM volume as well as other brain regions helps in distinguishing between normal and AD subjects.
Suggestions and future work	1. Apply the method to a standard database to evaluate the results. 2. Examine other medical modalities. 3. Utilize other types of classifiers as SVM and evaluate the results.
Study	Liu *et al* [67]
Contribution	Constructing hierarchal ensemble based classification algorithms to simplify and increase the accuracy of the classification through building a model that extracts and gradually assembles rich imaging features as well as decisions. Considering the spatial contiguity related to the local brain regions through applying hierarchical pyramid with a spatial structure from local patches of the brain to the large regions of the brain.

(Continued)

Table 13.2. (*Continued*)

Procedures and methods	1. Patch extraction:
	  • t-test.
	2. Classification:
	  a. Two low level classifiers for each patch where:
	    • Classifier 1: uses low level features (GM densities).
	    • Classifier 2: uses correlation-context features (Pearson correlation coefficient).
	  b. High level classification using: high level features obtained from the above steps.
	  c. Final ensemble classifier through applying weighted voting.

Subjects	198 AD, 229 NC
Modality	MRI
Availability of data	ADNI
Studied medical aspect	Structure of the brain.
Findings	1. The findings show better results when applying multiple classifiers than a single one. 2. The performance of the classification increases when applying a hierarchical ensemble of the multi-level classifiers.
Medical benefit	Assists in the diagnosing procedure.
Suggestions and future work	Evaluate the proposed method against other modalities.
Study	Hu *et al* [68]
Contribution	Adopting the feature selection method to deal with cortical thickness and use the results to train the adaptive boosting (AdaBoost) framework for final objective classification of AD.

Procedures and methods	1. Preprocessing:
	  • Freesurfer image analysis suite.
	2. Feature selection:
	  • Ensemble feature selection method.
	3. Classification:
	  • AdaBoost framework.

Subjects	16 AD, 16 NC
Modality	MRI
Availability of data	ADNI
Studied medical aspect	Structure of the brain.
Findings	Obtaining better results when combining features rather than selecting one feature subset.
Medical benefit	Assists in the diagnosis of AD.
Suggestions and future work	1. Apply the proposed feature selection method in other applications.
	2. Implement other classification methods to evaluate the classification results.
	3. Test the proposed work with other medical modalities.
	4. Enlarge the used data to evaluate and generalize the results.
Study	Yang *et al* [69]
Contribution	Proposing a CAD system that utilizes PCA for the selection of significant volumetric as well as shape features and uses the results to build a BP-ANN to discriminate between AD and normal subjects.
Procedures and methods	1. Spatial normalization:
	a. Image registration on a standard spatial coordinate system (the Talairach and Tournoux coordinate system).
	b. Spatial normalization using:
	• Optimum 12-parameter affine transformation.
	• A Bayesian framework.
	2. Feature extraction:
	a. Volumetric features:
	• Extract GM, WM, and CSF probability maps using clustering segmentation algorithm.
	• Extract binary ventricle volume data from MRI using region growing algorithm along with a threshold.
	b. Shape features:
	• Extract 2D and 3D features based on probability map.
	3. Feature selection:
	• PCA.
	4. Classification:
	• BP-ANN.

(Continued)

Table 13.2. (*Continued*)

Subjects	24 probable AD, 28 NC
Modality	MRI
Availability of data	ADNI
Studied medical aspect	Structure of the brain.
Findings	1. It is possible to extract volumetric and shape features due to their computational capability, which is low, in addition to the ability to discriminate them.
	2. Applying PCA to reduce the feature space dimensionality assists in achieving better classification accuracies than directly introducing volumetric and shape features to the ANN.
Medical benefit	Assists in diagnosing AD.
Suggestions and future work	Enlarge the tested dataset. Evaluate utilizing other classification methods.
Study Contribution	Patil and Yardi [70]
	Utilize a medical image processing, analysis, and visualization (MIPAV) tool to extract the statistical features of the hippocampus for further classification using BP-ANN for a purpose of discriminating between normal, MCI, and AD subjects.
Procedures and methods	1. Image preprocessing:
	• Statistical parametric mapping software (SPM5).
	• VBM tool.
	2. Feature extraction:
	• Statistical features using the MIPAV tool.
	3. Classification:
	• BP-ANN.
Subjects	30 AD, 30 NC
Modality	MRI
Availability of data	ADNI
Studied medical aspect	Structure of the brain.
Findings	1. Implementing a feed forward NN using the Trainlm training function shows better performance than other classifiers.
	2. Feed forward NN shows promising results that in turn reflect the role of utilizing ANN in medical image classification applications.

Medical benefit	Assists in diagnosing AD.
Suggestions and future work	Enlarge the tested dataset.
Study	Anand et al [71]
Contribution	Utilizing a fuzzy neural network (FNN) algorithm for the purpose of evaluating the anatomical changes that affects the brain and may related to the early analysis of AD.
Procedures and methods	1. Image preprocessing:
	a. Realignment and spatial normalization.
	b. Intensities normalization using:
	• Linear transformation.
	c. Brightness and contrast normalization using:
	• Mean and standard deviation adjustment.
	2. Feature extraction:
	• Statistical features using the MIPAV tool.
	3. Classification:
	• FNN classification.
Subjects	6 NC, 8 patients (MCI and AD patients)
Modality	MRI
Availability of data	Not available
Studied medical aspect	Structure of the brain.
Findings	1. The head of the hippocampus, hippocampus, parahippocampus, lateral ventricle, and frontal lobe structures are significant features that can help in discriminating between AD and normal controls.
	2. Measuring medial temporal lobe regions assists in differentiating between subjects with mild to moderate AD and MCI and in turn separating those subjects from normal ones.
	3. The methodology does not provide a tool of automatic diagnosis of diseases but rather assists in the diagnosing process.
Medical benefit	Assists in diagnosing AD.
Suggestions and future work	Evaluate the methodology against other medical modalities.

(Continued)

Table 13.2. (*Continued*)

Study	*Álvarez et al* [72]
Contribution	Utilizing PCA and SVM for solving the two main problems of using statistical classification along with the SPECT data:
	• The high dimensionality associated with the feature spaces.
	• The small size of the samples.
Procedures and methods	1. Preprocessing:
	a. Reconstructing the projected data using:
	• Filtered back-projection (FBP) algorithm.
	• Butterworth noise removal filter.
	b. Spatial normalization using:
	• SPM software.
	c. Image labeling:
	• Experienced physicians.
	2. Feature extraction:
	• PCA.
	3. Classification:
	• SVM.
Subjects	41 NC, 38 AD
Modality	SPECT
Availability of data	Not available
Studied medical aspect	Function of the brain.
Findings	1. The proposed method reaches 100% performance accuracy.
	2. PCA shows efficient results when produced with SVM.
Medical benefit	Assists in diagnosing AD.
Suggestions and future work	1. Evaluate the proposed method in other modalities.
	2. Enlarge the testing dataset and test the results.
Study	*Álvarez et al* [73]
Contribution	Eliminating the visual interpretation subjectivity that emerges in SPECT scans through implementing SVM over the independent component analysis (ICA) of such scans.

Procedures and methods	1. Preprocessing:
	a. Reconstructing the projected data using:
	• Filtered back-projection (FBP) algorithm.
	• Butterworth noise removal filter.
	b. Spatial normalization using:
	• SPM software.
	c. Image labeling:
	• Experienced physicians.
	2. Feature extraction:
	• PCA.
	3. Classification:
	• SVM.
Subjects	41 NC, 20 possible AD, 14 Proper AD, 4 certain AD
Modality	SPECT
Availability of data	Not available
Studied medical aspect	Function of the brain.
Findings	1. ICA is an effective tool for extracting salient features from SPECT images.
	2. Combining ICA and radial basis function (RBF) SVM shows high performance results.
Medical benefit	Assists in diagnosing the AD.
Suggestions and future work	1. Enlarge the tested dataset.
	2. Evaluate the performance of the proposed method against other modalities.
Study	Salas-Gonzalez et al [74]
Contribution	Utilizing Welch' t-test as a basis for selecting the important features that in turn assist in early diagnosis of AD.

(Continued)

Table 13.2. (*Continued*)

Procedures and methods	1. Preprocessing:
	  a. Reconstructing the projected data using:
	    • Filtered back-projection (FBP) algorithm.
	    • Butterworth noise removal filter.
	  b. Normalization and registration using:
	    • Affine normalization model.
	    • Non-rigid spatial transformation.
	    • Intensity normalization using maximum intensity.
	  c. Feature selection:
	    • Average of the brain image (mean images).
	    • Root-mean-square deviation of the brain images (standard deviation images).
	    • Welch's *t*-test.
	2. Classification:
	    • SVM.
	    • Classification trees.

Subjects	41 NC, 38 AD
Modality	SPECT
Availability of data	Not available
Studied medical aspect	Function of the brain.
Findings	SVM in general presents a higher correct rate than classification trees.
Medical benefit	Assists in diagnosing AD.
Suggestions and future work	1. Enlarging the tested dataset. 2. Evaluate the performance of the proposed method against other modalities. 3. Input data for MCI and evaluate the results.
Study	López *et al* [75]
Contribution	Facing the problem of the small sampling size through applying PCA with the aim of reducing the feature space.

Procedures and methods

1. Image preprocessing:
   a. Spatial registration using:
      - The algorithm mentioned in the paper.
   b. Selecting the voxels of interest using:
      - Mean intensity value.
2. Feature selection:
   - PCA.
3. Classification:
   - Bayesian classifier.

Subjects

PET: 18 NC, 42 AD
SPECT: 41 NC, 38 AD

Modality   PET, SPECT

Availability of data   Not available

Studied medical aspects   Function of the brain.

Findings   The obtained results outperform the reference work of this paper.

Medical benefit   Assists in diagnosing AD.

Suggestions and future work

1. Attempt to implement the proposed method with other modalities.
2. Test other classification methods.
3. Enlarge the tested dataset.

Study   Stoeckel et al [76]

Contribution   Study the influence related to the using of the SPECT perfusion images in the process of AD diagnosis through presenting a classifier that relies directly on the information embedded within the SPECT scans.

(Continued)

Table 13.2. (*Continued*)

Procedures and methods	1. Image preprocessing:
	  a. Spatial normalization using:
	    • Affine transformations where the similarity measure is the correlation ratio.
	  b. Intensity normalization:
	    • Dividing by the summed intensity in the cerebellum.
	2. Features: voxel intensities.
	3. Classification:
	    • Pseudo-Fisher linear discriminant classifier (PFLDC).
	    • Nearest mean classifier (NMC).

Subjects	Simulated data: 13 normal volumes, 60 AD volumes Real data: 31 NC, 99 AD
Modality	SPECT
Availability of data	Simulation: SimSET [77] Real data: The paper includes the reference for describing this data
Studied medical aspect	Function of the brain.
Findings	1. Using voxel values as features for classification is equal to or outperforms the results obtained by the other automatic methods. 2. Therefore, relying on the implicit knowledge of the data without any explicit knowledge is found to be possible.
Medical benefit	The implicit knowledge in the SPECT scan helps directly in diagnosing AD.
Suggestions and future work	1. Test other modalities to study their implicit knowledge. 2. Enlarge the tested dataset. 3. Input data for MCI and evaluate the results. 4. Evaluate feature extraction as well as other classifiers that suit the small sampling size.
Study	Stoeckel *et al* [78]
Contribution	Studying the interpretation of the SPECT scans for the subsequent diagnosis of AD.

Procedures and methods

1. Image preprocessing:
   a. Spatial normalization using:
      • Affine transformations with Powell's optimization method to maximize the correlation coefficients.
   b. Intensity normalization:
      • Dividing by the mean of the top one percent intensities.
      • Dividing by the sum of the brains' intensities.
   c. Subsampling using:
      • Mean filter.
2. Features: intensities of the images.
3. Classification:
   • PFLDC.
   • NMC.

Subjects 50 NC, 29 probable AD

Modality SPECT

Availability of data Dataset available through The Society of Nuclear Medicine Brain Imaging Council. The link is available in the paper.

Studied medical aspect Function of the brain.

Findings Promising results of classifying based on the implicit knowledge of SPECT scans.

Medical benefit The implicit knowledge in the SPECT scan helps directly in diagnosing AD.

Suggestions and future work
1. Evaluate feature extraction as well as other classifiers that suits the small sampling size.
2. Test other modalities to study their implicit knowledge.
3. Enlarge the tested dataset.
4. Input data for MCI and evaluate the results.

Study Fung and Stoeckel [79]

Contribution Study the influence of SPECT perfusion imaging in the AD diagnosis.

(Continued)

Table 13.2. (*Continued*)

Procedures and methods	1. Image preprocessing:
	  a. Spatial normalization using:
	    • Using affine transformations where the similarity measure is the correlation ratio.
	  b. Intensity normalization:
	    • Using affine transformations.
	2. Features: voxel intensities.
	3. Classification:
	    • Contiguous SVMC.

Subjects	31 NC, 99 AD
Modality	SPECT
Availability of data	The paper includes the reference for accessing this data.
Studied medical aspect	Function of the brain.
Findings	1. Contiguous SVM provides a performance that at least achieves that of human observers. 2. The classification without related pathological knowledge is possible.
Medical benefit	Selecting useful features that help in diagnosing AD.
Suggestions and future work	1. Enlarge the tested dataset. 2. Study how to obtain patient-specific information. 3. Evaluate the use of different methods of registration and normalization.
Study	Polikar *et al* [80]
Contribution	Analyzing event related potential (ERP) segments of EEG signals using DWT and utilizing the analysis results through ANN to classify subjects with AD or not.

Procedures and methods	1. Image preprocessing:
	    • Preparing the signals and using DWT.
	2. Classification:
	    • Multilayer perceptron.

Subjects	14 NC, 14 early AD
Modality	EEG signal
Availability of data	Not available

Studied medical aspect	Brain activity.
Findings	1. The multiresolution analysis is capable of analyzing EEG signals or in general biological signals that are nonstationary. 2. Training the ANN with adequate data shows assistance in diagnosing AD. 3. DWT can systematically and compactly represent ERPs and present them to the ANN for further classification tasks.
Medical benefit	Assists in diagnosing AD.
Suggestions and future work	1. Enlarge the tested dataset. 2. Try other classification methods and evaluate the results.
Study	Plocharski and Østergaard [81]
Contribution	Performing a morphological analysis of the cortical sulci, through computing the features of the sulcal medial surface, to serve classification purposes between NC and AD subjects.

Procedures and methods	1. Segmentation of sulci:
	    • BrainVISA 4.4.0 Morphologist 2013 pipeline.
	2. Feature extraction:
	    • Depth (mean and standard deviation).
	    • Sulcal length.
	    • Mean and Gaussian curvature.
	    • Surface area.
	3. Feature selection:
	    • Feature scaling.
	    • Forward feature selection.
	4. Classification:
	    • SVM.

Subjects	96 NC, 109 AD
Modality	MRI
Availability of data	ADNI
Studied medical aspects	Cortical sulci in the left hemisphere.

(Continued)

Table 13.2. (*Continued*)

Findings	The performance of the utilized features shows their applicability in addressing the differentiation task between the NC and the AD groups.
Medical benefit	1. The features obtained from the sulcal medial surface are found to be cortical neuroanatomical biomarkers of AD abnormalities.
	2. The utilized features were extracted from the left hemisphere due to the literature reports of it being severely influenced by AD as well as losing gray matter faster than the right hemisphere.
Suggestions and future work	Combine the extracted sulcal morphology with other promising measures that are found to be useful in addressing the same research problem.
Study	Beheshti *et al* [82]
Contribution	Constructing a computer aided diagnosis system to assist in the detection of AD relying on feature ranking and classification errors.
Procedures and methods	1. Preprocessing and GM analysis:
	• Using the VBM and DARTEL approach.
	2. Feature extraction:
	• Use the voxel intensity values of the VOIs as the raw features.
	3. Feature ranking using seven feature ranking methods: statistical dependency (SD), mutual information (MI), information gain (IG), Pearson's correlation coefficient (PCC), *t*-test score (TS), Fisher's criterion (FC), and the Gini index (GI). Then, calculating the top features using the error of the classification (resubstitution as well as cross-validation errors) of the NC and AD groups.
	4. Classification (using SVM) and data fusion between different methods of feature ranking.
Subjects	130 NC, 130 AD
Modality	MRI
Availability of data	ADNI
Studied medical aspect	GM
Findings	The presented method of feature selection showed its reliability for dealing with data of high dimensionality.
Medical benefit	Constructing the proposed CAD system with only MRI showed a comparable performance to the related literature.

Suggestions and future work	Apply the proposed feature selection methodology on other AD related modalities and examine the obtained performance.
Study	Garali et al [83]
Contribution	Develop an effectiveness based ranking method of the brain related ROIs for classification purposes of NC and AD groups.
Procedures and methods	1. Preprocessing: • Spatial normalization. • Image registration. • Smoothing. • Intensity normalization. 2. Brain image segmentation into 116 regions based on an anatomical atlas. 3. Feature selection and region ranking using a 'separation power factor' (SPF) in order to distinguish between the NC and the AD groups. 4. Classification using SVM.
Subjects	61 NC, 81 AD
Modality	^{18}FDG-PET scans
Availability of data	Not available
Studied medical aspects	Ranked regions among 116 anatomical brain regions.
Findings	1. Using the ranked regions showed either a similar or a slightly better performance compared to either using all 116 segmented regions or the whole voxels of the gray matter. 2. The computational time of the proposed work was found to be reduced compared to other related work.
Medical benefit	The performance using the ranked regions showed their discrimination power between the tested groups.
Suggestions and future work	1. Evaluate the performance of other features, such as the gradient, texture, etc. 2. Evaluate the proposed work on other AD related modalities.
Study	Hosseini-Asl et al [84]
Contribution	Utilizing a deep 3D convolutional neural network (3D-CNN) to assist in the prediction of AD based on the variations of the anatomical brain shape obtained from the MRI scans.

(Continued)

Table 13.2. (*Continued*)

Procedures and methods	1. Feature extraction using a three-dimensional convolutional autoencoder.
	2. Classification using a three-dimensional adaptive convolutional neural network. Note: The constructed system was applied in the source domain and target domain. For the source domain, the following additional steps were performed:
	   a. Preprocessing (i.e. spatial normalization) using a rigid registration approach.
	   b. Skull removal.
	   c. Intensity normalization.

Subjects	Source domain: 30 subjects Target domain: 70 NC, 70 MCI, 70 AD
Modality	MRI
Availability of data	CADDementia ADNI
Studied medical aspect	Cortical thickness and volume, brain size, ventricle size, and hippocampus model.
Findings	The proposed network showed a better performance compared to related work without performing a prior skull stripping.
Medical benefit	Shows the role of the utilized features in assisting the diagnosis process of AD.
Suggestions and future work	Implement the proposed analysis on other AD related modalities.
Study	Beheshti *et al* [85]
Contribution	Predicting the conversion from the MCI stage to AD, through presenting a CAD system that employs a feature ranking as well as a genetic algorithm, at between 1 to 3 years before being diagnosed clinically.

Procedures and methods	1. VOIs segmentation:
	   a. Utilizing a voxel based morphometry technique to determine the local as well as the global atrophies of the GM in the AD group compared to the NC group.
	   b. The corresponding regions of the GM volume reduction are segmented to represent the VOIs.
	2. Feature extraction:
	   Extract the voxel values of the atrophy based regions, in a feature vector, from all of the studied groups (i.e. NC, sMCI, pMCI, and AD) through utilizing the obtained VOIs.

3. Feature selection:
   a. Apply the *t*-test on the features then rank them according to their test score.
   b. Design a genetic algorithm to obtain the optimal subset of features where the Fisher criterion was included in the objective function of the genetic algorithm.

4. Classification using SVM.

Subjects	162 NC, 65 sMCI, 71 pMCI, 160 AD
Modality	MRI
Availability of data	ADNI
Studied medical aspects	GM
Findings	The classification performance reflects the ability of the proposed system to differentiate between the sMCI and the pMCI groups.
Medical benefit	The ability to implement the proposed system clinically due to its capability to perform the discrimination between the sMCI and pMCI groups.
Suggestions and future work	Apply the proposed framework on other related modalities of AD to assist in the analysis of those modalities in the context of the diagnosis of AD.
Study	Tong *et al* [86]
Contribution	Present a framework for multi-modality classification utilizing complementary information obtained from multimodal data to serve the diagnosis of AD.

Procedures and methods	1. Preprocessing and feature extraction:
	   a. Volumes in MRI.
	   b. Voxel intensities in FDG-PET scans.
	   c. A*β*42; T-Tau; P-tau in CSF.
	   d. The APOE allele in genetic data.
	2. Graph construction:
	   Use the random forest to calculate the similarity between two subjects, for each modality, to end up with a similarity matrix which in turn is used to construct the graph that represents the relation between the subjects.
	3. Feature selection.
	4. Nonlinear graph fusion and the calculation of the unified similarity matrix.
	5. Classification:
	   Random forest algorithm.

(Continued)

Table 13.2. (*Continued*)

Subjects	35 NC, 75 MCI, 37 AD
Modality	MRI, FDG-PET, CSF, genetic information
Availability of data	ADNI
Studied medical aspects	A total of 83 brain regions obtained from multi-atlas propagation with enhanced registration (MAPER), $A\beta42$, T-tau, and P-tau, as well as the APOE genotype information.
Findings	Using the unified graphs of the studied modalities showed better results than those using only single-modality biomarkers.
Medical benefit	Reflecting the role of the commentary information obtained from the studied biomarkers in addressing the differentiation between the AD groups.
Suggestions and future work	1. Incorporate imputation approaches for the purpose of enlarging the sample size through filling the missed data of the subjects that have been excluded from the study.
	2. Calculate the inter-modality based amount of complementary information thought estimating the inter-modality correlation.
	3. Due to the independence that can exist between the modality based features, the influence of incorporating the feature correlations on the classification task can be investigated in the future work.
	4. Investigate the role of different feature selection approaches on the classification performance.
	5. Investigate the integration of the nonlinear fusion technique to model the progress of the disease through using longitudinal data.
Study	Ben Ahmed *et al* [87]
Contribution	Assessing, in the diagnosis process of AD, a multimodal base of extracting local biomarkers from DTI and MRI images.
Procedures and methods	1. Preprocessing: a. Co-registration. b. Spatial normalization. 2. ROI selection: AAL atlas for hippocampus selection.

3. Feature extraction (signature generation):
   A multiresolution approach based on the Gauss–Laguerre circular harmonic function (GL-CHF) descriptors.
4. Feature fusion and classification:
   Multiple kernel learning where the combined features are:
   a. Mean diffusivity (MD) visual signature.
   b. MRI visual signature.
   c. The amount of CSF in the area of the hippocampus.

Subjects	52 NC, 85 MCI, and 45 AD
Modality	DTI, MRI
Availability of data	ADNI
Studied medical aspect	Hippocampus.
Findings	The proposed system shows better performance results compared to related work that uses different modalities but the same studied subjects.
Medical benefit	The utilized DTI could shows better results than related work that used PET scans, which in turn can prevent injecting the subjects with radioactive contrast agents, highly expensive scans, longer scanning time, and avoid the necessity of searching for specialized centers to perform PET scans. Generalizing the proposed system to other brain diseases such as brain tumors after adapting it to the studied problem.
Suggestions and future work	Evaluate the proposed system to predict the conversion and non-conversion of MCI to AD.
Study	El-Gamal *et al* [88]
Contribution	Utilizes PiB-PET scans to serve the personalized early diagnosis of AD through presenting a CAD system that visualizes the disease's effect in each of 116 brain regions separately.

Procedures and methods	1. Preprocessing:
	   a. Data standardization:
	      i. Re-orientation.
	      ii. Co-registration.
	      iii. Spatial normalization and reslicing.
	   b. Wavelet denoising.

(Continued)

Table 13.2. (*Continued*)

	2. Automatic brain labeling using an AAL atlas.
	3. Feature extraction:
	Blob detection using a Laplacian of Gaussian (LoG) with scale invariant.
	4. Classification: Probabilistic SVM (pSVM) for regional/local diagnosis followed by standard SVM for global/subject based diagnosis.
Subjects	19 NC, 65 MCI
Modality	^{11}C PiB-PET
Availability of data	ADNI
Studied medical aspect	116 anatomical brain regions (AAL atlas)
Findings	The proposed system showed superior results compared to related work that in turn reflects the power of the utilized features, as well as the classifier, in differentiating between the tested groups.
Medical benefit	Visualizing the effect of the disease in each of the brain regions separately and in turn determines the influence on the corresponding brain function that consequently can assist in the personalized treatment of the disease.
Suggestions and future work	1. Perform a detailed study of the significant brain regions affected by the disease at its early stage. 2. Perform the same detailed degree of analysis in other related brain modalities.
Study	Jha et al [89]
Contribution	Differentiation between NC and AD groups through proposing a dual-tree complex wavelet transform (DTCWT) for feature extraction, PCA for feature reduction, and a feed forward neural network (FNN) for classification.

Procedures and methods	1. Preprocessing and normalization.
	2. Feature extraction:
	   DTCWT.
	3. Feature selection:
	   PCA.
	4. Classification:
	   FNN.

Subjects	98 NC, 28 AD
Modality	MRI
Availability of data	OASIS
Studied medical aspects	Whole brain.
Findings	Better results compared to related work.

Medical benefit	Assists in the diagnosis of AD.
Suggestions and future work	1. Investigate other wavelet based variants (e.g. wavelet packet analysis), other feature selection methods (e.g. ICA), and other classifiers (e.g. SVM). 2. Test the proposed system on other modalities.
Study	Cevik et al [90]
Contribution	Utilize the multivariate adaptive regression splines (MARS) to present a fully automated voxel based procedure for assisting in the early diagnosis of AD.

Procedures and methods	1. Voxel-wise probability maps of the tissue:
	  a. Unified segmentation.
	  b. Nonlinear image registration.
	2. A three-step approach to determine the subset of the significant features:
	  a. Statistical analysis.
	  b. Tissue probability criteria.
	  c. Within-class norm thresholding.
	3. Classification with MARS.

Subjects	508 volume of MRI
Modality	MRI
Availability of data	ADNI
Studied medical aspects	WM, GM, CSF
Findings	1. Better sensitivity, NPV, and acceptable specificity results compared to other work in the context of early detection of AD and MCI. 2. Regarding the classification problem of converted and non-converted MCI, the system showed meaningful results compared to related work, with mostly better sensitivity compared to other work. 3. Regarding the classification problems of NC versus AD, and converted versus non-converted MCI groups, the proposed work showed lower average specificity results compared to other work. On the other hand, the amount of sensitivity gain was found to be greater than this loss of specificity in both of the studied problems.
Medical benefit	Assist in the diagnosis of AD.

(Continued)

Table 13.2. (*Continued*)

Suggestions and future work	Investigate other methods, such as CMARS, with mathematically more integration of the algorithmic part of MARS, with fewer heuristic elements and supported by the modern theory of optimization.
Study	Lahmiri and Shmuel [91]
Contribution	Investigate the role of certain of features, including fractals, in the classification of AD from NC groups using different classifiers.
Procedures and methods	Different features are extracted from the studied images, then four commonly known classifiers are used to investigate the most informative combination of features to serve the early diagnosis of AD.
Subjects	35 NC, 35 AD
Modality	MRI (MPRAGE images)
Availability of data	ADNI
Studied medical aspect	Cerebral cortex, cortical thickness, gyrification index, and Alzheimer's disease assessment scale (ADAS) test.
Findings	Applying SVM on the cortical thickness, gyrification index, and ADAS cognitive test scores shows better results other feature combinations as well as other classifiers.
Medical benefit	The neuroanatomy features as well as the cognitive assessment results are recommended to perform the classification task between the NC and the AD subjects.
Suggestions and future work	Test the system in a larger dataset.
Study	Kamathe and Joshi [92]
Contribution	Using clinical as well as other image based features from both the whole and segmented images of the brain to improve the final classification performance.
Procedures and methods	1. Preprocessing: a. Improve image appearance (using Gradwarp algorithm). b. Correct image color as well as intensity information (using B1 non-uniformity). c. Correct intensity distortion (using N3 bias field correction). d. GM, WM, and CSF segmentation (using SPM). 2. Feature extraction: a. Gray level co-occurrence matrix (GLCM). b. Scale invariant feature transform (SIFT). c. Local binary pattern (LBP). d. Histogram of oriented gradient (HOG). e. Bag of words (BOWs).

3. Classification:
   a. Support vector machine (SVM).
   b. K nearest neighbors (KNNs).
   c. Decision tree.
   d. Ensemble.

Subjects 90 NC, 105 MCI, 92 AD

Modality MRI as well as clinical data represented by the functional activities questionnaire (FAQ), neuropsychiatric inventory (NPI), and geriatric depression scale (GDS).

Availability of data ADNI

Studied medical aspect All brain regions in addition to the features obtained from the GM, WM, and CSF.

Findings
1. The evaluation of the visual features shows that the GLCM features produced superior results to the other texture based features.
2. Combining the imaging based biomarkers along with the clinical features helped in obtaining better results than not using them together.

Medical benefit Assists in identifying the stage of the subject in the AD cascade.

Suggestions and future work Construct a 3D feature model through the implementation of the proposed method on other MRI brain scan sections.

Study Luk *et al* [93]

Contribution Proposing a texture based analysis to predict MCI patients who are and who are not converted to the AD stage through studying the cerebral degeneration *in vivo* using the texture analysis of MRI scans.

Procedures and methods
1. Preprocessing: hippocampal as well as inferior lateral ventricle extraction using FreeSurfer.
2. Three-dimensional texture analysis: Using VGLCM-TOP-3D.
3. Statistical analysis of the subjects' texture maps using F-test. One-way analysis of variance, independent sample *t*-test, and chi-squared test were used for the determination of statistically based differences between various clinical based variables of the tested groups (i.e. NC, MCI, and AD) as well as converted and non-converted MCI groups.

Subjects 225 NC, 382 MCI, 183 AD. A subset of MCI patients was used to compare between the converted and non-converted MCI patients, 98 converted and 106 non-converted.

(Continued)

Table 13.2. (*Continued*)

Modality	MRI and clinical data
Availability of data	ADNI
Studied medical aspect	Hippocampal as well as inferior lateral ventricle.
Findings	A significant difference was obtained through the texture features between the NC, MCI, and AD subjects.
Medical benefit	The analysis indicates an early texture changes in MCI patients who proceed to AD.
Suggestions and future work	Test the proposed model on an independent dataset.
Study	El-Gamal *et al* [94]
Contribution	The personalized diagnosis of AD through utilizing the features detected from ^{11}C-PiB-PET scans as well as using statistical analysis provide precise diagnosis results of the MCI stage.

Procedures and methods	1. Preprocessing:
	  a. Data standardization:
	    i. Re-orientation.
	    ii. Co-registration.
	    iii. Spatial normalization and reslicing.
	  b. Wavelet denoising.
	2. Automatic brain labeling using AAL atlas.
	3. Feature extraction:
	  Blob detection using Laplacian of Gaussian (LoG) with scale invariant.
	4. Statistical analysis to determine the significant regions between the tested groups (NC versus MCI):
	  Using two-sample *t*-test.
	5. Classification:
	  Probabilistic SVM (pSVM) for regional/local diagnosis followed by standard SVM for global/subject based diagnosis.

Subjects	19 NC, 65 MCI
Modality	^{11}C PiB-PET
Availability of data	ADNI
Studied medical aspect	Cerebelum_3_L (left alar central lobule), Cerebelum_8_R (right biventer lobule), Cingulum_Post_L (left posterior cingulate gyrus), Olfactory_L and Olfactory_R (bilateral olfactory cortex), plus Vermis_1_2, Vermis_3, Vermis_8, and Vermis_9 (lobules I, II, III, VIII, and IX of the vermis).

Findings	1. The linear based SVM was found to exceed the performance of the other tested kernel based SVMs (radial basis function (RBF) and polynomial SVM). 2. Using the significant regions presented the best performance results compared to using all the brain regions, while excluding the significant regions represented the worst performance results.
Medical benefit	The high results through utilizing the significant brain regions reflects the influence of these regions on the disease, how they can assist in the diagnosis of the disease during its progression, and how they can assist in uncovering the ambiguity that surrounds the disease.
Suggestions and future work	1. Include all AD related stages in the study. 2. Apply the proposed analysis on other AD related modalities.
Study	El-Gamal et al [95]
Contribution	Supporting the personalized diagnosis of AD through utilizing the features detected from ^{11}C-PiB-PET scans as well as using statistical analysis in order to provide precise diagnosis results between the AD related groups (NC, MCI, and AD).

Procedures and methods	1. Preprocessing:
	  a. Data standardization:
	    i. Re-orientation.
	    ii. Co-registration.
	    iii. Spatial normalization and reslicing.
	  b. Wavelet denoising.
	2. Automatic brain labeling using the Brodmann area (BAs) atlas.
	3. Feature extraction:
	  Blob detection using the Laplacian of Gaussian (LoG) with scale invariant.
	4. Statistical analysis to determine the significant regions between the tested groups (NC versus the abnormal groups (MCI versus AD)):
	  a. Use the two-sample t-test in two statistical layers:
	    i. Between NC and MCI groups.
	    ii. Between NC and AD groups.
	  b. Treat all the obtained regions as significant regions between the NC and the remaining groups (MCI + AD) as the abnormal group.
	5. Classification:
	  PSVM for regional/local diagnosis followed by standard SVM for global/subject based diagnosis.

(Continued)

Table 13.2. (*Continued*)

Subjects	19 NC, 65 MCI, 19 AD
Modality	^{11}C PiB-PET
Availability of data	ADNI
Studied medical aspects	
Findings	Shown in figure 3 in the paper. 1. The linear based SVM was found to exceed the performance of the other tested kernel based SVMs (radial basis function (RBF) and polynomial SVM). 2. Achieving the same high performance results through using all the brain regions or the significant regions in the diagnosis process.
Medical benefit	The high performance through utilizing significant brain regions reflects the influence of these regions on the disease, how they can assist in the diagnosis of the disease during its progression, and how they can assist in uncovering the ambiguity that surrounds the disease.
Suggestions and future work	Apply the proposed analysis on other AD related modalities.
Study	El-Gamal *et al* [96]
Contribution	Utilizing the shape (Gaussian and mean curvature, sharpness, and curvedness) as well as the volume analysis of sMRI scans to serve the personalized early diagnosis of AD through visualizing the disease's effect is each of the anatomical brain regions.

Procedures and methods	1. Preprocessing and cortex extraction:
	  a. Skull stripping.
	  b. Data standardization.
	  c. Cortex extraction.
	2. Cortex reconstruction and regional analysis:
	  a. Marching cube method (cortex reconstruction).
	  b. Feature extraction (shape and volume features).
	  c. Anatomical brain labeling (76 region using AAL atlas).
	3. Feature fusion:
	  Using canonical correlation analysis (CCA).
	4. Diagnosis:
	  PSVM for regional/local diagnosis then standard SVM for global/subject based diagnosis.

Subjects	60 NC, 86 MCI
Modality	MRI
Availability of data	ADNI
Studied medical aspect	Cortical regions
Findings	1. Linear based SVM showed better results compared to other tested kernel based SVMs (RBF and polynomial kernels).
	2. Validating the obtained results using related work showed the promise of the results in solving the addressed problem of the early diagnosis of AD.
Medical benefit	Visualizing the disease's effects in each of the cortical regions, which can consequently assist in the diagnosis, treatment, and better understanding of the disease and its nature/behavior.
Suggestions and future work	Evaluate the proposed system in all the related stages of AD.

Study	El-Gamal *et al* [97]
Contribution	Utilizing the shape features (perimeter and bounding box) in the sMRI scans in addition to statistical analysis to present a precise personalized CAD system for the early diagnosis of AD.
Procedures and methods	1. Preprocessing:
	a. Skull stripping.
	b. Data standardization:
	i. Re-orientation.
	ii. Realignment.
	iii. Normalization, and reslicing.
	2. Anatomical brain labeling:
	Using the AAL atlas.
	3. Feature extraction and feature fusion:
	a. Extracting bounding box and perimeter feature.
	b. Using CCA technique to fuse them.
	4. Statistical analysis:
	Using two-sample *t*-test.
	5. Diagnosis:
	PSVM for regional/local diagnosis then standard SVM for global/subject based diagnosis.

(Continued)

Table 13.2. (*Continued*)

Subjects	60 NC, 87 MCI
Modality	MRI
Availability of data	ADNI
Studied medical aspect	116 anatomical region using the AAL atlas
Findings	Using the significant regions only, showed better results than using all the brain regions, which consequently both encourages utilizing those regions only in the diagnosis process, as well as reflecting the influence of the disease in these results due to the high performance obtained through utilizing them in the diagnosis.
Medical benefit	The regions obtained through the statistical analysis were: Frontal_Mid_Orb_R, Frontal_Sup_Orb_L, Rectus_L, Olfactory_R, Olfactory_L, Temporal_Mid_R, Heschl_L, Heschl_R, Temporal_Pole_Sup_L, Cingulum_Mid_R, Cingulum_Post_R, ParaHippocampal_R, Cerebelum_Crus1_R, Cerebelum_9_R, Cerebelum_6_L, and Cerebelum_8_L.
Suggestions and future work	1. Apply the analysis idea on other AD related modalities to assist in the analysis of AD from different perspectives. 2. Include all AD related groups in the evaluation step.
Study	Ju *et al* [98]
Contribution	Support the early diagnosis of AD through utilizing deep learning with brain networks as well as clinically relevant text information.

Procedures and methods	1. Preprocessing:
	  a. Motion correction.
	  b. Slice timing correction.
	  c. Skull stripping.
	  d. Normalization.
	  e. Smoothing.
	2. Identification of ROIs:
	  Using the AAL atlas.
	3. RS-fMRI classification:
	  Autoencoder based model.
	4. Brain network construction:
	  a. Quantile-quantile (QQ) plot and Shapiro–Wilk test (checks the Gaussianess of the time R-fMRI scans).
	  b. Pearson's correlation coefficient (measures the strength of functional connectivity).

Subjects	79 NC, 91 MCI
Modality	Resting-state fMRI (RS-fMRI) and clinically relevant text information on the subject that includes age, gender, as well as the ApoE gene.
Availability of data	ADNI
Studied medical aspect	Functional connectivity of different brain regions.
Findings	Combining the brain network and deep learning helped to improve the classification accuracy by about 20% compared to traditional methods.
Medical benefit	Assists in the early diagnosis of AD.
Suggestions and future work	1. Test the proposed work on larger dataset. 2. Apply the proposed method to diagnose other neurological diseases.

13.4.1.2 Studies related to CAD system stages

In addition to proposing complete CAD systems, various studies have addressed specific stages of CAD systems. Table 13.3 shows different approaches that accomplish specific tasks related to the stages of CAD systems.

13.4.1.3 Other AD related studies

In addition to CAD systems, different studies have emerged that serve the AD related research field. These studies include data analysis, developing database management systems, image retrieval, etc. Such studies help the experts in the field of AD in one way or another, as presented in table 13.4. The table presents different records related to different studies that benefit from AD data along with the medical benefits that can be obtained from such studies.

13.5 Discussion

Despite the availability of numerous scientific studies that aim to clarify the ambiguity that surrounds AD, there are limitations that challenge the scientific community working in this research area. Such limitations are related to different factors such as the available databases that are used to test the proposed methods, the available technologies that produce the medical images/biomarkers, the subjects who are the target of the research, and also the applied methods and techniques. This section aims to highlight some of the major obstacles that still stand in the way of further achievements in AD related research.

13.5.1 Databases

It is obvious from table 13.2 that different databases along with different objectives have emerged to serve the scientific community, related to the brain as well as the AD research field. Despite this, the existing repositories still face certain limitations that in turn affect the scientific development in this area. Such limitations, for example, include the absence of useful modalities which are not yet available within these repositories, such as EEG, SPECT, and microscopy images. Such an absence prevents performing further studies on these modalities alongside those already known to be useful, such as MRI for example, in particular with the emergence of powerful data mining and other techniques that can assist in such tasks.

In addition to the obstacle of the absence of various modalities, most databases offer MRI for use in studies. Of course, such availability of MRI is essential due to the significant role of this method in the context of AD as mentioned above. However, as implied in the review tables above, this trend directs researcher to work only in one direction (i.e. to deal with MRI only) and to avoid dealing with other modalities unless they have the opportunity to collect a few samples of other modalities to work with.

Also, most available data are unimodal with an absence in general of different modalities on the same subjects, which in turn hinders specific processes such as fusing the multimodal images/data to achieve maximum performance results. Additionally, most of the available records are for more elderly patients, 55 and

Table 13.3. Studies related to specific CAD system steps.

Preprocessing related studies

Study	Hajiesmaeili *et al* [99]
Contribution	Uses a wave atom shrinkage method as a preprocessing step to enhance noisy MR images to improve segmentation accuracy.

Procedures and methods	1. Preprocessing:
	  • Using wave atom shrinkage.
	2. Segmentation of hippocampus:
	  • Using a region-based active contour.

Subjects	9 scans (3 scans from ADNI; 6 scans including ground truth were obtained from the Department of Diagnostic Radiology at Henry Ford Hospital).
Modality	MRI
Availability of data	Part of the used data were acquired from ADNI.
Studied medical aspect	Hippocampus
Findings	The proposed wave atom method helped in noise removal and consequently increased the PSNR as a preprocessing step for hippocampus segmentation.
Medical benefit	Increasing the accuracy of hippocampus segmentation helps in the context of AD since hippocampal volume measurement is often used in detection and progression of AD.
Suggestions and future work	1. Evaluate the results against other modalities. 2. Enlarge the studied datasets to generalize the results. 3. Concerning the quantitative validation in addition to the accurate estimation regarding to the measurements of the hippocampus volume.
Study	Cattell *et al* [100]
Contribution	Proposing a novel approach to combine nonlinear PET-MR brain registration using a novel weighting between single-modality updates within the local correlation coefficient (LCC)-demons framework and investigate combining functional and anatomical information (the information of PET and MRI) during the registration of PET volumes to a template by comparing six registration methods, including their proposed approach.

(Continued)

Table 13.3. (*Continued*)

Procedures and methods	1. Rigidly register the PET volumes to their corresponding MR volumes.
	2. Apply the diffeomorphic demons registration algorithm with the implementation of the LCC as the similarity measure.
	3. Combine multiple update fields: combines functional and anatomical update fields.
Subjects	19 AD, 19 NC
Modality	PET-MRI
Availability of data	ADNI
Studied medical aspect	Functional and structural data of the brain.
Findings	All six tested registration methods resulted in significantly higher mean standardized uptake value ratios (SUVRs) for diseased subjects compared to healthy subjects.
Medical benefit	Higher SUVR in AD than normal controls in addition to its ability to separate between SUVRs and Dice overlaps, which in turn assists in distinguishing between AD and normal controls.
Suggestions and future work	1. Test the proposed method against images with sub-optimal quality.
	2. Evaluate the utilization of weighting parameters such as spatially varying weighting of the update combination when dealing with a larger validation dataset.
	3. Fuse the results using suitable methods to produce more informative images that include structural as well as functional information about the subjects.
	4. Enlarge the studied dataset.
Study	Leung *et al* [101]
Contribution	The proposal of identifying ROIs in the hippocampus based on the statistical differences in its shape between NCs and AD subjects using statistical shape models, and then quantify the atrophy rates within these local ROIs using the boundary shift integral (BSI).
Procedures and methods	1. Preprocessing the input images to perform various correction requirements.
	2. Segment hippocampal at baseline and 12 months using a technique known as SNT.
	3. Apply statistical shape models (SSM) using these baseline SNT hippocampal segmentations from the randomly chosen training set to produce *p*-maps that are then mapped onto 214 MCI subjects.
	4. Threshold the *p*-maps at different significance levels to generate different ROIs in the hippocampus of the MCI subjects.
	5. Align the hippocampus in the repeat images to the baseline image using a six-degree-of-freedom registration.
	6. Use the hippocampal boundary shift integral (HSBI) to quantify the atrophy rate between the baseline and repeat scans.

Subjects	60 NC, 214 MCI, 60 AD
Modality	MRI
Availability of data	ADNI
Studied medical aspect	Hippocampus
Findings	Demonstrated that the power to classify MCI converters and stable subjects using hippocampal atrophy rates can be increased by using local ROIs within the hippocampus identified using SSM on a training set of NC and AD.
Medical benefit	Assist in the early diagnosis of MCI subjects who will convert to clinical AD, which that is important for therapeutic decisions, patient counseling and clinical trials.
Suggestions and future work	Use an effective size map for the process of identifying the ROIs rather than relying on a *p*-map.

Study	Risser *et al* [102]
Contribution	Developing a theoretically well-justified simultaneous fine and coarse registration strategy in the LDDMM framework for 3D medical images.
Procedures and methods	1. Preprocess the MR images.
	2. Align all images with the MNI152 brain template using affine registration.
	3. Extract a region of interest of 1283 voxels around the hippocampus.
	4. Compare the baseline and follow-up images at large and small scales simultaneously and quantify the differences at the small scale.
	5. Register the baseline and follow-up MR images using several strategies.
Subjects	30 pairs of AD and healthy age-matched images
Modality	MRI
Availability of data	ADNI
Studied medical aspect	Structure of the brain
Findings	1. Estimating natural-looking deformations when registering images presenting variations at two different scales while retaining diffeomorphic properties.
	2. The ability to discriminate two groups of images by measuring the amplitudes of the deformations generated at a scale of interest.

(Continued)

Table 13.3. (*Continued*)

Medical benefit	Estimating large, smooth and invertible optimal deformations with a rich descriptive power for the quantification of temporal changes in the images.
Suggestions and future work	1. Validate the multiresolution approach in addition to investigating the geodesic properties of the obtained deformation. 2. Deep concern with characteristics of the apparent weights. 3. Enlarge the studied dataset. 4. Evaluate the proposed work against other modalities.
Study	Simpson *et al* [103]
Contribution	Deriving localized measurements of spatial uncertainty from a probabilistic registration framework, which provides a principled approach to image smoothing.
Procedures and methods	1. Preprocess the input images. 2. Register each of the scans to the MNI152 template using nine degrees of freedom to correct size and location differences. 3. Resample and process each scan to have 1 mm isotropic voxels and to remove non-brain tissue, respectively. 4. Create an initial atlas by averaging 40 healthy individual control scans after initial affine alignment to the MNI152 template. 5. Perform a non-rigid registration to the affine atlas using the probabilistic registration tool with 5 mm knot spacing. 6. The most recent follow-up scan is registered to the baseline scan of each individual using the probabilistic registration tool with 5 mm knot spacing where the probabilistic non-rigid registration algorithm is used to provide accurate spatial normalization, which allows measurement of spatial uncertainty.
Subjects	125 random control subjects and AD patients
Modality	MRI
Availability of data	ADNI
Studied medical aspect	Structure of the brain
Findings	Adaptive smoothing of spatially normalized image feature data, based on registration derived uncertainty, has been demonstrated to provide an increase in the ability to classify between patients with AD and NCs using longitudinal Jacobian maps as features.

Medical benefit	Increases the ability to distinguish between patients with AD and NCs.
Suggestions and future work	1. Add the MCI subjects in the case studies. 2. Investigate the use of the independent distribution of the uncertainty of each control point.
Study	Bossa et al [104]
Contribution	Evaluating the statistical power of TBM using stationary velocity field (SVF) diffeomorphic registration in a large population of subjects from the ADNI database.
Procedures and methods	1. MRI acquisition, image correction, and preprocessing. 2. Stationary velocity fields (SVF) diffeomorphic registration. 3. Unbiased average template. 4. Voxel-wise statistical analysis of brain atrophy. 5. Regions of interest statistical analysis.
Subjects	• For screening stage (from ADNI): 231 NC, 200 AD, 405 MCI (divided into converters (MCI-C) and non-converters (MCI-NC) because 127 MCI subjects were converted to the AD group during the clinical follow-up). • For comparison: 40 NC, 40 amnestic MCI, 40 probable AD.
Modality	MRI
Availability of data	ADNI
Studied medical aspect	Structure of the brain
Findings	The obtained brain atrophy patterns presented a higher spatial resolution with a larger statistical significance.
Medical benefit	Provides atrophy maps with very detailed anatomical resolution and with a high significance.
Suggestions and future work	Study the correlation between the atrophy measurements of the images against clinical variables (e.g. genotypes, biomarkers, etc).
Study	Martinez-Murcia et al [105]
Contribution	Proposing an algorithm for MRI projection from 3D to 2D using local binary patterns (LBPs) to provide a visual map that is assessable and that contains the test information required for distinguishing between AD and normal subjects.
Procedures and methods	1. Preprocess the images through spatially normalizing them, removing the skull, and segmenting them into GM and WM tissues. 2. Perform the 3D projection using LBPs.

(Continued)

Table 13.3. (*Continued*)

Subjects	180 NC, 180 AD
Modality	MRI
Availability of data	ADNI
Studied medical aspect	Structure of the brain
Findings	1. The presented descriptor that is used in the projection process is discriminative enough to accomplish such a goal.
	2. The results obtained through first order statistics (e.g. average, variance, etc) are improved.
	3. The discrimination power obtained through fusing the projections of GM and WM increased compared to using each projection separately.
Medical benefit	Produce visually assessable maps that present textural information which in turn enhance the process of diagnosing AD.
Suggestions and future work	1. Add MCI patients to the study and test the ability for early diagnosis of AD.
	2. Study the effect of using efficient fusion methods to fuse GM and WM descriptors to improve the obtained information.
Study	Meena and Raja [106]
Contribution	Comparing the obtained results of the K-means algorithm in different environments (i.e. the AForge.NET framework and MATLAB).
Procedures and methods	• In the AForge.NET framework:
	1. Convert the input images to byte stream.
	2. Store the result in a jagged array and text file format.
	3. Transform the data in the text file into the comma separated values (CSV) format and transform the result to a bitmap image.
	4. Use the K-means algorithm for image segmentation with a random number of clusters equal to five.
	• In the MATLAB environment:
	1. Load the image, determine the number of clusters, and initialize the clusters' centroid.
	2. Calculate the distance between the centroid of the clusters and the object.
	3. Update the clusters' centroid.
	4. Create a mask and finally segment the image.

Subjects	Not mentioned
Modality	PET
Availability of data	ADNI
Studied medical aspect	Function of the brain
Findings	The image segmentation through the AForge.NET Framework produces optimal results compared to the MATLAB output.
Medical benefit	More clarification of the PET scan and its embedded data.
Suggestions and future work	1. Evaluate the results in other modalities.
	2. Try other clustering algorithms to accomplish the segmentation goal.
	3. Use the optimal results for further processing steps related to AD.

Feature related studies

Study	Pagani et al [107]
Contribution	Analysis of the volume of interest (VOI) along with the PCA to study the differences of regional cerebral blood flow (rCBF) distribution between AD patients as well as NCs.
Procedures and methods	1. Preprocessing:
	• Spatial normalization to stereotactic space.
	2. VOI selection:
	• According to the literature mentioned in the paper.
	3. Statistical analysis using analysis of covariance (ANCOVA) in two steps:
	• Only consider VOIs.
	• Using PCA.
	4. Apply discriminant analysis (DA) for relationship estimation based on the clinical diagnosis as well as SPECT data.
Subjects	30 mild patients (mildAD), 27 moderate patients (modAD), 37 NC
Modality	SPECT, ^{99}mTc-hexamethylpropylene amine oxime (HMPAO)
Availability of data	Not available
Studied medical aspect	Regional cerebral blood flow (rCBF)
Findings	The role of PCA in illustrating a strong covariance between both the temporoparietal cortex and the limbic system.

(Continued)

Table 13.3. (*Continued*)

Medical benefit	1. Completing the information of the underlying changes in AD pathological networking. 2. Highlighting the connectivity of the functional points to another analysis approach of the neuroimaging data.
Suggestions and future work	1. Test the method against a standard database. 2. Enlarge the tested dataset. 3. Apply the results to build a CAD system.
Study	Aggarwal *et al* [108]
Contribution	Extracting the relevant features of the GM of five brain regions as different studies mention the relation between the degradation of GM and AD.
Procedures and methods	1. Image preprocessing by smoothly modulating 3D brain volumes. 2. Feature extraction through 3D discrete wavelet transform (DWT) (coiflet, daubechies, biorthogonal, and symmlet). 3. Feature selection by Fisher discriminant ratio (FDR) and minimum redundancy maximum relevance (mRMR).
Subjects	99 AD, 134 NC
Modality	MRI
Availability of data	OASIS database
Studied medical aspect	Hippocampus, amygdale, lateral ventricles, anterior cingulate, and posterior cingulate.
Findings	The better performance of the proposed approach when comparing sensitivity, specificity, classification accuracy, box plots, ROC curves, scoring measure, and computation time to other related works, in most cases.
Medical benefit	Helps in diagnosing the disease through showing the tissue degradation but with some prior knowledge related to the brain affected regions. The absence of such knowledge will cause such a method to be unhelpful.
Suggestions and future work	1. Test the method against other modalities. 2. Enhance the work to present a model that can accomplish the differentiation process without any prior knowledge.
Study	Ghorbanian *et al* [109]
Contribution	Extracting through DWT the discernment features from the EEG signals of a number of subjects while performing cognitive and auditory tasks.
Procedures and methods	1. Extract EEG features at each recording block using DWT. 2. Determine the statistically important features using the *t*-test and Kruskal–Wallis test. 3. The dominant features were finally determined through applying decision tree.

Subjects	10 AD, 14 NC
Modality	EEG signals
Availability of data	Not available
Studied medical aspect	Electrical activity of the brain
Findings	The applied steps help in obtaining the dominant discriminating features related to AD.
Medical benefit	Searching for discriminate features from EEG signals when the subjects perform tasks related to memory and cognition to simplify the diagnosis process. Such a search found that: 'the mean value of the β frequency band during 6 Hz auditory stimulation followed by the standard deviation of θ (4–8 Hz) frequency band of one card learning cognitive task are higher for AD patients compared to controls' (p 2937).
Suggestions and future work	1. Apply the proposed method against the standard database and evaluate the results. 2. Enlarge the tested data. 3. Input MCI subject data and evaluate the results. 4. Test the method against other modalities. 5. Apply the results to build a complete CAD system. 6. Compare the wavelet based results against complexity as well as spectral analysis.
Study	Ghorbanian et al [110]
Contribution	Extract through continuous WT (CWT) the discernment features from the EEG signals of a number of subjects while performing cognitive and auditory tasks.
Procedures and methods	1. Compute the absolute and relative geometric mean powers of Morlet wavelet coefficients at different scale ranges corresponding to the major brain frequency bands. 2. Determine the statistically important features of the cohort geometric means using the Kruskal–Wallis test. 3. The dominant features were finally classified through applying decision tree.
Subjects	10 AD, 14 age-matched NC
Modality	EEG signals
Availability of data	Not available
Studied medical aspect	Electrical activity of the brain
Findings	Identify a significant discriminating feature of EEG signal as well as providing superior performance results compared to short-time fast Fourier transform and DWT.

(Continued)

Table 13.3. (*Continued*)

Medical benefit	The absolute power of the θ frequency band during the second EO state was found to be higher for all AD patients when compared to control subjects.
Suggestions and future work	1. Apply the proposed method against a standard database and evaluate the results. 2. Enlarge the tested data. 3. Input MCI subject data and evaluate the results. 4. Test the method against other modalities. 5. Apply the results to build a complete CAD system. 6. Compare the wavelet based results against complexity as well as spectral analysis.
Study	Kontos *et al* [111]
Contribution	Discovering discriminative patterns of functional MRI activation to efficiently identify spatial regions that in turn help in decision making.
Procedures and methods	1. Apply statistical parametric mapping (SPM) [3] to analyze each fMRI voxel's changes independently and build a corresponding map of statistical values. 2. Use DRP on the 3D domain to discover highly informative 3D sub-regions with respect to the development of a disease.
Subjects	9 AD, 9 NC
Modality	fMRI
Availability of data	Not available
Studied medical aspect	Brain activation data
Findings	Identify discriminative spatial patterns arising from functional imaging information using DRP approach. Identify large hemispheric and lobar differences between Alzheimer's patients and controls.
Medical benefit	1. Apply the proposed method against a standard database and evaluate the results. 2. Enlarge the tested data. 3. Input MCI subject data and evaluate the results. 4. Test the method against other modalities. 5. Apply the results to build a CAD system.
Suggestions and future work	
Study	Kovalev *et al* [112]
Contribution	Comparing certain measures within pairs of anatomical regions to systematically investigate and compare the usefulness of new asymmetry features derived from SPECT perfusion scans.

Procedures and methods
1. Automatically register atlas SPECT volumes to each of the individual SPECT scans.
2. Extract the asymmetry features, where the symmetry was regarded as dissimilarity of 3D perfusion patterns within pairs of lobes in the left and right brain hemispheres.

Subjects 42 AD, 37 NC

Modality SPECT

Availability of data
- Reference data: Greitz Computerized Brain Atlas.
- Case study data: Not available.

Studied medical aspect Frontal, parietal, temporal, occipital, and cerebellar lobes of the brain.

Findings Obtain high-resolution asymmetry measures that provide very useful information for discriminating AD and normal controls.

Medical benefit Extracting the asymmetry measure gives additional information and should be used in conjunction with the more traditional methods for discriminating AD from controls and cannot alone suffice as a diagnostic feature for discrimination of AD from controls.

Suggestions and future work
1. Enlarge the tested data.
2. Test the method against other modalities.
3. Apply the results to build a CAD system.

Study Liu et al [113]

Contribution Producing a computer system that automatically explores very high dimensional image feature spaces in search of discriminative features.

Procedures and methods
1. 3D image alignment using an affine registration algorithm.
2. Defining regions of interest (ROIs) through medical experts.
3. Obtaining image features of two categories (statistical features, Law's texture features).
4. Image feature location: localize the potential discriminative features in the ROI
5. Discriminative feature evaluation and screening to mainly avoid redundancy.
6. Separability analysis to output a set of image feature subspaces with the highest classification rates indicating the best separation among image classes.
7. Prediction of an unseen test sample to evaluate the prediction accuracy of the learned classifier.

Subjects 20 NC, 20 MCI, 20 AD

Modality MRI

Availability of data Not available

(Continued)

Table 13.3. (*Continued*)

Studied medical aspect	Hippocampus
Findings	Regional image features contain highly discriminative information to separate different CNS diseases, e.g. SZ or AD, from normal brains on several limited (15–27 subjects in each disease class) but well-chosen image sets.
Medical benefit	The study encourages the study of hippocampus asymmetry as a highly discriminative feature for studying AD.
Suggestions and future work	1. Enlarge the tested data and attempt to use a standard database for comparison and generalization. 2. Test the method against other modalities.
Study	Zhu *et al* [114]
Contribution	Propose a novel multi-modality canonical feature selection method to select canonical-cross-modality features that are useful for the tasks of clinical score regression and multi-class disease identification.
Procedures and methods	1. Preprocessing and feature extraction. 2. Apply canonical correlation analysis (CCA). 3. Perform canonical feature selection.
Subjects	51 AD, 43 MCI-C, 56 MCI-NC, 52 NC
Modality	MRI, PET
Availability of data	ADNI
Studied medical aspect	Focusing on gray matter of the brain.
Findings	When comparing MRI and PET, MRI showed better performance than PET. When comparing applying MRI and PET separately against combining them, the performance of such bi-modality was better than each of them separately.
Medical benefit	1. Achieves the best performance for the joint clinical score regression and multi-class clinical status identification. 2. Encourages the testing of the ability of more than bi-modality fusion to further enhance diagnostic accuracy.
Suggestions and future work	1. Evaluate the results against CT in place of MR images due to the promising results of CT with PET as indicated by different studies. 2. Fuse more than two modalities and evaluate the accuracy of the diagnosis.
Study	Trojacanec *et al* [115]
Contribution	Proposing a new representation of the information extracted from the MRI volumes as quantitative measurements that are highlighted in the literature as valuable markers for distinguishing AD from healthy controls.

Procedures and methods	1. Segmentation. 2. Quantitative measurements estimation.
Subjects	Not mentioned
Modality	MRI
Availability of data	ADNI
Studied medical aspect	Ventricle volume: left and lateral ventricle, third and fourth ventricle, the volume of the left and right hippocampus, and left and right amygdala, as well as the cortical thickness of the separate cortical structures.
Findings	The method is suitable for the study of single-scan studies or multiple-scan studies (provided at certain time points) since it performs a representation that includes the measures of the ventricle volumes
Medical benefit	Comprehensiveness, the suitability to reflect the patient's condition changes over time and in efficiency.
Suggestions and future work	1. Follow the same approach with other modalities and search for new representations of their features. 2. Incorporate the proposed method in the context of content based image retrieval and evaluate the results.
Study	Fiot et al [116]
Contribution	Evaluation of global, semi-local, and local descriptors of the hippocampus evolutions between two time points to assist in separating stable MCI (MCI patients not converting to AD) and progressive MCI.
Procedures and methods	1. Computation of local deformation descriptors. 2. Computation of population template. 3. Transporting deformation descriptors to the temple. 4. SVM classification.
Subjects	103 patients were screened and 'month 12' was the second time point selected. At screening point: all patients were MCI. By month 12: 19 became AD and the remaining 84 stayed MCI.
Modality	Binary segmentations of hippocampus
Availability of data	ADNI
Studied medical aspect	Hippocampus
Findings	In general and based on the proposed method, the local descriptors outperform the global descriptors on the studied dataset.
Medical benefit	Helps in predicting AD conversion for MCI patients.
Suggestions and future work	Apply the results to build a CAD system.

(Continued)

Table 13.3. (*Continued*)

Study	Rueda *et al* [117]
Contribution	Evaluate the suitability of a bag-of-feature (BOF) representation for automatic classification (distinguishing) between normal control and Alzheimer's disease.
Procedures and methods	1. BOF image representation (represents an image as a frequency histogram of an unordered collection of individual regions (patches or blocks)) by local feature detection and description, dictionary construction, and histogram image representation. 2. Binary SVM classification.
Subjects	Two groups Group 1: 86 subjects, mild AD; 20 AD, 66 NC. Group 2: 136 subjects, mild and very mild AD; 70 AD, 66 NC.
Modality	MRI
Availability of data	OASIS
Studied medical aspect	Structural data of the brain
Findings	The proposed method shows a competitive performance that improves the results of distinguishing between NCs and AD patients.
Medical benefit	Encouraging and suggesting that the BOF representation has the ability to capture visual patterns useful for discriminating healthy MR brain volumes from those exhibiting AD.
Suggestions and future work	1. Evaluate the method against other modalities. 2. Evaluate the method against other datasets. 3. Explore feature descriptors (e.g. SIFT, HOG, and feature combinations, etc). 4. Analyze the visual dictionary searching for characteristic discriminant visual words for AD.
Study	Coupé *et al* [118]
Contribution	Proposing a method for simultaneous segmentation and grading of the hippocampus to better capture the patterns of pathology occurring during AD.
Procedures and methods	1. Image preprocessing. 2. Segmentation and grading: using nonlocal means-based approaches and the new concept of patch-based grading, respectively.

Subjects	60 NC, 60 AD
Modality	MRI
Availability of data	ADNI
Studied medical aspect	Hippocampus
Findings	When the proposed novel grading measure is used alone it leads to a success rate of 89% and 90% when combined with HC volume.
Medical benefit	Grading the atrophy degree in the context of AD enables an accurate distinction between NC subjects and patients with AD.
Suggestions and future work	1. Evaluate the method against other modalities. 2. Input MCI in the case studies.

Table 13.4. AD related studies unrelated to CAD systems.

	Analysis
Study	Hamou *et al* [119]
Contribution	Utilizing clustering analysis along with decision trees to investigate the influence of AD related data (i.e. MRI data, demographic data, and clinical data) on the occurrence of AD/MCI.

Procedures and methods	1. Preprocessing:
	   • Intensity correction.
	   • Brain tissue segmentation.
	   • Regional brain separation.
	2. Cortical thickness measurements: to detect the differences in the pre-symptomatic individuals who have variant frontotemporal dementia in their family.
	3. Cluster analysis:
	   • Modified *k*-means clustering algorithm.
	   Note: apply PCI (for variable selection) to deal with grouping overlap and produce better defined clusters.
	   • Decision trees (refining the decision making process).

Subjects	112 NC, 122 MCI, 120 AD
Modality	MRI
Availability of data	Not available
Studied medical aspect	Structure of the brain
Findings	Adequate separation results of the groups as well as the possibility of decision making.
Medical benefit	1. Assists in the early detection of AD/MCI. 2. Aids in the production of new drug therapies that assist in inhibiting AD progression.
Suggestions and future work	1. Analyze the influence of each included variable separately through measuring the distance between them and the cluster center. 2. Investigate the results against other modalities. 3. Evaluate the method through using other database.
Study	Ortiz *et al* [120]
Contribution	Assessing the discriminative regions of the brain in diagnosing AD through statistically testing the tissue density in different brain regions as well as utilizing ApoE genetic information to increase diagnosis accuracy.

Procedures and methods	1. Preprocessing:
	• Spatial normalization (registration) according to the VBM-T1 template.
	• Brain tissue segmentation through SPM/VBM to gather information regarding to the distribution of GM and WM tissues.
	2. Feature extraction (density computation):
	• Calculate the volume of GM and WM in certain brain regions (116 regions according to the AAL brain atlas).
	• Calculate tissue density.
	Note: these computations involve using a threshold that is optimally determined using SVM.
	3. Feature selection:
	• Select the most distinctive brain regions for AD using the t-statistic.
	4. Include ApoE genetic data of the subjects obtained from the ADNI database to the feature space.
	5. Classification:
	Assess the distinctiveness of the tissues density capabilities using SVM through testing:
	a. Only use density information as features.
	b. Fuse the WM and GM densities and use them as features.
	c. Fuse ApoE data with WM/GM densities.
Subjects	68 NC, 111 MCI, 70 AD
Modality	MRI
Availability of data	ADNI
Studied medical aspect	Structure of the brain
Findings	1. Although GM contains the most information, WM also introduces discriminate information that when fused with GM helps in improving the classification performance for NC and AD subjects.
	2. ApoE data by itself does not introduce sufficient information to accomplish the classification task but fusing these data with WM/GM densities helps in increasing the classification performance.
Medical benefit	1. For distinguishing between AD and NC subjects, GM contains the most information,
	2. WM also introduces discriminate information, but does not introduce high discriminative capabilities, therefore it cannot be used alone to accomplish the distinguishing task.

(Continued)

Table 13.4. (*Continued*)

Suggestions and future work	1. Investigate the fusion of the obtained results with other sources of information (i.e. function information) to improve the diagnosing process. 2. Test the ability to analyze other discriminative brain features for further assessment in diagnosing AD.
Study	Eblenkamp *et al* [121]
Contribution	Relying on the fact that changes in the neurons of the dendritic spine produce equivalent changes to AD neurons, counting the large numbers of spines (25 000 or more) is considered a major task in AD research. Therefore, this study presents an image analysis tool for the purpose of dendritic spine evaluation to minimize both classification error and changes in the classification procedure due to examiner dependence.

Procedures and methods	1. Preprocessing:
	  a. Image standardization:
	    • A consistent size of 1300 × 1300 pixels.
	    • Transformation to 8-bit contrast resolution.
	  b. Qualitative adaptation:
	    • Histogram equalization.
	    • Background signal minimization through darkening homogeneous regions.
	    • Contrast the objects of interest by enhancing bright image elements.
	2. Object recognition:
	  a. Object of interest (dendrites) identification.
	  b. Pinching off the dendritic spine candidates from the dendrite.
	  c. Applying the cutting process of the second step in limited cases for the separation of spine candidates.
	  d. Assign the identified spines to three classes, namely mushroom, stubby, and thin.

Subjects	Not mentioned
Modality	Fluorescence image *z*-stacks (10–20 slices) of *in vivo* two-photon laser scan microscopy images.
Availability of data	Not available
Studied medical aspect	Dendritic spines
Findings	Robust separation of spines from the corresponding dendrites.
Medical benefit	Extending and speeding up the evaluation of large amounts of dendrites regarding their dendritic spines and therefore speeding up AD related research.

Suggestions and future work	Use the obtained results along with other AD related data for the AD diagnosis process.
Study	Rusina *et al* [122]
Contribution	Proposing a SPECT based analysis method through utilizing a fuzzy edge detector, watershed transform, and orientation toward activity separation of parietal lobe domains with the goal of counting the total number of regions in both AD and NC subjects to finally assist in the AD diagnosing procedure.
Procedures and methods	1. Image smoothing:
	• Using a Gaussian 3D filter.
	2. Intensity normalization:
	• Using a (0,1) interval.
	3. Background elimination:
	• Using threshold.
	4. Fuzzy edge detection:
	• Using Lukasiewicz BL-algebra.
	5. Segmentation:
	• Standard 3D watershed transform.
	6. Counting the segmented regions:
	• Using Student's two-sample *t*-test.
Subjects	10 NC (with amyotrophic lateral sclerosis (ALS)), 17 AD
Modality	SPECT
Availability of data	Not available
Studied medical aspect	Function of the brain
Findings	AD patients are characterized by a significant reduction in the total number of watershed regions compared to the number of regions in NC subjects.
Medical benefit	Assists in diagnosing AD.
Suggestions and future work	1. Test the method against generalized database and more conditions.
	2. Input the MCI in the analysis process.
	3. Analyze the proposed method against other AD related modalities.

(Continued)

Table 13.4. (*Continued*)

Study	Kodama et al [123]
Contribution	Propose texture analysis to assist the possibility of a CAD system to differentiate between two types of dementia (dementia with Lewy bodies (DLB) and AD).

Procedures and methods	1. Extract image information:
	• Using four nearest neighbors based erosion operation twice.
	2. Extract cerebral parenchyma image:
	• Using four nearest neighbors based dilation operation twice.
	3. Texture analysis: 76 texture feature were determined:
	• 14 features from the co-occurrence matrix.
	• Five features from run-length matrix.
	Note:
	  a. These features were determined at 0°, 45°, 90°, and 135° for each subject.
	  b. The stepwise method was used to select seven features.

Subjects	10 DLB, 21 AD, 20 NC
Modality	MRI
Availability of data	Not available
Studied medical aspect	Structure of the brain
Findings	1. CAD systems can be constructed for diagnosing these dementia types. 2. Texture features can accurately detect both morphological changes as well as local gray level distribution changes.
Medical benefit	Assists in diagnosing DLB and AD.
Suggestions and future work	1. Evaluate other texture features. 2. Enlarge the studied conditions.

Database management systems

Study	Stanchev and Fotouhi [124]
Contribution	Presenting MEDIMAGE (a multimedia database for AD patients) that contains imaging, text, and voice data and that can be used to find some correlations of brain atrophy in Alzheimer's patients with different demographic factors.

Details

- The MEDIMAGE system has four databases:
  1. MEDIMAGE MR database.
  2. MEDIMAGE segmented and 3D reconstructed database.
  3. MEDIMAGE test database (patient results from the standard tests for AD and related disorders).
  4. MEDIMAGE radiologist comments database (text and voice radiologist findings).
- The MEDIMAGE system has three main tools for image processing:
  1. MEDIMAGE MR image segmentation tools.
  2. MEDIMAGE MR 3D reconstruction tools.
  3. MEDIMAGE MR measurement tools.
- In the MEDIMAGE database management system there are definition, storage, manipulation, and viewing tools.

Findings

1. Generality: The system could easily be modified for other medical image collections.
2. Practical applicability: The results obtained with the system define essential medical findings.

Medical consequences

1. Managing and accessing multimedia information on the analysis of brain data.
2. The database links MR images to patient data in a way that permits the user to view and query medical information using alphanumeric and feature-based predicates.
3. The visualization results support the wide variety of data types and presentation methods required by neuroradiologists.

Suggestions and future work

1. Apply content based image retrieval methods to simplify working with such a system.
2. Utilize data mining algorithms to assist in extracting essential information from the included data.

Data mining analysis for studying risk factors and identifiers of AD

Study

Contribution

Ertek *et al* [125]

Examining the effects of age, social and economic status, gender, medical tests, and other factors on demographics and test statistics of the examined subjects through data visualization and mining methods.

Details

The process of data mining includes three main types of analysis: classification tree (decision tree) analysis, hierarchical clustering, and classification analysis. The data mining process steps are:

1. Reading the data from file, validating it through displaying it in a data table and observing its histogram, scatter plot, and attribute statistics.
2. Each of the attributes is specified either as the class attribute or one of the predictor attributes.
3. Use C4.5 as the classification tree algorithm, the hierarchical algorithm as the clustering analysis, and four classification algorithms (k nearest neighbors, C4.5, SVM, and classification tree) for classification analysis.

(Continued)

Table 13.4. (*Continued*)

Findings	Table 2 in the paper summarizes the insights found based on the given factors.
Medical consequences	1. Helps individual people as well as medical experts to estimate the related risks through the obtained results and consequently take the required prevention.
	2. In addition, health institutions, pharmaceutical companies, insurance companies, and government institutions can benefit from the obtained insights in their current and future plans.
Suggestions and future work	Study the effect of implementing the results on medical AD imaging modalities to verify the obtained results in addition to helping in studying the corresponding variations in the images related to the obtained results which in turn assist in extracting new insights on the medical imaging modalities.

Quantifying metabolic asymmetry modulo structures in AD

Study	Fletcher *et al* [126]
Contribution	1. Presenting a quantifying method that relies on the construction of an atlas with large deformation diffeomorphic metric mapping (LDDMM) for analyzing and quantifying metabolic asymmetry that in turn assists in diagnosing AD. The study derived from the fact that in elderly people, glucose metabolism is considered symmetric in average while individual dementia patients can sometimes face metabolic asymmetry.
	2. Construct an atlas of symmetric and normative MR images along with the corresponding metabolic asymmetry of these images.
Details	The steps of the proposed method are:
	1. Preprocess the input PET and MR images that were obtained from ADNI (correct the inhomogeneity of the image intensity, and orient the images into a standard coordinate system through rigid registration of PET and MRI images).
	2. Apply LDDMM to perform the following steps:
	a. Quantify the structural asymmetry of individuals.
	b. Quantify the metabolic asymmetry modulo structural asymmetry of the individuals.
	c. Build statistical atlas of the normal cross-sectional metabolic asymmetry.
	d. Test the asymmetry of the individuals statistically against the atlas data.
Findings	The study was only applied to ten normative subjects and the figures in the paper present the obtained results visually.

Medical consequences

Quantifying and comparing the normal range of the individuals' metabolic asymmetry against the possible variations associated with AD to help in the diagnosis process.

Suggestions and future work

1. Extend the evaluation process to a larger dataset.
2. Evaluate the obtained results against the clinical symptoms to search for new insights that can assist in further diagnosis and treatment processes.

Connectivity analysis of the hippocampus in AD subjects

Study

Hasan et al [127]

Contribution

Use DTI to study the connectivity between both the hippocampus (i.e. left and right) and the other brain regions through tracking the tracts of the inter-region fibers using probabilistic tractography.

Details

- The study was performed on 12 AD patients and 24 NCs obtained from the ADNI database.

The steps of the method are as follows steps:

1. Convert the data into a suitable format.
2. Construct time series from the 3D DTI data.
3. Apply a correction for the results of step 2.
4. Extract a free of diffusion image for a reference image.
5. Extract the brain in the reference image.
6. Track the results for crossing fibers through computing the fractional anisotropy and Bayesian tractography and registering with standard weighted images MNI152-T1.
7. Streamline the fibers from seed ROIs along with the ROI segmentation in MNI152-T1 through a normalization process.
8. Finally, perform a comparison of the AD patients against the CTR subjects through statistical analysis.

Findings

- The non-zero voxels in AD patients are more than those in NC.
- The non-zero voxels in AD patients are more scattered and dispersed than those in NC.
- The paths of the fibers are much lower in intensity in AD patients than in NC.

Medical consequences

The scattered fiber paths through the hippocampus cause communication interruptions and consequently impair the functionality of the memory in all brain regions.

Suggestions and future work

1. Perform further analysis of the brain's connectivity in Alzheimer's patients.
2. Search for other imaging biomarkers that can assist in differentiating between NCs and Alzheimer's patients.

(Continued)

Table 13.4. (*Continued*)

	Network and content analysis of AD related research
Study	Song *et al* [128]
Contribution	Utilize the graphs and topic modeling to present a comprehensive approach to perform productivity, network, and content analysis of AD literature collected from PubMed.
Details	The study adopted Dirichlet multinomial regression (DMR) as a topical model for tracking related trends through time.
Findings	1. In productivity analysis: The results show that AD research is dynamic with several experts directing the work in the field, in particular in the last ten years where the most research productivity was founded to be in 2013. 2. In network analysis: The results show that although the top bio-entities are more concentrated than MeSH terms in topics such as diagnostic imaging and neoplasm, there are partial similarities between these entities and the frequently appearing MeSH terms. 3. In content analysis: The results identify the notable topics related to the causes of AD and rising trends in advanced science and technology (e.g. brain imaging techniques).
Medical consequences	Provides a clearer view of the recent findings relating to AD to help in turn in improving the quality of diagnosis.
Suggestions and future work	1. Collect articles from resources other than PubMed to help in improving the quality of the results and generalizing the findings. 2. Utilize the results of such analysis to build a relationship between the findings of the studied articles to help in enhancing the diagnosis process and curing the disease.
	Image retrieval for AD detection
Study	Agarwal and Mostafa [129]
Contribution	Propose an AD related system for retrieving as well as diagnosing the disease through multi-level architecture with: segmentation, feature extraction, classification, and ranking.
Details	• The system was focused on the MR images and went through the following steps: 1. Image segmentation: using histogram. 2. Feature extraction: using DCT. 3. Reduce the obtained features using 'normalization average' to facilitate the retrieval procedure. 4. Classify the images to one of three classes (NC, MCI, and AD) using SVM.

5. Perform ranking operation using Euclidean distance weighted by a probability value that refers in this paper to the decision values of the classifier.

6. The system offers a user interface to enter a textual keyword.

• The evaluation of the system was derived based on the pre-classified MRI data obtained from the ADNI database.

Findings

Medical consequences

Using classification along with the ranking procedures improves the final results.

1. Helps in finding information regarding other subjects with similar conditions.

2. Assists in diagnosing AD.

Suggestions and future work

Further investigations of time efficiency, measure representation, ranking performance, and usability are needed.

older, preventing the analysis of younger people who also suffer from AD. Even if these cases are limited, the possibility still exists, and such cases need to be considered.

13.5.2 Modalities

Despite the large sources of information that can be gathered through different imaging technologies, the disadvantages that are associated with these modalities prevent maximum utilization of these technologies. For example, despite its high quality results, PET technology suffers from high radiation that can subsequently affect human health. MRI also provides another example where, despite its powerful role in the AD diagnosis process, it cannot be used in the presence of metallic devices, in addition to its high cost. To avoid the limitations associated with certain modalities, other modalities can be used, but of course only by compromising, for example, the quality of the obtained output or the financial investments that some modalities require.

In addition, some modalities fail in the process of differential diagnosis between AD and other disorders, which in turn affects the decision making processes. In addition, some modalities fail to assist in the early diagnosis of the disease, although they can assist in subsequent diagnosis stages. Therefore, advances in medical imaging technologies continue to overcome the current limitations and also to serve the medical field more and more.

13.5.3 Applied techniques

Various methods and techniques have been and still are used to decipher the complexity of AD and consequently assist experts in their decision making. Over time, various studies have been proposed that used different techniques and achieved a lot of useful findings that serve the field of AD research. However, various limitations are associated with these techniques and prevent them from being generalized and applied in real clinical life.

As is known, the performance quality and accuracy are essential factors, in particular when dealing with medical applications since any mistake may put human life in danger. Starting from such a fact, although various methods achieve high quality and accuracy, they still need much more improvement to be applied in real life. In addition to performance quality and accuracy, computation time is also a vital factor in real applications, and is an obstacle for the implementation of some techniques which, although providing acceptable results, cannot be implemented due to their relatively high computation time.

Dependence on the included data also represents a source of limitations in the context of applying computerized methods/techniques in AD. In other words, some implementations require adaptation of specific parameters depending on case studies, which in turn prevent such results from being generalized unless its performance is otherwise verified. Therefore, in general, further advances are still needed to help in enhancing the quality of the applied methods/techniques to achieve maximum benefits in their implementation.

13.5.4 Subjects

The subjects themselves can also represent another obstacle in the further development of AD research. For example, the movement of patients during the acquisition process introduces a source of noise that can be detected either by preprocessing steps or by taking such movements into consideration during further work (e.g. slow eye movements during the earliest phase of drowsiness when recording EEG signals).

Additionally, suffering from other disorders in addition to AD can lead to misclassification. In other words, this prevents accurate decision making regarding the stage of AD due to the overlapping that may occur between the symptoms of the disorders. In addition, the subjects may suffer from a specific obstacle (e.g. using metallic devices) that prevents the acquisition of images using specific technologies (e.g. MRI).

13.6 Conclusion

Despite the availability of many scientific studies that aim to clarify the ambiguity that surrounds AD, the door is still open for further achievements and the scientific community is still searching for an accurate description of AD and its associated characteristics. This chapter aimed to illustrate the numerous achievements in dealing with AD from different perspectives, which ultimately share the same goal of removing, as much as possible, the ambiguity surrounding AD. This chapter also discussed the current considerations and limitations that challenge researchers and their further achievements regarding AD.

This work could also be applied to various other applications in medical imaging, such as the kidney, heart, prostate, lung, and retina, as well as several non-medical applications [130–133].

One application is renal transplant functional assessment, in particular in terms of developing noninvasive CAD systems, utilizing different image modalities (e.g. ultrasound, computed tomography (CT), MRI, etc). Accurate assessment of renal transplant function is critically important for graft survival. Although transplantation can improve a patient's wellbeing, there is a potential post-transplantation risk of kidney dysfunction that, if not treated in a timely manner, can lead to the loss of the entire graft, and even patient death. In particular, dynamic and diffusion MRI-based systems have been clinically used to assess transplanted kidneys with the advantage of providing information on each kidney separately. For more details about renal transplant functional assessment, see [134–151, 151–161].

The heart is also an important application for this work. The clinical assessment of myocardial perfusion plays a major role in the diagnosis, management, and prognosis of ischemic heart disease patients. Thus, there have been ongoing efforts to develop automated systems for accurate analysis of myocardial perfusion using first-pass images [162–178].

Another application for this work could be the detection of retinal abnormalities. The majority of ophthalmologists depend on visual interpretation for the identification of diseases types. However, inaccurate diagnosis will affect the treatment procedure which may lead to fatal results. Hence, there is a crucial need for computer automated diagnosis systems that yield highly accurate results. Optical coherence tomography (OCT) has become a powerful modality for the non invasive diagnosis of various retinal abnormalities such as glaucoma, diabetic macular edema, and macular degeneration. The problem with diabetic retinopathy (DR) is that the patient is not aware of the disease until the changes in the retina have progressed to a level that treatment tends to be less effective. Therefore, automated early detection could limit the severity of the disease and assist ophthalmologists in investigating and treating it more efficiently [179–181].

Abnormalities of the lung could also be another promising area of research and a related application to this work. Radiation-induced lung injury is the main side effect of radiation therapy for lung cancer patients. Although higher radiation doses increase the effectiveness of radiation therapy for tumor control, this can lead to lung injury as a greater quantity of normal lung tissues is included in the treated area. Almost 1/3 of patients who undergo radiation therapy develop lung injury following radiation treatment. The severity of radiation-induced lung injury ranges from ground-glass opacities and consolidation at the early phase to fibrosis and traction bronchiectasis in the late phase. Early detection of lung injury will thus help to improve management of the treatment [182–224].

This work can also be applied to other brain abnormalities, such as dyslexia and autism. Dyslexia is one of the most complicated developmental brain disorders that affect children's learning abilities. Dyslexia leads to the failure to develop age-appropriate reading skills despite a normal intelligence level and adequate reading instruction. Neuropathological studies have revealed the abnormal anatomy of some structures, such as the corpus callosum in dyslexic brains. There has been a lot of work in the literature that aims to develop CAD systems for diagnosing such disorders, along with other brain disorders [225–246].

For the vascular system [247], this work could also be applied for the extraction of blood vessels, e.g. from phase contrast (PC) magnetic resonance angiography (MRA). Accurate cerebrovascular segmentation using non invasive MRA is crucial for the early diagnosis and timely treatment of intracranial vascular disease [230, 231, 248–253].

References

[1] Feldman H 2007 *Atlas of Alzheimer's Disease* (Boca Raton, FL: CRC Press)
[2] Turkington C 2003 *The Encyclopedia of Alzheimer's DiseaseFacts on File Library of Health and Living* (New York: Facts On File)
[3] Gauthier S 2006 *Clinical Diagnosis and Management of Alzheimer's Disease* (Boca Raton, FL: CRC Press)
[4] NLM 2019 *National Library of Medicine—National Institutes of Health* https://www.nlm.nih.gov/ (Accessed: 14 March 2019)
[5] Chamberlin S, Narins B and Group G 2005 *The Gale Encyclopedia of Neurological Disorders* (Farmington Hills, MI: Thomson Gale)
[6] World Health Organisation 2019 http://www.who.int/ (Accessed: 12 February 2019)

[7] Minagar A and Alezander J 2005 *Inflammatory Disorders of the Nervous System: Pathogenesis, Immunology, and Clinical Management* (New York: Humana)

[8] Starkey C 2013 *Therapeutic Modalities* (Philadelphia, PA: F A Davis Company)

[9] Neighbors M and Tannehill-Jones R 2014 *Human Diseases* (Boston, MA: Cengage Learning)

[10] Myers J, Neighbors M and Tannehill-Jones R 2002 *Principles of Pathophysiology and Emergency Medical Care* (Clifton Park, NY: Delmar/Thomson Learning)

[11] Hardiman O and Doherty C 2011 *Neurodegenerative Disorders: A Clinical Guide* (London: Springer)

[12] Brown D 2012 *Brain Diseases and Metalloproteins* (Singapore: Pan Stanford)

[13] IASP 2019 International Association for the Study of Pain http://www.iasp-pain.org/ (Accessed: 12 February 2019)

[14] World Health Organization 2006 *Neurological Disorders: Public Health Challenges* (Geneva: World Health Organization)

[15] Smith H 2009 *Current Therapy in Pain* (Amsterdam: Saunders/Elsevier)

[16] Rosner L and Ross S 2008 *Multiple Sclerosis: New Hope and Practical Advice for People with MS and Their Families* (New York: Atria)

[17] Forbes J, Tweedie A, Conolly J and Dunglison R 1859 *The Cyclopaedia of Practical Medicine: Comprising Treatises on the Nature and Treatment of Diseases, Materia Medica and Therapeutics, Medical Jurisprudence, Etc., Etc* (Philadelphia, PA: Lea and Blanchard)

[18] Moore K L, Agur A M R and Dalley A F 2015 *Essential Clinical Anatomy* (Wolters Kluwer Health)

[19] Naidich T, Duvernoy H, Delman B, Sorensen A, Kollias S and Haacke E 2009 *Duvernoy's Atlas of the Human Brain Stem and Cerebellum: High-Field MRI, Surface Anatomy, Internal Structure, Vascularization and 3D Sectional Anatomy* (Vienna: Springer)

[20] ICD9Data 2019 The Web's Free ICD-9-CM Medical Coding Reference http://www.icd9data.com/ (Accessed: 13 February 2019)

[21] Hodler J, Schulthess G and Zollikofer C 2008 *Diseases of the Brain, Head and Neck, Spine: Diagnostic Imaging and Interventional Techniques* (Milan: Springer)

[22] Fritz S, Chaitow L and Hymel G 2008 *Clinical Massage in the Healthcare Setting* (Amsterdam: Elsevier Health Sciences)

[23] Hodler J, von Schulthess G and Zollikofer C 2012 *Diseases of the Brain, Head and Neck, Spine 2012-2015: Diagnostic Imaging and Interventional Techniques* (Milan: Springer)

[24] Morris J, Galvin J and Holtzman D 2006 *Handbook of Dementing Illnesses* (London: Taylor and Francis)

[25] Wang K, Zhang Z and Kobeissy F 2014 *Biomarkers of Brain Injury and Neurological Disorders* (London: Taylor and Francis)

[26] Alzheimer's Association 2019 Alzheimer's Disease and Dementia https://www.alz.org/alzheimer_s_dementia (Accessed: 13 February 2019)

[27] Lechtenberg R and Schutta H 1997 *Neurology Practice Guidelines* (London: Taylor and Francis)

[28] Davis L, King M and Schultz J 2005 *Fundamentals of Neurologic Disease* (New York: Demos Medical)

[29] Zatta P 2003 *Metal Ions and Neurodegenerative Disorders* (Singapore: World Scientific)

[30] Lu L and D J 2011 *Alzheimer's Disease* (Santa Barbara, CA: ABC-CLIO)

[31] ARMS 2019 Alzheimer Research Management System http://www.alzheimer-research.eu/ (Accessed: 13 February 2019)

[32] NIH 2019 National Institute on Aging, http://www.nia.nih.gov/ (Accessed: 12 February 2019)

[33] Ali N 2012 *Understanding Alzheimer's: An Introduction for Patients and Caregivers* (Lanham, MD: Rowman and Littlefield)

[34] Blickman J, Parker B and Barnes P 2009 *Pediatric Radiology: The Requisites* (Amsterdam: Mosby/Elsevier)

[35] Acharya R, Wasserman R, Stevens J and Hinojosa C 1995 Biomedical imaging modalities: a tutorial *Computer. Med. Imaging Graph.* **19** 3–25

[36] Varghese T, Sheelakumari R, James J S and Mathuranath P S 2013 A review of neuroimaging biomarkers of Alzheimer's disease *Neurol. Asia* **18** 239

[37] El-Gamal F E-Z A, Elmogy M and Atwan A 2016 Current trends in medical image registration and fusion *Egypt. Inf. J.* **17** 99–124

[38] Bigler E and Clement P 1997 *Diagnostic Clinical Neuropsychology* (Austin, TX: University of Texas Press)

[39] Smith N and Webb A 2010 *Introduction to Medical Imaging: Physics, Engineering and Clinical Applications* (Cambridge: Cambridge University Press)

[40] Farncombe T and Iniewski K 2013 *Medical Imaging: Technology and Applications* (Boca Raton, FL: CRC Press)

[41] Haidekker M 2013 *Medical Imaging Technology* (New York: Springer)

[42] Exarchos T 2009 *Handbook of Research on Advanced Techniques in Diagnostic Imaging and Biomedical Applications* (Hershey, PA: IGI Global)

[43] Aiyagari V and Gorelick P 2010 *Hypertension and Stroke: Pathophysiology and Management* (New York: Humana)

[44] Van Hecke W, Emsell L and Sunaert S 2015 *Diffusion Tensor Imaging: A Practical Handbook* (New York: Springer)

[45] Analoui M, Bronzino J and Peterson D 2012 *Medical Imaging: Principles and Practices* (London: Taylor and Francis)

[46] Kruger L 2001 *Methods in Pain Research* (Boca Raton, FL: CRC Press)

[47] Elgazzar A 2006 *The Pathophysiologic Basis of Nuclear Medicine* (Berlin: Springer)

[48] Van Heertum R, Tikofsky R and Ichise M 2013 *Functional Cerebral SPECT and PET Imaging* (Philadelphia, PA: Wolters Kluwer Health)

[49] Saha G B, MacIntyre W J and Go R T 1994 Radiopharmaceuticals for brain imaging *Seminars in Nuclear Medicine* vol 24 (Amsterdam: Elsevier) pp 324–49

[50] Reilly R B and Lee T C 2010 Electrograms (ECG, EEG, EMG, EOG) *Technol. Health Care* **18** 443–58

[51] Voros N S *et al* 2015 *Cyberphysical Systems for Epilepsy and Related Brain Disorders* (Berlin: Springer)

[52] Lee T C and Niederer P 2010 *Basic Engineering for Medics and Biologists: An ESEM Primer* vol 152 (Amsterdam: IOS Press)

[53] The Free Dictionary 2019 http://medical-dictionary.thefreedictionary.com/operating+micro-scope/ (Accessed: 13 February 2019)

[54] Toole G and Toole S 1997 *Advanced Human and Social Biology* (Cheltenham: Stanley Thornes)

[55] Perry G 2006 *Alzheimer's Disease: A Century of Scientific and Clinical Research* (Amsterdam: IOS Press)

[56] Panjarathinam R 2007 *Medical Microbiology* (New Delhi: New Age International)

[57] Su S-S, Chen K-W and Huang Q 2014 Discriminant analysis in the study of Alzheimer's disease using feature extractions and support vector machines in positron emission tomography with 18 F-FDG *J. Shanghai Jiaotong Univ. (Science)* **19** 555–60

[58] Varghese T, Kumari R S, Mathuranath P and Singh N A 2014 Discrimination between Alzheimer's disease, mild cognitive impairment and normal aging using ANN based MR brain image segmentation *Proc. of the Int. Conf. on Frontiers of Intelligent Computing: Theory and Applications (FICTA) 2013* (Berlin: Springer) pp 129–36

[59] Martnez F, Salas-Gonzalez D, Górriz J, Ramrez J, Puntonet C G and Gómez-Ro M 2011 Analysis of SPECT brain images using Wilcoxon and relative entropy criteria and quadratic multivariate classifiers for the diagnosis of Alzheimer's disease *Int. Work-Conf. on the Interplay Between Natural and Artificial Computation* (Berlin: Springer) pp 41–8

[60] Jie B, Zhang D, Cheng B and Shen D 2013 Manifold regularized multi-task feature selection for multi-modality classification in Alzheimer's disease *Int. Conf. on Medical Image Computing and Computer-Assisted Intervention* (Berlin: Springer) pp 275–83

[61] Simoes R, van Walsum A-M v C and Slump C H 2014 Classification and localization of early-stage Alzheimer's disease in magnetic resonance images using a patch-based classifier ensemble *Neuroradiology* **56** 709–21

[62] Khedher L, Ramrez J, Górriz J M, Brahim A and Illán I 2015 Independent component analysis-based classification of Alzheimer's disease from segmented MRI data *Int. Work-Conf. on the Interplay between Natural and Artificial Computation* (Berlin: Springer) pp 78–87

[63] Ahmed O B *et al* 2015 Classification of Alzheimer's disease subjects from MRI using hippocampal visual features *Multimedia Tools Appl.* **74** 1249–66

[64] Dyrba M *et al* 2012 Combining DTI and MRI for the automated detection of Alzheimer's disease using a large European multicenter dataset *Int. Workshop on Multimodal Brain Image Analysis* (Berlin: Springer) pp 18–28

[65] Fan Y 2011 Ordinal ranking for detecting mild cognitive impairment and Alzheimer's disease based on multimodal neuroimages and CSF biomarkers *Int. Workshop on Multimodal Brain Image Analysis* (Berlin: Springer) pp 44–51

[66] Varghese T, Kumari R S, Mathuranath P and Singh N A 2012 Performance evaluation of bacterial foraging optimization algorithm for the early diagnosis and tracking of Alzheimer's disease *Int. Conf. on Swarm, Evolutionary, and Memetic Computing* (Berlin: Springer) pp 41–8

[67] Liu M, Zhang D, Yap P-T and Shen D 2012 Hierarchical ensemble of multi-level classifiers for diagnosis of Alzheimer's disease *Int. Workshop on Machine Learning in Medical Imaging* (Berlin: Springer) pp 27–35

[68] Hu Z, Pan Z, Lu H and Li W 2011 Classification of Alzheimer's disease based on cortical thickness using AdaBoost and combination feature selection method *Int. Conf. on Information and Management Engineering* (Berlin: Springer) pp 392–401

[69] Yang S-T, Lee J-D, Huang C-H, Wang J-J, Hsu W-C and Wai Y-Y 2010 Computer-aided diagnosis of Alzheimer's disease using multiple features with artificial neural network *Pacific Rim International Conference on Artificial Intelligence* (Berlin: Springer) pp 699–705

[70] Patil M and Yardi A 2011 Diagnosis of Alzheimer's disease from 3D MR images with statistical features of hippocampus *Comput. Intell. Inf. Technol.* (Berlin: Springer) pp 744–9

[71] Rao M M *et al* 2009 Automated diagnosis of early Alzheimer's disease using fuzzy neural network *4th Eur. Conf. of the Int. Federation for Medical and Biological Engineering* (Berlin: Springer) pp 1455–8

[72] Álvarez I, Górriz J M, Ramrez J, Salas-Gonzalez D, López M, Puntonet C and Segovia F 2009 Alzheimer's diagnosis using eigenbrains and support vector machines *Electron. Lett.* **45** 342–3

[73] Álvarez I, Górriz J M, Ramrez J, Salas-Gonzalez D, López M, Puntonet C G and Segovia F 2009 Independent component analysis of SPECT images to assist the Alzheimer's disease diagnosis *The Sixth Int. Symp. on Neural Networks (ISNN 2009)* (Berlin: Springer) pp 411–9

[74] Salas-Gonzalez D, Górriz J M, Ramrez J, López M, Álvarez I, Segovia F and Puntonet C G 2009 Selecting regions of interest for the diagnosis of Alzheimer's disease in brain SPECT images using Welch's *t*-test *Int. Work-Conf. on Artificial Neural Networks* (Berlin: Springer) pp 965–72

[75] López M, Ramrez J, Górriz J M, Álvarez I, Salas-Gonzalez D, Segovia F and Puntonet C G 2009 Computer aided diagnosis of Alzheimer's disease using principal component analysis and Bayesian classifiers *The Sixth Int. Symp. on Neural Networks (ISNN 2009)* (Berlin: Springer) pp 213–21

[76] Stoeckel J, Ayache N, Malandain G, Koulibaly P M, Ebmeier K P and Darcourt J 2004 Automatic classification of SPECT images of Alzheimer's disease patients and control subjects *Int. Conf. on Medical Image Computing and Computer-Assisted Intervention* (Berlin: Springer) pp 654–62

[77] University of Washington Division of Nuclear Medicine 2019 Simulation System for Emission Tomography http://depts.washington.edu/simset/html/simset_main.html/ (Accessed: 13 February 2019)

[78] Stoeckel J, Malandain G, Migneco O, Koulibaly P M, Robert P, Ayache N and Darcourt J 2001 Classification of SPECT images of normal subjects versus images of Alzheimer's disease patients *Int. Conf. on Medical Image Computing and Computer-Assisted Intervention* (Berlin: Springer) pp 666–74

[79] Fung G and Stoeckel J 2007 SVM feature selection for classification of SPECT images of Alzheimer's disease using spatial information *Knowl. Inf. Syst.* **11** 243–58

[80] Polikar R, Keinert F and Greer M H 2001 Wavelet analysis of event related potentials for early diagnosis of Alzheimer's disease *Wavelets in Signal and Image Analysis* (Berlin: Springer) pp 453–78

[81] Plocharski M *et al* 2016 Extraction of sulcal medial surface and classification of Alzheimer's disease using sulcal features *Comput. Methods Programs Biomed.* **133** 35–44

[82] Beheshti I *et al* 2016 Structural MRI-based detection of Alzheimer's disease using feature ranking and classification error *Comput. Methods Programs Biomed.* **137** 177–93

[83] Garali I, Adel M, Bourennane S and Guedj E 2016 Brain region ranking for 18FDG-PET computer-aided diagnosis of Alzheimer's disease *Biomed. Signal Process. Control* **27** 15–23

[84] Hosseini-Asl E, Keynton R and El-Baz A 2016 Alzheimer's disease diagnostics by adaptation of 3D convolutional network *2016 IEEE Int. Conf. on Image Processing (ICIP)* pp 126–30

[85] Beheshti I, Demirel H and Matsuda H 2017 Classification of Alzheimer's disease and prediction of mild cognitive impairment-to-Alzheimer's conversion from structural magnetic resource imaging using feature ranking and a genetic algorithm *Comput. Biol. Med.* **83** 109–19

[86] Tong T, Gray K, Gao Q, Chen L and Rueckert D 2017 Multi-modal classification of Alzheimer's disease using nonlinear graph fusion *Pattern Recogn.* **63** 171–81

[87] Ahmed O B, Benois-Pineau J, Allard M, Catheline G and Amar C B 2017 Recognition of Alzheimer's disease and mild cognitive impairment with multimodal image-derived biomarkers and multiple kernel learning *Neurocomputing* **220** 98–110

[88] El-Gamal F E A, Elmogy M M, Ghazal M, AtwanA, Barnes G N, Casanova M F, Keynton R and El-Baz A S 2017 A novel CAD system for local and global early diagnosis of Alzheimer's disease based on PiB-PET scans *2017 IEEE Int. Conf. on Image Processing (ICIP)* pp 3270–4

[89] Jha D, Kim J-I and Kwon G-R 2017 Diagnosis of Alzheimer's disease using dual-tree complex wavelet transform, PCA, and feed-forward neural network *J. Healthc. Eng.* **2017** 1–13

[90] Çevik A, Weber G-W, Eyüboğlu B M and Oğuz K K A 2017 Voxel-Mars: a method for early detection of Alzheimer's disease by classification of structural brain MRI *Ann. Oper. Res.* **258** 31–57

[91] Lahmiri S and Shmuel A 2018 Performance of machine learning methods applied to structural MRI and ADAS cognitive scores in diagnosing Alzheimer's disease *Biomed. Signal Process. Control* **52** 414–9

[92] Kamathe R S and Joshi K R 2018 A novel method based on independent component analysis for brain MR image tissue classification into CSF, WM and GM for atrophy detection in Alzheimer's disease *Biomed. Signal Process. Control* **40** 41–8

[93] Luk C C, Ishaque A, Khan M, Ta D, Chenji S, Yang Y-H, Eurich D and Kalra S 2018 Alzheimer's disease: 3-dimensional MRI texture for prediction of conversion from mild cognitive impairment *Alzheimer's Dement.* **10** 755–63

[94] El-Gamal F E A, Elmogy M M, Ghazal M, Atwan A, Casanova M F, Barnes G N, Keynton R, El-Baz A S and Khalil A 2018 A novel early diagnosis system for mild cognitive impairment based on local region analysis: a pilot study *Front. Hum. Neurosci.* **11** 643

[95] El-Gamal F E A, Elmogy M M, Atwan A, Ghazal M, Barnes G N, Hajjdiab H, Keynton R and El-Baz A S 2018 Significant region-based framework for early diagnosis of Alzheimer's disease using ^{11}C PiB-PET scans *2018 24th Int. Conf. on Pattern Recognition (ICPR)* pp 2989–94

[96] El-Gamal F E A, Elmogy M M, Hajjdiab H, Ghazal M, Soliman H, Atwan A, Keynton R, El-Baz A S and Barnes G N 2018 A cortical based diagnosis system for MCI based on SMRI features fusion *2018 IEEE Int. Conf. on Imaging Systems and Techniques (IST)* pp 1–6

[97] El-Gamal F E A, Elmogy M M, Khalil A, Ghazal M, Soliman H, Atwan A, Keynton R, Barnes G N and El-Baz A S 2018 A significant regional-based diagnosis system for early detection of Alzheimer's disease using SMRI scans *2018 IEEE Int. Symp. on Signal Processing and Information Technology (ISSPIT)* pp 407–12

[98] Ju R, Hu C, Zhou P and Li Q 2019 Early diagnosis of Alzheimer's disease based on resting-state brain networks and deep learning *IEEE/ACM Trans. Comput. Biol. Bioinform.* **16** 244–57

[99] Hajiesmaeili M, Bagherinakhjavanlo B, Dehmeshki J and Ellis T 2012 Segmentation of the hippocampus for detection of Alzheimer's disease *Int. Symp. on Visual Computing* (Berlin: Springer) pp 42–50

[100] Cattell L, Schnabel J A, Declerck J and Hutton C 2014 Combined PET-MR brain registration to discriminate between Alzheimer's disease and healthy controls *Int. Workshop on Biomedical Image Registration* (Berlin: Springer) pp 134–43

[101] Leung K K *et al* 2010 Increasing power to predict mild cognitive impairment conversion to Alzheimer's disease using hippocampal atrophy rate and statistical shape models *Int. Conf. on Medical Image Computing and Computer-Assisted Intervention* (Berlin: Springer) pp 125–32

[102] Risser L, Vialard F-X, Wolz R, Holm D D and Rueckert D 2010 Simultaneous fine and coarse diffeomorphic registration: application to atrophy measurement in Alzheimer's disease *Int. Conf. on Medical Image Computing and Computer-Assisted Intervention* (Berlin: Springer) pp 610–7

[103] Simpson I J, Woolrich M, Groves A R and Schnabel J A 2011 Longitudinal brain MRI analysis with uncertain registration *Int. Conf. on Medical Image Computing and Computer-Assisted Intervention* (Berlin: Springer) pp 647–54

[104] Bossa M N, Zacur E and Olmos S 2009 Tensor-based morphometry with mappings parameterized by stationary velocity fields in Alzheimer's disease neuroimaging initiative *Int. Conf. on Medical Image Computing and Computer-Assisted Intervention* (Berlin: Springer) pp 240–7

[105] Martinez-Murcia F J, Ortiz A, Górriz J M, Ramrez J and Illán I 2015 A volumetric radial LBP projection of MRI brain images for the diagnosis of Alzheimer's disease *Int. Work-Conf. on the Interplay Between Natural and Artificial Computation* (Berlin: Springer) pp 19–28

[106] Meena A and Raja K 2014 K-means segmentation of Alzheimer's disease in PET scan datasets—an implementation *International Joint Conference on Advances in Signal Processing and Information Technology* (Cham: Springer) pp 168–72

[107] Pagani M, Salmaso D, Rodriguez G, Nardo D and Nobili F 2009 Principal component analysis in mild and moderate Alzheimer's disease—a novel approach to clinical diagnosis *Psychiatry Res.: Neuroimaging* **173** 8–14

[108] Aggarwal N, Rana B and Agrawal R 2015 3D discrete wavelet transform for computer aided diagnosis of Alzheimer's disease using T1-weighted brain MRI *Int. J. Imaging Syst. Technol.* **25** 179–90

[109] Ghorbanian P, Devilbiss D M, Simon A J, Bernstein A, Hess T and Ashrafiuon H 2012 Discrete wavelet transform EEG features of Alzheimer's disease in activated states *2012 Annu. Int. Conf. of the IEEE Engineering in Medicine and Biology Society* (Piscataway, NJ: IEEE) pp 2937–40

[110] Ghorbanian P, Devilbiss D M, Hess T, Bernstein A, Simon A J and Ashrafiuon H 2015 Exploration of EEG features of Alzheimer's disease using continuous wavelet transform *Med. Biol. Eng. Comput.* **53** 843–55

[111] Kontos D, Megalooikonomou V, Pokrajac D, Lazarevic A, Obradovic Z, Boyko O B, Ford J, Makedon F and Saykin A J 2004 Extraction of discriminative functional MRI activation patterns and an application to Alzheimer's disease *Int. Conf. on Medical Image Computing and Computer-Assisted Intervention* (Berlin: Springer) pp 727–35

[112] Kovalev V A, Thurfjell L, Lundqvist R and Pagani M 2006 Asymmetry of SPECT perfusion image patterns as a diagnostic feature for Alzheimer's disease *Int. Conf. on Medical Image Computing and Computer-Assisted Intervention* (Berlin: Springer) pp 421–8

[113] Liu Y *et al* 2004 Discriminative MR image feature analysis for automatic schizophrenia and Alzheimer's disease classification *Int. Conf. on Medical Image Computing and Computer-assisted Intervention* (Berlin: Springer) pp 393–401

[114] Zhu X, Suk H-I and Shen D 2014 Multi-modality canonical feature selection for Alzheimer's disease diagnosis *Int. Conf. on Medical Image Computing and Computer-Assisted Intervention* (Berlin: Springer) pp 162–9

[115] Trojacanec K, Kitanovski I, Dimitrovski I and Loshkovska S 2015 New representation of information extracted from MRI volumes applied to Alzheimer's disease *ICT Innovations 2014* (Berlin: Springer) pp 249–58

[116] Fiot J-B, Risser L, Cohen L D, Fripp J and Vialard F-X 2012 Local vs global descriptors of hippocampus shape evolution for Alzheimer's longitudinal population analysis *Int. Workshop on Spatio-temporal Image Analysis for Longitudinal and Time-Series Image Data* (Berlin: Springer) pp 13–24

[117] Rueda A, Arevalo J, Cruz A, Romero E and González F A 2012 Bag of features for automatic classification of Alzheimer's disease in magnetic resonance images *Iberoamerican Congress on Pattern Recognition* (Berlin: Springer) pp 559–66

[118] Coupé P, Eskildsen S F, Manjón J V, Fonov V and Collins D L 2011 Simultaneous segmentation and grading of hippocampus for patient classification with Alzheimer's disease *Int. Conf. on Medical Image Computing and Computer-Assisted Intervention* (Berlin: Springer) pp 149–57

[119] Hamou A *et al* 2010 Cluster analysis and decision trees of MR imaging in patients suffering Alzheimer's *Trends in Practical Applications of Agents and Multiagent Systems* (Berlin: Springer) pp 477–84

[120] Ortiz A, Moreno-Estévez M, Górriz J M, Ramrez J, Garca-Tarifa M J, Munilla J and Haba N 2015 Automated diagnosis of Alzheimer's disease by integrating genetic bio-markers and tissue density information *Int. Work-Conf. on the Interplay Between Natural and Artificial Computation* (Berlin: Springer) pp 1–8

[121] Eblenkamp M, Haag L, Pfeifer S and Wintermantel E 2015 Software-based system for automatic 3D dendritic spine evaluation for research on Alzheimers disease *6th Eur. Conf. of the Int. Federation for Medical and Biological Engineering* (Berlin: Springer) pp 138–41

[122] Rusina R, Kukal J, Bělíček T, Buncová M and Matěj R 2010 Use of fuzzy edge single-photon emission computed tomography analysis in definite Alzheimer's disease—a retrospective study *BMC Med. Imaging* **10** 20

[123] Kodama N, Kawase Y and Okamoto K 2007 Application of texture analysis to differentiation of dementia with lewy bodies from Alzheimer's disease on magnetic resonance images *World Congress on Medical Physics and Biomedical Engineering 2006* (Berlin: Springer) pp 1444–6

[124] Stanchev P L and Fotouhi F 2002 Medimage—a multimedia database management system for Alzheimer's disease patients *Int. Conf. on Advances in Visual Information Systems* (Berlin: Springer) pp 187–93

[125] Ertek G, Tokdil B and Günaydn İ 2014 Risk factors and identifiers for Alzheimer's disease: a data mining analysis *Industrial Conf. on Data Mining* (Berlin: Springer) pp 1–11

[126] Fletcher P T, Powell S, Foster N L and Joshi S C 2007 Quantifying metabolic asymmetry modulo structure in Alzheimer's disease *Biennial Int. Conf. on Information Processing in Medical Imaging* (Berlin: Springer) pp 446–57

[127] Hasan M K, Lee W, Park B and Han K 2011 Connectivity analysis of hippocampus in Alzheimer's brain using probabilistic tractography *Int. Conf. on Intelligent Computing* (Berlin: Springer) pp 521–8

[128] Song M, Heo G E and Lee D 2015 Identifying the landscape of Alzheimer's disease research with network and content analysis *Scientometrics* **102** 905–27

[129] Agarwal M and Mostafa J 2009 Image retrieval for Alzheimer's disease detection *MICCAI Int. Workshop on Medical Content-Based Retrieval for Clinical Decision Support* (Berlin: Springer) pp 49–60

[130] Mahmoud A H 2014 Utilizing radiation for smart robotic applications using visible, thermal, and polarization images *PhD Dissertation* University of Louisville, KY

[131] Mahmoud A, El-Barkouky A, Graham J and Farag A 2014 Pedestrian detection using mixed partial derivative based histogram of oriented gradients *2014 IEEE Int. Conf. on Image Processing (ICIP)* (Piscataway, NJ: IEEE) pp 2334–7

[132] El-Barkouky A, Mahmoud A, Graham J and Farag A 2013 An interactive educational drawing system using a humanoid robot and light polarization *2013 IEEE Int. Conf. on Image Processing* (Piscataway, NJ: IEEE) pp 3407–11

[133] Mahmoud A H, El-Melegy M T and Farag A A 2012 Direct method for shape recovery from polarization and shading *2012 19th IEEE Int. Conf. on Image Processing* (Piscataway, NJ: IEEE) pp 1769–72

[134] Ali A M, Farag A A and El-Baz A 2007 Graph cuts framework for kidney segmentation with prior shape constraints *Proc. of Int. Conf. on Medical Image Computing and Computer-Assisted Intervention, (MICCAI'07) (Brisbane, Australia, 29 October–2 November 2007)* **vol 1** pp 384–92

[135] Chowdhury A S, Roy R, Bose S, Elnakib F K A and El-Baz A 2012 Non-rigid biomedical image registration using graph cuts with a novel data term *Proc. of IEEE Int. Symp. on Biomedical Imaging: From Nano to Macro, (ISBI'12) (Barcelona, Spain, 2–5 May 2012)* pp 446–9

[136] El-Baz A, Farag A A, Yuksel S E, El-Ghar M E, Eldiasty T A and Ghoneim M A 2007 Application of deformable models for the detection of acute renal rejection *Deformable Models* (New York: Springer) pp 293–333

[137] El-Baz A, Farag A, Fahmi R, Yuksel S, El-Ghar M A and Eldiasty T 2006 Image analysis of renal DCE MRI for the detection of acute renal rejection *Proc. of IAPR Int. Conf. on Pattern Recognition (ICPR'06) (Hong Kong, 20–24 August)* pp 822–82

[138] El-Baz A, Farag A, Fahmi R, Yuksel S, Miller W, El-Ghar M A, El-Diasty T and Ghoneim M 2006 A new CAD system for the evaluation of kidney diseases using DCE-MRI *Proc. of Int. Conf. on Medical Image Computing and Computer-Assisted Intervention, (MICCAI'08) (Copenhagen, Denmark, 1–6 October 2006)* pp 446–53

[139] El-Baz A, Gimel'farb G and El-Ghar M A 2008 A novel image analysis approach for accurate identification of acute renal rejection *Proc. of IEEE Int. Conf. on Image Processing, (ICIP'08) (San Diego, California, USA, 12–15 October 2008)* pp 1812–5

[140] El-Baz A, Gimel'farb G and El-Ghar M A 2008 Image analysis approach for identification of renal transplant rejection *Proc. of IAPR Int. Conf. on Pattern Recognition, (ICPR'08) (Tampa, Florida, USA, 8–11 December 2008)* pp 1–4

[141] El-Baz A, Gimel'farb G and El-Ghar M A 2007 New motion correction models for automatic identification of renal transplant rejection *Proc. of Int. Conf. on Medical Image Computing and Computer-Assisted Intervention, (MICCAI'07) (Brisbane, Australia, 29 October–2 November, 2007)* pp 235–43

[142] Farag A, El-Baz A, Yuksel S, El-Ghar M A and Eldiasty T 2006 A framework for the detection of acute rejection with Dynamic Contrast Enhanced Magnetic Resonance

Imaging *Proc. of IEEE Int. Symp. on Biomedical Imaging: From Nano to Macro, (ISBI'06)* *(Arlington, Virginia, USA, 6–9 April 2006)* pp 418–21

[143] Khalifa F, Beache G M, El-Ghar M A, El-Diasty T, Gimel'farb G, Kong M and El-Baz A 2013 Dynamic contrast-enhanced MRI-based early detection of acute renal transplant rejection *IEEE Trans. Med. Imaging* **32** 1910–27

[144] Khalifa F, El-Baz A, Gimel'farb G and El-Ghar M A 2010 Non-invasive image-based approach for early detection of acute renal rejection *Proc. of Int. Conf. Medical Image Computing and Computer-Assisted Intervention, (MICCAI'10) (Beijing, China, 20–24 September 2010)* pp 10–8

[145] Khalifa F, El-Baz A, Gimel'farb G, Ouseph R and El-Ghar M A 2010 Shape-appearance guided level-set deformable model for image segmentation *Proc. of IAPR Int. Conf. on Pattern Recognition, (ICPR'10) (Istanbul, Turkey, 23–26 August 2010)* pp 4581–4

[146] Khalifa F, El-Ghar M A, Abdollahi B, Frieboes H, El-Diasty T and El-Baz A 2013 A comprehensive non-invasive framework for automated evaluation of acute renal transplant rejection using DCE-MRI *NMR Biomed.* **26** 1460–70

[147] Khalifa F, El-Ghar M A, Abdollahi B, Frieboes H B, El-Diasty T and El-Baz A 2014 Dynamic contrast-enhanced MRI-based early detection of acute renal transplant rejection *2014 Annu. Scientific Meeting and Educational Course Brochure of the Society of Abdominal Radiology, (SAR'14) Boca Raton, Florida* (23–28 March 2014) CID: 1855912

[148] Khalifa F, Elnakib A, Beache G M, Gimel'farb G, El-Ghar M A, Sokhadze G, Manning S, McClure P and El-Baz A 2011 3D kidney segmentation from CT images using a level set approach guided by a novel stochastic speed function *Proc. of Int. Conf. Medical Image Computing and Computer-Assisted Intervention, (MICCAI'11) (Toronto, Canada, 18–22 September 2011)* pp 587–94

[149] Khalifa F, Gimel'farb G, El-Ghar M A, Sokhadze G, Manning S, McClure P, Ouseph R and El-Baz A 2011 A new deformable model-based segmentation approach for accurate extraction of the kidney from abdominal CT images *Proc. of IEEE Int. Conf. on Image Processing, (ICIP'11) (Brussels, Belgium, 11–14 September 2011)* pp 3393–6

[150] Mostapha M, Khalifa F, Alansary A, Soliman A, Suri J and El-Baz A 2014 Computer-aided diagnosis systems for acute renal transplant rejection: challenges and methodologies *Abdomen and Thoracic Imaging* ed A El-Baz, L saba and J Suri (Berlin: Springer) pp 1–35

[151] Shehata M, Khalifa F, Hollis E, Soliman A, Hosseini-Asl E, El-Ghar M A, El-Baz M, Dwyer A C, El-Baz A and Keynton R 2016 A new non-invasive approach for early classification of renal rejection types using diffusion-weighted MRI *IEEE Int. Conf. on Image Processing (ICIP), 2016* (Piscataway, NJ: IEEE) pp 136–40

[152] Khalifa F, Soliman A, Takieldeen A, Shehata M, Mostapha M, Shaffie A, Ouseph R, Elmaghraby A and El-Baz A 2016 Kidney segmentation from CT images using a 3D NMF-guided active contour model *IEEE 13th Int. Symp. on Biomedical Imaging (ISBI), 2016* (Piscataway, NJ: IEEE) pp 432–5

[153] Shehata M, Khalifa F, Soliman A, Takieldeen A, El-Ghar M A, Shaffie A, Dwyer A C, Ouseph R, El-Baz A and Keynton R 2016 3D diffusion MRI-based CAD system for early diagnosis of acute renal rejection *2016 IEEE 13th Int. Symp. on Biomedical Imaging (ISBI)* (Piscataway, NJ: IEEE) pp 1177–80

[154] Shehata M, Khalifa F, Soliman A, Alrefai R, El-Ghar M A, Dwyer A C, Ouseph R and El-Baz A 2015 A level set-based framework for 3D kidney segmentation from diffusion MR images *IEEE Int. Conf. on Image Processing (ICIP), 2015* (Piscataway, NJ: IEEE) pp 4441–5

[155] Shehata M, Khalifa F, Soliman A, El-Ghar M A, Dwyer A C, Gimel'farb G, Keynton R and El-Baz A 2016 A promising non-invasive CAD system for kidney function assessment *Int. Conf. on Medical Image Computing and Computer-Assisted Intervention* (Berlin: Springer) pp 613–21

[156] Khalifa F, Soliman A, Elmaghraby A, Gimel'farb G and El-Baz A 2017 3D kidney segmentation from abdominal images using spatial-appearance models *Comput. Math. Methods Med.* **2017** 1–10

[157] Hollis E, Shehata M, Khalifa F, El-Ghar M A, El-Diasty T and El-Baz A 2016 Towards non-invasive diagnostic techniques for early detection of acute renal transplant rejection: a review *Egypt. J. Radiol. Nucl. Med.* **48** 257–69

[158] Shehata M, Khalifa F, Soliman A, El-Ghar M A, Dwyer A C and El-Baz A 2017 Assessment of renal transplant using image and clinical-based biomarkers *Proc. of 13th Annu. Scientific Meeting of American Society for Diagnostics and Interventional Nephrology (ASDIN'17) (New Orleans, LA, USA, 10–12 February 2017)*

[159] Shehata M, Khalifa F, Soliman A, El-Ghar M A, Dwyer A C and El-Baz A 2017 Early assessment of acute renal rejection *Proc. of 12th Annual Scientific Meeting of American Society for Diagnostics and Interventional Nephrology (ASDIN'16) (Pheonix, AZ, USA, 19–21 February 2016)*

[160] Eltanboly A, Ghazal M, Hajjdiab H, Shalaby A, Switala A, Mahmoud A, Sahoo P, El-Azab M and El-Baz A 2019 Level sets-based image segmentation approach using statistical shape priors *Appl. Math. Comput.* **340** 164–79

[161] Shehata M, Mahmoud A, Soliman A, Khalifa F, Ghazal M, El-Ghar M A, El-Melegy M and El-Baz A 2018 3D kidney segmentation from abdominal diffusion MRI using an appearance-guided deformable boundary *PLoS One* **13** e0200082

[162] Khalifa F, Beache G, El-Baz A and Gimel'farb G 2010 Deformable model guided by stochastic speed with application in cine images segmentation *Proc. of IEEE Int. Conf. on Image Processing, (ICIP'10) (Hong Kong, 26–29 September 2010)* pp 1725–8

[163] Khalifa F, Beache G M, Elnakib A, Sliman H, Gimel'farb G, Welch K C and El-Baz A 2013 A new shape-based framework for the left ventricle wall segmentation from cardiac first-pass perfusion MRI *Proc. of IEEE Int. Symp. on Biomedical Imaging: From Nano to Macro, (ISBI'13) (San Francisco, CA, 7–11 April 2013)* pp 41–4

[164] Khalifa F, Beache G M, Elnakib A, Sliman H, Gimel'farb G, Welch K C and El-Baz A 2012 A new nonrigid registration framework for improved visualization of transmural perfusion gradients on cardiac first–pass perfusion MRI *Proceedings of IEEE International Symposium on Biomedical Imaging: From Nano to Macro, (ISBI'12) (Barcelona, Spain, 2–5 May 2012)* pp 828–31

[165] Khalifa F, Beache G M, Firjani A, Welch K C, Gimel'farb G and El-Baz A 2012 A new nonrigid registration approach for motion correction of cardiac first-pass perfusion MRI *Proc. of IEEE Int. Conf. on Image Processing, (ICIP'12) (Lake Buena Vista, Florida, 30 September–3 October 2012)* pp 1665–8

[166] Khalifa F, Beache G M, Gimel'farb G and El-Baz A 2012 A novel CAD system for analyzing cardiac first-pass MR images *Proc. of IAPR Int. Conf. on Pattern Recognition (ICPR'12) (Tsukuba Science City, Japan, 11–15 November 2012)* pp 77–80

[167] Khalifa F, Beache G M, Gimel'farb G and El-Baz A 2011 A novel approach for accurate estimation of left ventricle global indexes from short-axis cine MRI *Proc. of IEEE Int. Conf. on Image Processing, (ICIP'11) (Brussels, Belgium, 11–14 September 2011)* pp 2645–9

[168] Khalifa F, Beache G M, Gimel'farb G, Giridharan G A and El-Baz A 2011 A new image-based framework for analyzing cine images *Handbook of Multi Modality State-of-the-Art Medical Image Segmentation and Registration Methodologies* vol 2 ed A El-Baz, U R Acharya, M Mirmedhdi and J S Suri (New York: Springer) ch 3, pp 69–98

[169] Khalifa F, Beache G M, Gimel'farb G, Giridharan G A and El-Baz A 2012 Accurate automatic analysis of cardiac cine images *IEEE Trans. Biomed. Eng.* **59** 445–55

[170] Khalifa F, Beache G M, Nitzken M, Gimel'farb G, Giridharan G A and El-Baz A 2011 Automatic analysis of left ventricle wall thickness using short-axis cine CMR images *Proc. of IEEE Int. Symp. on Biomedical Imaging: From Nano to Macro, (ISBI'11) (Chicago, Illinois, 30 March–2 April 2011)* pp 1306–9

[171] Nitzken M, Beache G, Elnakib A, Khalifa F, Gimel'farb G and El-Baz A 2012 Accurate modeling of tagged CMR 3D image appearance characteristics to improve cardiac cycle strain estimation *2012 19th IEEE Int. Conf. on Image Processing (ICIP)* (Piscataway, NJ: IEEE) pp 521–4

[172] Nitzken M, Beache G, Elnakib A, Khalifa F, Gimel'farb G and El-Baz A 2012 Improving full-cardiac cycle strain estimation from tagged CMR by accurate modeling of 3D image appearance characteristics *2012 9th IEEE Int. Symp. on Biomedical Imaging (ISBI)* (Piscataway, NJ: IEEE) pp 462–5

[173] Nitzken M J, El-Baz A S and Beache G M 2012 Markov–Gibbs random field model for improved full-cardiac cycle strain estimation from tagged CMR *J. Cardiovasc. Magn. Reson.* **14** 1–2

[174] Sliman H, Elnakib A, Beache G, Elmaghraby A and El-Baz A 2014 Assessment of myocardial function from cine cardiac MRI using a novel 4D tracking approach *J. Comput. Sci. Syst. Biol.* **7** 169–73

[175] Sliman H, Elnakib A, Beache G M, Soliman A, Khalifa F, Gimel'farb G, Elmaghraby A and El-Baz A 2014 A novel 4D PDE-based approach for accurate assessment of myocardium function using cine cardiac magnetic resonance images *Proc. of IEEE Int. Conf. on Image Processing (ICIP'14) (Paris, France, 27–30 October 2014)* pp 3537–41

[176] Sliman H, Khalifa F, Elnakib A, Beache G M, Elmaghraby A and El-Baz A 2013 A new segmentation-based tracking framework for extracting the left ventricle cavity from cine cardiac MRI *Proc. of IEEE Int. Conf. on Image Processing, (ICIP'13) (Melbourne, Australia, 15–18 September 2013)* pp 685–9

[177] Sliman H, Khalifa F, Elnakib A, Soliman A, Beache G M, Elmaghraby A, Gimel'farb G and El-Baz A 2013 Myocardial borders segmentation from cine MR images using bi-directional coupled parametric deformable models *Med. Phys.* **40** 1–13

[178] Sliman H, Khalifa F, Elnakib A, Soliman A, Beache G M, Gimel'farb G, Emam A, Elmaghraby A and El-Baz A 2013 Accurate segmentation framework for the left ventricle wall from cardiac cine MRI *Proc. of Int. Symp. on Computational Models for Life Science, (CMLS'13) (Sydney, Australia, 27–29 November 2013)* **vol 1559** pp 287–96

[179] Eladawi N, Elmogy M, Ghazal M, Helmy O, Aboelfetouh A, Riad A, Schaal S and El-Baz A 2018 Classification of retinal diseases based on OCT images *Front. Biosci.* **23** 247–64

[180] ElTanboly A, Ismail M, Shalaby A, Switala A, El-Baz A, Schaal S, Gimel'farb G and El-Azab M 2017 A computer-aided diagnostic system for detecting diabetic retinopathy in optical coherence tomography images *Med. Phys.* **44** 914–23

[181] Sandhu H S, El-Baz A and Seddon J M 2018 Progress in automated deep learning for macular degeneration *JAMA Ophthalmol.* **136** 1366–7

[182] Abdollahi B, Civelek A C, Li X-F, Suri J and El-Baz A 2014 PET/CT nodule segmentation and diagnosis: a survey *Multi Detector CT Imaging* ed L Saba and J S Suri (London: Taylor and Francis) ch 30, pp 639–51

[183] Abdollahi B, El-Baz A and Amini A A 2011 A multi-scale non-linear vessel enhancement technique *2011 Annu. Int. Conf. of the IEEE Engineering in Medicine and Biology Society, EMBC* (Piscataway, NJ: IEEE) pp 3925–9

[184] Abdollahi B, Soliman A, Civelek A, Li X-F, Gimel'farb G and El-Baz A 2012 A novel Gaussian scale space-based joint MGRF framework for precise lung segmentation *Proc. of IEEE Int. Conf. on Image Processing, (ICIP'12)* (Piscataway, NJ: IEEE) pp 2029–32

[185] Abdollahi B, Soliman A, Civelek A, Li X-F, Gimel'farb G and El-Baz A 2012 A novel 3D joint MGRF framework for precise lung segmentation *Machine Learning in Medical Imaging* (Berlin: Springer) pp 86–93

[186] Ali A M, El-Baz A S and Farag A A 2007 A novel framework for accurate lung segmentation using graph cuts *Proc. of IEEE Int. Symp. on Biomedical Imaging: From Nano to Macro, (ISBI'07)* (Piscataway, NJ: IEEE) pp 908–11

[187] El-Baz A, Beache G M, Gimel'farb G, Suzuki K and Okada K 2013 Lung imaging data analysis *Int. J. Biomed. Imaging* **2013** 1–2

[188] El-Baz A, Beache G M, Gimel'farb G, Suzuki K, Okada K, Elnakib A, Soliman A and Abdollahi B 2013 Computer-aided diagnosis systems for lung cancer: challenges and methodologies *Int. J. Biomed. Imaging* **2013** 1–46

[189] El-Baz A, Elnakib A, Abou El-Ghar M, Gimel'farb G, Falk R and Farag A 2013 Automatic detection of 2D and 3D lung nodules in chest spiral CT scans *Int. J. Biomed. Imaging* **2013** 1–11

[190] El-Baz A, Farag A A, Falk R and La Rocca R 2003 A unified approach for detection, visualization, and identification of lung abnormalities in chest spiral CT scans *International Congress Series* vol 1256 (Amsterdam: Elsevier) pp 998–1004

[191] El-Baz A, Farag A A, Falk R and La Rocca R 2002 Detection, visualization and identification of lung abnormalities in chest spiral CT scan: Phase-I *Proc. of Int. conf. on Biomedical Engineering (Cairo, Egypt,* **vol 12**

[192] El-Baz A, Farag A, Gimel'farb G, Falk R, El-Ghar M A and Eldiasty T 2006 A framework for automatic segmentation of lung nodules from low dose chest CT scans *Proc. of Int. Conf. on Pattern Recognition, (ICPR'06)* **vol 3** (Piscataway, NJ: IEEE) pp 611–4

[193] El-Baz A, Farag A, Gimel'farb G, Falk R and El-Ghar M A 2011 A novel level set-based computer-aided detection system for automatic detection of lung nodules in low dose chest computed tomography scans *Lung Imaging and Computer Aided Diagnosis* vol 10 pp 221–38

[194] El-Baz A, Gimel'farb G, Abou El-Ghar M and Falk R 2012 Appearance-based diagnostic system for early assessment of malignant lung nodules *Proc. of IEEE Int. Conf. on Image Processing, (ICIP'12)* (Piscataway, NJ: IEEE) pp 533–6

[195] El-Baz A, Gimel'farb G and Falk R 2011 A novel 3D framework for automatic lung segmentation from low dose CT images *Lung Imaging and Computer Aided Diagnosis*; El-Baz A and Suri J S (London: Taylor and Francis) ch 1, pp 1–16

[196] El-Baz A, Gimel'farb G, Falk R and El-Ghar M 2010 Appearance analysis for diagnosing malignant lung nodules *Proc. of IEEE Int. Symp. on Biomedical Imaging: From Nano to Macro (ISBI'10)* (Piscataway, NJ: IEEE) pp 193–6

[197] El-Baz A, Gimel'farb G, Falk R and El-Ghar M A 2011 A novel level set-based CAD system for automatic detection of lung nodules in low dose chest CT scans *Lung Imaging*

and Computer Aided Diagnosis **vol 1**; El-Baz A and Suri J S (London: Taylor and Francis) ch 10, pp 221–38

[198] El-Baz A, Gimel'farb G, Falk R and El-Ghar M A 2008 A new approach for automatic analysis of 3D low dose CT images for accurate monitoring the detected lung nodules *Proc. of Int. Conf. on Pattern Recognition, (ICPR'08)* (Piscataway, NJ: IEEE) pp 1–4

[199] El-Baz A, Gimel'farb G, Falk R and El-Ghar M A 2007 A novel approach for automatic follow-up of detected lung nodules *Proc. of IEEE Int. Conf. on Image Processing, (ICIP'07)* **vol 5** (Piscataway, NJ: IEEE) pp V–501

[200] El-Baz A, Gimel'farb G, Falk R and El-Ghar M A 2007 A new CAD system for early diagnosis of detected lung nodules *ICIP 2007 . IEEE Int. Conf. on Image Processing, 2007* **vol 2** (Piscataway, NJ: IEEE) pp II–461

[201] El-Baz A, Gimel'farb G, Falk R, El-Ghar M A and Refaie H 2008 Promising results for early diagnosis of lung cancer *Proc. of IEEE Int. Symp. on Biomedical Imaging: From Nano to Macro, (ISBI'08)* (Piscataway, NJ: IEEE) pp 1151–4

[202] El-Baz A, Gimel'farb G L, Falk R, Abou El-Ghar M, Holland T and Shaffer T 2008 A new stochastic framework for accurate lung segmentation *Proc. Medical Image Computing and Computer-Assisted Intervention, (MICCAI'08)* pp 322–30

[203] El-Baz A, Gimel'farb G L, Falk R, Heredis D and Abou El-Ghar M 2008 A novel approach for accurate estimation of the growth rate of the detected lung nodules *Proc. of Int. Workshop on Pulmonary Image Analysis* pp 33–42

[204] El-Baz A, Gimel'farb G L, Falk R, Holland T and Shaffer T 2008 A framework for unsupervised segmentation of lung tissues from low dose computed tomography images *Proc. of British Machine Vision, (BMVC'08)* pp 1–10

[205] El-Baz A, Gimel'farb G, Falk R and El-Ghar M A 2011 3D MGRF-based appearance modeling for robust segmentation of pulmonary nodules in 3D LDCT chest images *Lung Imaging and Computer Aided Diagnosis* (Boca Raton, FL: CRC Press) ch 3, pp 51–63

[206] El-Baz A, Gimel'farb G, Falk R and El-Ghar M A 2009 Automatic analysis of 3D low dose CT images for early diagnosis of lung cancer *Pattern Recogn.* **42** 1041–51

[207] El-Baz A, Gimel'farb G, Falk R, El-Ghar M A, Rainey S, Heredia D and Shaffer T 2009 Toward early diagnosis of lung cancer *Proc. of Medical Image Computing and Computer-Assisted Intervention, (MICCAI'09)* (Berlin: Springer) pp 682–9

[208] El-Baz A, Gimel'farb G, Falk R, El-Ghar M A and Suri J 2011 Appearance analysis for the early assessment of detected lung nodules *Lung Imaging and Computer Aided Diagnosis* (Boca Raton, FL: CRC Press) ch 17, pp 395–404

[209] El-Baz A, Khalifa F, Elnakib A, Nitkzen M, Soliman A, McClure P, Gimel'farb G and El-Ghar M A 2012 A novel approach for global lung registration using 3D Markov Gibbs appearance model *Proc. of Int. Conf. Medical Image Computing and Computer-Assisted Intervention, (MICCAI'12) (Nice, France, 1–5 October 2012)* pp 114–21

[210] El-Baz A, Nitzken M, Elnakib A, Khalifa F, Gimel'farb G, Falk R and El-Ghar M A 2011 3D shape analysis for early diagnosis of malignant lung nodules *Proc. of Int. Conf. Medical Image Computing and Computer-Assisted Intervention, (MICCAI'11) (Toronto, Canada, 18–22 September 2011)* pp 175–82

[211] El-Baz A, Nitzken M, Gimel'farb G, Van Bogaert E, Falk R, El-Ghar M A and Suri J 2011 Three-dimensional shape analysis using spherical harmonics for early assessment of

detected lung nodules *Lung Imaging and Computer Aided Diagnosis* (Boca Raton, FL: CRC Press) ch 19, pp 421–38

[212] El-Baz A, Nitzken M, Khalifa F, Elnakib A, Gimel'farb G, Falk R and El-Ghar M A 2011 3D shape analysis for early diagnosis of malignant lung nodules *Proc. of Int. Conf. on Information Processing in Medical Imaging, (IPMI'11) (Monastery Irsee, Germany (Bavaria), 3–8 July 2011)* pp 772–83

[213] El-Baz A, Nitzken M, Vanbogaert E, Gimel'Farb G, Falk R and Abo El-Ghar M 2011 A novel shape-based diagnostic approach for early diagnosis of lung nodules *2011 IEEE Int. Symp. on Biomedical Imaging: From Nano to Macro* (Piscataway, NJ: IEEE) pp 137–40

[214] El-Baz A, Sethu P, Gimel'farb G, Khalifa F, Elnakib A, Falk R and El-Ghar M A 2011 Elastic phantoms generated by microfluidics technology: validation of an imaged-based approach for accurate measurement of the growth rate of lung nodules *Biotechnol. J.* **6** 195–203

[215] El-Baz A, Sethu P, Gimel'farb G, Khalifa F, Elnakib A, Falk R and El-Ghar M A 2010 A new validation approach for the growth rate measurement using elastic phantoms generated by state-of-the-art microfluidics technology *Proc. of IEEE Int. Conf. on Image Processing, (ICIP'10) (Hong Kong, 26–29 September 2010)* pp 4381–3

[216] El-Baz A, Sethu P, Gimel'farb G, Khalifa F, Elnakib A, Falk R and Suri M A E-G J 2011 Validation of a new imaged-based approach for the accurate estimating of the growth rate of detected lung nodules using real CT images and elastic phantoms generated by state-of-the-art microfluidics technology *Handbook of Lung Imaging and Computer Aided Diagnosis* **vol 1**; El-Baz A and Suri J S (New York: Taylor and Francis) ch 18, pp 405–20

[217] El-Baz A, Soliman A, McClure P, Gimel'farb G, El-Ghar M A and Falk R 2012 Early assessment of malignant lung nodules based on the spatial analysis of detected lung nodules *Proc. of IEEE Int. Symp. on Biomedical Imaging: From Nano to Macro, (ISBI'12)* (Piscataway, NJ: IEEE) pp 1463–6

[218] El-Baz A, Yuksel S E, Elshazly S and Farag A A 2005 Non-rigid registration techniques for automatic follow-up of lung nodules *Proc. of Computer Assisted Radiology and Surgery, (CARS'05)* **vol 1281** (Amsterdam: Elsevier) pp 1115–20

[219] El-Baz A S and Suri J S 2011 *Lung Imaging and Computer Aided Diagnosis* (Boca Raton, FL: CRC Press)

[220] Soliman A, Khalifa F, Dunlap N, Wang B, El-Ghar M and El-Baz A 2016 An iso-surfaces based local deformation handling framework of lung tissues *2016 IEEE 13th Int. Symp. on Biomedical Imaging (ISBI)* (Piscataway, NJ: IEEE) pp 1253–9

[221] Soliman A, Khalifa F, Shaffie A, Dunlap N, Wang B, Elmaghraby A and El-Baz A 2016 Detection of lung injury using 4D-CT chest images *2016 IEEE 13th Int. Symp. on Biomedical Imaging (ISBI)* (Piscataway, NJ: IEEE) pp 1274–7

[222] Soliman A, Khalifa F, Shaffie A, Dunlap N, Wang B, Elmaghraby A, Gimel'farb G, Ghazal M and El-Baz A 2017 A comprehensive framework for early assessment of lung injury *2017 IEEE Int. Conf. on Image Processing (ICIP)* (Piscataway, NJ: IEEE) pp 3275–9

[223] Shaffie A, Soliman A, Ghazal M, Taher F, Dunlap N, Wang B, Elmaghraby A, Gimel'farb G and El-Baz A 2017 A new framework for incorporating appearance and shape features of lung nodules for precise diagnosis of lung cancer *2017 IEEE Int. Conf. on Image Processing (ICIP)* (Piscataway, NJ: IEEE) pp 1372–76

[224] Soliman A, Khalifa F, Shaffie A, Liu N, Dunlap N, Wang B, Elmaghraby A, Gimel'farb G and El-Baz A 2016 Image-based CAD system for accurate identification of lung injury *2016 IEEE Int. Conf. on Image Processing (ICIP)* (Piscataway, NJ: IEEE) pp 121–5

[225] Dombroski B, Nitzken M, Elnakib A, Khalifa F, El-Baz A and Casanova M F 2014 Cortical surface complexity in a population-based normative sample *Transl. Neurosci.* **5** 17–24

[226] El-Baz A, Casanova M, Gimel'farb G, Mott M and Switala A 2008 An MRI-based diagnostic framework for early diagnosis of dyslexia *Int. J. Comput. Assist. Radiol. Surg.* **3** 181–9

[227] El-Baz A, Casanova M, Gimel'farb G, Mott M, Switala A, Vanbogaert E and McCracken R 2008 A new CAD system for early diagnosis of dyslexic brains *Proc. Int. Conf. on Image Processing (ICIP'2008)* (Piscataway, NJ: IEEE) pp 1820–3

[228] El-Baz A, Casanova M F, Gimel'farb G, Mott M and Switwala A E 2007 A new image analysis approach for automatic classification of autistic brains *Proc. IEEE Int. Symp. on Biomedical Imaging: From Nano to Macro (ISBI'2007)* (Piscataway, NJ: IEEE) pp 352–5

[229] El-Baz A, Elnakib A, Khalifa F, El-Ghar M A, McClure P, Soliman A and Gimel'farb G 2012 Precise segmentation of 3-D magnetic resonance angiography *IEEE Trans. Biomed. Eng.* **59** 2019–29

[230] El-Baz A, Farag A, Gimel'farb G, El-Ghar M A and Eldiasty T 2006 Probabilistic modeling of blood vessels for segmenting MRA images *18th Int. Conf. on Pattern Recognition (ICPR'06)* **vol 3** (Piscataway, NJ: IEEE) pp 917–20

[231] El-Baz A, Farag A A, Gimel'farb G, El-Ghar M A and Eldiasty T 2006 A new adaptive probabilistic model of blood vessels for segmenting MRA images *Medical Image Computing and Computer-Assisted Intervention–MICCAI 2006* vol 4191 (Berlin: Springer) pp 799–806

[232] El-Baz A, Farag A A, Gimel'farb G and Hushek S G 2005 Automatic cerebrovascular segmentation by accurate probabilistic modeling of TOF-MRA images *Medical Image Computing and Computer-Assisted Intervention–MICCAI 2005* (Berlin: Springer) pp 34–42

[233] El-Baz A, Farag A, Elnakib A, Casanova M F, Gimel'farb G, Switala A E, Jordan D and Rainey S 2011 Accurate automated detection of autism related corpus callosum abnormalities *J. Med. Syst.* **35** 929–39

[234] El-Baz A, Farag A and Gimelfarb G 2005 Cerebrovascular segmentation by accurate probabilistic modeling of TOF-MRA images *Image Analysis* vol 3540 (Berlin: Springer) pp 1128–37

[235] El-Baz A, Gimel'farb G, Falk R, El-Ghar M A, Kumar V and Heredia D 2009 A novel 3D joint Markov–Gibbs model for extracting blood vessels from PC–MRA images *Medical Image Computing and Computer-Assisted Intervention—MICCAI 2009* vol 5762 (Berlin: Springer) pp 943–50

[236] Elnakib A, El-Baz A, Casanova M F, Gimel'farb G and Switala A E 2010 Image-based detection of corpus callosum variability for more accurate discrimination between dyslexic and normal brains *Proc. IEEE Int. Symp. on Biomedical Imaging: From Nano to Macro (ISBI'2010)* (Piscataway, NJ: IEEE) pp 109–12

[237] Elnakib A, Casanova M F, Gimel'farb G, Switala A E and El-Baz A 2011 Autism diagnostics by centerline-based shape analysis of the corpus callosum *Proc. IEEE Int. Symp. on Biomedical Imaging: From Nano to Macro (ISBI'2011)* (Piscataway, NJ: IEEE) pp 1843–6

[238] Elnakib A, Nitzken M, Casanova M, Park H, Gimel'farb G and El-Baz A 2012 Quantification of age-related brain cortex change using 3D shape analysis *2012 21st Int. Conf. on Pattern Recognition (ICPR)* (Piscataway, NJ: IEEE) pp 41–4

[239] Mostapha M, Soliman A, Khalifa F, Elnakib A, Alansary A, Nitzken M, Casanova M F and El-Baz A 2014 A statistical framework for the classification of infant DT images *2014 IEEE Int. Conf. on Image Processing (ICIP)* (Piscataway, NJ: IEEE) pp 2222–6

[240] Nitzken M, Casanova M, Gimel'farb G, Elnakib A, Khalifa F, Switala A and El-Baz A 2011 3D shape analysis of the brain cortex with application to dyslexia *2011 18th IEEE Int. Conf. on Image Processing (ICIP)* (Piscataway NJ: IEEE) pp 2657–60

[241] Ismail M, Soliman A, Ghazal M, Switala A E, Gimel'farb G, Barnes G N, Khalil A and El-Baz A 2017 A fast stochastic framework for automatic MR brain images segmentation *PloS one* **12** e0187391

[242] Ismail M M, Keynton R S, Mostapha M M, ElTanboly A H, Casanova M F, Gimel'farb G L and El-Baz A 2016 Studying autism spectrum disorder with structural and diffusion magnetic resonance imaging: a survey *Front. Hum. Neurosci.* **10** 211

[243] Alansary A *et al* 2016 Infant brain extraction in T1-weighted MR images using BET and refinement using LCDG and MGRF models *IEEE J. Biomed. Health Inform.* **20** 925–35

[244] Ismail M, Soliman A, ElTanboly A, Switala A, Mahmoud M, Khalifa F, Gimel'farb G, Casanova M F, Keynton R and El-Baz A 2016 Detection of white matter abnormalities in MR brain images for diagnosis of autism in children *2016 IEEE 13th Int. Symp. on Biomedical Imaging (ISBI)* pp 6–9

[245] Ismail M, Mostapha M, Soliman A, Nitzken M, Khalifa F, Elnakib A, Gimel'farb G, Casanova M and El-Baz A 2015 Segmentation of infant brain MR images based on adaptive shape prior and higher-order MGRF *2015 IEEE International Conference on Image Processing (ICIP)* pp 4327–31

[246] Asl E H, Ghazal M, Mahmoud A, Aslantas A, Shalaby A, Casanova M, Barnes G, Gimel'farb G, Keynton R and El-Baz A 2018 Alzheimer's disease diagnostics by a 3D deeply supervised adaptable convolutional network *Front. Biosci.* **23** 584–96

[247] Mahmoud A, El-Barkouky A, Farag H, Graham J and Farag A 2013 A non-invasive method for measuring blood flow rate in superficial veins from a single thermal image *Proc. of the IEEE Conf. on Computer Vision and Pattern Recognition Workshops* pp 354–9

[248] El-baz A, Shalaby A, Taher F, El-Baz M, Ghazal M, El-Ghar M A, Takieldeen A and Suri J 2017 Probabilistic modeling of blood vessels for segmenting magnetic resonance angiography images *Med. Res. Arch.* **5**

[249] Chowdhury A S, Rudra A K, Sen M, Elnakib A and El-Baz A 2010 Cerebral white matter segmentation from MRI using probabilistic graph cuts and geometric shape priors *ICIP* pp 3649–52

[250] Gebru Y, Giridharan G, Ghazal M, Mahmoud A, Shalaby A and El-Baz A 2018 Detection of cerebrovascular changes using magnetic resonance angiography *Cardiovascular Imaging and Image Analysis* (Boca Raton, FL: CRC Press) pp 1–22

[251] Mahmoud A, Shalaby A, Taher F, El-Baz M, Suri J S and El-Baz A 2018 Vascular tree segmentation from different image modalities *Cardiovascular Imaging and Image Analysis* (Boca Raton, FL: CRC Press) pp 43–70

[252] Taher F, Mahmoud A, Shalaby A and El-Baz A 2018 A review on the cerebrovascular segmentation methods *2018 IEEE Int. Symp. on Signal Processing and Information Technology (ISSPIT)* (Piscataway, NJ: IEEE) pp 359–64

[253] Kandil H, Soliman A, Fraiwan L, Shalaby A, Mahmoud A, ElTanboly A, Elmaghraby A, Giridharan G and El-Baz A 2018 A novel MRA framework based on integrated global and local analysis for accurate segmentation of the cerebral vascular system *2018 IEEE 15th Int. Symp. on Biomedical Imaging (ISBI 2018)* (Piscataway, NJ: IEEE) pp 1365–8

[254] Zhang Y W, Thompson R, Zhang H and Xu H 2011 APP processing in Alzheimer's disease *Mol. Brain* **4** 3

IOP Publishing

Neurological Disorders and Imaging Physics, Volume 3
Application to autism spectrum disorders and Alzheimer's
Ayman El-Baz and Jasjit S Suri

Chapter 14

A noninvasive image-based approach toward an early diagnosis of autism

Yaser ElNakieb, Ahmed Shalaby, Ali Mahmoud, Hassan Hajjdiab, Ashraf Khalil, Mohammed Ghazal, Fatma Taher, Ahmed Soliman, Robert Keynton, Gregory Barnes and Ayman El-Baz

Autism spectrum disorder (ASD) denotes a significant growing public health concern. Currently, one in 68 children has been diagnosed with ASD in the United States, and most children are diagnosed after the age of four, despite the fact that ASD can be identified as early as age two. The ultimate goal of this chapter is to develop a computer-aided diagnostic (CAD) system for the accurate and early diagnosis of ASD using diffusion tensor imaging (DTI). This CAD system consists of three main steps. First, the brain tissues are segmented based on three image descriptors: a visual appearance model that has the ability to model a large dimensional feature space, a shape model that is adapted during the segmentation process using first- and second-order visual appearance features, and a spatially invariant second-order homogeneity descriptor. Second, discriminatory features are extracted from the segmented brains. Cortex shape variability is assessed using shape construction methods, and white matter integrity is further examined through connectivity analysis. Finally, the diagnostic capabilities of these extracted features are investigated. The accuracy of the presented CAD system has been tested on 38 infants with a high risk of developing ASD. The statistical analysis using the unpaired t-test and the diagnostic results using the random forest classifier (87% accuracy and an AUC of 0.96) confirm the high performance and the efficiency of the proposed CAD system.

14.1 Introduction

Autism spectrum disorder (ASD) is a group of life-long developmental disabilities that are defined by significant social, communication, and behavioral challenges. Currently, ASD is a significant growing public health concern. According to the report issued by the US Centers for Disease Control and Prevention (CDC) in 2014, one in 68 children has been diagnosed with ASD in the United States, which is approximately 30% greater than previous estimates in 2012 of one in 88 children [1]. Moreover, this last report indicated that most children with ASD are currently diagnosed after the age of four, despite the fact that ASD can be identified as early as age two [1]. Thus, there is an urgent need for a noninvasive technology with the capability of providing new laboratory-based measures that confer an early and accurate diagnosis of ASD, as an early detection could lead to the reduction or the elimination of the manifestation of the disorder through effective early intervention [2].

Recent molecular and functional connectivity studies have indicated that brain connectivity and the underlying white matter tracts might be impaired in patients with ASD, but these studies failed to provide sufficient information about the morphological characteristics of these white matter tracts [3]. Structure based studies are based on extracting volumetric [4] or shape [5] information from the brain to detect differences between control and autistic patients. However, most of these studies were age sensitive and failed to detect abnormalities in infant brains [6]. Fortunately, recent advances in diffusion tensor imaging (DTI) have allowed researchers to study, in a noninvasive manner, both the macrostructure and microstructure of the white matter tracts of the brain. These new advances have allowed the growth of DTI-based studies investigating ASD in the last decade [3]. However, developing a noninvasive computer-aided diagnostic (CAD) system for the early detection of ASDs from DTI is still challenging. One of the major existing challenges is the segmentation of anatomical structures such as white matter (WM) and gray matter (GM) regions from infant DTI images. This is due to the fact that most of the current magnetic resonance (MR) infant segmentation techniques are dedicated to segmenting infant brains either in the early infantile stage ($\leqslant 5$ months) or early adult-like stage ($\geqslant 12$ months) by using a T1 or T2 scan or the combination of both [7, 8]. However, these methods would fail in the case of infants in the isointense stage (6–12 months), which is the primary focus of this work, because both WM and GM have roughly the same intensity levels [8]. Therefore, segmentation frameworks that integrate DTI image contrasts (i.e. fractional anistropy (FA) maps) would result in a better differentiation between WM and GM in the isointense stage [9–12]. However, none of these frameworks has tried to integrate other DTI image contrasts in their segmentation procedure.

To overcome the limitations mentioned above, we propose a new CAD system that integrates both shape and connectivity extracted features in the classification process. One major contribution in the segmentation step is the use of nonnegative matrix factorization (NMF) to find the most discriminating attributes in high-dimensional feature spaces and combine them in order to create a lower-dimensional space where classes are better separated and training data are encoded. Before describing the proposed methods, the basic notation used throughout this chapter is presented:

- $\mathbf{R} = \{(x, y, z): 0 \leqslant x \leqslant X - 1, 0 \leqslant y \leqslant Y - 1, 0 \leqslant z \leqslant Z - 1\}$—a finite arithmetic lattice supporting digital images and their region maps, and a voxel

$s = (x, y, z)$ is associated with its neighbors, $\{(x + \xi, y + \eta, z + \zeta)$: $(x + \xi, y + \eta, z + \zeta) \in \mathbf{R}; (\xi, \eta, \zeta) \in \nu_s\}$, where ν_s is the 26-neighborhood defined by $\xi \in \{-1, 0, 1\}$, $\eta \in \{-1, 0, 1\}$, and $\zeta \in \{-1, 0, 1\}$.

- $\mathbf{g} = \{g_{x,y,z}: (x, y, z) \in \mathbf{R}; g_{x,y,z} \in \mathbf{Q}\}$—a gray scale image taking values from a finite set of gray levels $\mathbf{Q} = \{0, 1, \ldots, Q - 1\}$.
- $\mathbf{m} = \{m_{x,y,z}: (x, y, z) \in \mathbf{R}; m_{x,y,z} \in \mathbf{L}\}$—a region map taking values from from a finite set $\mathbf{L} = \{0, \ldots, L\}$.
- $\mathbf{A} = \{a_{i,n}: i = 1, \ldots, I, n = 1, \ldots, XYZ; a_{i,n} \in \mathbf{Q}\}$—a 2D matrix containing image features of all n voxels in a vector form, where i is the dimension size of image features.
- $\mathbf{W} = \{w_{i,j}: i = 1, \ldots, I, j = 1, \ldots, J; w_{i,j} \in \mathbb{R}^+\}$—a 2D matrix contains J basis image features.
- $\mathbf{H} = \{h_{j,n}: j = 1, \ldots, J, n = 1, \ldots, XYZ; h_{i,j} \in \mathbb{R}^+\}$—a 2D matrix contains n voxels in new feature space of size J.

14.2 Methods

The proposed CAD system consists of three main steps: (i) segmentation of the brain cortex and the WM tracts from DTI, (ii) extraction of discriminatory features from the segmented brain tissues, i.e. shape features from the brain cortex and connectivity features from the WM tracts, and (iii) classification of the input infant brain to autistic or control based on analyzing the extracted shape and connectivity features for both control and autistic brains. Figure 14.1 shows the entire proposed CAD system framework.

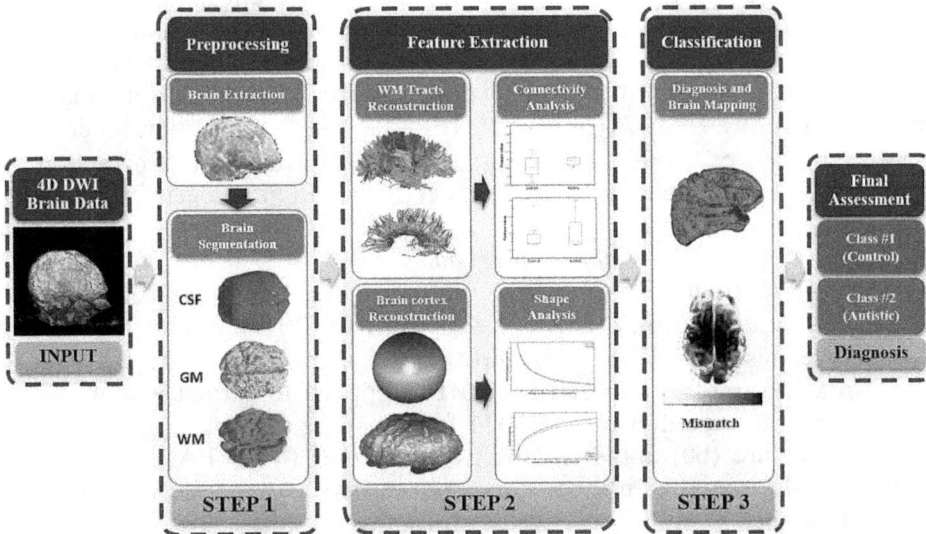

Figure 14.1. Proposed computer-aided diagnostic system framework.

14.2.1 Segmentation

In order to increase the segmentation accuracy, the brain is first extracted to minimize the overall intensity overlap and decrease the segmentation search space. To achieve this step, the following new stochastic brain extraction algorithm was developed.

Algorithm 1. Steps of the proposed brain extraction approach.

1. Approximate the marginal intensity distribution of the input image **g** using a linear combination of discrete Gaussians (LCDG) model with two dominant modes for the brain and the non-brain classes [13].
2. Determine the discriminant threshold, τ, which ensures the best separation between the brain and the non-brain voxel signals.
3. Reduce the inhomogeneity of **g** by applying the 3D generalized Gauss–Markov random field (GGMRF) model to each voxel signal (q_s; $s \in \mathbf{R}$, $q \in \mathbf{Q}$) to obtain the modified image $\hat{\mathbf{g}}$ [14].
4. Nudge pixels that may fall close to the boundary between the brain and the non-brain classes by comparing the signal of modified voxels $\hat{q}_s$ to the threshold τ obtained in step 2, and add or subtract a small fixed bias ϵ to or from these modified signals.
5. Improve $\hat{\mathbf{g}}$ using gradient descent optimization algorithm.
6. Apply 3D region growing and connected component analysis to obtain the final results.

Infant brain segmentation. A novel 3D joint Markov–Gibbs random field (MGRF) model is presented in this chapter. An input brain image, **g**, co-aligned to the training database, and its map, **m**, are described with a joint probability model, $P(\mathbf{g}, \mathbf{m}) = P(\mathbf{g}|\mathbf{m})P(\mathbf{m})$, which combines a conditional distribution of the images given the map $P(\mathbf{g}|\mathbf{m})$, and an unconditional probability distribution of maps $P(\mathbf{m}) = P_{sp}(\mathbf{m})P_V(\mathbf{m})$. Here, $P(\mathbf{g}|\mathbf{m})$ is a NMF-based visual appearance model, $P_{sp}(\mathbf{m})$ is an adaptive shape model, and $P_V(\mathbf{m})$ is a Gibbs probability distribution with potentials **V**, which specifies a MGRF model of spatially homogeneous maps **m**.

14.2.1.1 NMF-based visual appearance model

Following DTI estimation, five different anisotropy features were calculated using the 3D Slicer software [15], namely, mean diffusivity (MD), fractional anisotropy (FA), relative anisotropy (RA), axial diffusivity ($\lambda_{\parallel}$), and radial diffusivity ($\lambda_{\perp}$) [16]. In this work, feature fusion based on NMF is applied to extract new meaningful features from the large dimensional DTI feature space (**A**), which consists of one appearance feature (b0) and five anisotropy features (MD, FA, RA, $\lambda_{\parallel}$, $\lambda_{\perp}$). In particular, NMF is used in this segmentation framework to find the weights for each input DTI feature (**W**) in order to create a new feature space (**H**). Using NMF, the input data matrix $\mathbf{A} \in \mathbb{R}^{+I \times XYZ}$ can be factorized into two matrices: $\mathbf{A} \approx \mathbf{WH}$, where

$\mathbf{W} \in \mathbb{R}^{+I \times J}$ contains the basis voxels as its columns, and $\mathbf{H} \in \mathbb{R}^{+J \times XYZ}$ is the new feature encoding of voxels [17]. To process the 3D DTI features using NMF, each DTI feature for each voxel (x, y, z) in the 3D brain is converted to vectors in the input data matrix $\mathbf{A}$. $\mathbf{W}$ and $\mathbf{H}$ were calculated by minimizing the Euclidean distance between $\mathbf{A}$ and $\mathbf{WH}$ with the constraint that $\mathbf{W}$ and $\mathbf{H}$ contained only nonnegative values. Mathematically, this can be expressed by the following constrained optimization problem:

$$\underset{\mathbf{W},\mathbf{H}}{\text{minimize}} \; \frac{1}{2} \, \|\mathbf{A} - \mathbf{WH}\|^2 \; \text{subject to} \; \mathbf{W}, \mathbf{H} \geqslant 0, \tag{14.1}$$

and since the advent of NMF, several methods have been used to optimize equation (14.1). In this chapter, the alternating least square (ALS) algorithm [18] was used because of its high speed and flexibility compared to other competing algorithms.

In the proposed framework, NMF was performed on an input data matrix that was Ith-dimensional, one dimension for each calculated feature, and a column vector for each voxel in the training volumes. The resulting $\mathbf{W}$ was used as the basis vectors to transform new feature vectors into the new Jth-dimensional space. The resulting $\mathbf{H}$ was used to find the J-dimensional centroids corresponding to each brain label (C_l; $l \in \mathbf{L}$). In the testing phase, a new input data matrix was created ($\mathbf{B}$) with the same Ith-dimensional for each feature as in the training phase, and a column vector for each voxel in the test volume. The new Jth-dimensional vectors corresponding to these new input vectors were calculated by multiplying them by the psuedo-inverse of $\mathbf{W}$. Mathematically, this is given by

$$\mathbf{H}_B = \mathbf{W}^\dagger \mathbf{B}. \tag{14.2}$$

To model the visual appearance of the new NMF fused features, a K-means classifier was used with the classes J-dimensional centroids (C_l; $l \in \mathbf{L}$) that were calculated in the H-space during the training phase. For each voxel $(x, y, z) \in \mathbf{R}$ in the testing volume $\mathbf{B}$, a new J-dimensional test vector $H_{B:x,y,z}$ is formed using equation (14.2). The labels' probabilities associated with $H_{B:x,y,z}$ were calculated based on the Euclidean distance $d_l(T_{B:x,y,z})$ from the test vector $H_{B:x,y,z}$ to each of the centroids C_l; $l \in \mathbf{L}$. The NMF-based probabilities for brain label $l \in \mathbf{L}$ and voxel $(x, y, z) \in \mathbf{R}$ are defined as

$$P_{x,y,z}(\mathbf{g}|\mathbf{m} = l) = (1/d_l(H_{B:x,y,z})) \left/ \left(\sum_{l=1}^{L} 1/d_l(H_{B:x,y,z}) \right) \right. . \tag{14.3}$$

14.2.1.2 Adaptive shape model

To enhance the segmentation accuracy, the expected shapes of each brain label were constrained with an adaptive probabilistic shape prior. To create the shape (atlas) database, a training set of images was collected for different subjects (ten datasets) with their new NMF fused features. They were co-aligned by 3D affine transformations with 12 degrees of freedom in a way that maximized their mutual information. The shape priors are spatially variant independent random fields of region labels for the co-aligned data:

$$P_{sp}(\mathbf{m}) = \prod_{(x,y,z) \in R} p_{sp:x,y,z}(m_{x,y,z}), \tag{14.4}$$

where $p_{sp:x,y,z}(l)$ refers to the voxel-wise empirical probabilities for each brain label $l \in \mathbf{L}$. To generate the ground truth labels for the atlas, the infant tissue probability maps provided by IDEA lab [19] were used with the unified segmentation algorithm, implemented in the statistical parametric mapping (SPM) software, to segment the non-diffusion (b0) scans of the training datasets [19]. Then, an MR expert refined the generated initial segmentation to produce the final brain labels. For each input subject data to be segmented, the shape prior was constructed by an adaptive process guided by the visual appearance of its NMF fused features, as described in [20].

14.2.1.3 Spatial interaction MGRF model

In order to overcome noise effects and to ensure segmentation homogeneity, spatially homogeneous 3D pair-wise interactions between the region labels are additionally incorporated in the proposed segmentation model. The utilized second-order 3D MGRF model of the region map $\mathbf{m}$ is defined as [13]:

$$P_V(\mathbf{m}) = \frac{1}{Z_{\nu_s}} \exp \sum_{(x,y,z) \in \mathbf{R}} \sum_{(\xi,\eta,\zeta) \in \nu_s} \mathbf{V}(m_{x,y,z}, m_{x+\xi,y+\eta,z+\zeta}), \tag{14.5}$$

where Z_{ν_s} is the normalization factor and $\mathbf{V}$ is the bi-valued Gibbs potential, which depends on whether the nearest pair of labels are equal or not. The initial region map results in an approximation with the following analytical maximum likelihood estimates of the potentials [13]

$$v_{eq} = -v_{ne} \approx 2f_{eq}(\mathbf{m}) - 1, \tag{14.6}$$

where $f_{eq}(\mathbf{m})$ is the relative frequency of equal labels in the voxel pairs $\{((x, y, z), (x + \xi, y + \eta, z + \zeta)): (x, y, z) \in \mathbf{R}; \quad (x + \xi, y + \eta, z + \zeta) \in \mathbf{R}; (\xi, \eta, \zeta) \in \nu_s\}$. These estimates allow for computing the voxel-wise probabilities $p_{x,y,z}(m_{x,y,z} = l)$ of each brain label; $l \in \mathbf{L}$. In total, the complete segmentation steps of the proposed framework are presented in figure 14.2 and are summarized in algorithm 2.

Algorithm 2 Steps for the proposed segmentation approach.

1. Detect artifacts, correct motion and eddy current distortions using the DTIprep software [21].
2. Derive DTI from DWI using the weighted linear least square (WLLS) method, and extract five DTI anisotropy features (MD, FA, RA, $\lambda_{\parallel}$, $\lambda_{\perp}$) using the 3D Slicer software [15].

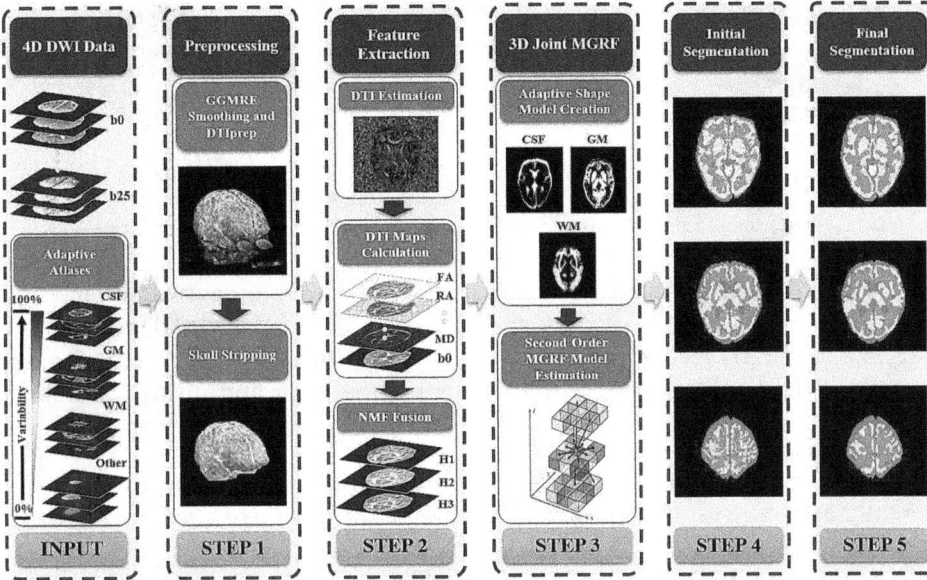

Figure 14.2. Details of the proposed segmentation framework.

3. Remove any non-brain tissues from DWI brain images using the proposed automated brain extraction approach (algorithm 1).

4. Create a new NMF input data matrix **B** from the obtained DTI features and the b0 volume.

5. Use the weight matrix **W** (calculated in the training phase) to transform **B** from the original Ith-dimensional space to the new Jth-dimensional space, according to equation (14.2).

6. Create the subject-specific shape prior model according to the approach described in [20].

7. Form an initial region map **m** using the NMF-based probabilities and prior adaptive shapes.

8. Find the Gibbs potentials for the MGRF model from the initial map **m** [13].

9. Improve the region map **m** using voxel-wise Bayes classifier after integrating the three descriptors in the proposed joint MGRF model.

14.2.2 Feature extraction

14.2.2.1 Shape features

The shape analysis was based on spherical harmonic reconstruction [22], which considers 3D surface data (i.e. the brain cortex) as a linear combination of specific basis functions, namely spherical harmonics (SHs). First, a 3D mesh model of the segmented brain cortex surface is generated using a Delaunay triangulated 3D mesh. Second, a smoothed version of the 3D mesh is created to ensure the uniqueness of each point in the given dataset. Then, the smoothed brain mesh is mapped to a unit sphere utilizing the attraction–repulsion mapping approach [23]. The SHs are

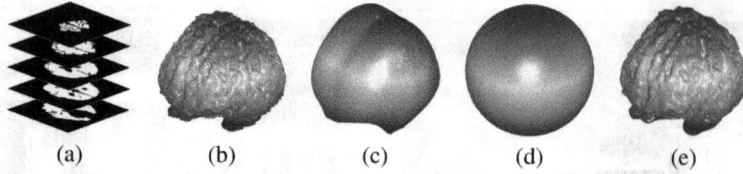

Figure 14.3. Illustration of the spherical harmonics shape analysis steps: (a) segmented brain cortex, (b) original mesh, (c) smoothed mesh, (d) unit sphere mesh, and (e) the SHs reconstructed mesh.

produced by solving an isotropic heat equation for the brain cortex surface on the unit sphere [23]. A step-by-step of the proposed SH analysis is demonstrated in figure 14.3. After SH reconstruction, two techniques for measuring the complexity of the cerebral cortex are proposed, SH reconstruction error and surface complexity.

SH reconstruction error

Due to the unit sphere mapping, the original cortex mesh for each subject is inherently aligned with the SH approximated mesh. The error between the original cortex mesh nodes and the SH approximated cortex mesh nodes can be calculated in terms of the Euclidean distance. This error generates a reconstruction error curve that is unique to each subject, where the area under the curve is examined to provide a representative metric for the brain cortex [23].

Surface complexity

A new metric for examining the complexity of the brain using the SH coefficients is also proposed. For a unit sphere f, having an SH expansion, the surface complexity metric $S(f)$ is defined as

$$S(f) = \sum_{N=0}^{\infty} \varepsilon_N^2 = \sum_{N=0}^{\infty} N B_N^2, \tag{14.7}$$

where N is the number of harmonics, and B are the previously calculated SH coefficients. The squared residual ε_N^2 is defined as

$$\varepsilon_N^2 = \| f - f_N \|^2 = \left\| \sum_{n=N+1}^{\infty} \sum_{m=-n}^{n} b_{nm} Y_n^m \right\|^2$$
$$= \sum_{n=N+1}^{\infty} \sum_{m=-n}^{n} |b_{nm}|^2 = \sum_{n=N+1}^{\infty} B_n^2. \tag{14.8}$$

For use in 3D SH analysis there are three sets of coefficients for each direction, x, y, and z. Therefore the surface complexity is expanded from equation (14.7) to be defined as

$$S(f) = \frac{\sum_{N=0}^{\infty} N(B_{N,x}^2 + B_{N,y}^2 + B_{N,z}^2)}{\| f_x \|^2 + \| f_y \|^2 + \| f_z \|^2}. \tag{14.9}$$

This metric generates a unique curve for each subject similar to the SH reconstruction error curves. Some of the advantages to this calculation are that it depends only on the SH coefficients, making it a self-contained metric, and it serves to represent the average degree of SH expansion. Also, it is considered as a convergent metric, and can be computed over the range of harmonics of interest. Figure 14.4(b) shows the surface complexity error versus number of harmonics used.

14.2.2.2 Connectivity features

In this chapter, the obtained WM label maps were used as seeds to generate the required WM tracts according to existing tractography methods built into the 3D slicer software [15]. After WM fiber tracts were extracted, FA values were generated for each fiber tract to measure the degree of anisotropy of local diffusivity. In addition to the FA values, axial ($\lambda_{\parallel}$) and radial ($\lambda_{\perp}$) diffusivity values, which represent diffusion parallel and transverse to axonal directions, were also produced [16]. Figure 14.5 illustrates different features extracted from gray matter and from white matter.

14.2.2.3 NMF feature fusion

As in the segmentation step, the ALS algorithm [18] was used to approximate a weight matrix **W** for an input matrix **A**. The columns of the input matrix **A** corresponded to all the features extracted from both the shape and the connectivity analysis. In the proposed approach, the feature vectors of both training and testing data were included in **A**, and the dimensionality of the transformation space was set to 2. Once NMF was performed, the resulting **H** matrix was used as the input to the classification step, which was performed using the k_n-nearest neighbor algorithm.

14.3 Experimental results and conclusions

This chapter includes data from the Infant Brain Imaging Study (IBIS) [24]. The study participants are six-month-old infants with a high risk of developing ASD. Based on an ASD cutoff threshold, the high-risk infants were divided into two groups: ASD-negative (control) and ASD-positive (autistic). From the 280 subjects

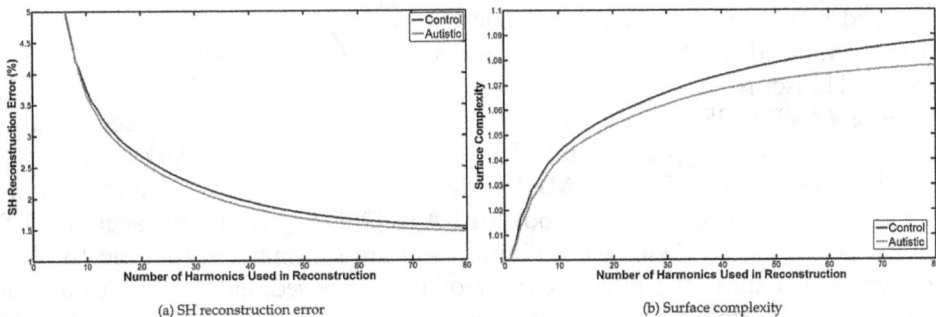

(a) SH reconstruction error   (b) Surface complexity

Figure 14.4. Reconstruction error and surface complexity metrics versus the number of harmonics used in reconstruction for autistic versus typically developed subjects.

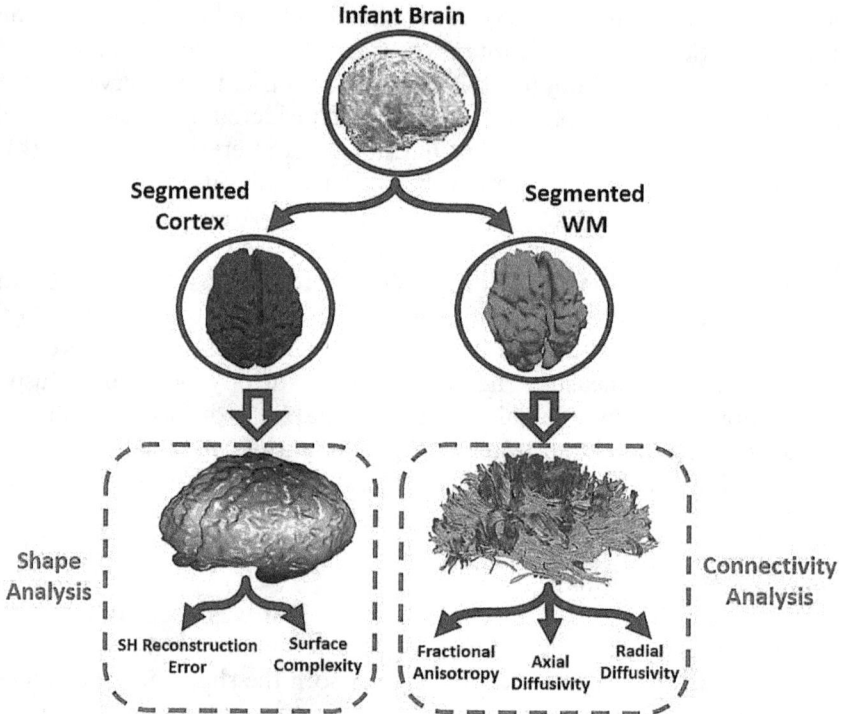

Figure 14.5. Feature extraction framework.

available, there were only 38 subjects with an available final diagnosis provided to our group (19 autistic and 19 control). Diffusion weighted MR brain scans were acquired using the following parameters: field of view = 190 mm, number of slices = 75–81, voxel resolution = $2 \times 2 \times 2$ mm^3, variable b values between 0 and 1000 s mm^{-2}, and 25 gradient directions [24].

Since the segmentation accuracy significantly affects the diagnostic results, performance assessment of the segmentation results is necessary. This is achieved by applying the proposed segmentation techniques on the 38 diffusion weighted infant MR brain datasets, and evaluated on ten training datasets with a manually segmented ground truth, using leave-one-out cross validation. Three performance metrics were used: (i) the Dice similarity coefficient (DSC), (ii) the 95-percentile modified Hausdorff distance (MHD), and (iii) the percentage absolute volume difference (AVD) [28]. The proposed brain extraction and brain segmentation frameworks were compared with the available software packages: the brain extraction tool (BET) [25], the FMRIB software library (FSL) [27], and the infant brain extraction and analysis toolbox (iBEAT) [26]. Segmentation accuracies for each method using the DSC, MHD, and AVD are summarized in table 14.1. The obtained results show the high accuracy of the proposed segmentation approach compared to the state-of-the-art-software packages as confirmed from the DSC paired t-test results (p-value < 0.05). In addition, the proposed segmentation

Table 14.1. Accuracy of our segmentation approach using DSC (%), MHD (mm), and AVD (%).

	Infant brain extraction			Infant brain segmentation					
	Brain mask			Brain cortex			White matter		
Metric	Proposed	BET [25]	iBEAT [26]	Proposed	FSL [27]	iBEAT [26]	Proposed	FSL [27]	iBEAT [26]
DSC	**96.96 ± 1.42**	91.16 ± 0.77	93.15 ± 2.40	**97.58 ± 3.60**	93.60 ± 5.36	92.52 ± 1.73	**95.11 ± 5.07**	79.49 ± 4.69	88.85 ± 0.84
p-value		**0.0003**	**0.0001**		**0.0191**	**0.0002**		**0.0002**	**0.0001**
MHD	**6.57 ± 4.08**	10.01 ± 1.05	9.49 ± 5.41	**2.71 ± 0.84**	4.66 ± 1.91	6.46 ± 2.06	**1.98 ± 0.55**	5.84 ± 0.90	5.05 ± 1.05
p-value		0.0067	0.0302		0.0094	0.0001		0.0009	0.0001
AVD	**2.46 ± 3.34**	4.09 ± 2.10	3.96 ± 2.97	**2.03 ± 0.82**	13.78 ± 13.99	16.19 ± 3.93	**5.19 ± 1.98**	51.76 ± 16.68	25.10 ± 2.12
p-value		0.3116	0.0480		0.0001	0.0027		0.0004	0.0008

Table 14.2. Feature extraction results.

		Control	Autistic	p-value
Shape features	SH reconstruction error	261.42 ± 37.53	233.26 ± 30.14	0.0251
	Surface complexity	87.29 ± 2.52	85.59 ± 1.36	0.0139
Connectivity features	Mean fractional anisotropy	0.66 ± 0.08	0.70 ± 0.02	0.0408
	Mean axial diffusivity	4.55 ± 4.01	5.87 ± 3.07	0.2607
	Mean radial diffusivity	1.09 ± 0.54	1.77 ± 1.06	0.0166

techniques show a particular advantage in terms of processing speed (<1 min) when compared to the competing techniques (iBeat >75 min and FSL >15 min).

The statistical analysis results of the extracted shape and connectivity features are shown in table 14.2. According to the unpaired t-test results, all shape features and two of the three connectivity features (FA, $\lambda_\perp$) show statistically significant differences between the control and autistic infant brains (p-value < 0.05). These results encouraged us to explore the classification potential of these four statistically significant features. In this chapter, four different classifiers are explored to find a suitable classifier for such diagnosis application, namely, (i) naïve Bayes, (ii) K-nearest neighbor (K-NN), (iii) support vector machine (SVM), and (iv) random forest. Table 14.3 provides a comparison between the performance of these classifiers in terms of the classification accuracy (ACC) and the area under the ROC curve (AUC) using leave-one-out cross validation. It is clear that the random forest classifier provides the best diagnostic results (an ACC of 86.84% and AUC of 0.96), which confirms both the robustness and the accuracy of the random forest classifier (figure 14.6). Currently, we are investigating with our medical collaborator the clinical explanation behind the subjects' misclassification.

To conclude, this chapter presents a novel CAD system for diagnosing ASD. A major advantage of the proposed CAD system is its ability to extract both shape and connectivity features using a single image modality, which eliminates errors associated with multi-modality CAD systems. The preliminary diagnostic results, based on the available limited dataset, are promising in identifying autistic from control patients. Our future work will include collecting more data to build a more powerful classifier, and to extend our work to involve advanced identification of

Table 14.3. Classification results.

Classifier	ACC	AUC
Naive Bayes	76.31%	0.864
K-nearest neighbor	86.84%	0.906
Support vector machine	86.84%	0.868
Random forest	**86.84%**	**0.960**

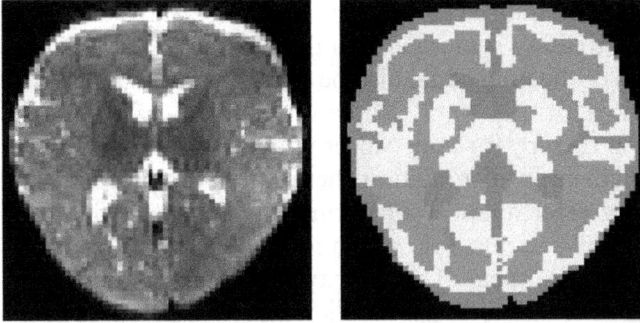

Figure 14.6. Segmentation example.

brain regions that have significant differences between autistic and control subjects using constructed brain maps.

This work could also be applied to various other applications in medical imaging, such as the kidney, the heart, the prostate, the lung, and the retina, as well as several non-medical applications [29–32].

One application is renal transplant functional assessment, in particular with developing noninvasive CAD systems for renal transplant function assessment, utilizing different image modalities (e.g. ultrasound, computed tomography (CT), MRI, etc). Accurate assessment of renal transplant function is critically important for graft survival. Although transplantation can improve a patient's wellbeing, there is a potential post-transplantation risk of kidney dysfunction that, if not treated in a timely manner, can lead to the loss of the entire graft, and even patient death. In particular, dynamic and diffusion MRI-based systems have been clinically used to assess transplanted kidneys with the advantage of providing information on each kidney separately. For more details about renal transplant functional assessment, see [33–50, 50–60].

The heart is also an important application for this work. The clinical assessment of myocardial perfusion plays a major role in the diagnosis, management, and prognosis of ischemic heart disease patients. Thus, there have been ongoing efforts to develop automated systems for accurate analysis of myocardial perfusion using first-pass images [61–77].

Another application for this work could be the detection of retinal abnormalities. The majority of ophthalmologists depend on visual interpretation for the identification of diseases types. However, inaccurate diagnosis will affect the treatment procedure which may lead to fatal results. Hence, there is a crucial need for computer automated diagnosis systems that yield highly accurate results. Optical coherence tomography (OCT) has become a powerful modality for the noninvasive diagnosis of various retinal abnormalities such as glaucoma, diabetic macular edema, and macular degeneration. The problem with diabetic retinopathy (DR) is that the patient is not aware of the disease until the changes in the retina have progressed to a level that treatment tends to be less effective. Therefore, automated early detection could limit the severity of the disease and assist ophthalmologists in investigating and treating it more efficiently [78–80].

Abnormalities of the lung could also be another promising area of research and a related application for this work. Radiation-induced lung injury is the main side effect of radiation therapy for lung cancer patients. Although higher radiation doses increase the effectiveness of radiation therapy for tumor control, this can lead to lung injury as a greater quantity of normal lung tissues is included in the treated area. Almost 1/3 of patients who undergo radiation therapy develop lung injury following radiation treatment. The severity of radiation-induced lung injury ranges from ground-glass opacities and consolidation at the early phase to fibrosis and traction bronchiectasis in the late phase. Early detection of lung injury will thus help to improve management of the treatment [81–123].

This work can also be applied to other brain abnormalities such as dyslexia, in addition to autism. Dyslexia is one of the most complicated developmental brain disorders that affects children's learning abilities. Dyslexia leads to the failure to develop age-appropriate reading skills in spite of a normal intelligence level and adequate reading instruction. Neuropathological studies have revealed the abnormal anatomy of some structures, such as the corpus callosum in dyslexic brains. A lot of work has been published in the literature that aims at developing CAD systems for diagnosing such disorders, along with other brain disorders [124–146].

For the vascular system [147], this work could also be applied for the extraction of blood vessels, e.g. from phase contrast (PC) magnetic resonance angiography (MRA). Accurate cerebrovascular segmentation using noninvasive MRA is crucial for the early diagnosis and timely treatment of intracranial vascular diseases [129, 130, 148–153].

References

[1] Wingate M *et al* 2014 Prevalence of autism spectrum disorder among children aged 8 years—autism and developmental disabilities monitoring network, 11 sites, United States, 2010 *MMWR Surveill. Summ.* **63** 1–21

[2] Dawson G 2008 Early behavioral intervention, brain plasticity, and the prevention of autism spectrum disorder *Dev. Psychopathol.* **20** 775–803

[3] Travers B G *et al* 2012 Diffusion tensor imaging in autism spectrum disorder: a review *Autism Res.* **5** 289–313

[4] Aylward E H *et al* 2002 Effects of age on brain volume and head circumference in autism *Neurology* **59** 175–83

[5] El-Baz A *et al* 2007 Autism diagnostics by 3D texture analysis of cerebral white matter gyrifications *MICCAI* (Berlin: Springer) pp 882–90

[6] Hazlett H C *et al* 2012 Brain volume findings in 6-month-old infants at high familial risk for autism *Am. J. Psychiatry* **169** 601–8

[7] Shi F *et al* 2010 Neonatal brain image segmentation in longitudinal MRI studies *Neuroimage* **49** 391–400

[8] Xue H *et al* 2007 Automatic segmentation and reconstruction of the cortex from neonatal MRI *Neuroimage* **38** 461–77

[9] Wang L *et al* 2013 Integration of sparse multi-modality representation and geometrical constraint for isointense infant brain segmentation *MICCAI* (Berlin: Springer) pp 703–10

[10] Wang L *et al* 2012 4D multi-modality tissue segmentation of serial infant images *PLoS One* **7** e44596

[11] Wang X *et al* 2014 Online discriminative multi-atlas learning for isointense infant brain segmentation *Machine Learning in Medical Imaging* (Berlin: Springer) pp 297–305

[12] Wang L *et al* 2015 Links: learning-based multi-source integration framework for segmentation of infant brain images *NeuroImage* **108** 160–72

[13] Farag A *et al* 2006 Precise segmentation of multimodal images *IEEE Trans. Image Process.* **15** 952–68

[14] Bouman C and Sauer K 1993 A generalized Gaussian image model for edge-preserving map estimation *IEEE Trans. Image Process.* **2** 296–310

[15] Fedorov A *et al* 2012 3D slicer as an image computing platform for the quantitative imaging network *Multidiscip. Respir. Med.* **30** 1323–41

[16] Mori S and Tournier J-D 2013 *Introduction to Diffusion Tensor Imaging: And Higher Order Models* (New York: Academic)

[17] Lee D D and Seung H S 1999 Learning the parts of objects by non-negative matrix factorization *Nature* **401** 788–91

[18] Berry M W *et al* 2007 Algorithms and applications for approximate nonnegative matrix factorization *Comput. Stat. Data Anal.* **52** 155–73

[19] Shi F *et al* 2011 Infant brain atlases from neonates to 1-and 2-year-olds *PLoS One* **6** e18746

[20] Mostapha M *et al* 2014 A statistical framework for the classification of Infant DT Images *ICIP (Paris, France, 27–30 October 2014)* pp 2222–6

[21] Liu Z *et al* 2010 Quality control of diffusion weighted images *Proc. SPIE* **7628** 76280J

[22] Gerig G *et al* 2001 Shape analysis of brain ventricles using SPHARM *MMBIA* (Piscataway, NJ: IEEE) pp 171–8

[23] Nitzken M *et al* 2011 3D shape analysis of the brain cortex with application to autism *ISBI* (Piscataway, NJ: IEEE) pp 1847–50

[24] Infant Brain Imaging Study (IBIS) http://ibisnetwork.org/

[25] Smith S M 2002 Fast robust automated brain extraction *Hum. Brain Mapp.* **17** 143–55

[26] Yakang D *et al* 2013 iBEAT: a toolbox for infant brain magnetic resonance image processing *Neuroinformatics* **11** 211–5

[27] Jenkinson M *et al* 2012 Fsl *NeuroImage* **62** 782–90

[28] Babalola K O *et al* 2009 An evaluation of four automatic methods of segmenting the subcortical structures in the brain *Neuroimage* **47** 1435–47

[29] Mahmoud A H 2014 Utilizing radiation for smart robotic applications using visible, thermal, and polarization images *PhD Dissertation* University of Louisville

[30] Mahmoud A, El-Barkouky A, Graham J and Farag A 2014 Pedestrian detection using mixed partial derivative based histogram of oriented gradients *2014 IEEE Int. Conf. on Image Processing (ICIP)* (Piscataway, NJ: IEEE) pp 2334–7

[31] El-Barkouky A, Mahmoud A, Graham J and Farag A 2013 An interactive educational drawing system using a humanoid robot and light polarization *2013 IEEE Int. Conf. on Image Processing* (Piscataway, NJ: IEEE) pp 3407–11

[32] Mahmoud A H, El-Melegy M T and Farag A A 2012 Direct method for shape recovery from polarization and shading *2012 19th IEEE Int. Conf. on Image Processing* (Piscataway, NJ: IEEE) pp 1769–72

[33] Ali A M, Farag A A and El-Baz A 2007 Graph cuts framework for kidney segmentation with prior shape constraints *Proc. of Int. Conf. on Medical Image Computing and Computer-Assisted Intervention, (MICCAI'07) (Brisbane, Australia, 29 October–2 November 2007)* **vol 1** pp 384–92

[34] Chowdhury A S, Roy R, Bose S, Elnakib F K A and El-Baz A 2012 Non-rigid biomedical image registration using graph cuts with a novel data term *Proc. of IEEE Int. Symp. on Biomedical Imaging: From Nano to Macro, (ISBI'12)* (Barcelona, Spain 2–5 May 2012) *pp 446–9*

[35] El-Baz A, Farag A A, Yuksel S E, El-Ghar M E, Eldiasty T A and Ghoneim M A 2007 Application of deformable models for the detection of acute renal rejection *Deformable Models* (New York: Springer) pp 293–333

[36] El-Baz A, Farag A, Fahmi R, Yuksel S, El-Ghar M A and Eldiasty T 2006 Image analysis of renal DCE MRI for the detection of acute renal rejection *Proc. of IAPR Int. Conf. on Pattern Recognition (ICPR'06) (Hong Kong, 20–24 August 2006)* pp 822–5

[37] El-Baz A, Farag A, Fahmi R, Yuksel S, Miller W, El-Ghar M A, El-Diasty T and Ghoneim M 2006 A new CAD system for the evaluation of kidney diseases using DCE-MRI *Proc. of Int. Conf. on Medical Image Computing and Computer-Assisted Intervention, (MICCAI'08) (Copenhagen, Denmark, 1–6 October 2006)* pp 446–53

[38] El-Baz A, Gimel'farb G and El-Ghar M A 2008 A novel image analysis approach for accurate identification of acute renal rejection *Proc. of IEEE Int. Conf. on Image Processing, (ICIP'08) (San Diego, California, USA, 12–15 October 2008)* pp 1812–5

[39] El-Baz A, Gimel'farb G and El-Ghar M A 2008 Image analysis approach for identification of renal transplant rejection *Proc. of IAPR Int. Conf. on Pattern Recognition, (ICPR'08) (Tampa, Florida, USA, 8–11 December 2008)* pp 1–4

[40] El-Baz A, Gimel'farb G and El-Ghar M A 2007 New motion correction models for automatic identification of renal transplant rejection *Proc. of Int. Conf. on Medical Image Computing and Computer-Assisted Intervention, (MICCAI'07) (Brisbane, Australia, 29 October–2 November 2007)* pp 235–43

[41] Farag A, El-Baz A, Yuksel S, El-Ghar M A and Eldiasty T 2006 A framework for the detection of acute rejection with dynamic contrast enhanced magnetic resonance imaging *Proc. of IEEE Int. Symp. on Biomedical Imaging: From Nano to Macro, (ISBI'06)* (Arlington, Virginia, USA 6–9 April 2006) *pp 418–21*

[42] Khalifa F, Beache G M, El-Ghar M A, El-Diasty T, Gimel'farb G, Kong M and El-Baz A 2013 Dynamic contrast-enhanced MRI-based early detection of acute renal transplant rejection *IEEE Trans. Med. Imaging* **32** 1910–27

[43] Khalifa F, El-Baz A, Gimel'farb G and El-Ghar M A 2010 Non-invasive image-based approach for early detection of acute renal rejection *Proc. of Int. Conf. Medical Image Computing and Computer-Assisted Intervention, (MICCAI'10) (Beijing, China, 20–24 September 2010)* pp 10–8

[44] Khalifa F, El-Baz A, Gimel'farb G, Ouseph R and El-Ghar M A 2010 Shape-appearance guided level-set deformable model for image segmentation *Proc. of IAPR Int. Conf. on Pattern Recognition, (ICPR'10) (Istanbul, Turkey, 23–26 August 2010)* pp 4581–4

[45] Khalifa F, El-Ghar M A, Abdollahi B, Frieboes H, El-Diasty T and El-Baz A 2013 A comprehensive non-invasive framework for automated evaluation of acute renal transplant rejection using DCE-MRI *NMR Biomed.* **26** 1460–70

[46] Khalifa F, El-Ghar M A, Abdollahi B, Frieboes H B, El-Diasty T and El-Baz A 2014 Dynamic contrast-enhanced MRI-based early detection of acute renal transplant rejection *2014 Annu. Scientific Meeting and Educational Course Brochure of the Society of Abdominal Radiology, (SAR'14) (Boca Raton, FL, 23–28 March 2014)* CID: 1855 912

[47] Khalifa F, Elnakib A, Beache G M, Gimel'farb G, El-Ghar M A, Sokhadze G, Manning S, McClure P and El-Baz A 2011 3D kidney segmentation from CT images using a level set approach guided by a novel stochastic speed function *Proc. of Int. Conf. Medical Image Computing and Computer-Assisted Intervention, (MICCAI'11) (Toronto, Canada, 18–22 September 2011)* pp 587–94

[48] Khalifa F, Gimel'farb G, El-Ghar M A, Sokhadze G, Manning S, McClure P, Ouseph R and El-Baz A 2011 A new deformable model-based segmentation approach for accurate extraction of the kidney from abdominal CT images *Proc. of IEEE Int. Conf. on Image Processing, (ICIP'11) (Brussels, Belgium, 11–14 September 2011)* pp 3393–6

[49] Mostapha M, Khalifa F, Alansary A, Soliman A, Suri J and El-Baz A 2014 Computer-aided diagnosis systems for acute renal transplant rejection: challenges and methodologies *Abdomen and Thoracic Imaging* ed A El-Baz, L saba and J Suri (Berlin: Springer) pp 1–35

[50] Shehata M, Khalifa F, Hollis E, Soliman A, Hosseini-Asl E, El-Ghar M A, El-Baz M, Dwyer A C, El-Baz A and Keynton R 2016 A new non-invasive approach for early classification of renal rejection types using diffusion-weighted MRI *IEEE Int. Conf. on Image Processing (ICIP), 2016* (Piscataway, NJ: IEEE) pp 136–40

[51] Khalifa F, Soliman A, Takieldeen A, Shehata M, Mostapha M, Shaffie A, Ouseph R, Elmaghraby A and El-Baz A 2016 Kidney segmentation from CT images using a 3D NMF-guided active contour model *IEEE 13th Int. Symp. on Biomedical Imaging (ISBI), 2016* (Piscataway, NJ: IEEE) pp 432–5

[52] Shehata M, Khalifa F, Soliman A, Takieldeen A, El-Ghar M A, Shaffie A, Dwyer A C, Ouseph R, El-Baz A and Keynton R 2016 3D diffusion MRI-based cad system for early diagnosis of acute renal rejection *2016 IEEE 13th Int. Symp. on Biomedical Imaging (ISBI)* (Piscataway, NJ: IEEE) pp 1177–80

[53] Shehata M, Khalifa F, Soliman A, Alrefai R, El-Ghar M A, Dwyer A C, Ouseph R and El-Baz A 2015 A level set-based framework for 3D kidney segmentation from diffusion MR images *IEEE Int. Conf. on Image Processing (ICIP), 2015* (Piscataway, NJ: IEEE) pp 4441–5

[54] Shehata M, Khalifa F, Soliman A, El-Ghar M A, Dwyer A C, Gimel'farb G, Keynton R and El-Baz A 2016 A promising non-invasive CAD system for kidney function assessment *Int. Conf. on Medical Image Computing and Computer-Assisted Intervention* (Berlin: Springer) pp 613–21

[55] Khalifa F, Soliman A, Elmaghraby A, Gimel'farb G and El-Baz A 2017 3D kidney segmentation from abdominal images using spatial-appearance models *Comput. Math. Methods Med.* **2017** 1–10

[56] Hollis E, Shehata M, Khalifa F, El-Ghar M A, El-Diasty T and El-Baz A 2016 Towards non-invasive diagnostic techniques for early detection of acute renal transplant rejection: a review *Egypt. J. Radiol. Nucl. Med.* **48** 257–69

[57] Shehata M, Khalifa F, Soliman A, El-Ghar M A, Dwyer A C and El-Baz A 2017 Assessment of renal transplant using image and clinical-based biomarkers *Proc. of 13th Annu. Scientific Meeting of American Society for Diagnostics and Interventional Nephrology (ASDIN'17) (New Orleans, LA, USA 10–12 February 2017)*

[58] Shehata M, Khalifa F, Soliman A, El-Ghar M A, Dwyer A C and El-Baz A 2017 Early assessment of acute renal rejection *Proc. of 12th Annu. Scientific Meeting of American Society for Diagnostics and Interventional Nephrology (ASDIN'16)* (*Phoenix, AZ, USA* 19–21 February 2016)

[59] Eltanboly A, Ghazal M, Hajjdiab H, Shalaby A, Switala A, Mahmoud A, Sahoo P, El-Azab M and El-Baz A 2019 Level sets-based image segmentation approach using statistical shape priors *Appl. Math. Comput.* **340** 164–79

[60] Shehata M, Mahmoud A, Soliman A, Khalifa F, Ghazal M, El-Ghar M A, El-Melegy M and El-Baz A 2018 3D kidney segmentation from abdominal diffusion MRI using an appearance-guided deformable boundary *PLoS One* **13** e0200082

[61] Khalifa F, Beache G, El-Baz A and Gimel'farb G 2010 Deformable model guided by stochastic speed with application in cine images segmentation *Proc. of IEEE Int. Conf. on Image Processing, (ICIP'10)* (*Hong Kong, 26–29 September 2010*) pp 1725–8

[62] Khalifa F, Beache G M, Elnakib A, Sliman H, Gimel'farb G, Welch K C and El-Baz A 2013 A new shape-based framework for the left ventricle wall segmentation from cardiac first-pass perfusion MRI *Proc. of IEEE Int. Symp. on Biomedical Imaging: From Nano to Macro, (ISBI'13)* (*San Francisco, CA* 7–11 April 2013) *pp 41–4*

[63] Khalifa F, Beache G M, Elnakib A, Sliman H, Gimel'farb G, Welch K C and El-Baz A 2012 A new nonrigid registration framework for improved visualization of transmural perfusion gradients on cardiac first–pass perfusion MRI *Proc. of IEEE Int. Symp. on Biomedical Imaging: From Nano to Macro, (ISBI'12)* (*Barcelona, Spain* 2–5 May) *pp 828–31*

[64] Khalifa F, Beache G M, Firjani A, Welch K C, Gimel'farb G and El-Baz A 2012 A new nonrigid registration approach for motion correction of cardiac first-pass perfusion MRI *Proc. of IEEE Int. Conf. on Image Processing, (ICIP'12)* (*Lake Buena Vista, Florida, 30 September–3 October 2012*) pp 1665–8

[65] Khalifa F, Beache G M, Gimel'farb G and El-Baz A 2012 A novel CAD system for analyzing cardiac first-pass MR images *Proc. of IAPR Int. Conf. on Pattern Recognition (ICPR'12)* (*Tsukuba Science City, Japan, 11–15 November 2012*) pp 77–80

[66] Khalifa F, Beache G M, Gimel'farb G and El-Baz A 2011 A novel approach for accurate estimation of left ventricle global indexes from short-axis cine MRI *Proc. of IEEE Int. Conf. on Image Processing, (ICIP'11)* (*Brussels, Belgium, 11–14 September 2011*) pp 2645–9

[67] Khalifa F, Beache G M, Gimel'farb G, Giridharan G A and El-Baz A 2011 A new image-based framework for analyzing cine images *Handbook of Multi Modality State-of-the-Art Medical Image Segmentation and Registration Methodologies* vol 2 ed A El-Baz, U R Acharya, M Mirmedhdi and J S Suri (New York: Springer) ch 3, pp 69–98

[68] Khalifa F, Beache G M, Gimel'farb G, Giridharan G A and El-Baz A 2012 Accurate automatic analysis of cardiac cine images *IEEE Trans. Biomed. Eng.* **59** 445–55

[69] Khalifa F, Beache G M, Nitzken M, Gimel'farb G, Giridharan G A and El-Baz A 2011 Automatic analysis of left ventricle wall thickness using short-axis cine CMR images *Proc. of IEEE Int. Symp. on Biomedical Imaging: From Nano to Macro, (ISBI'11)* (*Chicago, Illinois* 30 March–2 April 2011) *pp 1306–9*

[70] Nitzken M, Beache G, Elnakib A, Khalifa F, Gimel'farb G and El-Baz A 2012 Accurate modeling of tagged CMR 3D image appearance characteristics to improve cardiac cycle strain estimation *2012 19th IEEE Int. Conf. on Image Processing (ICIP)* (Piscataway, NJ: IEEE) pp 521–4

[71] Nitzken M, Beache G, Elnakib A, Khalifa F, Gimel'farb G and El-Baz A 2012 Improving full-cardiac cycle strain estimation from tagged CMR by accurate modeling of 3D image appearance characteristics *2012 9th IEEE Int. Symp. on Biomedical Imaging (ISBI)* (Piscataway, NJ: IEEE) pp 462–5

[72] Nitzken M J, El-Baz A S and Beache G M 2012 Markov–Gibbs random field model for improved full-cardiac cycle strain estimation from tagged CMR *J. Cardiovasc. Magn. Reson.* **14** 1–2

[73] Sliman H, Elnakib A, Beache G, Elmaghraby A and El-Baz A 2014 Assessment of myocardial function from cine cardiac MRI using a novel 4D tracking approach *J. Comput. Sci. Syst. Biol.* **7** 169–73

[74] Sliman H, Elnakib A, Beache G M, Soliman A, Khalifa F, Gimel'farb G, Elmaghraby A and El-Baz A 2014 A novel 4D PDE-based approach for accurate assessment of myocardium function using cine cardiac magnetic resonance images *Proc. of IEEE Int. Conf. on Image Processing (ICIP'14) (Paris, France, 27–30 October 2014)* pp 3537–41

[75] Sliman H, Khalifa F, Elnakib A, Beache G M, Elmaghraby A and El-Baz A 2013 A new segmentation-based tracking framework for extracting the left ventricle cavity from cine cardiac MRI *Proc. of IEEE Int. Conf. on Image Processing, (ICIP'13) (Melbourne, Australia, 15–18 September 2013)* pp 685–9

[76] Sliman H, Khalifa F, Elnakib A, Soliman A, Beache G M, Elmaghraby A, Gimel'farb G and El-Baz A 2013 Myocardial borders segmentation from cine MR images using bi-directional coupled parametric deformable models *Med. Phys.* **40** 1–13

[77] Sliman H, Khalifa F, Elnakib A, Soliman A, Beache G M, Gimel'farb G, Emam A, Elmaghraby A and El-Baz A 2013 Accurate segmentation framework for the left ventricle wall from cardiac cine MRI *Proc. of Int. Symp. on Computational Models for Life Science, (CMLS'13) (Sydney, Australia 27–29 November 2013)* **vol 1559** *pp 287–96*

[78] Eladawi N, Elmogy M, Ghazal M, Helmy O, Aboelfetouh A, Riad A, Schaal S and El-Baz A 2018 Classification of retinal diseases based on OCT images *Front. Biosci.* **23** 247–64

[79] ElTanboly A, Ismail M, Shalaby A, Switala A, El-Baz A, Schaal S, Gimelfarb G and El-Azab M 2017 A computer-aided diagnostic system for detecting diabetic retinopathy in optical coherence tomography images *Med. Phys.* **44** 914–23

[80] Sandhu H S, El-Baz A and Seddon J M 2018 Progress in automated deep learning for macular degeneration *JAMA Ophthalmol.* **136** 1366–7

[81] Abdollahi B, Civelek A C, Li X-F, Suri J and El-Baz A 2014 PET/CT nodule segmentation and diagnosis: a survey *Multi Detector CT Imaging* ed L Saba and J S Suri (London: Taylor and Francis) ch 30, pp 639–51

[82] Abdollahi B, El-Baz A and Amini A A 2011 A multi-scale non-linear vessel enhancement technique *2011 Annu. Int. Conf. of the IEEE Engineering in Medicine and Biology Society, EMBC* (Piscataway, NJ: IEEE) pp 3925–9

[83] Abdollahi B, Soliman A, Civelek A, Li X-F, Gimel'farb G and El-Baz A 2012 A novel Gaussian scale space-based joint MGRF framework for precise lung segmentation *Proc. of IEEE Int. Conf. on Image Processing, (ICIP'12)* (Piscataway, NJ: IEEE) pp 2029–32

[84] Abdollahi B, Soliman A, Civelek A, Li X-F, Gimel'farb G and El-Baz A 2012 A novel 3D joint MGRF framework for precise lung segmentation *Machine Learning in Medical Imaging* (Berlin: Springer) pp 86–93

[85] Ali A M, El-Baz A S and Farag A A 2007 A novel framework for accurate lung segmentation using graph cuts *Proc. of IEEE Int. Symp. on Biomedical Imaging: From Nano to Macro, (ISBI'07)* (Piscataway, NJ: IEEE) pp 908–11

[86] El-Baz A, Beache G M, Gimel'farb G, Suzuki K and Okada K 2013 Lung imaging data analysis *Int. J. Biomed. Imaging* **2013** 1–2

[87] El-Baz A, Beache G M, Gimel'farb G, Suzuki K, Okada K, Elnakib A, Soliman A and Abdollahi B 2013 Computer-aided diagnosis systems for lung cancer: challenges and methodologies *Int. J. Biomed. Imaging* **2013** 1–46

[88] El-Baz A, Elnakib A, Abou El-Ghar M, Gimel'farb G, Falk R and Farag A 2013 Automatic detection of 2D and 3D lung nodules in chest spiral CT scans *Int. J. Biomed. Imaging* **2013** 1–11

[89] El-Baz A, Farag A A, Falk R and La Rocca R 2003 A unified approach for detection, visualization, and identification of lung abnormalities in chest spiral CT scans *International Congress Series* vol 1256 (Amsterdam: Elsevier) pp 998–1004

[90] El-Baz A, Farag A A, Falk R and La Rocca R 2002 Detection, visualization and identification of lung abnormalities in chest spiral CT scan: Phase-I *Proc. of Int. Conf. on Biomedical Engineering Cairo, Egypt,* **vol 12**

[91] El-Baz A, Farag A, Gimel'farb G, Falk R, El-Ghar M A and Eldiasty T 2006 A framework for automatic segmentation of lung nodules from low dose chest CT scans *Proc. of Int. Conf. on Pattern Recognition, (ICPR'06)* **vol 3** (Piscataway, NJ: IEEE) pp 611–4

[92] El-Baz A, Farag A, Gimel'farb G, Falk R and El-Ghar M A 2011 A novel level set-based computer-aided detection system for automatic detection of lung nodules in low dose chest computed tomography scans *Lung Imaging and Computer Aided Diagnosis* (London: Taylor and Francis) ch 10, pp 221–38

[93] El-Baz A, Gimel'farb G, Abou El-Ghar M and Falk R 2012 Appearance-based diagnostic system for early assessment of malignant lung nodules *Proc. of IEEE Int. Conf. on Image Processing, (ICIP'12)* (Piscataway, NJ: IEEE) pp 533–6

[94] El-Baz A, Gimel'farb G and Falk R 2011 A novel 3D framework for automatic lung segmentation from low dose CT images *Lung Imaging and Computer Aided Diagnosis* ed A El-Baz and J S Suri (London: Taylor and Francis) ch 1, pp 1–16

[95] El-Baz A, Gimel'farb G, Falk R and El-Ghar M 2010 Appearance analysis for diagnosing malignant lung nodules *Proc. of IEEE Int. Symp. on Biomedical Imaging: From Nano to Macro (ISBI'10)* (Piscataway, NJ: IEEE) pp 193–6

[96] El-Baz A, Gimel'farb G, Falk R and El-Ghar M A 2011 A novel level set-based CAD system for automatic detection of lung nodules in low dose chest CT scans *Lung Imaging and Computer Aided Diagnosis* vol 1 ed A El-Baz and J S Suri (London: Taylor and Francis) ch 10, pp 221–38

[97] El-Baz A, Gimel'farb G, Falk R and El-Ghar M A 2008 A new approach for automatic analysis of 3D low dose CT images for accurate monitoring the detected lung nodules *Proc. of Int. Conf. on Pattern Recognition, (ICPR'08)* (Piscataway, NJ: IEEE) pp 1–4

[98] El-Baz A, Gimel'farb G, Falk R and El-Ghar M A 2007 A novel approach for automatic follow-up of detected lung nodules *Proc. of IEEE Int. Conf. on Image Processing, (ICIP'07)* **vol 5** (Piscataway, NJ: IEEE) pp V–501

[99] El-Baz A, Gimel'farb G, Falk R and El-Ghar M A 2007 A new CAD system for early diagnosis of detected lung nodules *ICIP 2007 IEEE Int. Conf. on Image Processing, 2007* **vol 2** (Piscataway, NJ: IEEE) pp II–461

[100] El-Baz A, Gimel'farb G, Falk R, El-Ghar M A and Refaie H 2008 Promising results for early diagnosis of lung cancer *Proc. of IEEE Int. Symp. on Biomedical Imaging: From Nano to Macro, (ISBI'08)* (Piscataway, NJ: IEEE) pp 1151–4

[101] El-Baz A, Gimel'farb G L, Falk R, Abou El-Ghar M, Holland T and Shaffer T 2008 A new stochastic framework for accurate lung segmentation *Proc. of Medical Image Computing and Computer-Assisted Intervention, (MICCAI'08)* pp 322–30

[102] El-Baz A, Gimel'farb G L, Falk R, Heredis D and Abou El-Ghar M 2008 A novel approach for accurate estimation of the growth rate of the detected lung nodules *Proc. of Int. Workshop on Pulmonary Image Analysis* pp 33–42

[103] El-Baz A, Gimel'farb G L, Falk R, Holland T and Shaffer T 2008 A framework for unsupervised segmentation of lung tissues from low dose computed tomography images *Proc. of British Machine Vision, (BMVC'08)* pp 1–10

[104] El-Baz A, Gimel'farb G, Falk R and El-Ghar M A 2011 3D MGRF-based appearance modeling for robust segmentation of pulmonary nodules in 3D LDCT chest images *Lung Imaging and Computer Aided Diagnosis* (London: Taylor and Francis) ch 3, pp 51–63

[105] El-Baz A, Gimel'farb G, Falk R and El-Ghar M A 2009 Automatic analysis of 3D low dose CT images for early diagnosis of lung cancer *Pattern Recogn.* **42** 1041–51

[106] El-Baz A, Gimel'farb G, Falk R, El-Ghar M A, Rainey S, Heredia D and Shaffer T 2009 Toward early diagnosis of lung cancer *Proc. of Medical Image Computing and Computer-Assisted Intervention, (MICCAI'09)* (Berlin: Springer) pp 682–9

[107] El-Baz A, Gimel'farb G, Falk R, El-Ghar M A and Suri J 2011 Appearance analysis for the early assessment of detected lung nodules *Lung Imaging and Computer Aided Diagnosis* (London: Taylor and Francis) ch 17, pp 395–404

[108] El-Baz A, Khalifa F, Elnakib A, Nitkzen M, Soliman A, McClure P, Gimel'farb G and El-Ghar M A 2012 A novel approach for global lung registration using 3D Markov Gibbs appearance model *Proc. of Int. Conf. Medical Image Computing and Computer-Assisted Intervention, (MICCAI'12) (Nice, France, 1–5 October 2012)* pp 114–21

[109] El-Baz A, Nitzken M, Elnakib A, Khalifa F, Gimel'farb G, Falk R and El-Ghar M A 2011 3D shape analysis for early diagnosis of malignant lung nodules *Proc. of Int. Conf. Medical Image Computing and Computer-Assisted Intervention, (MICCAI'11) (Toronto, Canada, 18–22 September 2011)* pp 175–82

[110] El-Baz A, Nitzken M, Gimel'farb G, Van Bogaert E, Falk R, El-Ghar M A and Suri J 2011 Three-dimensional shape analysis using spherical harmonics for early assessment of detected lung nodules *Lung Imaging and Computer Aided Diagnosis* (London: Taylor and Francis) ch 19, pp 421–38

[111] El-Baz A, Nitzken M, Khalifa F, Elnakib A, Gimel'farb G, Falk R and El-Ghar M A 2011 3D shape analysis for early diagnosis of malignant lung nodules *Proc. of Int. Conf. on Information Processing in Medical Imaging, (IPMI'11) (Monastery Irsee, Germany (Bavaria), 3–8 July 2011)* pp 772–83

[112] El-Baz A, Nitzken M, Vanbogaert E, Gimel'Farb G, Falk R and Abo El-Ghar M 2011 A novel shape-based diagnostic approach for early diagnosis of lung nodules *2011 IEEE Int. Symp. on Biomedical Imaging: From Nano to Macro* (Piscataway, NJ: IEEE) pp 137–40

[113] El-Baz A, Sethu P, Gimel'farb G, Khalifa F, Elnakib A, Falk R and El-Ghar M A 2011 Elastic phantoms generated by microfluidics technology: validation of an imaged-based approach for accurate measurement of the growth rate of lung nodules *Biotechnol. J.* **6** 195–203

[114] El-Baz A, Sethu P, Gimel'farb G, Khalifa F, Elnakib A, Falk R and El-Ghar M A 2010 A new validation approach for the growth rate measurement using elastic phantoms generated by state-of-the-art microfluidics technology *Proc. of IEEE Int. Conf. on Image Processing, (ICIP'10) (Hong Kong, 26–29 September 2010)* pp 4381–3

[115] El-Baz A, Sethu P, Gimel'farb G, Khalifa F, Elnakib A, Falk R and Suri M A E-G J 2011 Validation of a new imaged-based approach for the accurate estimating of the growth rate of detected lung nodules using real CT images and elastic phantoms generated by state-of-the-art microfluidics technology *Handbook of Lung Imaging and Computer Aided Diagnosis* vol 1 ed A El-Baz and J S Suri (New York: Taylor and Francis) ch 18, pp 405–20

[116] El-Baz A, Soliman A, McClure P, Gimel'farb G, El-Ghar M A and Falk R 2012 Early assessment of malignant lung nodules based on the spatial analysis of detected lung nodules *Proc. of IEEE Int. Symp. on Biomedical Imaging: From Nano to Macro, (ISBI'12)* (Piscataway, NJ: IEEE) pp 1463–6

[117] El-Baz A, Yuksel S E, Elshazly S and Farag A A 2005 Non-rigid registration techniques for automatic follow-up of lung nodules *Proc. of Computer Assisted Radiology and Surgery, (CARS'05)* **vol 1281** (Amsterdam: Elsevier) pp 1115–20

[118] El-Baz A S and Suri J S 2011 *Lung Imaging and Computer Aided Diagnosis* (Boca Raton, FL: CRC Press)

[119] Soliman A, Khalifa F, Dunlap N, Wang B, El-Ghar M and El-Baz A 2016 An iso-surfaces based local deformation handling framework of lung tissues *2016 IEEE 13th Int. Symp. on Biomedical Imaging (ISBI)* (Piscataway, NJ: IEEE) pp 1253–9

[120] Soliman A, Khalifa F, Shaffie A, Dunlap N, Wang B, Elmaghraby A and El-Baz A 2016 Detection of lung injury using 4D-CT chest images *2016 IEEE 13th Int. Symp. on Biomedical Imaging (ISBI)* (Piscataway, NJ: IEEE) pp 1274–7

[121] Soliman A, Khalifa F, Shaffie A, Dunlap N, Wang B, Elmaghraby A, Gimel'farb G, Ghazal M and El-Baz A 2017 A comprehensive framework for early assessment of lung injury *2017 IEEE Int. Conf. on Image Processing (ICIP)* (Piscataway, NJ: IEEE) pp 3275–9

[122] Shaffie A, Soliman A, Ghazal M, Taher F, Dunlap N, Wang B, Elmaghraby A, Gimel'farb G and El-Baz A 2017 A new framework for incorporating appearance and shape features of lung nodules for precise diagnosis of lung cancer *2017 IEEE Int. Conf. on Image Processing (ICIP)* (Piscataway, NJ: IEEE) pp 1372–6

[123] Soliman A, Khalifa F, Shaffie A, Liu N, Dunlap N, Wang B, Elmaghraby A, Gimel'farb G and El-Baz A 2016 Image-based CAD system for accurate identification of lung injury *2016 IEEE Int. Conf. on Image Processing (ICIP)* (Piscataway, NJ: IEEE) pp 121–5

[124] Dombroski B, Nitzken M, Elnakib A, Khalifa F, El-Baz A and Casanova M F 2014 Cortical surface complexity in a population-based normative sample *Transl. Neurosci.* **5** 17–24

[125] El-Baz A, Casanova M, Gimel'farb G, Mott M and Switala A 2008 An MRI-based diagnostic framework for early diagnosis of dyslexia *Int. J. Comput. Assist. Radiol. Surg.* **3** 181–9

[126] El-Baz A, Casanova M, Gimel'farb G, Mott M, Switala A, Vanbogaert E and McCracken R 2008 A new CAD system for early diagnosis of dyslexic brains *Proc. Int. Conf. on Image Processing (ICIP'2008)* (Piscataway, NJ: IEEE) p 1820–3

[127] El-Baz A, Casanova M F, Gimel'farb G, Mott M and Switwala A E 2007 A new image analysis approach for automatic classification of autistic brains *Proc. IEEE Int. Symp. on Biomedical Imaging: From Nano to Macro (ISBI'2007)* (Piscataway, NJ: IEEE) pp 352–5

[128] El-Baz A, Elnakib A, Khalifa F, El-Ghar M A, McClure P, Soliman A and Gimel'farb G 2012 Precise segmentation of 3-D magnetic resonance angiography *IEEE Trans. Biomed. Eng.* **59** 2019–29

[129] El-Baz A, Farag A, Gimel'farb G, El-Ghar M A and Eldiasty T 2006 Probabilistic modeling of blood vessels for segmenting MRA images *18th Int. Conf. on Pattern Recognition (ICPR'06)* **vol 3** (Piscataway, NJ: IEEE) pp 917–20

[130] El-Baz A, Farag A A, Gimel'farb G, El-Ghar M A and Eldiasty T 2006 A new adaptive probabilistic model of blood vessels for segmenting MRA images *Medical Image Computing and Computer-Assisted Intervention–MICCAI 2006* vol 4191 (Berlin: Springer) pp 799–806

[131] El-Baz A, Farag A A, Gimel'farb G and Hushek S G 2005 Automatic cerebrovascular segmentation by accurate probabilistic modeling of TOF-MRA images *Medical Image Computing and Computer-Assisted Intervention–MICCAI 2005* (Berlin: Springer) pp 34–42

[132] El-Baz A, Farag A, Elnakib A, Casanova M F, Gimel'farb G, Switala A E, Jordan D and Rainey S 2011 Accurate automated detection of autism related corpus callosum abnormalities *J. Med. Syst.* **35** 929–39

[133] El-Baz A, Farag A and Gimelfarb G 2005 Cerebrovascular segmentation by accurate probabilistic modeling of TOF-MRA images *Image Analysis* vol 3540 (Berlin: Springer) pp 1128–37

[134] El-Baz A, Gimel'farb G, Falk R, El-Ghar M A, Kumar V and Heredia D 2009 A novel 3D joint Markov–Gibbs model for extracting blood vessels from PC–MRA images *Medical Image Computing and Computer-Assisted Intervention–MICCAI 2009* vol 5762 (Berlin: Springer) pp 943–50

[135] Elnakib A, El-Baz A, Casanova M F, Gimel'farb G and Switala A E 2010 Image-based detection of corpus callosum variability for more accurate discrimination between dyslexic and normal brains *Proc. IEEE Int. Symp. on Biomedical Imaging: From Nano to Macro (ISBI'2010)* (Piscataway, NJ: IEEE) pp 109–12

[136] Elnakib A, Casanova M F, Gimel'farb G, Switala A E and El-Baz A 2011 Autism diagnostics by centerline-based shape analysis of the corpus callosum *Proc. IEEE Int. Symp. on Biomedical Imaging: From Nano to Macro (ISBI'2011)* (Piscataway, NJ: IEEE) pp 1843–6

[137] Elnakib A, Nitzken M, Casanova M, Park H, Gimel'farb G and El-Baz A 2012 Quantification of age-related brain cortex change using 3D shape analysis *2012 21st Int. Conf. on Pattern Recognition (ICPR)* (Piscataway, NJ: IEEE) pp 41–4

[138] Mostapha M, Soliman A, Khalifa F, Elnakib A, Alansary A, Nitzken M, Casanova M F and El-Baz A 2014 A statistical framework for the classification of infant DT images *2014 IEEE Int. Conf. on Image Processing (ICIP)* (Piscataway, NJ: IEEE) pp 2222–6

[139] Nitzken M, Casanova M, Gimel'farb G, Elnakib A, Khalifa F, Switala A and El-Baz A 2011 3D shape analysis of the brain cortex with application to dyslexia *2011 18th IEEE Int. Conf. on Image Processing (ICIP)* (Piscataway, NJ: IEEE) pp 2657–60

[140] El-Gamal F E-Z A, Elmogy M M, Ghazal M, Atwan A, Barnes G N, Casanova M F, Keynton R and El-Baz A S 2017 A novel CAD system for local and global early diagnosis of Alzheimer's disease based on PiB-PET scans *2017 IEEE Int. Conf. on Image Processing (ICIP)* (Piscataway, NJ: IEEE) pp 3270–4

[141] Ismail M, Soliman A, Ghazal M, Switala A E, Gimelfarb G, Barnes G N, Khalil A and El-Baz A 2017 A fast stochastic framework for automatic MR brain images segmentation *PLOS One*

[142] Ismail M M, Keynton R S, Mostapha M M, ElTanboly A H, Casanova M F, Gimel'farb G L and El-Baz A 2016 Studying autism spectrum disorder with structural and diffusion magnetic resonance imaging: a survey *Front. Human Neurosci.* **10** 211

[143] Alansary A *et al* 2016 Infant brain extraction in T1-weighted MR images using BET and refinement using LCDG and MGRF models *IEEE J. Biomed. Health Inform.* **20** 925–35

[144] Ismail M, Soliman A, ElTanboly A, Switala A, Mahmoud M, Khalifa F, Gimel'farb G, Casanova M F, Keynton R and El-Baz A 2016 Detection of white matter abnormalities in MR brain images for diagnosis of autism in children *2016 IEEE 13th International Symposium on Biomedical Imaging (ISBI)* pp 6–9

[145] Ismail M, Mostapha M, Soliman A, Nitzken M, Khalifa F, Elnakib A, Gimel'farb G, Casanova M and El-Baz A 2015 Segmentation of infant brain MR images based on adaptive shape prior and higher-order MGRF *2015 IEEE Int. Conf. on Image Processing (ICIP)* pp 4327–31

[146] Asl E H, Ghazal M, Mahmoud A, Aslantas A, Shalaby A, Casanova M, Barnes G, Gimelfarb G, Keynton R and El-Baz A 2018 Alzheimers disease diagnostics by a 3D deeply supervised adaptable convolutional network *Front. Biosci.* **23** 584–96

[147] Mahmoud A, El-Barkouky A, Farag H, Graham J and Farag A 2013 A non-invasive method for measuring blood flow rate in superficial veins from a single thermal image *Proc. of the IEEE Conf. on Computer Vision and Pattern Recognition Workshops* pp 354–9

[148] El-baz A, Shalaby A, Taher F, El-Baz M, Ghazal M, El-Ghar M A, Takieldeen A and Suri J 2017 Probabilistic modeling of blood vessels for segmenting magnetic resonance angiography images *Med. Res. Arch.* **5**

[149] Chowdhury A S, Rudra A K, Sen M, Elnakib A and El-Baz A 2010 Cerebral white matter segmentation from MRI using probabilistic graph cuts and geometric shape priors *ICIP* 3649–52

[150] Gebru Y, Giridharan G, Ghazal M, Mahmoud A, Shalaby A and El-Baz A 2018 Detection of cerebrovascular changes using magnetic resonance angiography *Cardiovascular Imaging and Image Analysis* (Boca Raton, FL: CRC Press) pp 1–22

[151] Mahmoud A, Shalaby A, Taher F, El-Baz M, Suri J S and El-Baz A 2018 Vascular tree segmentation from different image modalities *Cardiovascular Imaging and Image Analysis* (Boca Raton, FL: CRC Press) pp 43–70

[152] Taher F, Mahmoud A, Shalaby A and El-Baz A 2018 A review on the cerebrovascular segmentation methods *2018 IEEE Int. Symp. on Signal Processing and Information Technology (ISSPIT)* (Piscataway, NJ: IEEE) pp 359–64

[153] Kandil H, Soliman A, Fraiwan L, Shalaby A, Mahmoud A, ElTanboly A, Elmaghraby A, Giridharan G and El-Baz A 2018 A novel MRA framework based on integrated global and local analysis for accurate segmentation of the cerebral vascular system *2018 IEEE 15th Int. Symp. on Biomedical Imaging (ISBI 2018)* (Piscataway, NJ: IEEE) pp 1365–8

IOP Publishing

Neurological Disorders and Imaging Physics, Volume 3
Application to autism spectrum disorders and Alzheimer's
Ayman El-Baz and Jasjit S Suri

Chapter 15

Towards a robust CAD system for early diagnosis of autism using structural MRI

Ali Mahmoud, Yaser ElNakieb, Ahmed Shalaby, Ahmed Soliman, Fatma Taher, Hassan Hajjdiab, Ashraf Khalil, Mohammed Ghazal, Robert Keynton, Jasjit S Suri, Gregory N Barnes and Ayman El-Baz

This chapter discusses a promising computer-aided diagnosis (CAD) system, devised by our research team, for diagnosing autism at various stages of life, making use of the shape information in brain magnetic resonance imaging (MRI). Our system integrates the shape features extracted from both the cerebral white matter (CWM) and the cerebral cortex (Cx). At the first stage of the CAD system, a 3D probability model that jointly includes shape, intensity, and spatial information is used to segment the CWM and the Cx. The features of the Cx 3D shape are then precisely extracted by applying spherical harmonics (SPHARMs) to the reconstructed meshes, where four values are obtained for each mesh point: Cx surface reconstruction error, mean curvature, normal curvature, and Gaussian curvature. Regarding the CWM, its shape features are the distance maps calculated for its gyri after being segmented by the 3D fast marching technique, in addition to the mean curvature, sharpness, and curvedness of the surfaces. The last stage of the CAD system includes a multi-level deep network that integrates all the shape features for the purpose of diagnosis. To evaluate our CAD system, we tested it on three databases: (1) the ABIDE database for subjects of ages ranging from 8 to 12.8 years, which resulted in an accuracy of 92.4%; (2) NDAR/Pitt database for subjects of ages from 16 to 51 years, which gave an accuracy of 96%; and (3) NDAR/IBIS database for infant subjects, which achieved an accuracy of 85%. This indicates how promising our CAD system is to be used robustly for the early diagnosis of autism from structural MRI.

15.1 Introduction

Autism spectrum disorder (ASD) is one of the most crucial neuro-developmental disorders that can significantly affect the social, linguistic, and behavioral skills of individuals [1], which has made it one of the most active research areas. The literature contains many works that have studied the changes linked to autism using different types of magnetic resonance imaging (MRI), e.g. diffusion tensor imaging (DTI), structural MRI (sMRI), and functional MRI (fMRI). The method discussed in this chapter is limited to using sMRI data for autism diagnosis, however, we will discuss integrating other modalities in the system in the future.

The United States has many sites for acquiring sMRI data, which has led to establishing several databases in a comprehensive manner, such as ABIDE [2], NDAR/IBIS, and NDAR/Pitt [3]. Exploring the literature, sMRI studies on ASD can be categorized into either ROI-based volumetry (RBV), or shape-based morphometry (SBM). RBV approaches focus on the total volume for a particular region, e.g. the cerebral gray matter (CGM) for toddlers with autism was found to be enlarged by the study [4] that split the Cx into 34 parts to compute the lobe volume measures. An RBV study [5] was conducted on infants aged 6 months from the NDAR/IBIS database for comparing the cases that are at low risk with those at high risk. The study used Fourier harmonics for parameterizing the contour of the head in order to compute some volumetric measures, however, the results revealed no significant apparent differences between the groups. Also, the study on infants under 2 years of age [6] reported an increase in cortical volume. The RBV approaches are gender and age-sensitive, which requires involving age-correction coefficients in the system, which does not enable having robust CAD systems. On the other hand, SBM studies mainly address topological shape features, such as cortical thickness. For instance, the study performed in [7] on children with autism revealed the decrease of the cortical thickness in the cerebral lobes for autistic subjects, when compared to control subjects. The study applied the BRAINS package for surface analysis. Moreover, the cortical surface of adolescents was studied in [8] and indicated a significant cortical thinning in two areas of the left hemisphere for autistic subjects, when compared to control subjects. Although SBM approaches are capable of accounting for the topology features inherent in the brain, most of the current work directly deals with the meshes developed from raw data without parameterizing the brain or aligning the meshes, which makes the computations sensitive to pre-processing and makes it difficult to compare information. In addition, using SBM to study autistic infants scarcely exists in the literature. To our knowledge, there are no CAD systems addressing infants and children. To handle such limitations, this chapter discusses a novel shape-based automated approach for autism diagnosis by our research team [9, 10]. What significantly motivated our research team to investigate integrating the Cx and the CWM features is the suggestion made recently to consider the self magnifications of minicolumnopathy and dysplastic Cx via the CWM abnormal connections [11].

15.2 Methods

The first stage of the presented CAD system is to segment the Cx and CWM from the brain. This is followed by performing shape analysis on the reconstructed meshes, in order to extract eight shape-based features that are finally integrated for the purpose of classifying the brain as autistic or control. The stages of the presented method are illustrated in figure 15.1.

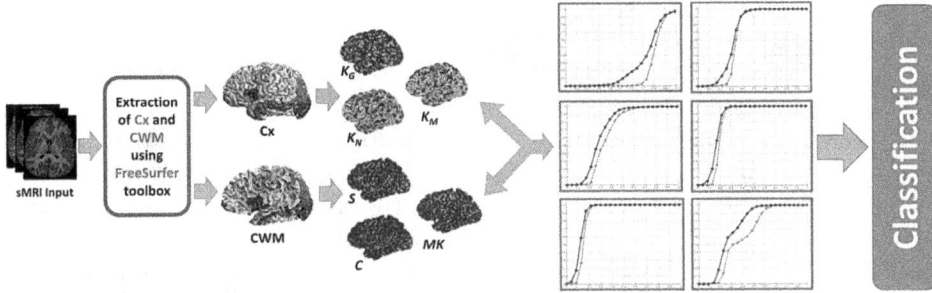

Figure 15.1. The framework of our CAD system.

15.2.1 Cx and CWM segmentation from MRI scans

The segmentation framework consists mainly of a 3D model that jointly combines shape, spatial, and intensity information. The joint probability model $P(\mathbf{g}, \mathbf{m}) = P(\mathbf{g}|\mathbf{m})P(\mathbf{m})$ is used to describe the brain input image, $\mathbf{g}$, which is co-aligned to the training database, and its map, $\mathbf{m}$, where $P(\mathbf{m}) = P_{sp}(\mathbf{m})P_V(\mathbf{m})$. Here, the weighted shape prior is denoted by $P_{sp}(\mathbf{m})$, and the Gibbs probability distribution with the potentials $\mathbf{V}$ is denoted by $P_V(\mathbf{m})$. Details of the three components are given below.

Adaptive shape model, $P_{sp}(\mathbf{m})$: The segmentation starts by creating an atlas, where the expected shapes of each brain label are constrained with an adaptive shape prior. A training set collected for 15 data sets (not included as test subjects), are co-aligned by 3D affine transformations by maximizing their mutual information [12]. The 15 subjects covered different ages and included control subjects in addition to autistic subjects (to avoid being biased to a certain group). To segment a subject, we exploit the normalized cross correlation similarity coefficient to pick the subject from the shape database for which the similarity with the input is the highest. The selected subject is then considered as a reference prototype to co-align the input subject via the 3D affine transformation.

First-order intensity model, $P(\mathbf{g}|\mathbf{m})$: Intensity information is also incorporated. The first-order visual appearance of each brain label is modeled by separating a mixed distribution of voxel intensities into individual components associated with the dominant modes of the mixture, and approximated with the linear combinations of discrete Gaussians (LCDG) method [13].

Higher-order clique MGRF model, $P_V(\mathbf{m})$: In order to handle inhomogeneities, in particular for infant scans that have low contrast between different tissue types, the spatial information is considered [14] using a novel higher-order Markov–Gibbs random field (MGRF) model, which spatially includes the triple and quad cliques families in addition to the pairwise cliques. Let $\mathbf{C}_a$ denote a family of a-order cliques of an interaction graph with nodes in the 3D lattice sites $p = (x, y, z)$ and edges connecting the interacting sites. An MGRF homogeneously models the label interactions over the 26 neighborhoods of voxels:

$$P_{\mathbf{V}}(\mathbf{m}) = \frac{1}{Z_{\mathbf{V}}} \exp\left(\sum_{a=1}^{A} \sum_{\mathbf{c} \in \mathbf{C}_a} V_a(\mathbf{m}(x, y, z): (x, y, z) \in \mathbf{c})\right), \qquad (15.1)$$

where A clique families specify the geometry of the graph interactions, $\mathbf{V} = [V_a: \{0, \ldots, L\} \rightarrow (-\infty, \infty): a = 1, \ldots, A]$ is a collection of Gibbs potential functions V_a for the families $\mathbf{C}_a$, and the partition function $Z_{\mathbf{V}}$ normalizes the probabilities. An initial region map $\mathbf{m}$ allows for analytically approximating the maximum likelihood estimates of the potentials and computing the voxel-wise probabilities of the region labels. The second-, third-, and forth-order potentials for the region map label $m_{\mathbf{p}_i}$ are given by

$$V_a(m_{\mathbf{p}_1}, m_{\mathbf{p}_2}) = \begin{cases} V_{2:a:\text{eq}} & \text{if } m_{\mathbf{p}_1} = m_{\mathbf{p}_2} \\ -V_{2:a:\text{eq}} & \text{otherwise} \end{cases}, \qquad (15.2)$$

where $V_{2:a:\text{eq}} = -V_{2:a:\text{ne}} = 4(F_{a:\text{eq}}(\mathbf{m}°) - \frac{1}{2})$, and $\mathbf{F}(\mathbf{m}°) = [\rho_a F_a(\mu_1, \ldots, \mu_s|\mathbf{m}°) : (\mu_1, \ldots, \mu_s) \in \{0, \ldots, L\}^s; a = 1, \ldots, A]$ is the collection of scaled relative frequencies of co-occurrences of configurations $(\mu_1, \ldots, \mu_s)$ of the labels in the cliques of each family $\mathbf{C}_a$ over a given training map $\mathbf{m}°$.

$$V_a(m_{\mathbf{p}_1}, m_{\mathbf{p}_2}, m_{\mathbf{p}_3}) = \begin{cases} V_{3:a:\text{eq}_3} & \text{if } m_{\mathbf{p}_1} = m_{\mathbf{p}_2} = m_{\mathbf{p}_3} \\ -V_{3:a:\text{eq}_3} & \text{otherwise} \end{cases}, \qquad (15.3)$$

where $V_{3:a:\text{eq}_3} = -V_{3:a:\text{eq}_2} = \frac{16}{3}(F_{a:\text{eq}_3}(\mathbf{m}°) - \frac{1}{4})$

$$V_a(m_{\mathbf{p}_1}, m_{\mathbf{p}_2}, m_{\mathbf{p}_3}, m_{\mathbf{p}_4}) = \begin{cases} V_{4:a:\text{eq}_4} & \text{if 4 equal labels} \\ V_{4:a:\text{eq}_3} & \text{if 3 equal labels}, \\ -(V_{4:a:\text{eq}_3} + V_{4:a:\text{eq}_4}) & \text{otherwise} \end{cases} \qquad (15.4)$$

where

$$V_{4:a:\text{eq}_4} = \lambda^*\left(F_{a:\text{eq}_4}(\mathbf{m}°) - \frac{1}{8}\right)$$

$$V_{4:a:\text{eq}_3} = \lambda^*\left(F_{a:\text{eq}_3}(\mathbf{m}°) - \frac{1}{2}\right)$$

$$V_{4:a:\text{eq}_2} = \lambda^*\left(F_{a:\text{eq}_2}(\mathbf{m}°) - \frac{3}{8}\right) = -\left(V_{4:a:\text{eq}_4} + V_{4:a:\text{eq}_3}\right)$$

and

$$\lambda^* = \frac{\displaystyle\sum_{a=1}^{A}\left(\left(F_{a:\text{eq}_4}(\mathbf{m}°) - \frac{1}{8}\right)^2 + \left(F_{a:\text{eq}_3}(\mathbf{m}°) - \frac{1}{2}\right)^2 + \left(F_{a:\text{eq}_2}(\mathbf{m}°) - \frac{3}{8}\right)^2\right)}{\displaystyle\sum_{a=1}^{A}\left(\frac{7}{64}\left(F_{a:\text{eq}_4}(\mathbf{m}°) - \frac{1}{8}\right)^2 + \frac{1}{4}\left(F_{a:\text{eq}_3}(\mathbf{m}°) - \frac{1}{2}\right)^2 + \frac{15}{64}\left(F_{a:\text{eq}_2}(\mathbf{m}°) - \frac{3}{8}\right)^2\right)}.$$

To prove the analytic estimation of potentials, let $\mathbf{L} = \{0, \ldots, L\}$ and $\mathbf{R} = \{(x, y, z): 0 \leqslant x \leqslant X - 1, 0 \leqslant y \leqslant Y - 1, 0 \leqslant z \leqslant Z - 1, \}$ denote a set of region labels L and a finite arithmetic lattice supporting binary maps $\mathbf{m}: \mathbf{R} \to \mathbf{L}$, respectively. Let $\mathbf{C}_a$ be a family of s-order cliques of the interaction graph with nodes in the lattice sites (x, y, z) and edges connecting the interacting (interdependent) sites. Let A clique families describe the spatial geometry of interactions between the region labels of region maps for an MGRF model:

$$P_\mathbf{V}(\mathbf{m}) = \frac{1}{Z_\mathbf{V}} \exp\left(\sum_{a=1}^{A} \sum_{c \in \mathbf{C}_a} V_a(\mathbf{m}(x, y, z): (x, y, z) \in \mathbf{c}) \right),$$

where $\mathbf{V} = [V_a: \{0, \ldots, L\} \to (-\infty, \infty): a = 1, \ldots, A]$ is a collection of potential functions for the families $\mathbf{C}_a$ and $Z_\mathbf{V}$ is the partition function

$$Z_\mathbf{V} = \sum_{\mathbf{m} \in \mathsf{M}} \exp\left(\sum_{a=1}^{A} \sum_{c \in \mathbf{C}_a} V_a(\mathbf{m}(x, y, z): (x, y, z) \in \mathbf{c}) \right)$$

normalizing the probabilities over the population $\mathsf{M} = \{0, \ldots, L\}^{XYZ}$ of the maps. Let $\mathbf{F}(\mathbf{m}^\circ) = [\rho_a F_a(\mu_1, \ldots, \mu_s | \mathbf{m}^\circ): (\mu_1, \ldots, \mu_s) \in \{0, \ldots, L\}^s; a = 1, \ldots, A]$ and $\mathbf{P}_\mathbf{V} = [\rho_a P_a(\mu_1, \ldots, \mu_s | \mathbf{V}): (\mu_1, \ldots, \mu_s) \in \{0, \ldots, L\}^s; a = 1, \ldots, A]$, where $\rho_a = \frac{1}{XYZ} |\mathbf{C}_a|$ is the relative cardinality of the clique family with regard to the lattice cardinality, describe the collection of scaled relative frequencies $F_a(\mu_1, \ldots, \mu_s | \mathbf{m}^\circ)$ of co-occurrences of configurations $(\mu_1, \ldots, \mu_s)$ of the labels in the cliques for every family $\mathbf{C}_a$ across a given training map $\mathbf{m}^\circ$ and the corresponding scaled marginal probabilities $P_a(\mu_1, \ldots, \mu_s | \mathbf{V})$ of the MGRF model configurations, respectively.

The normalized log-likelihood $L(\mathbf{V}|\mathbf{m}^\circ) = \frac{1}{XYZ} \log \mathbf{P}_\mathbf{V}(\mathbf{m}^\circ)$ of the map $\mathbf{m}^\circ$ has the gradient

$$\nabla L(\mathbf{V}|\mathbf{m}^\circ) \equiv \frac{\partial}{\partial \mathbf{V}} L(\mathbf{V}|\mathbf{m}^\circ) = \mathbf{F}(\mathbf{m}^\circ) - \mathbf{P}_\mathbf{V}$$

and its Hessian matrix of the second derivatives is equal to the negated covariance matrix of the marginal probabilities of signal co-occurrences in the cliques (due to non-negative definiteness of the covariance matrix, the log-likelihood is unimodal over the potential space).

The analytical potential estimate (i.e. the approximate maximum likelihood estimate) is specified as a scaled gradient vector at the origin in the potential space,

$$\mathbf{V}^* = \lambda^*(\mathbf{F}(\mathbf{m}^\circ) - \mathbf{P}_0),$$

where the factor λ^* maximizes the truncated (to the first three terms) Taylor series decomposition of the log-likelihood $L(\mathbf{V}|\mathbf{m}^\circ)$ about the origin:

$$L(\mathbf{V}|\mathbf{m}^\circ) \approx L(0|\mathbf{m}^\circ) + \lambda(\mathbf{F}(\mathbf{m}^\circ) - \mathbf{P}_0)^\mathsf{T}(\mathbf{F}(\mathbf{m}^\circ) - \mathbf{P}_0)$$
$$- \frac{\lambda^2}{2}(\mathbf{F}(\mathbf{m}^\circ) - \mathbf{P}_0)^\mathsf{T} \mathbf{D}_0 (\mathbf{F}(\mathbf{m}^\circ) - \mathbf{P}_0),$$

where $\mathbf{D}_0$ denotes the covariance matrix for marginal clique-wise probabilities at the origin. Generally,

$$\lambda^* = \frac{(\mathbf{F}(\mathbf{m}^\circ) - \mathbf{P}_0)^\mathsf{T}(\mathbf{F}(\mathbf{m}^\circ) - \mathbf{P}_0)}{(\mathbf{F}(\mathbf{m}^\circ) - \mathbf{P}_0)^\mathsf{T}\mathbf{D}_0(\mathbf{F}(\mathbf{m}^\circ) - \mathbf{P}_0)}.$$

The origin, $\mathbf{V} = \mathbf{0}$ (zero potentials), corresponds to an independent random field (IRF) of equiprobable labels, so that the covariance matrix is closely approximated with the diagonal matrix of variances.

For the labels, $m(x, y, z) \in \{0, \dots, L\}$, the marginal co-occurrence probabilities over cliques of second, third, and fourth order are $\frac{1}{4}$, $\frac{1}{8}$, and $\frac{1}{16}$, respectively. If for the sake of symmetry, only equality and inequality of the labels are taken into account, then the combinations of co-occurrences and their probabilities are summarized in table 15.1.

Provided the cardinalities of the clique families are close to the lattice cardinality (so that $\rho_a \approx 1$) for all the families $a = 1, \dots, A$, the resulting potential estimates for the second- and third-order models are independent of the number A of the clique families:

Second-order: $V_{2:a:\text{eq}} = -V_{2:a:\text{ne}} = 4\left(F_{a:\text{eq}}(\mathbf{m}^\circ) - \frac{1}{2}\right)$

Third-order: $V_{3:a:\text{eq}_3} = -V_{3:a:\text{eq}_2} = \frac{16}{3}\left(F_{a:\text{eq}_3}(\mathbf{m}^\circ) - \frac{1}{4}\right).$

But for the fourth-order model the factor depends on the number of clique families and the marginal probabilities of label combinations for these families on the training map $\mathbf{m}^\circ$:

$$V_{4:a:\text{eq}_4} = \lambda^*\left(F_{a:\text{eq}_4}(\mathbf{m}^\circ) - \frac{1}{8}\right)$$

$$V_{4:a:\text{eq}_3} = \lambda^*\left(F_{a:\text{eq}_3}(\mathbf{m}^\circ) - \frac{1}{2}\right)$$

$$V_{4:a:\text{eq}_2} = \lambda^*\left(F_{a:\text{eq}_2}(\mathbf{m}^\circ) - \frac{3}{8}\right) = -\left(V_{4:a:\text{eq}_4} + V_{4:a:\text{eq}_3}\right),$$

Table 15.1. Label combinations and their marginal probabilities for second-, third-, and fourth-order cliques. Here, p denotes the probability, 'eq' and 'ne' denote two equal or non-equal labels for a second-order clique, and 'eq$_i$' denote i equal labels for a third- or fourth-order clique.

Clique order	2		3		4		
Label combinations	eq	ne	eq$_3$	eq$_2$	eq$_4$	eq$_3$	eq$_2$
Marginal p for the IRF	1/2	1/2	1/4	3/4	1/8	1/2	3/8
Its variance $p(1 - p)$	1/4	1/4	3/16	3/16	7/64	1/4	15/64

where

$$
\lambda^* = \frac{\displaystyle\sum_{a=1}^{A}\left(\left(F_{a:\mathrm{eq}_4}(\mathbf{m}^\circ) - \frac{1}{8}\right)^2 + \left(F_{a:\mathrm{eq}_3}(\mathbf{m}^\circ) - \frac{1}{2}\right)^2 + \left(F_{a:\mathrm{eq}_2}(\mathbf{m}^\circ) - \frac{3}{8}\right)^2\right)}{\displaystyle\sum_{a=1}^{A}\left(\frac{7}{64}\left(F_{a:\mathrm{eq}_4}(\mathbf{m}^\circ) - \frac{1}{8}\right)^2 + \frac{1}{4}\left(F_{a:\mathrm{eq}_3}(\mathbf{m}^\circ) - \frac{1}{2}\right)^2 + \frac{15}{64}\left(F_{a:\mathrm{eq}_2}(\mathbf{m}^\circ) - \frac{3}{8}\right)^2\right)}.
$$

Finally, the region map $\mathbf{m}$ is improved using an iterative conditional mode (ICM) algorithm [15] that maximizes the probabilities of the 3D joint model.

15.2.2 Extraction of shape features

After the segmentation of the CWM and the Cx, shape analysis is performed to acquire the CAD system. In particular, we extracted four shape features from four lobes of the Cx (temporal, frontal, parietal, and occipital) per hemisphere. In addition, we extracted four other shape features from the CWM, which resulted in having 64 features in total per scan (eight per lobe per hemisphere). Shape features are discussed below in more details

Cx shape features: For extracting the shape features from the Cx, it is required to accurately approximate its 3D shape, in addition to being capable of comparing different brain subjects. Among the ways to achieve this, is to perform SPHARM analysis [16, 17] that approximates the cortex surface so that the sensitivity to any errors that might have resulted from the segmentation is greatly reduced. Moreover, it makes the reconstruction independent to acquiring data from different sites. For SPHARM to be applied, a modified version of the TETGEN algorithm [18] should first be used to generate a mesh manifold from the segmented scans. During the construction of the meshes, we are restricted to using the same number of nodes for all subjects. The reason behind this is that although the shape of each mesh differs from the other, having a consistent number of nodes makes it possible to align the meshes with each other. Having the alignment done using SPAHRM-based registration [17], each node from the lobe mesh (four per hemisphere) is then described by four features. Specifically, such values are the reconstruction error of the Cx by computing the Euclidean distance from the origin D_{Euc}; the Gaussian curvature K_G (the result of multiplying the principal curvatures); the mean curvature K_M (the result of averaging the principal curvatures); and the normal surface curvature K_N (the largest of the principal curvatures).

CWM shape features: For analyzing the shape of the CWM, we first apply a level-set based method to extract its gyri. The computation of the distance map is performed inside the 3D segmented CWM by the fast marching level-set method [19], accompanied by the propagation of two waves. The first wave is responsible for determining the voxels whose location from the CWM boundary is less than or equal to T, while the second propagating wave is to determine the voxels which have a distance less than or equal to T from voxels located at a distance T from the boundary of the CWM, as illustrated in figure 15.2(c). The voxels hit by the second

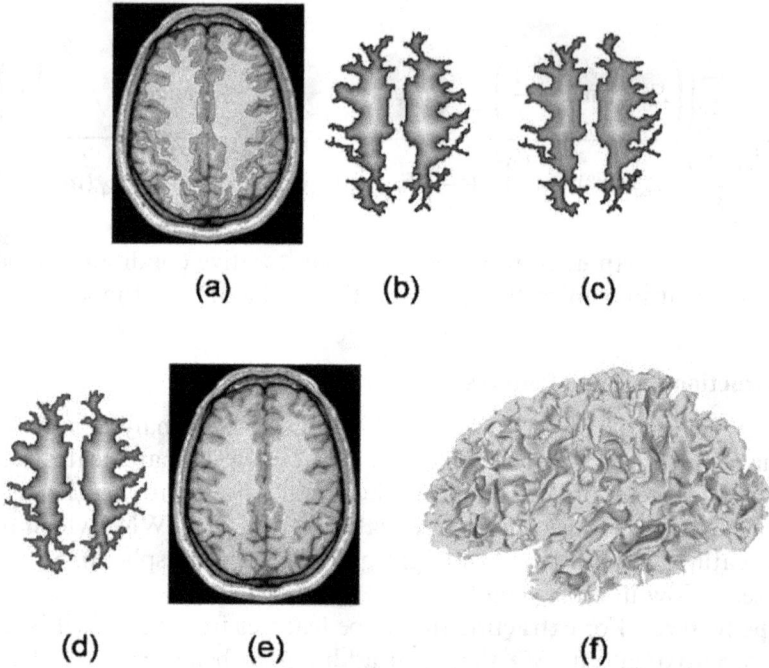

Figure 15.2. (a) The segmented CWM boundary. (b) The first distance map. (c) The second distance map. (d) and (e) The extracted CWM gyri. (f) Visualizing the extracted gyri in 3D.

wave are then excluded from those hit by the first wave, where the remainder is the gyri of the CWM as illustrated in figure 15.2(d) and (e).

The distance maps, DM, were computed from the extracted gyral CWM, in order to be used as a shape feature. Also, three more features are obtained through the curvature-based analysis of the 3D brain CWM meshes: the sharpness S, which is used to describe the folding sharpness; the curvedness C, which discriminates the highly folded parts from those that are less folded; and the mean curvature MK [20].

15.2.3 Deep fusion classification network (DFCN)

The main objective of this stage is to globally achieve a diagnosis by making use of the shape features that were collected, which are eight features per lobe per hemisphere, i.e. 64 features per scan. The traditional way to achieve this is the vertex-wise utilization of the raw features, however having 48K nodes in each mesh makes this inefficient, as a long time is needed to handle about 400K points for each subject. This problem is avoided by representing the features using the values of their cumulative distribution functions (CDFs), which keeps all the data while reducing its dimensionality.

To include all of the information in the distribution, the CDFs are computed using the minimum increment that results from sorting all the measures and then choosing the minimum difference among consecutive points. Having 64 features per subject would require a deep network scheme to account for all data. The model is based mainly on having a stack of auto-encoders (AEs) for each feature in addition to an output layer of softmax regression [21]. It is composed of two stages as illustrated in figure 15.3. During the first stage, the CDF of each of the 64 features is learned separately with an AE using sparsity (KLdivergence) and non-negative constraints, in order to capture most prominent variations and thus obtain discriminatory features. The scheme employed here uses AEs with non-negativity constraint (NCAE) [22] that decompose the data, along with sparsity (KL divergence), to extract the most uncorrelated and definitive features based on non-negative matrix factorization. In the second stage, the total loss, or in other words the negative log-likelihood for a given labeled training data is minimized using a supervised back-propagation technique. The last step is to concatenate the 64 features that were extracted from each stacked NCAE (SNCAE) into a new combined high-level feature, which is input to another SNCAE to perform non-linear fusion and final diagnosis, by having two probabilities in the output layer that correspond to autism and control. This is where the term 'deep fusion' comes from.

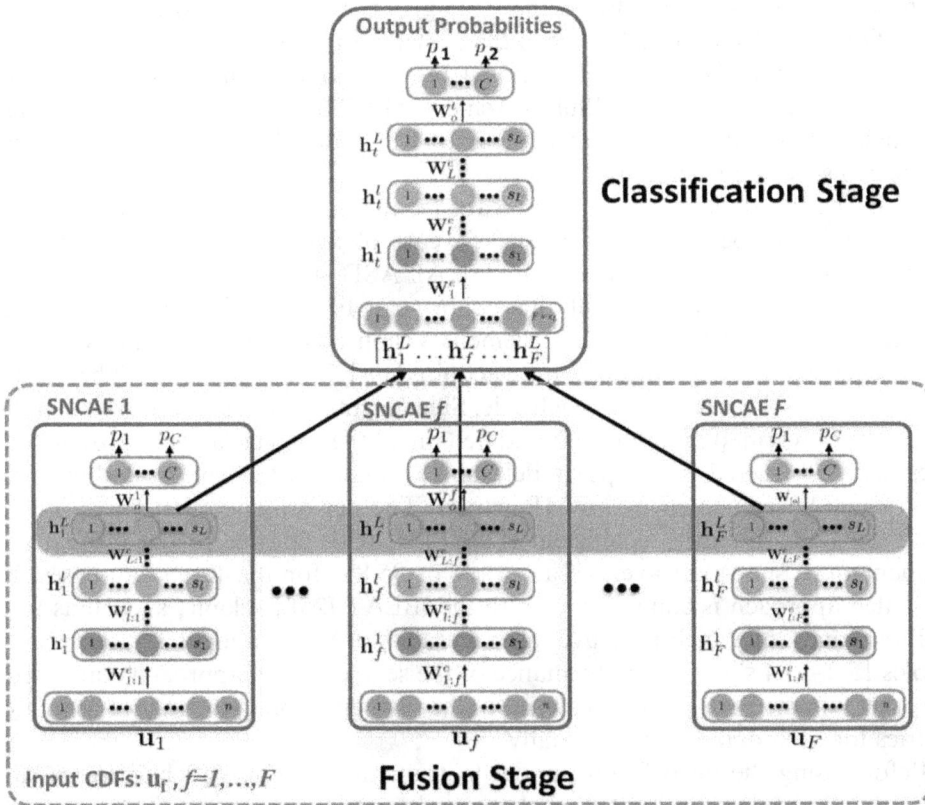

Figure 15.3. Structure of DFCN.

15.3 Experimental results and conclusions

The presented work was tested on three databases at different ages to show its generality and robustness. T1-weighted, whole brain MRI were obtained from three sources, as follows. (1) the Infant Brain Imaging Study (hereafter, IBIS), a multisite, longitudinal study of infants at high risk of developing autism. These data are available from the National Database for Autism Research (NDAR). (2) Data from the Kennedy Krieger Institute (hereafter, KKI), made available through the Autism Brain Imaging Data Exchange [2]. (3) Scans obtained as part of the Collaborative Program for Excellence in Autism at the University of Pittsburgh (hereafter, Pitt), again available from NDAR.

The IBIS data comprised MRI from 20 individuals at 6 or 7 months of age, ten of whom (three girls) were later diagnosed with ASD, while the other ten (five girls) were not. T1-weighted images were acquired on a 3T scanner with TR = 2400 ms, TE = 3.16 ms, TI = 1200 ms, and 8° flip angle. 160 sagittal slices were acquired at 1 mm thickness, with each slice being 224 × 256 pixels with 1 mm resolution.

The KKI data include 21 children with ASD and 21 controls [2]. Participants were between 8 and 13 years of age when scanned; four of those with ASD and seven of those without were girls. From the ASD group, three were low functioning (full scale IQ below 80), and 12 had been diagnosed with at least one other psychiatric condition: ADHD (9), ODD (6), phobia (4), or mood disorder (2). T1-weighted images were acquired on a 3T Philips Achieva with TR = 8 ms, TE = 3.7 ms, and 8° flip angle. 128 sagittal slices were acquired at 1.33 mm thickness. The pixels of each 256 × 256 slice were 1 mm each side.

From the Pitt dataset we analyzed scans from 16 individuals with ASD, three of them female, all between 16.5 and 53.5 years of age, and 16 controls matched pairwise for age, sex, and full scale IQ [3]. All ASD subjects in the Pitt study were high-functioning. Full details of this cohort can be found in the literature [23]. Participants were scanned on 1.5T Siemens Vision equipment using an MPRAGE pulse sequence. 128 sagittal slices were acquired at 1.25 mm thickness. The pixels of each 256 × 256 slice were 1 mm each side. Since segmentation is one key in the CAD system, the presented method was validated on all the subjects above (different from ones that constructed the shape model), and evaluated with their ground truth that was manually segmented by an MR expert. The performance that was evaluated using the Dice similarity coefficient (DSC) [24] with relative to the ground truth segmentation, shows an average accuracy of 95.8% for the three databases. The presented approach is compared also to the iBEAT [25] package, as well as to the FSL package [26], which have average accuracies of 84.1% and 89.3%, respectively. Tables 15.2–15.4 show the performance of the segmentation approach compared to the other two methods, and also measures the segmentation performance using three metrics for each database individually.

Before using the deep fusion model to train the features, we had to test their significance and determine whether they are discriminatory or not. Thus, statistical analysis was conducted for each dataset and each measurement individually. Mixed linear models were used, with the fixed effects diagnosis (ASD/control), lobe, and

Table 15.2. Illustrating the accuracy of our segmentation approach using the Dice similarity coefficient (DSC) (%), the modified Hausdorff distance (MHD), and absolute brain volume difference (ABVD) (%) for the IBIS database. Metrics are represented as mean ± standard deviation.

Metric	WM			Cx		
	Ours	iBEAT	FSL	Ours	iBEAT	FSL
DSC	94.7 ± 1.53	73.3 ± 1.27	80.4 ± 1.57	93.86 ± 0.13	81.6 ± 3.5	89.5 ± 0.65
p-value			0.000 1			0.000 1
MHD	7.3 ± 1.23	18.27 ± 1.53	13.6 ± 1.28	3.5 ± 0.24	23.3 ± 0.52	15.68 ± 0.86
p-value			0.000 1			0.000 1
ABVD	3.17 ± 1.73	37.94 ± 0.61	15.8 ± 0.6	1.62 ± 1.24	34.46 ± 0.18	14.56 ± 1.17
p-value			0.000 1			0.000 1

Table 15.3. Illustrating the accuracy of our segmentation approach using the Dice similarity coefficient DSC (%), the modified Hausdorff distance (MHD), and absolute brain volume difference ABVD (%) for the KKI database. Metrics are represented as mean ± standard deviation.

Metric	WM			Cx		
	Ours	iBEAT	FSL	Ours	iBEAT	FSL
DSC	95.8 ± 1.5	85 ± 3.7	88.4 ± 2.5	96.7 ± 1.2	88 ± 2.7	92.3 ± 1.85
p-value			0.000 1			0.000 1
MHD	5.2 ± 1.5	10.2 ± 3.4	9 ± 1.2	3.1 ± 1.7	15.1 ± 3.5	11.8 ± 1.2
p-value			0.000 1			0.000 1
ABVD	2.5 ± 1.6	23.3 ± 1.1	9.8 ± 1.6	1.12 ± 1.1	18.4 ± 1.8	8.3 ± 2.7
p-value			0.000 1			0.000 1

Table 15.4. Illustrating the accuracy of our segmentation approach using Dice similarity coefficient DSC (%), the modified Hausdorff distance MHD, and absolute brain volume difference ABVD (%) for the University of Pittsburgh database. Metrics are represented as mean ± standard deviation.

Metric	WM			Cx		
	Ours	iBEAT	FSL	Ours	iBEAT	FSL
DSC	96.1 ± 1.5	87.3 ± 1.2	92.4 ± 1.5	97.8 ± 0.13	89.6 ± 2.5	93.2 ± 1.3
p-value			0.000 1			0.000 1
MHD	2.3 ± 1.2	15.6 ± 1.09	11.8 ± 1.8	0.9 ± 0.2	13.1 ± 1.2	9.1 ± 1.6
p-value			0.000 1			0.000 1
ABVD	1.5 ± 1.7	15.4 ± 1.1	10.7 ± 1.6	1.2 ± 1.4	12.6 ± 1.8	10.6 ± 1.73
p-value			0.000 1			0.000 1

hemisphere, along with their factor interactions. Age at time of scan was incorporated as a covariate when available (i.e. for KKI and NDAR/Pitt data). A random intercept term was included for the individual subject, nested within sex.

The feature mean value was the dependent variable, and it was calculated from the corresponding CDF. Statistically significant effects of interest were, for the NDAR/IBIS dataset, diagnosis–lobe interaction for curvedness and sharpness. Post hoc testing of individual lobes suggested that curvedness differed between ASD and control in the right parietal and left occipital lobes, while sharpness differed in the right parietal and right occipital lobes. The KKI data showed a significant diagnosis–hemisphere interaction for Gaussian curvature, with differences in the left temporal and left parietal lobes, and diagnosis–lobe interaction for CWM mean curvature, with differences in the left temporal lobe. Moreover, all the four of the CWM metrics had significant dependence on the diagnosis–lobe interaction in the NDAR/Pitt dataset. Furthermore, significant post hoc findings were obtained in the right frontal and both of the of occipital lobes for mean gyral white matter depth, in the left temporal, both parietal, and right occipital lobes for CWM mean curvature, and in the left temporal lobe for sharpness, as illustrated in figure 15.4.

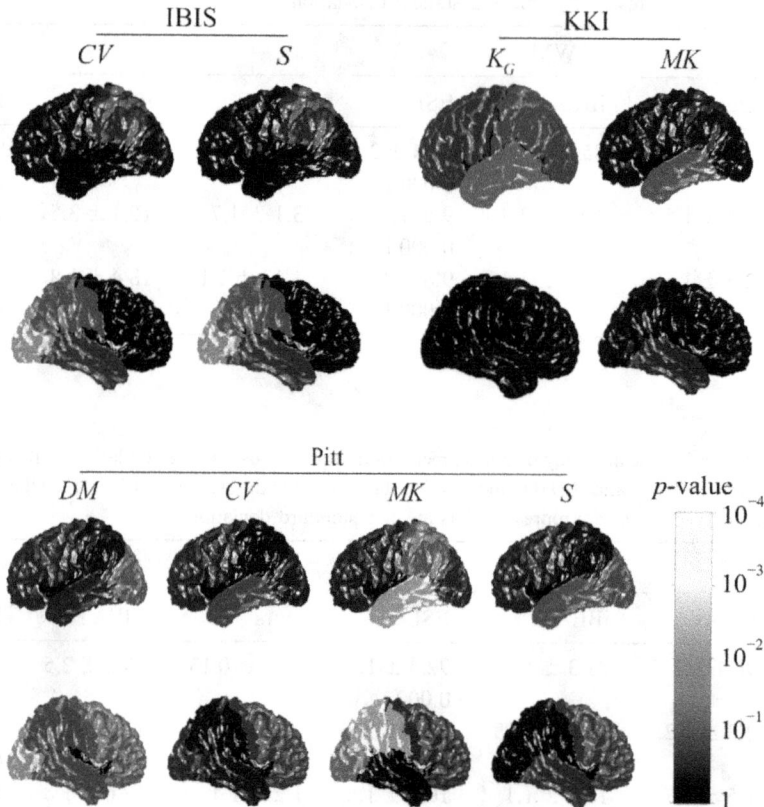

Figure 15.4. Individual lobes showing strong evidence against ASD and a control with the same mean value for each feature with a significant diagnosis–lobe or diagnosis–hemisphere interaction.

The presence of some features for each database with statistical significance was satisfactory enough to move towards the diagnosing stage. The deep network was first trained and tested only with the discriminatory features resulting from statistical analysis using the leave-one-subject-out technique. The classification accuracy was calculated for the three databases for each significant feature for each lobe in each hemisphere as illustrated in table 15.5, where there was a consistency between the high accuracies and the statistical analysis. The next step was to investigate fusing the discriminatory features from all eight lobes for each database for the purpose of achieving higher classification accuracy. The fusion of features from KKI, NDAR/IBIS, and NDAR/Pitt databases achieved classification accuracies of 81.25%, 90.47%, and 93.75%, respectively. Such results guided us to train and test the deep fusion network with all of the 64 extracted features. The integration of all features and testing with the leave-one-subject-out technique yielded the following accuracies: 92.8% for KKI, 85% for NDAR/IBIS, and 96.88% for NDAR/Pitt. While attempting to improve the accuracies, it was found that some features other than those listed in table 15.5 were close to being significant, which have eventually improved accuracies when integrated with significant features.

In conclusion, this chapter presented an approach for diagnosing autism at different life stages. It started with brain segmentation into Cx and CWM, followed by shape feature extraction from both. Local areas of the brain were analyzed to find any correlation with autism, using a deep fusion network. The work presented in this chapter could lead us to a better understanding of the autism spectrum by finding the local areas that correlate to the disease, as seen in figure 15.5. In the future, we will discuss integrating other modalities in the system.

This work could also be applied to various other applications in medical imaging, such as the kidney, the heart, the prostate, the lung, and the retina, as well as several non-medical applications [27–30].

One application is renal transplant functional assessment, in particular developing noninvasive CAD systems, utilizing different image modalities (e.g. ultrasound, computed tomography (CT), MRI, etc). Accurate assessment of renal transplant function is critically important for transplant survival. Although transplantation can improve a patient's wellbeing, there is a potential post-transplantation risk of kidney

Table 15.5. The accuracy of classification per lobe per hemisphere of statistically significant features with NCAE. 'I' indicates NDAR/IBIS, while 'P' stands for NDAR/Pitt, and 'K' for KKI.

Metric	Frontal L	Frontal R	Temporal L	Temporal R	Parietal L	Parietal R	Occipital L	Occipital R
C	NA	NA	NA	NA	NA	75% (I)	85% (I)	NA
DM	NA	92%(P)	NA	NA	NA	NA	91.5%(P)	94%(P)
MK	NA	NA	93%(P), 85.7%(K)	NA	93.5%(P)	92.5%(P)	NA	91%(P)
S	NA	NA	94%(P)	NA	NA	80%(I)	NA	85%(I)
K_G	NA	NA	88%(K)	NA	90%(K)	NA	NA	NA

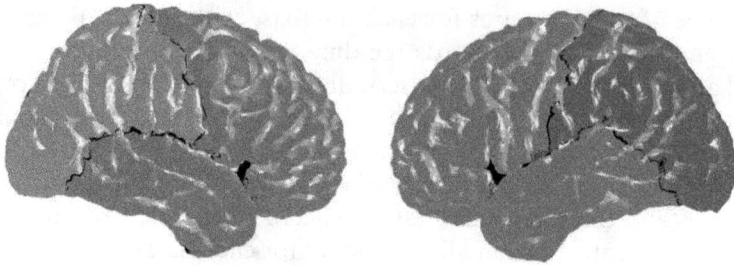

Figure 15.5. Cerebral lobes are color-coded according to their relative contribution to the classification of ASD versus control. Blue: highly informative; red: uninformative; orange: moderately informative.

dysfunction that, if not treated in a timely manner, can lead to the loss of the entire transplant, and even patient death. In particular, dynamic and diffusion MRI-based systems have been clinically used to assess transplanted kidneys with the advantage of providing information on each kidney separately. For more details about renal transplant functional assessment, see [10, 31–47, 47–57].

The heart is also an important application to this work. The clinical assessment of myocardial perfusion plays a major role in the diagnosis, management, and prognosis of ischemic heart disease patients. Thus, there have been ongoing efforts to develop automated systems for accurate analysis of myocardial perfusion using first-pass images [58–74].

Another application for this work could be the detection of retinal abnormalities. The majority of ophthalmologists depend on visual interpretation for the identification of diseases types. However, inaccurate diagnosis will affect the treatment procedure which may lead to fatal results. Hence, there is a crucial need for computer automated diagnosis systems that yield highly accurate results. Optical coherence tomography (OCT) has become a powerful modality for the noninvasive diagnosis of various retinal abnormalities such as glaucoma, diabetic macular edema, and macular degeneration. The problem with diabetic retinopathy (DR) is that the patient is not aware of the disease until the changes in the retina have progressed to a level that treatment tends to be less effective. Therefore, automated early detection could limit the severity of the disease and assist ophthalmologists in investigating and treating it more efficiently [75–77].

Abnormalities of the lung could also be another promising area of research and a related application to this work. Radiation-induced lung injury is the main side effect of radiation therapy for lung cancer patients. Although higher radiation doses increase the effectiveness of radiation therapy for tumor control, this can lead to lung injury as a greater quantity of normal lung tissue is included in the treated area. Almost 1/3 of patients who undergo radiation therapy develop lung injury following radiation treatment. The severity of radiation-induced lung injury ranges from ground-glass opacities and consolidation at the early phase to fibrosis and traction bronchiectasis in the late phase. Early detection of lung injury will thus help to improve management of the treatment [78–120].

This work can also be applied to other brain abnormalities, such as dyslexia in addition to autism. Dyslexia is one of the most complicated developmental brain disorders that affect children's learning abilities. Dyslexia leads to the failure to develop age-appropriate reading skills in spite of a normal intelligence level and adequate reading instruction. Neuropathological studies have revealed an abnormal anatomy of some structures, such as the corpus callosum in dyslexic brains. There has been a lot of work in the literature that aims at developing CAD systems for diagnosing such disorder, along with other brain disorders [121–142].

For the vascular system [143], this work could also be applied for the extraction of blood vessels, e.g. from phase contrast (PC) magnetic resonance angiography (MRA). Accurate cerebrovascular segmentation using noninvasive MRA is crucial for the early diagnosis and timely treatment of intracranial vascular diseases [125, 126, 144–149].

References

[1] NIMH 2015 *Autism Spectrum Disorder* (Bethesda, MD: Office of Science Policy)

[2] Di Martino A *et al* 2014 The autism brain imaging data exchange: towards a large-scale evaluation of the intrinsic brain architecture in autism *Mol. Psychiatry* **19** 659–67

[3] Hall D *et al* 2012 Sharing heterogeneous data: the National Database for Autism Research *Neuroinformatics* **10** 3319

[4] Schumann C M *et al* 2010 Longitudinal magnetic resonance imaging study of cortical development through early childhood in autism *J. Neurosci.* **30** 4419–27

[5] Hazlett H C *et al* 2012 Brain volume findings in 6-month-old infants at high familial risk for autism *Am. J. Psychiatry* **169** 601–8

[6] Hazlett H *et al* 2011 Early brain overgrowth in autism associated with an increase in cortical surface area before age 2 years *Arch. Gen. Psychiatry* **68** 467–76

[7] Hardan A *et al* 2009 A preliminary longitudinal magnetic resonance imaging study of brain volume and cortical thickness in autism *Biol. Psychiatry* **66** 320–6

[8] Wallace G L *et al* 2015 Longitudinal cortical development during adolescence and young adulthood in autism spectrum disorders: Increased cortical thinning but comparable surface area changes *J. Am. Acad. Child Adolesc. Psychiatry* **54** 464–9

[9] Ismail M, Barnes G, Nitzken M, Switala A, Shalaby A, Hosseini-Asl E, Casanova M, Keynton R, Khalil A and El-Baz A 2017 A new deep-learning approach for early detection of shape variations in autism using structural MRI *2017 IEEE Int. Conf. on Image Processing (ICIP) (Piscataway, NJ:,* (IEEE) *pp 1057–61*

[10] Khalifa F, Beache G M, El-Ghar M A, El-Diasty T, Gimel'farb G, Kong M and El-Baz A 2013 Dynamic contrast-enhanced MRI-based early detection of acute renal transplant rejection *IEEE Trans. Med. Imaging* **32** 1910–27

[11] Casanova M F *et al* 2006 Minicolumnar abnormalities in autism *Acta Neuropathol.* **112** 287–303

[12] Viola P A and Wells W M III 1997 Alignment by maximization of mutual information *Int. J. Comput. Vis.* **24** 137–54

[13] El-Baz A, Elnakib A, Khalifa F, El-Ghar M A, McClure P, Soliman A and Gimel'farb G 2012 Precise segmentation of 3-D magnetic resonance angiography *IEEE Trans. Biomed. Eng.* **59** 2019–29

[14] Alansary A, Soliman A, Khalifa F, Elnakib A, Mostapha M, Nitzken M, Casanova M and El-Baz A 2013 MAP-based framework for segmentation of MR brain images based on visual appearance and prior shape *MIDAS J.* **1** 1–13

[15] Besag J 1986 On the statistical analysis of dirty pictures *J. R. Stat. Soc. B* **48** 259–302

[16] Gerig G *et al* 2001 *Shape Analysis of Brain Ventricles using SPHARM MMBIA* (Piscataway, NJ: IEEE) 171–8

[17] Chung M K *et al* 2008 Tensor-based cortical surface morphometry via weighted spherical harmonic representation *IEEE Trans. Med. Imaging* **27** 1143–51

[18] The CGAL Project 2015, CGAL User and Reference Manual, 4.7 edn. CGAL Editorial Board, available at: http://doc.cgal.org/4.7/Manual/packages.html

[19] Adalsteinsson D *et al* 1995 A fast level set method for propagating interfaces *J. Comput. Phys.* **118** 269–77

[20] Pienaar R *et al* 2008 A methodology for analyzing curvature in the developing brain from preterm to adult *IJIST* **18** 42–68

[21] Bengio Y, Lamblin P, Popovici D and Larochelle H 2007 Greedy layer-wise training of deep networks *Adv. Neural Inf. Process. Syst.* **19** 153–60

[22] Anonymous.

[23] Conturo T E, Williams D L, Smith C D, Gultepe E, Akbudak E and Minshew N J 2008 Neuronal fiber pathway abnormalities in autism: an initial MRI diffusion tensor tracking study of hippocampo-fusiform and amygdalo-fusiform pathways *J. Int. Neuropsychol. Soc.* **14** 933–46

[24] Dice L R 1945 Measures of the amount of ecologic association between species *Ecology* **26** 297–302

[25] Yakang D *et al* 2013 iBEAT: a toolbox for infant brain magnetic resonance image processing *Neuroinformatics* **11** 211–25

[26] Jenkinson M *et al* 2012 FSL *Neuroimage* **62** 782–90

[27] Mahmoud A H 2014 Utilizing radiation for smart robotic applications using visible, thermal, and polarization images *PhD Dissertation* (University of Louisville KY)

[28] Mahmoud A, El-Barkouky A, Graham J and Farag A 2014 Pedestrian detection using mixed partial derivative based histogram of oriented gradients *2014 IEEE Int. Conf. on Image Processing (ICIP)* (Piscataway, NJ: IEEE) pp 2334–7

[29] El-Barkouky A, Mahmoud A, Graham J and Farag A 2013 An interactive educational drawing system using a humanoid robot and light polarization *2013 IEEE Int. Conf. on Image Processing* (Piscataway, NJ: IEEE) pp 3407–11

[30] Mahmoud A H, El-Melegy M T and Farag A A 2012 Direct method for shape recovery from polarization and shading *2012 19th IEEE Int. Conf. on Image Processing* (Piscataway, NJ: IEEE) pp 1769–72

[31] Ali A M, Farag A A and El-Baz A 2007 Graph cuts framework for kidney segmentation with prior shape constraints *Proc. of Int. Conf. on Medical Image Computing and Computer-Assisted Intervention, (MICCAI'07) (Brisbane, Australia, 29 October–2 November 2007)* **vol 1** pp 384–92

[32] Chowdhury A S, Roy R, Bose S, Elnakib F K A and El-Baz A 2012 Non-rigid biomedical image registration using graph cuts with a novel data term *Proc. of IEEE Int. Symp. on Biomedical Imaging: From Nano to Macro, (ISBI'12) (Barcelona, Spain, 2–5 May 2012)* pp 446–9

[33] El-Baz A, Farag A A, Yuksel S E, El-Ghar M E, Eldiasty T A and Ghoneim M A 2007 Application of deformable models for the detection of acute renal rejection *Deformable Models* (New York: Springer) pp 293–333

[34] El-Baz A, Farag A, Fahmi R, Yuksel S, El-Ghar M A and Eldiasty T 2006 Image analysis of renal DCE MRI for the detection of acute renal rejection *Proc. of IAPR Int. Conf. on Pattern Recognition (ICPR'06) (Hong Kong, 20–24 August 2006)* pp 822–5

[35] El-Baz A, Farag A, Fahmi R, Yuksel S, Miller W, El-Ghar M A, El-Diasty T and Ghoneim M 2006 A new CAD system for the evaluation of kidney diseases using DCE-MRI *Proc. of Int. Conf. on Medical Image Computing and Computer-Assisted Intervention, (MICCAI'08) (Copenhagen, Denmark, 1–6 October 2006)* pp 446–53

[36] El-Baz A, Gimel'farb G and El-Ghar M A 2008 A novel image analysis approach for accurate identification of acute renal rejection *Proc. of IEEE Int. Conf. on Image Processing, (ICIP'08) (San Diego, California, USA, 12–15 October 2008)* pp 1812–5

[37] El-Baz A, Gimel'farb G and El-Ghar M A 2008 Image analysis approach for identification of renal transplant rejection *Proc. of IAPR Int. Conf. on Pattern Recognition, (ICPR'08) (Tampa, Florida, USA, 8–11 December 2008)* pp 1–4

[38] El-Baz A, Gimel'farb G and El-Ghar M A 2007 New motion correction models for automatic identification of renal transplant rejection *Proc. of Int. Conf. on Medical Image Computing and Computer-Assisted Intervention, (MICCAI'07) (Brisbane, Australia, 29 October–2 November 2007)* pp 235–43

[39] Farag A, El-Baz A, Yuksel S, El-Ghar M A and Eldiasty T 2006 A framework for the detection of acute rejection with Dynamic Contrast Enhanced Magnetic Resonance Imaging *Proc. of IEEE Int. Symp. on Biomedical Imaging: From Nano to Macro, (ISBI'06) (Arlington, Virginia, USA, 6–9 April 2006)* pp 418–21

[40] Khalifa F, El-Baz A, Gimel'farb G and El-Ghar M A 2010 Non-invasive image-based approach for early detection of acute renal rejection *Proc. of Int. Conf. Medical Image Computing and Computer-Assisted Intervention, (MICCAI'10) (Beijing, China, 20–24 September 2010)* pp 10–8

[41] Khalifa F, El-Baz A, Gimel'farb G, Ouseph R and El-Ghar M A 2010 Shape-appearance guided level-set deformable model for image segmentation *Proc. of IAPR Int. Conf. on Pattern Recognition, (ICPR'10) (Istanbul, Turkey, 23–26 August 2010)* pp 4581–4

[42] Khalifa F, El-Ghar M A, Abdollahi B, Frieboes H, El-Diasty T and El-Baz A 2013 A comprehensive non-invasive framework for automated evaluation of acute renal transplant rejection using DCE-MRI *NMR Biomed.* **26** 1460–70

[43] Khalifa F, El-Ghar M A, Abdollahi B, Frieboes H B, El-Diasty T and El-Baz A 2014 Dynamic contrast-enhanced MRI-based early detection of acute renal transplant rejection *2014 Annu. Scientific Meeting and Educational Course Brochure of the Society of Abdominal Radiology, (SAR'14) (Boca Raton, FL, 23–28 March 2014)* CID: 1855912

[44] Khalifa F, Elnakib A, Beache G M, Gimel'farb G, El-Ghar M A, Sokhadze G, Manning S, McClure P and El-Baz A 2011 3D kidney segmentation from CT images using a level set approach guided by a novel stochastic speed function *Proc. of Int. Conf. Medical Image Computing and Computer-Assisted Intervention, (MICCAI'11) (Toronto, Canada, 18–22 September 2011)* pp 587–94

[45] Khalifa F, Gimel'farb G, El-Ghar M A, Sokhadze G, Manning S, McClure P, Ouseph R and El-Baz A 2011 A new deformable model-based segmentation approach for accurate extraction of the kidney from abdominal CT images *Proc. of IEEE Int. Conf. on Image Processing, (ICIP'11) (Brussels, Belgium, 11–14 September 2011)* pp 3393–6

[46] Mostapha M, Khalifa F, Alansary A, Soliman A, Suri J and El-Baz A 2014 Computer-aided diagnosis systems for acute renal transplant rejection: challenges and methodologies *Abdomen and Thoracic Imaging* ed A El-Baz and L S J Suri (Berlin: Springer) pp 1–35

[47] Shehata M, Khalifa F, Hollis E, Soliman A, Hosseini-Asl E, El-Ghar M A, El-Baz M, Dwyer A C, El-Baz A and Keynton R 2016 A new non-invasive approach for early classification of renal rejection types using diffusion-weighted MRI *IEEE Int. Conf. on Image Processing (ICIP), 2016* (Piscataway, NJ: IEEE) pp 136–40

[48] Khalifa F, Soliman A, Takieldeen A, Shehata M, Mostapha M, Shaffie A, Ouseph R, Elmaghraby A and El-Baz A 2016 Kidney segmentation from CT images using a 3D NMF-guided active contour model *IEEE 13th Int. Symp. on Biomedical Imaging (ISBI), 2016* (Piscataway, NJ: IEEE) 432–5

[49] Shehata M, Khalifa F, Soliman A, Takieldeen A, El-Ghar M A, Shaffie A, Dwyer A C, Ouseph R, El-Baz A and Keynton R 2016 3D diffusion MRI-based CAD system for early diagnosis of acute renal rejection *2016 IEEE 13th Int. Symp. on Biomedical Imaging (ISBI)* (Piscataway, NJ: IEEE) pp 1177–80

[50] Shehata M, Khalifa F, Soliman A, Alrefai R, El-Ghar M A, Dwyer A C, Ouseph R and El-Baz A 2015 A level set-based framework for 3D kidney segmentation from diffusion MR images *IEEE Int. Conf. on Image Processing (ICIP), 2015* (Piscataway, NJ: IEEE) pp 4441–5

[51] Shehata M, Khalifa F, Soliman A, El-Ghar M A, Dwyer A C, Gimel'farb G, Keynton R and El-Baz A 2016 A promising non-invasive CAD system for kidney function assessment *Int. Conf. on Medical Image Computing and Computer-Assisted Intervention* (Berlin: Springer) pp 613–21

[52] Khalifa F, Soliman A, Elmaghraby A, Gimel'farb G and El-Baz A 2017 3D kidney segmentation from abdominal images using spatial-appearance models *Comput. Math. Methods Med.* **2017** 1–10

[53] Hollis E, Shehata M, Khalifa F, El-Ghar M A, El-Diasty T and El-Baz A 2016 Towards non-invasive diagnostic techniques for early detection of acute renal transplant rejection: a review *Egypt. J. Radiol. Nucl. Med.* **48** 257–69

[54] Shehata M, Khalifa F, Soliman A, El-Ghar M A, Dwyer A C and El-Baz A 2017 Assessment of renal transplant using image and clinical-based biomarkers *Proc. of 13th Annu. Scientific Meeting of American Society for Diagnostics and Interventional Nephrology (ASDIN'17) (New Orleans, LA, USA, 10–12 February 2017)*

[55] Shehata M, Khalifa F, Soliman A, El-Ghar M A, Dwyer A C and El-Baz A 2017 Early assessment of acute renal rejection *Proc. of 12th Annu. Scientific Meeting of American Society for Diagnostics and Interventional Nephrology (ASDIN'16) (Phoenix, AZ, USA, 19–21 February 2016)*

[56] Eltanboly A, Ghazal M, Hajjdiab H, Shalaby A, Switala A, Mahmoud A, Sahoo P, El-Azab M and El-Baz A 2019 Level sets-based image segmentation approach using statistical shape priors *Appl. Math. Comput.* **340** 164–79

[57] Shehata M, Mahmoud A, Soliman A, Khalifa F, Ghazal M, El-Ghar M A, El-Melegy M and El-Baz A 2018 3D kidney segmentation from abdominal diffusion MRI using an appearance-guided deformable boundary *PLoS One* **13** p e0200082

[58] Khalifa F, Beache G, El-Baz A and Gimel'farb G 2010 Deformable model guided by stochastic speed with application in cine images segmentation *Proc. of IEEE Int. Conf. on Image Processing, (ICIP'10) (Hong Kong, 26–29 September 2010)* pp 1725–8

[59] Khalifa F, Beache G M, Elnakib A, Sliman H, Gimel'farb G, Welch K C and El-Baz A 2013 A new shape-based framework for the left ventricle wall segmentation from cardiac first-pass perfusion MRI *Proc. of IEEE Int. Symp. on Biomedical Imaging: From Nano to Macro, (ISBI'13) (San Francisco, CA, 7–11 April 2013)* pp 41–4

[60] Khalifa F, Beache G M, Elnakib A, Sliman H, Gimel'farb G, Welch K C and El-Baz A 2012 A new nonrigid registration framework for improved visualization of transmural perfusion gradients on cardiac first-pass perfusion MRI *Proc. of IEEE Int. Symp. on Biomedical Imaging: From Nano to Macro, (ISBI'12) (Barcelona, Spain, 2–5 May 2012)* pp 828–31

[61] Khalifa F, Beache G M, Firjani A, Welch K C, Gimel'farb G and El-Baz A 2012 A new nonrigid registration approach for motion correction of cardiac first-pass perfusion MRI *Proc. of IEEE Int. Conf. on Image Processing, (ICIP'12) (Lake Buena Vista, FL, 30 September–3 October 2012)* pp 1665–8

[62] Khalifa F, Beache G M, Gimel'farb G and El-Baz A 2012 A novel CAD system for analyzing cardiac first-pass MR images *Proc. of IAPR Int. Conf. on Pattern Recognition (ICPR'12) (Tsukuba Science City, Japan, 11–15 November 2012)* pp 77–80

[63] Khalifa F, Beache G M, Gimel'farb G and El-Baz A 2011 A novel approach for accurate estimation of left ventricle global indexes from short-axis cine MRI *Proc. of IEEE Int. Conf. on Image Processing, (ICIP'11) (Brussels, Belgium, 11–14 September 2011)* pp 2645–9

[64] Khalifa F, Beache G M, Gimel'farb G, Giridharan G A and El-Baz A 2011 A new image-based framework for analyzing cine images *Handbook of Multi Modality State-of-the-Art Medical Image Segmentation and Registration Methodologies* vol 2 ed A El-Baz, U R Acharya, M Mirmedhdi and J S Suri (New York: Springer) ch 3 pp 69–98

[65] Khalifa F, Beache G M, Gimel'farb G, Giridharan G A and El-Baz A 2012 Accurate automatic analysis of cardiac cine images *IEEE Trans. Biomed. Eng.* **59** 445–55

[66] Khalifa F, Beache G M, Nitzken M, Gimel'farb G, Giridharan G A and El-Baz A 2011 Automatic analysis of left ventricle wall thickness using short-axis cine CMR images *Proc. of IEEE Int. Symp. on Biomedical Imaging: From Nano to Macro, (ISBI'11) (Chicago, IL, 30 March–2 April 2011)* pp 1306–9

[67] Nitzken M, Beache G, Elnakib A, Khalifa F, Gimel'farb G and El-Baz A 2012 Accurate modeling of tagged CMR 3D image appearance characteristics to improve cardiac cycle strain estimation *2012 19th IEEE Int. Conf. on Image Processing (ICIP) (Orlando, FL, USA,* (Piscataway, NJ: IEEE) 521–4

[68] Nitzken M, Beache G, Elnakib A, Khalifa F, Gimel'farb G and El-Baz A 2012 Improving full-cardiac cycle strain estimation from tagged CMR by accurate modeling of 3D image appearance characteristics *2012 9th IEEE Int. Symp. on Biomedical Imaging (ISBI) (Barcelona, Spain,* (Piscataway, NJ: IEEE) 462–5

[69] Nitzken M J, El-Baz A S and Beache G M 2012 Markov–Gibbs random field model for improved full-cardiac cycle strain estimation from tagged CMR *J. Cardiovasc. Magn. Reson.* **14** 1–2

[70] Sliman H, Elnakib A, Beache G, Elmaghraby A and El-Baz A 2014 Assessment of myocardial function from cine cardiac MRI using a novel 4D tracking approach *J. Comput. Sci. Syst. Biol.* **7** 169–73

[71] Sliman H, Elnakib A, Beache G M, Soliman A, Khalifa F, Gimel'farb G, Elmaghraby A and El-Baz A 2014 A novel 4D PDE-based approach for accurate assessment of

myocardium function using cine cardiac magnetic resonance images *Proc. of IEEE Int. Conf. on Image Processing (ICIP'14) (Paris, France, 27–30 October 2014)* pp 3537–41

[72] Sliman H, Khalifa F, Elnakib A, Beache G M, Elmaghraby A and El-Baz A 2013 A new segmentation-based tracking framework for extracting the left ventricle cavity from cine cardiac MRI *Proc. of IEEE Int. Conf. on Image Processing, (ICIP'13) (Melbourne, Australia, 15–18 September 2013)* pp 685–9

[73] Sliman H, Khalifa F, Elnakib A, Soliman A, Beache G M, Elmaghraby A, Gimel'farb G and El-Baz A 2013 Myocardial borders segmentation from cine MR images using bi-directional coupled parametric deformable models *Med. Phys.* **40** 1–13

[74] Sliman H, Khalifa F, Elnakib A, Soliman A, Beache G M, Gimel'farb G, Emam A, Elmaghraby A and El-Baz A 2013 Accurate segmentation framework for the left ventricle wall from cardiac cine MRI *Proc. of Int. Symp. on Computational Models for Life Science, (CMLS'13) (Sydney, Australia, 27–29 November 2013)* **vol 1559** pp 287–96

[75] Eladawi N, Elmogy M, Ghazal M, Helmy O, Aboelfetouh A, Riad A, Schaal S and El-Baz A 2018 Classification of retinal diseases based on OCT images *Front. Biosci.* **23** 247–64

[76] ElTanboly A, Ismail M, Shalaby A, Switala A, El-Baz A, Schaal S, Gimelfarb G and El-Azab M 2017 A computer-aided diagnostic system for detecting diabetic retinopathy in optical coherence tomography images *Med. Phys.* **44** 914–23

[77] Sandhu H S, El-Baz A and Seddon J M 2018 Progress in automated deep learning for macular degeneration *JAMA Ophthalmol.* **36** 1366–7

[78] Abdollahi B, Civelek A C, Li X-F, Suri J and El-Baz A 2014 PET/CT nodule segmentation and diagnosis: a survey *Multi Detector CT Imaging* ed L Saba and J S Suri (London: Taylor and Francis) ch 30 pp 639–51

[79] Abdollahi B, El-Baz A and Amini A A 2011 A multi-scale non-linear vessel enhancement technique *2011 Annu. Int. Conf. of the IEEE Engineering in Medicine and Biology Society, EMBC* (Piscataway, NJ: IEEE) pp 3925–9

[80] Abdollahi B, Soliman A, Civelek A, Li X-F, Gimel'farb G and El-Baz A 2012 A novel Gaussian scale space-based joint MGRF framework for precise lung segmentation *Proc. of IEEE Int. Conf. on Image Processing, (ICIP'12)* (Piscataway, NJ: IEEE) pp 2029–32

[81] Abdollahi B, Soliman A, Civelek A, Li X-F, Gimel'farb G and El-Baz A 2012 A novel 3D joint MGRF framework for precise lung segmentation *Machine Learning in Medical Imaging* (Berlin: Springer) pp 86–93

[82] Ali A M, El-Baz A S and Farag A A 2007 A novel framework for accurate lung segmentation using graph cuts *Proc. of IEEE Int. Symp. on Biomedical Imaging: From Nano to Macro, (ISBI'07)* (Piscataway, NJ: IEEE) pp 908–11

[83] El-Baz A, Beache G M, Gimel'farb G, Suzuki K and Okada K 2013 Lung imaging data analysis *Int. J. Biomed. Imaging* **2013** 1–2

[84] El-Baz A, Beache G M, Gimel'farb G, Suzuki K, Okada K, Elnakib A, Soliman A and Abdollahi B 2013 Computer-aided diagnosis systems for lung cancer: challenges and methodologies *Int. J. Biomed. Imaging* **2013** 1–46

[85] El-Baz A, Elnakib A, Abou El-Ghar M, Gimel'farb G, Falk R and Farag A 2013 Automatic detection of 2D and 3D lung nodules in chest spiral CT scans *Int. J. Biomed. Imaging* **2013** 1–11

[86] El-Baz A, Farag A A, Falk R and La Rocca R 2003 A unified approach for detection, visualization, and identification of lung abnormalities in chest spiral CT scans *International Congress Series* vol 1256 (Amsterdam: Elsevier) pp 998–1004

[87] El-Baz A, Farag A A, Falk R and La Rocca R 2002 Detection, visualization and identification of lung abnormalities in chest spiral CT scan: Phase-I *Proc. of Int. Conf. on Biomedical Engineering (Cairo, Egypt)* **vol 12**

[88] El-Baz A, Farag A, Gimel'farb G, Falk R, El-Ghar M A and Eldiasty T 2006 A framework for automatic segmentation of lung nodules from low dose chest CT scans *Proc. of Int. Conf. on Pattern Recognition, (ICPR'06)* **vol 3** (Piscataway, NJ: IEEE) 611–4

[89] El-Baz A, Farag A, Gimel'farb G, Falk R and El-Ghar M A 2011 A novel level set-based computer-aided detection system for automatic detection of lung nodules in low dose chest computed tomography scans *Lung Imaging and Computer Aided Diagnosis* (London: Taylor and Francis) ch 10, pp 221–38

[90] El-Baz A, Gimel'farb G, Abou El-Ghar M and Falk R 2012 Appearance-based diagnostic system for early assessment of malignant lung nodules *Proc. of IEEE Int. Conf on Image Processing, (ICIP'12)* (Piscataway, NJ: IEEE) pp 533–6

[91] El-Baz A, Gimel'farb G and Falk R 2011 A novel 3D framework for automatic lung segmentation from low dose CT images *Lung Imaging and Computer Aided Diagnosis* ed A El-Baz and J S Suri (London: Taylor and Francis) ch 1 pp 1–16

[92] El-Baz A, Gimel'farb G, Falk R and El-Ghar M 2010 Appearance analysis for diagnosing malignant lung nodules *Proc. of IEEE Int. Symp. on Biomedical Imaging: From Nano to Macro (ISBI'10)* (Piscataway, NJ: IEEE) pp 193–6

[93] El-Baz A, Gimel'farb G, Falk R and El-Ghar M A 2011 A novel level set-based CAD system for automatic detection of lung nodules in low dose chest CT scans *Lung Imaging and Computer Aided Diagnosis* vol 1 ed A El-Baz and J S Suri (London: Taylor and Francis) ch 10, pp 221–38

[94] El-Baz A, Gimel'farb G, Falk R and El-Ghar M A 2008 A new approach for automatic analysis of 3D low dose CT images for accurate monitoring the detected lung nodules *Proc. of Int. Conf. on Pattern Recognition, (ICPR'08)* (Piscataway, NJ: IEEE) pp 1–4

[95] El-Baz A, Gimel'farb G, Falk R and El-Ghar M A 2007 A novel approach for automatic follow-up of detected lung nodules *Proc. of IEEE Int. Conf. on Image Processing, (ICIP'07)* **vol 5** (Piscataway, NJ: IEEE) pp V–501

[96] El-Baz A, Gimel'farb G, Falk R and El-Ghar M A 2007 A new CAD system for early diagnosis of detected lung nodules *ICIP 2007. IEEE Int. Conf. on Image Processing, 2007* **vol 2** (Piscataway, NJ: IEEE) pp II–461

[97] El-Baz A, Gimel'farb G, Falk R, El-Ghar M A and Refaie H 2008 Promising results for early diagnosis of lung cancer *Proc. of IEEE Int. Symp. Biomedical Imaging: From Nano to Macro, (ISBI'08)* (Piscataway, NJ: IEEE) pp 1151–4

[98] El-Baz A, Gimel'farb G L, Falk R, Abou El-Ghar M, Holland T and Shaffer T 2008 A new stochastic framework for accurate lung segmentation *Proc. of Medical Image Computing and Computer-Assisted Intervention, (MICCAI'08)* pp 322–30

[99] El-Baz A, Gimel'farb G L, Falk R, Heredis D and Abou El-Ghar M 2008 A novel approach for accurate estimation of the growth rate of the detected lung nodules *Proc. of Int. Workshop on Pulmonary Image Analysis* pp 33–42

[100] El-Baz A, Gimel'farb G L, Falk R, Holland T and Shaffer T 2008 A framework for unsupervised segmentation of lung tissues from low dose computed tomography images *Proc. of British Machine Vision, (BMVC'08)* pp 1–10

[101] El-Baz A, Gimel'farb G, Falk R and El-Ghar M A 2011 3D MGRF-based appearance modeling for robust segmentation of pulmonary nodules in 3D LDCT chest images *Lung Imaging and Computer Aided Diagnosis* (London: Taylor and Francis) ch 3, pp 51–63

[102] El-Baz A, Gimel'farb G, Falk R and El-Ghar M A 2009 Automatic analysis of 3D low dose CT images for early diagnosis of lung cancer *Pattern Recogn.* **42** 1041–51

[103] El-Baz A, Gimel'farb G, Falk R, El-Ghar M A, Rainey S, Heredia D and Shaffer T 2009 Toward early diagnosis of lung cancer *Proc. of Medical Image Computing and Computer-Assisted Intervention, (MICCAI'09)* (Berlin: Springer) pp 682–9

[104] El-Baz A, Gimel'farb G, Falk R, El-Ghar M A and Suri J 2011 Appearance analysis for the early assessment of detected lung nodules *Lung Imaging and Computer Aided Diagnosis* (London: Taylor and Francis) ch 17, pp 395–404

[105] El-Baz A, Khalifa F, Elnakib A, Nitkzen M, Soliman A, McClure P, Gimel'farb G and El-Ghar M A 2012 A novel approach for global lung registration using 3D Markov Gibbs appearance model *Proc. of Int. Conf. Medical Image Computing and Computer-Assisted Intervention, (MICCAI'12) (Nice, France, 1–5 October 2012)* pp 114–21

[106] El-Baz A, Nitzken M, Elnakib A, Khalifa F, Gimel'farb G, Falk R and El-Ghar M A 2011 3D shape analysis for early diagnosis of malignant lung nodules *Proceedings of International Conference Medical Image Computing and Computer-Assisted Intervention, (MICCAI'11) (Toronto, Canada, 18–22 September 2011)* pp 175–82

[107] El-Baz A, Nitzken M, Gimel'farb G, Van Bogaert E, Falk R, El-Ghar M A and Suri J 2011 Three-dimensional shape analysis using spherical harmonics for early assessment of detected lung nodules *Lung Imaging and Computer Aided Diagnosis* (London: Taylor and Francis) ch 19, pp 421–38

[108] El-Baz A, Nitzken M, Khalifa F, Elnakib A, Gimel'farb G, Falk R and El-Ghar M A 2011 3D shape analysis for early diagnosis of malignant lung nodules *Proc. of Int. Conf. on Information Processing in Medical Imaging, (IPMI'11) (Monastery Irsee, Germany (Bavaria), 3–8 July 2011)* pp 772–83

[109] El-Baz A, Nitzken M, Vanbogaert E, Gimel'Farb G, Falk R and Abo El-Ghar M 2011 A novel shape-based diagnostic approach for early diagnosis of lung nodules *2011 IEEE Int. Symp. on Biomedical Imaging: From Nano to Macro* (Piscataway, NJ: IEEE) pp 137–40

[110] El-Baz A, Sethu P, Gimel'farb G, Khalifa F, Elnakib A, Falk R and El-Ghar M A 2011 Elastic phantoms generated by microfluidics technology: validation of an imaged-based approach for accurate measurement of the growth rate of lung nodules *Biotechnol. J.* **6** 195–203

[111] El-Baz A, Sethu P, Gimel'farb G, Khalifa F, Elnakib A, Falk R and El-Ghar M A 2010 A new validation approach for the growth rate measurement using elastic phantoms generated by state-of-the-art microfluidics technology *Proc. of IEEE Int. Conf. on Image Processing, (ICIP'10) (Hong Kong, 26–29 September 2010)* pp 4381–3

[112] El-Baz A, Sethu P, Gimel'farb G, Khalifa F, Elnakib A, Falk R and Suri M A E -G J 2011 Validation of a new imaged-based approach for the accurate estimating of the growth rate of detected lung nodules using real CT images and elastic phantoms generated by state-of-the-art microfluidics technology *Handbook of Lung Imaging and Computer Aided Diagnosis* vol 1 ed A El-Baz and J S Suri (New York: Taylor and Francis) ch 18 pp 405–20

[113] El-Baz A, Soliman A, McClure P, Gimel'farb G, El-Ghar M A and Falk R 2012 Early assessment of malignant lung nodules based on the spatial analysis of detected lung nodules *Proc. of IEEE Int. Symp. on Biomedical Imaging: From Nano to Macro, (ISBI'12)* (Piscataway, NJ: IEEE) pp 1463–6

[114] El-Baz A, Yuksel S E, Elshazly S and Farag A A 2005 Non-rigid registration techniques for automatic follow-up of lung nodules *Proc. of Computer Assisted Radiology and Surgery, (CARS'05)* **vol 1281** (Amsterdam: Elsevier) pp 1115–20

[115] El-Baz A S and Suri J S 2011 *Lung Imaging and Computer Aided Diagnosis* (Boca Raton, FL: CRC Press)

[116] Soliman A, Khalifa F, Dunlap N, Wang B, El-Ghar M and El-Baz A 2016 An iso-surfaces based local deformation handling framework of lung tissues *2016 IEEE 13th Int. Symp. on Biomedical Imaging (ISBI)* (Piscataway, NJ: IEEE) pp 1253–9

[117] Soliman A, Khalifa F, Shaffie A, Dunlap N, Wang B, Elmaghraby A and El-Baz A 2016 Detection of lung injury using 4D-CT chest images *2016 IEEE 13th Int. Symposium on Biomedical Imaging (ISBI)* (Piscataway, NJ: IEEE) pp 1274–7

[118] Soliman A, Khalifa F, Shaffie A, Dunlap N, Wang B, Elmaghraby A, Gimel'farb G, Ghazal M and El-Baz A 2017 A comprehensive framework for early assessment of lung injury *2017 IEEE Int. Conf. on Image Processing (ICIP)* (Piscataway, NJ: IEEE) pp 3275–9

[119] Shaffie A, Soliman A, Ghazal M, Taher F, Dunlap N, Wang B, Elmaghraby A, Gimel'farb G and El-Baz A 2017 A new framework for incorporating appearance and shape features of lung nodules for precise diagnosis of lung cancer *2017 IEEE Int. Conf. on Image Processing (ICIP)* (Piscataway, NJ: IEEE) pp 1372–6

[120] Soliman A, Khalifa F, Shaffie A, Liu N, Dunlap N, Wang B, Elmaghraby A, Gimel'farb G and El-Baz A 2016 Image-based CAD system for accurate identification of lung injury *2016 IEEE Int. Conf. on Image Processing (ICIP)* (Piscataway, NJ: IEEE) pp 121–5

[121] Dombroski B, Nitzken M, Elnakib A, Khalifa F, El-Baz A and Casanova M F 2014 Cortical surface complexity in a population-based normative sample *Transl. Neurosci.* **5** 17–24

[122] El-Baz A, Casanova M, Gimel'farb G, Mott M and Switala A 2008 An MRI-based diagnostic framework for early diagnosis of dyslexia *Int. J. Comput. Assist. Radiol. Surg.* **3** 181–9

[123] El-Baz A, Casanova M, Gimel'farb G, Mott M, Switala A, Vanbogaert E and McCracken R 2008 A new CAD system for early diagnosis of dyslexic brains *Proc. Int. Conf. on Image Processing (ICIP'2008)* (Piscataway, NJ: IEEE) pp 1820–3

[124] El-Baz A, Casanova M F, Gimel'farb G, Mott M and Switwala A E 2007 A new image analysis approach for automatic classification of autistic brains *Proc. IEEE Int. Symp. on Biomedical Imaging: From Nano to Macro (ISBI'2007)* (Piscataway, NJ: IEEE) pp 352–5

[125] El-Baz A, Farag A, Gimel'farb G, El-Ghar M A and Eldiasty T 2006 Probabilistic modeling of blood vessels for segmenting MRA images *18th Int. Conf. on Pattern Recognition (ICPR'06)* vol 3 (Piscataway, NJ: IEEE) pp 917–20

[126] El-Baz A, Farag A A, Gimel'farb G, El-Ghar M A and Eldiasty T 2006 A new adaptive probabilistic model of blood vessels for segmenting MRA images *Medical Image Computing and Computer-Assisted Intervention-MICCAI 2006* vol 4191 (Berlin: Springer) pp 799–806

[127] El-Baz A, Farag A A, Gimel'farb G and Hushek S G 2005 Automatic cerebrovascular segmentation by accurate probabilistic modeling of TOF-MRA images *Medical Image Computing and Computer-Assisted Intervention—MICCAI 2005* (Berlin: Springer) pp 34–42

[128] El-Baz A, Farag A, Elnakib A, Casanova M F, Gimel'farb G, Switala A E, Jordan D and Rainey S 2011 Accurate automated detection of autism related corpus callosum abnormalities *J. Med. Syst.* **35** 929–39

[129] El-Baz A, Farag A and Gimelfarb G 2005 Cerebrovascular segmentation by accurate probabilistic modeling of TOF-MRA images *Image Analysis* vol 3540 (Berlin: Springer) pp 1128–37

[130] El-Baz A, Gimel'farb G, Falk R, El-Ghar M A, Kumar V and Heredia D 2009 A novel 3D joint Markov–Gibbs model for extracting blood vessels from PC-MRA images *Medical Image Computing and Computer-Assisted Intervention—MICCAI 2009* (Berlin: Springer) pp 943–50

[131] Elnakib A, El-Baz A, Casanova M F, Gimel'farb G and Switala A E 2010 Image-based detection of corpus callosum variability for more accurate discrimination between dyslexic and normal brains *Proc. IEEE Int. Symp. on Biomedical Imaging: From Nano to Macro (ISBI'2010)* (Piscataway, NJ: IEEE) pp 109–12

[132] Elnakib A, Casanova M F, Gimel'farb G, Switala A E and El-Baz A 2011 Autism diagnostics by centerline-based shape analysis of the corpus callosum *Proc. IEEE Int. Symp. on Biomedical Imaging: From Nano to Macro (ISBI'2011)* (Piscataway, NJ: IEEE) pp 1843–6

[133] Elnakib A, Nitzken M, Casanova M, Park H, Gimel'farb G and El-Baz A 2012 Quantification of age-related brain cortex change using 3D shape analysis *2012 21st Int. Conf. on Pattern Recognition (ICPR)* (Piscataway, NJ: IEEE) pp 41–4

[134] Mostapha M, Soliman A, Khalifa F, Elnakib A, Alansary A, Nitzken M, Casanova M F and El-Baz A 2014 A statistical framework for the classification of infant DT images *2014 IEEE Int. Conf. on Image Processing (ICIP)* (Piscataway, NJ: IEEE) pp 2222–6

[135] Nitzken M, Casanova M, Gimel'farb G, Elnakib A, Khalifa F, Switala A and El-Baz A 2011 3D shape analysis of the brain cortex with application to dyslexia *2011 18th IEEE Int. Conf. on Image Processing (ICIP)* (Brussels: IEEE) pp 2657–60

[136] El-Gamal F E-Z A, Elmogy M M, Ghazal M, Atwan A, Barnes G N, Casanova M F, Keynton R and El-Baz A S 2017 A novel CAD system for local and global early diagnosis of Alzheimer's disease based on PiB-PET scans *2017 IEEE Int. Conf. on Image Processing (ICIP)* (Piscataway, NJ: IEEE) pp 3270–4

[137] Ismail M, Soliman A, Ghazal M, Switala A E, Gimelfarb G, Barnes G N, Khalil A and El-Baz A 2017 A fast stochastic framework for automatic MR brain images segmentation *PloS one* **12** e0187391

[138] Ismail M M, Keynton R S, Mostapha M M, ElTanboly A H, Casanova M F, Gimel'farb G L and El-Baz A 2016 Studying autism spectrum disorder with structural and diffusion magnetic resonance imaging: a survey *Front. Hum. Neurosci.* **10** 211

[139] Alansary A *et al* 2016 Infant brain extraction in T1-weighted MR images using BET and refinement using LCDG and MGRF models *IEEE J. Biomed. Health Inform.* **20** pp 925–35

[140] Ismail M, Soliman A, ElTanboly A, Switala A, Mahmoud M, Khalifa F, Gimel'farb G, Casanova M F, Keynton R and El-Baz A 2016 Detection of white matter abnormalities in MR brain images for diagnosis of autism in children *2016 IEEE 13th International Symposium on Biomedical Imaging (ISBI)* pp 6–9

[141] Ismail M, Mostapha M, Soliman A, Nitzken M, Khalifa F, Elnakib A, Gimel'farb G, Casanova M and El-Baz A 2015 Segmentation of infant brain MR images based on adaptive shape prior and higher-order MGRF *2015 IEEE Int. Conf. on Image Processing (ICIP)* pp 4327–31

[142] Asl E H, Ghazal M, Mahmoud A, Aslantas A, Shalaby A, Casanova M, Barnes G, Gimelfarb G, Keynton R and El-Baz A 2018 Alzheimers disease diagnostics by a 3D deeply supervised adaptable convolutional network *Front. Biosci.* **23** 584–96

[143] Mahmoud A, El-Barkouky A, Farag H, Graham J and Farag A 2013 A non-invasive method for measuring blood flow rate in superficial veins from a single thermal image *Proc. of the IEEE Conf. on Computer Vision and Pattern Recognition Workshops* pp 354–9

[144] El-baz A, Shalaby A, Taher F, El-Baz M, Ghazal M, El-Ghar M A, Takieldeen A and Suri J 2017 Probabilistic modeling of blood vessels for segmenting magnetic resonance angiography images *Med. Res. Arch.* **5**

[145] Chowdhury A S, Rudra A K, Sen M, Elnakib A and El-Baz A 2010 Cerebral white matter segmentation from MRI using probabilistic graph cuts and geometric shape priors *ICIP* pp 3649–52

[146] Gebru Y, Giridharan G, Ghazal M, Mahmoud A, Shalaby A and El-Baz A 2018 Detection of cerebrovascular changes using magnetic resonance angiography *Cardiovascular Imaging and Image Analysis* (Boca Raton, FL: CRC Press) pp 1–22

[147] Mahmoud A, Shalaby A, Taher F, El-Baz M, Suri J S and El-Baz A 2018 Vascular tree segmentation from different image modalities *Cardiovascular Imaging and Image Analysis* (Boca Raton, FL: CRC Press) pp 43–70

[148] Taher F, Mahmoud A, Shalaby A and El-Baz A 2018 A review on the cerebrovascular segmentation methods *2018 IEEE Int. Symp. on Signal Processing and Information Technology (ISSPIT)* (Piscataway, NJ: IEEE) pp 359–64

[149] Kandil H, Soliman A, Fraiwan L, Shalaby A, Mahmoud A, ElTanboly A, Elmaghraby A, Giridharan G and El-Baz A 2018 A novel MRA framework based on integrated global and local analysis for accurate segmentation of the cerebral vascular system *2018 IEEE 15th Int. Symp. on Biomedical Imaging (ISBI 2018)* (Piscataway, NJ: IEEE) pp 1365–8

IOP Publishing

Neurological Disorders and Imaging Physics, Volume 3
Application to autism spectrum disorders and Alzheimer's
Ayman El-Baz and Jasjit S Suri

Chapter 16

Computational analysis techniques: a case study on fMRI for autism spectrum disorder

Omar Dekhil, Ali Mahmoud, Ahmed Shalaby, Ahmed Soliman, Fatma Taher, Hassan Hajjdiab, Ashraf Khalil, Mohammed Ghazal, Robert Keynton, Gregory Barnes and Ayman El-Baz

Functional MRI is one of the most promising techniques in neuro-imaging. It is mainly used to either record the response of a subject to a certain task (task based fMRI) or to assess the functional connectivity while at rest (resting state fMRI). In this survey, both fMRI techniques are explained and the most commonly used techniques in each of them are discussed and criticized. For each technique the hypothesis used and the mathematical background is discussed.

After discussing the analytical methods, a case study of both task based fMRI and resting state fMRI in autism spectrum disorder is discussed. In this case study, examples from previous studies are provided and summarized. For each study the population, experimental design, and major findings are summarized. Using this review could be a useful reference for future work about fMRI analysis techniques and its application to ASD.

16.1 Introduction

Functional MRI (fMRI) is a trending research and diagnosis technique [1–4]. In fMRI, the brain activity is measured in terms of blood oxygenated level dependent (BOLD) signal, which reflects the associated activity [5].

There are two major fMRI study types: (i) task based fMRI and (ii) resting state fMRI (RfMRI) [6]. In task based fMRI, the patient is asked to do a certain task and his/her response to this task is recorded; while in RfMRI, the patient is asked to stay in a nonmoving position, with their eyes open, and not to fall asleep during the scan.

In task based fMRI, too many experiments are used. Some examples of task based fMRI experiments and studies include: (i) motor tasks [7–9], where the subjects are asked to either move their hands or fingers for a certain amount of time

doi:10.1088/978-0-7503-1793-1ch16

(task condition) followed by a rest time; (ii) face tasks [10–12], in which the activation in response to both familiar and non-familiar faces to the subject is recorded; and (iii) reward tasks [13–15], in which the response of the subject to either monetary rewards or social rewards is measured. Usually in these studies, subjects are asked to complete a certain task, and accordingly they are awarded or denied either a social reward (a smiling face for example), or monetary reward (coins or gift cards for example). The mentioned studies are just few examples out of an abundance of experimental studies with different stimuli and different conditions.

In the next sections the computational methods for fMRI preprocessing and analysis in both task based fMRI and RfMRI are discussed.

16.2 Task based fMRI analysis

In task based fMRI it is required to check to what extent each voxel follows the set of conditions in experiment design, where these conditions are listed in the form a design matrix. Figure 16.1 shows an example of how voxel activity over time might be represented in terms of design matrix conditions. In order to model different conditions during task based fMRI analysis, a general linear model (GLM) is used, then to determine the most significant voxels a statistical analysis is applied to each voxel according to the desired conditions under test. The fMRI analysis is usually done on two levels, where the first level is concerned with mapping subject time series to the activation patterns, and the second level is concerned with group or session variations and it allows estimating group specific statistical parameters [16].

16.2.1 Building the design matrix

For task based fMRI, the design matrix is what expresses the experimental protocol, and it consists of a set of regressors corresponding to each experimental condition. In addition to the experimental conditions, some nuisance variable could be modeled and embedded in the design matrix, for example motion and motion drifts [4].

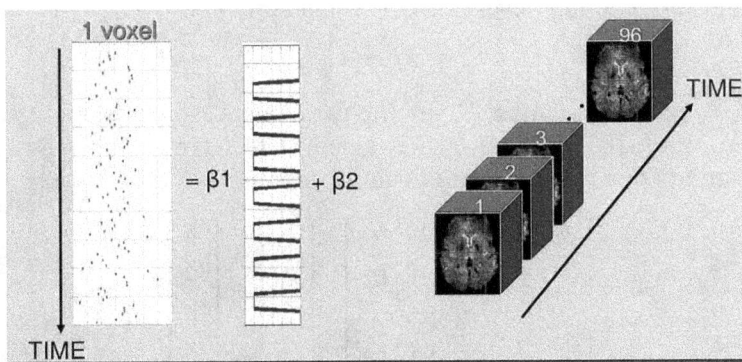

Figure 16.1. An illustrating example shows how the actual BOLD response in a certain voxel is modeled in terms of design matrix conditions, where β_1 and β_2 are the parameters to be estimated.

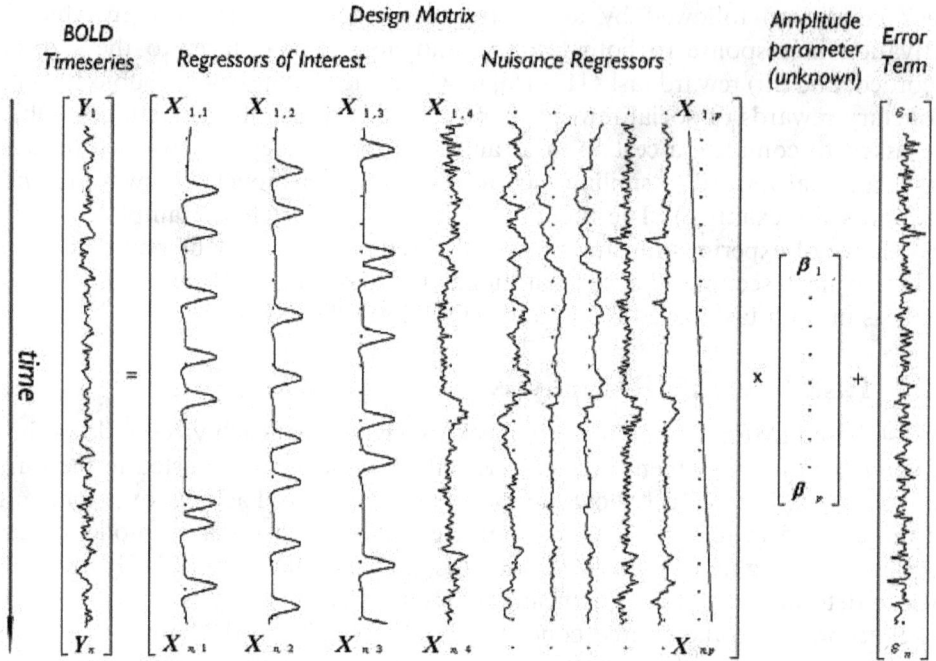

Figure 16.2. An illustrative example shows how to express the BOLD time series in terms of set regressors. The ten regressors here are an example of both regressors of interest for three different tasks and nuisance regressors (six for motion modeling and one for linear drift).

Figure 16.2 shows an illustrating example of how a design matrix is built with three task conditions (regressors of interest), six motion parameters, and one linear drift parameter (nuisance parameter).

16.2.2 Multi level general linear model (GLM) in fMRI

Consider an experiment with N subjects, and for each subject K there is a vector of T time points, Y_k. The first level GLM is defined as

$$Y = X\beta + \epsilon, \tag{16.1}$$

where X_k is the design matrix, β_k are the parameters estimates, and ϵ_k is the error term (subject residuals). The residual is assumed to have a zero mean $E(\epsilon_k) = 0$. Putting equation (16.1) in its matrix form it becomes

$$Y = \begin{bmatrix} X_1 & 0 & \dots & 0 & 0 \\ 0 & X_2 & & \dots & 0 \\ \dots & \dots & \dots & \dots & \dots \\ 0 & 0 & & \dots & x_N \end{bmatrix} * \begin{bmatrix} \beta_1 \\ \beta_2 \\ \vdots \\ \beta_N \end{bmatrix} + \begin{bmatrix} \epsilon_1 \\ \epsilon_2 \\ \vdots \\ \epsilon_N \end{bmatrix}. \tag{16.2}$$

By adding the second level of analysis on the groups level we have

$$\beta = X_G B_G + \eta, \tag{16.3}$$

where X_G is the group level design matrix, and it separates the two groups (controls and patients), β_G is the vector of the group level parameters, and η is the group level residuals. Also η is considered to have zero mean, $E(\eta) = 0$. Substituting with equation (16.3) in (16.1) we obtain

$$Y = XX_G\beta_G + \gamma, \tag{16.4}$$

where

$$\gamma = X\eta + \epsilon. \tag{16.5}$$

Also γ has zero mean, $E(\gamma) = 0$. The covariance of γ is $\text{cov}(\gamma) = w = XV_GX^T + V$, where V is the covariance of ϵ and V_G is the covariance of η.

16.2.3 GLM parameter estimates

Using the general least squares approach [17], the first level parameters could be estimated as

$$\hat{\beta} = (X^TV^{-1}X)X^TV^{-1}Y \tag{16.6}$$

and

$$\text{cov}(\hat{\beta}) = (X^TV^{-1}X)^{-1}. \tag{16.7}$$

Using the same approach but on the group level, the group parameters are given by

$$\hat{\beta}_G = \left(X_G^TV_G^{-1}X_G\right)^{-1}X^TV^{-1}Y \tag{16.8}$$

and

$$\text{cov}(\hat{\beta}_G) = \left(X_G^TV_G^{-1}X_G\right)^{-1}. \tag{16.9}$$

After estimating the parameters for each voxel, it is required to define some contrasts and check which voxels are significant with respect to these conditions. The most common techniques for testing the voxel significance are the paired t-test and paired f-test [16, 18]. The statistical analysis result could be expressed as a map of corrected P-values at each voxel. Figure 16.3 show an example of the most significant result obtained from a task based fMRI experiment, where the used test is a paired t-test and the threshold was P-value $= 0.05$.

16.2.4 Limitations of GLM for parameter estimation in fMRI analysis

Although it is a widely used approach and it has achieved very good results in many studies and applications, there are some underlying assumptions that should be satisfied for the GLM model to generate unbiased accurate results [4].

As stated above the method used to estimate the model parameters is the least squares methods. According to Gauss–Markov theorem [19], there are some

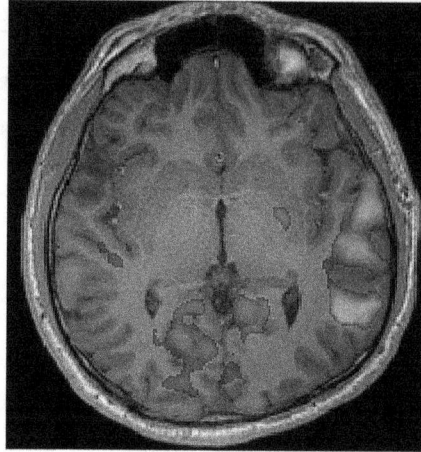

Figure 16.3. An illustrating example shows the most significant voxels for a task based fMRI using a paired *t*-test and a threshold at *P*-value = 0.05.

assumptions that should be verified prior to using the least squares method. The five main assumptions to estimate the best linear unbiased estimator are:

1. The residual errors have zero mean.
2. The residual errors are uncorrelated.
3. The residual errors have the equal variance.
4. The regressors in the design matrix X are fully deterministic and independent of the errors.
5. No regressor in the design matrix X is a linear transformation or combination of one or more regressors.

When trying to apply these assumptions, some potential violations might exist. For example, the second assumption might be violated due to some hardware related noise, such as low frequency drifts, and also due to some movement artifacts [3, 20]. To overcome these assumption violations, there are three suggested approaches in [4]:

1. *Precoloring the data*: In precoloring, a modification is suggested to equation (16.1), being:

$$Y = X\beta + \Sigma\epsilon, \tag{16.10}$$

where Σ is a latent process that describes the serial correlation and ϵ is the error term with desired distribution, $\epsilon \sim N(0, \sigma^2 I)$. By finding a linear transformation S and applying it to equation (16.10),

$$SY = SX\beta + S\Sigma\epsilon, \tag{16.11}$$

if the transformation S is robust enough such that $S\Sigma S^T \sim SS^T$, the colored noise will have the typically desired distribution $\sim N(0, \sigma^2 S^T)$ [21]. Although this solution does not cause the parameters to be biased, it decreases the power and efficiency of the model, and it might cause low frequency signals of interest to be attenuated [22].

2. *Prewhitening the data*: Another approach to have the noise following the Gauss–Markov theorem is noise prewhitening. In this approach, a GLM is first run with noise violating the first three assumptions and then the autocorrelation is modeled using an auto-regressive model. Once the autocorrelation is estimated, it is then removed and a second round of GLM is applied for model parameter estimation. In this approach, equation (16.1) is changed to

$$K^{-1}Y = K^{-1}X\beta + K^{-1}\Sigma\epsilon, \tag{16.12}$$

where Σ is the autocorrelation of the data, $k \approx \Sigma$. Accordingly, equations (16.6) and (16.7) become

$$\hat{\beta} = (X^T K^{-1} X) X^T k^{-1} Y \tag{16.13}$$

and

$$\mathrm{cov}(\hat{\beta}) = (X^T K^{-1} X)^{-1}. \tag{16.14}$$

This approach is more efficient than precoloring [23] for rapid task based designs where many of the signals of interest are concentrated in high temporal frequencies. The main challenge in this approach is to model autocorrelation correctly; non-optimal modeling may lead to a drop in the efficiency [24].

3. *Explicit noise modeling*: Another approach proposed in [25] suggests direct modeling of the noise correlation sources within the design matrix. Some of these nuisance sources are low frequency drifts, unhandled motion artifacts, respiration, and any other factors believed to cause noise correlation. The main challenge in this approach is the ability to determine the noise correlation sources accurately, and hence model them in an appropriate way in the design matrix.

16.3 RfMRI analysis

In RfMRI analysis, it is necessary to capture the spatio-temporal features and decompose the recorded signal into a set of activation areas and corresponding time courses for each subject with very limited to no information about how spatial and temporal components interact. This problem is analogous to the blind source separation problem [26], where there are set of individual independent sources mixed together and with no prior information about any of these sources it is required to restore the each of the sources again from the mixture. Unlike task based fMRI, it is very difficult to use the model based solutions such as GLM here. In the GLM the actual task timing and any other related parameters are modeled in the form of a design matrix, but in the case of RfMRI, there is no prior information to build such matrix, that is why the majority of reported methods in the literature use model-free methods [27]. Figure 16.4 shows an example of blind source separation,

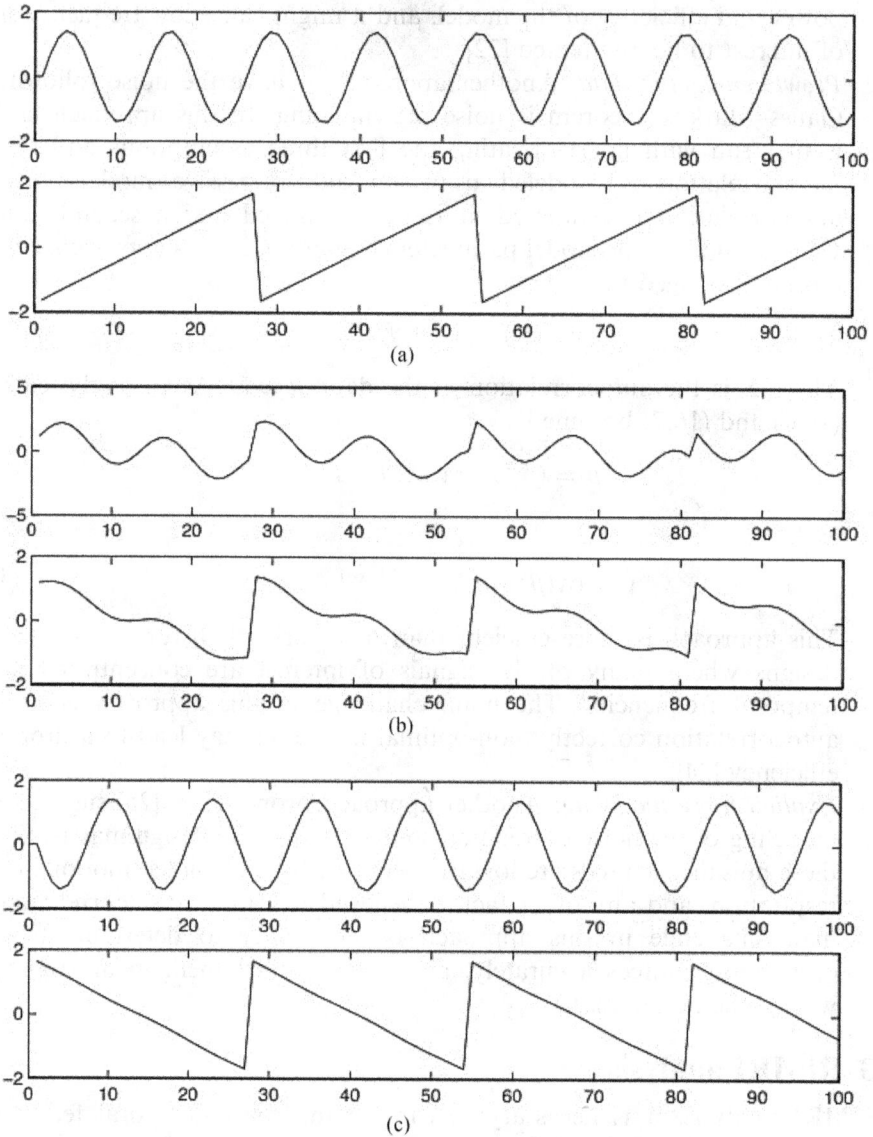

Figure 16.4. An example of (a) two original signals, (b) two recorded mixtures, and (c) the restored signals from the mixtures.

where two recorded mixtures of two signals, the original two signals, and the restored two signals are shown. In the next subsections some of the most common model-free techniques used in RfMRI, such as principle components analysis (PCA), non-negative matrix factorization (NFM), independent components analysis (ICA), and restricted Boltzmann machines (RBMs) will be explained and for each model the algorithm, challenges, and limitations will be discussed.

16.3.1 Principle components analysis in RfMRI analysis

The intention behind using PCA is to decompose the RfMRI data X with T time points and n voxels into set of orthogonal components, each of these components is a product of a temporal pattern and a spatial map. The selected components to use are those with the highest variance when data are projected on them [28].

To achieve this, singular value decomposition (SVD) is used [29] for the matrix X as follows:

$$X = USV^T = \sum_{i=1}^{P} S_i U_I V_i^T,$$ (16.15)

where P is the number of components, U are the temporal components, V is the corresponding spatial pattern, and S is the singular value of X.

For a better illustration of how PCA and SVD work in RfMRI decomposition, figure 16.5 shows a signal Y of 128 time points and 40 voxels and their SVD decomposition result using only three components. The reconstructed signal Y^* is close to the initial Y signal.

There are some limitations that make PCA unable to perform optimally in RfMRI analysis. The first limitation that it is limited to linear correlation, hence it cannot capture any non-linear relation between the data. PCA also is limited to finding orthogonal components. In addition PCA uses only up to second order statistics (covariance matrix and correlation matrix) [30].

16.3.2 Independent components analysis in RfMRI analysis

Another model-free approach for RfMRI analysis is independent components analysis (ICA) [31, 32]. In ICA the data matrix X is assumed to be the product of matrices A and S:

$$X = AS,$$ (16.16)

where in the RfMRI context, S is the independent components matrix and A is the time courses matrix. After solving for A, simply calculate W, the inverse of A and solve to obtain s using

$$S = Wx.$$ (16.17)

It is impossible to solve such a problem without having some constraints on the solution. The basic two constraints on the ICA solution are: (i) the components are statistically independent and (ii) the components are non-Gaussian.

To understand why the components should be non-Gaussian, we can consider the matrix A as a rotation matrix that transfers the data X to the component space S. In the case of Gaussian components, the distribution will be symmetric along each direction, which makes it impossible to find a rotation matrix A that could transform the data X to the components S. Figure 16.6 shows an illustrative case where there are two components s_1 and s_2 and the corresponding data in two dimensions. It is obvious that just by rotating the data, the independent components could be

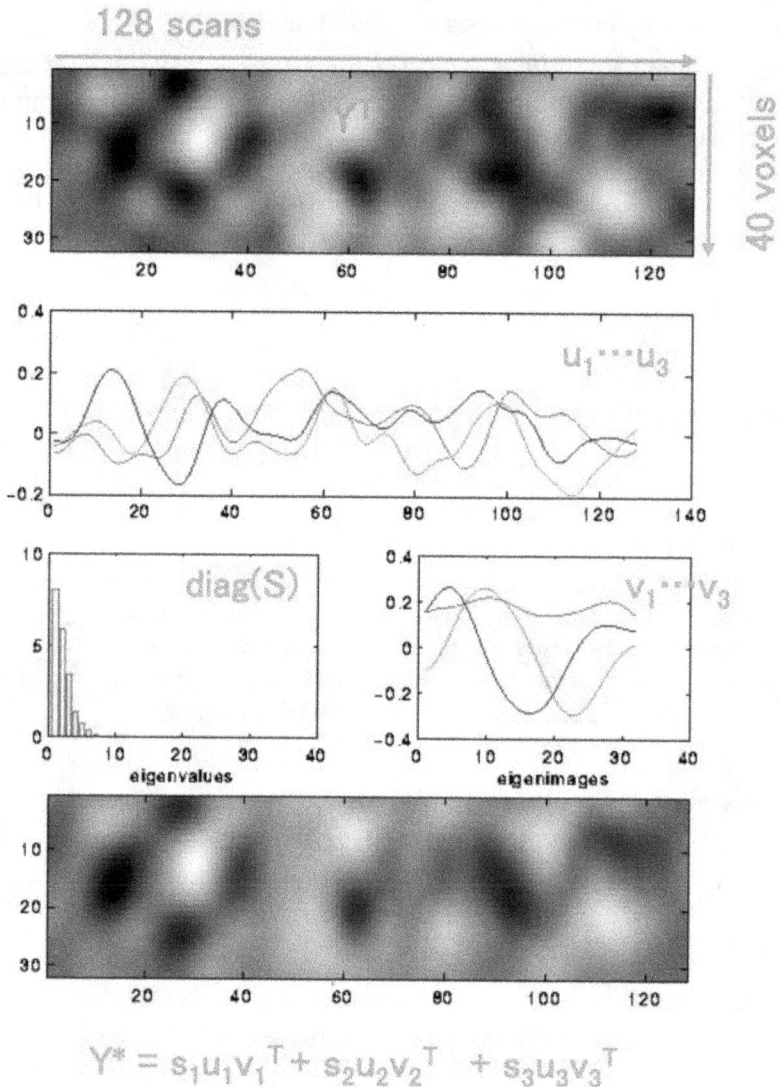

Figure 16.5. The SVD decomposition of the signal Y using only three components, and the reconstructed signal y^*.

obtained. Also figure 16.7 shows a case where the components are Gaussian, and how they are symmetric, thus no rotation matrix can help in finding a solution.

In order to solve for the mixing (rotation) matrix A, let us first denote

$$Y = W^T X = \sum_i w_i x_i. \tag{16.18}$$

In equation (16.18), if w is one row of the A^{-1}, then y is one of the independent components. From equations (16.16) and (16.18) we have

$$Y = W^T X = W^T A S = Z^T S, \tag{16.19}$$

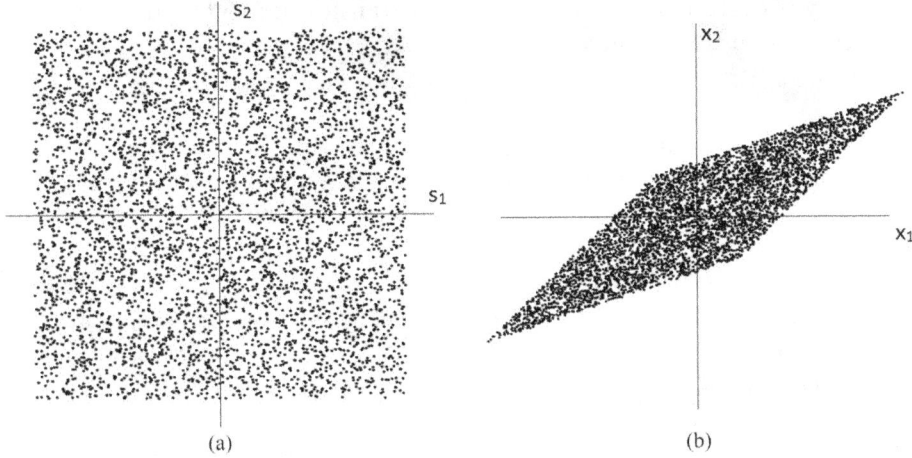

Figure 16.6. An illustrative figure of (a) two independent components s_1 and s_2 and (b) the corresponding mixture data in two dimensions x_1 and x_2. In this case, it is obvious that using a rotation matrix can transfer the data into independent components space.

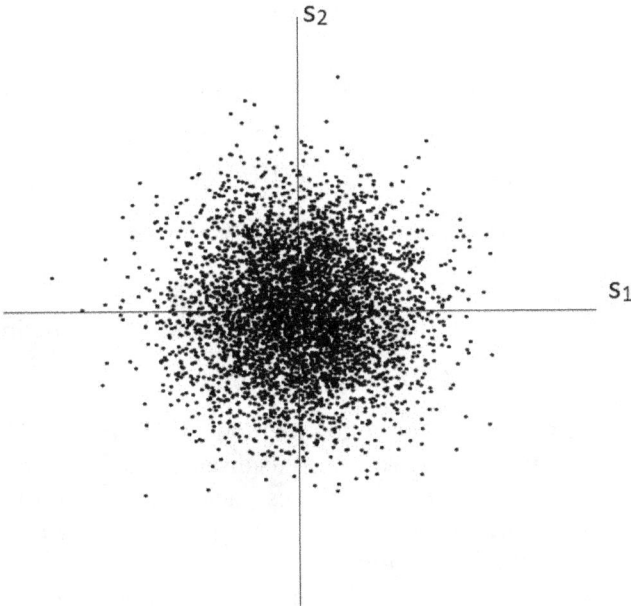

Figure 16.7. The multivariate distribution of two independent Gaussian variables. It is obvious that using any rotation matrix cannot lead to a solution due to the symmetry.

where $Z^T = A^T W$. According to the central limit theorem (CLT), the sum of any two independent random variables is more Gaussian than any of them. Based on this fact, and from equation (16.19), the linear combination Z^{TS} is more Gaussian than any of the independent components s, and Z^{TS} becomes least Gaussian, when it is equal to exactly one of the components. In this case Z^T is all zeros except for one element only. Accordingly the problem is now transformed to find the vector w that maximizes the non-Gaussianity of

$$w^T x = z^T s, \tag{16.20}$$

where both sides in equation (16.20) give exactly independent components.

There are several measures of non-Gaussianity that could be used to optimize this problem. One of these measures is kurtosis, and it is defined as

$$\text{Kurt}(y) = E\{y^4\} - 3(E\{y^2\})^2. \tag{16.21}$$

One of the major drawbacks of kurtosis is its sensitivity to outliers [33], that is why it is not commonly used in practice. Another measure of non-Gaussianity is neg-entropy, where the neg-entropy is denoted by

$$j(y)(H(y_{\text{Gaussian}}) - H(y)), \tag{16.22}$$

where $H(y_{\text{Gaussian}})$ is the entropy of Gaussian distribution, and $H(y)$ is the entropy of the distribution y. To calculate the entropy of a distribution it is necessary to know the probability distribution function (PDF) of this distribution:

$$H(y) - \sum_i p_i(y)\log_2 p_i(y). \tag{16.23}$$

In practical applications, the PDF is not necessarily known, that is why the approximation suggested in [34] is used for the neg-entropy:

$$j(y) \propto [EG(y) - E\{(G(\nu))\}]^2, \tag{16.24}$$

where G is a non-quadratic function and ν is a Gaussian variable with zero mean and unit variance.

To solve for the independent components, each row r of the matrix Y is solved in an iterative way using the first and second derivatives of the function G as follows:

$$r \leftarrow \langle XG'(y) - \langle G''(y)\rangle r\rangle. \tag{16.25}$$

Although ICA is one of the most common methods used for RfMRI analysis [6, 32, 35], its solution is still limited by the constraints discussed above. The ICA solved the problem of orthogonality, but it still assumes linearity, in addition it requires that the components should be non-Gaussian. Figure 16.8 shows a sample of brain activation components obtained using ICA.

16.3.3 Restricted Boltzmann machines in RfMRI analysis

Unlike ICA and PCA and any another matrix factorization technique, restricted Boltzmann machines (RBMs) are probabilistic energy-based graphical models [36]. The intuition behind using RBMs is to fit a probability distribution model over a set of visible random variables ν to the observed data. In general energy-based models the probability $p(\nu)$ is defined in terms of energy $E(\nu)$ as

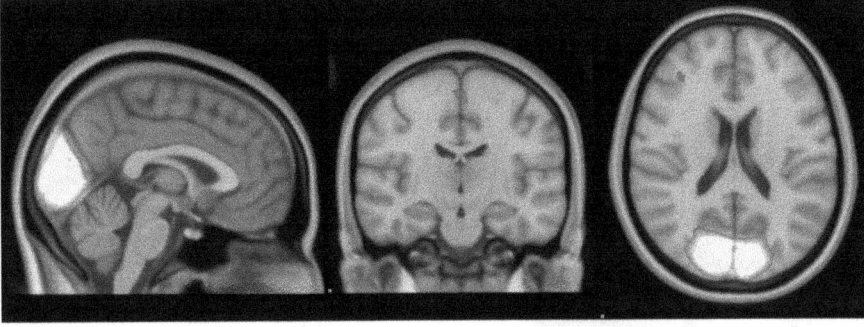

Figure 16.8. A sample of one of the spatially independent components of activation obtained using ICA decomposition.

$$p(v) = \frac{1}{Z}e^{\{E(v)\}}, \tag{16.26}$$

where $Z = \sum_v e^{-E(v)}$ is a normalization constant. By introducing a latent variable h, the visible variable probability is given as the marginal distribution over the joint distribution of both hidden and visible variables:

$$p(v) = \sum_h p(v, h) = \sum_h \frac{1}{Z}e^{E(v, h)}. \tag{16.27}$$

By introducing conditionally independent random hidden variables h and when the visible variables v are conditionally independent, RBMs model the data through latent factors obtained from interactions between visible and hidden variables.

For RBM data modeling, the interactions between the visible and latent variables are defined by the following energy function:

$$E(v, h) = -\sum_{i,j} v_i h_j w_{i,j} - \sum_i a_i v_i - \sum_j b_j h_j. \tag{16.28}$$

The first term in the energy function describes the interaction between the visible variables and the latent variables through the weights W_{ij}, where $W_{i,j}$ is the weight between the visible unit v_i and the latent variable h_j. The second and the third terms describe the visible and hidden units' binary states and their biases a_i and b_i.

The probability of assigning the network to a certain training example could be achieved by assigning weights and biases that lower the energy function. To achieve this, the derivative of the log probability with respect to the weights is used,

$$\frac{\partial \log P(v)}{\partial w_{ij}} = \langle v_i h_j \rangle_{\text{data}} - \langle v_i h_j \rangle_{\text{model}}, \tag{16.29}$$

where the angle brackets denote the expectation. Using this derivative leads to the following weight updating rule:

$$\Delta w_{ij} = \epsilon(\langle v_i h_j \rangle_{\text{data}} - \langle v_i h_j \rangle_{\text{model}}), \tag{16.30}$$

where ϵ is the learning rate. This learning rule is a good approximation to a learning function called contrastive divergence [37], and it approximates the difference between two Kullback–Liebler divergences. It is easy to calculate $\langle v_i h_j \rangle_{\text{data}}$ by calculating $p(h_j = 1|v)$ and $p(v_i = 1|h)$. These conditional probabilities are given by

$$p(h_j = 1|v) = \sigma\left(b_j + \sum_i v_i w_{ij}\right) \tag{16.31}$$

and

$$p(v_i = 1|h) = \sigma\left(a_i + \sum_j h_j w_{ij}\right), \tag{16.32}$$

where $\sigma(x) = \frac{1}{1+e^{-x}}$ is the sigmoid activation function. Although it is more difficult to calculate $\langle v_i h_j \rangle_{\text{model}}$, it could be calculated using Gibbs sampling and Markov chain Monte Carlo (MCMC) iteratively. An approach to calculate an approximation of $\langle v_i h_j \rangle_{\text{model}}$ was suggested in [37]. In this approach a training vector is used to set the binary states of the visible units then, using equation (16.31), the binary states of the hidden variables are calculated. Using equation (16.32), a reconstructed version of the training vector is calculated and equation (16.30) is changed to

$$\Delta w_{ij} = \epsilon(\langle v_i h_j \rangle_{\text{data}} - \langle v_i h_j \rangle_{\text{recon}}). \tag{16.33}$$

The model described works for binary vectors. To generalize it to work with real values, which is the case for RfMRI, Gaussian visible units with variance σ_i are used. In this case, the energy function in equation (16.28) is changed to

$$E(v, h) = -\sum_{i,j} \frac{v_i}{\sigma_i} h_j w_{i,j} - \sum_i \frac{(v_i - a_i)^2}{\sigma_i} v_i - \sum_j b_j h_j. \tag{16.34}$$

To map this approach to the RfMRI context [38], the visible units v represent the input data and each visible unit v_i corresponds to a single voxel, the hidden units represent the latent representation and their number is the number of the independent components, and the weight matrix represents the spatial maps. It is obvious that the RBM does not generate the time courses explicitly. However, by formulating the problem in the form of a matrix decomposition as in ICA, the time courses could be easily calculated as linear combination of the input and spatial maps:

$$S = W^T X, \tag{16.35}$$

where the columns of S are the TCs of each spatial map in W. Figure 16.9 shows a graphical representation of how RBMs are used for fMRI analysis.

One of the major drawbacks of RBMs is using MCMC in the training. One of the major drawbacks of MCMC is the convergence time [39]. In addition using MCMC is not stable enough and it is not guaranteed to converge [40, 41].

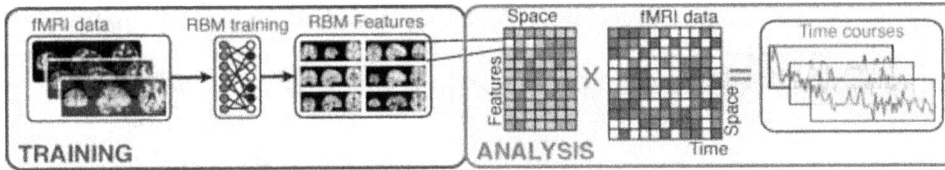

Figure 16.9. A graphical representation that illustrates how RBMs are used for fMRI analysis.

16.4 A case study of fMRI in autism

Autism spectrum disorder (ASD) is a neuro-developmental disorder which includes difficulties in social communication, deficits in both social and verbal communication, and repetitive behaviors. A causative factor has not been found for idiopathic autism. There is an overabundance of theories and hypotheses as to putative causative agents as well as risk factors. The available neuropathological and structural imaging data suggest that autism is the result of a developmental disorder capable of affecting brain growth and differentiation [42].

fMRI has been a major research technique introduced for studying autism by comparing functional connectivity measurements between both groups (autistic and typically developed (TD)) in response to a certain task [43]. The authors of [44] reported the 'under-connectivity theory', according to which ASD is both a cognitive and a neurobiological disorder. The cognitive disorder in ASD is noticeable in the form of less synchronized activity across the brain areas for tasks that need integrative processing (e.g. integrating a set of words together to form a sentence) [43].

16.4.1 Task based fMRI findings in autism

Many studies investigate a variety of tasks to address functional connectivity abnormalities between groups using fMRI. For example, the authors of [45] conducted a figure task experiment which showed that autistic subjects displayed less activation in the left dorsolateral prefrontal and inferior parietal areas while more activation was recorded in the right occipital (visuospatial) areas and bilateral superior parietal regions.

These results support the under-connectivity hypothesis proposed by [44]. A hypothesis of [46] is that ASD is highly related to the abnormal processing of unexpected stimuli. To verify this hypothesis, [46] conducted an experiment to test the functional response to odd audio frequencies in both autistic and typically developed children. In autistic children activation in the anterior cingulate cortex was noted to be less than that in typically developed individuals for new or odd frequency detection. Another experiment by [47] studied the response of autistic and typically developed individuals to an *n*-back working memory task with letters. In this task, several letters are displayed in each trial and participants are asked to report the letter presented at the beginning of the trial. This experiment showed that typically developed individuals tend to use verbal codes to remember the letter, which was reflected by increased activity in the left parietal regions compared to the right parietal region. Autistic individuals tend to use visual codes with more activation recorded in the right parietal regions compared to left parietal regions.

In order to examine the different responses between autistic and typically developed individuals to face recognition and social cognition tasks, [48] used some cartoon character faces. This experiment showed that the amygdalae and fusiform gyrus are activated in autistic people in a different manner than in typically developed subjects. The response to facial expressions was addressed in the study by [49], where various facial expressions were displayed to both groups and activation was monitored. In this experiment autistic individuals showed higher activation in the amygdala, vPFC, and striatum specifically for sad facial expressions. It was also noticed that age and activation are negatively correlated.

ASD affects both cognitive and motor function in the cerebellum. Allen and Courchesne [50] investigated the cerebellar activation in response to both attention and motor tasks compared to rest state activation. According to this experiment, it was found that autistic subjects showed higher activation in response to motor tasks than attention tasks when compared to typically developed subjects.

Another experimental task used to investigate functional connectivity is the reward-processing task. For example, the authors of [51] investigated both autistic and typically developed individual responses towards social and monetary rewards. Autistic subjects showed less activation in the right nucleus accumbens and more activation in left midfrontal and anterior cingulate gyrus than typically developed individuals in response to monetary rewards. For the social rewards, autistic subjects showed more activation in the bilateral amygdala and bilateral insular cortex. The less activation hypothesis in autism as a response towards social compared to nonsocial rewards was supported by the study [52]. In the study [53], where both autistic and typically developed individuals participated in a social and monetary rewards reception experiment, autistic subjects showed less activation in the left dorsal striatum (DS). Minshew and Keller [54] found a correlation between increased activation in the left rostral anterior cingulate cortex during reward-processing and social impairment in autistic individuals. Another monetary and social rewards experiment was conducted in [15], where autistic subjects showed less response towards both monetary and social stimuli. Moreover, less response was noticed in frontostriatal areas in autistic individuals in response to social rewards than that to monetary rewards. Table 16.1 summarizes the task based fMRI findings in autism in each of the experimental types discussed above.

16.4.2 Resting state fMRI findings in autism

In addition to task based experiments, many studies addressed the resting state connectivity for both ASD and typically developing subjects. The main hypothesis in these studies is the functional under-connectivity in ASD. In [55], the functional connectivity (FC) was compared in both ASDs and TDs using the default mode network (DMN). The DMN is a network that is active during passive resting states and cognitive processes related to the social deficits seen in ASD. Using a dataset of

Table 16.1. Summary of some task based fMRI experiments, where the study reference, experiment type, experiment summary, study population, and main findings are listed.

Study	Experiment type	Experiment summary	Population	Findings
Damarla et al [45]	Figures experiment	Participants decided if the target figure was embedded in the simultaneously presented more complex figure and indicated their decision, pressing one of two response buttons.	13 high functioning ASDs and 13 TDs matched in age, gender, and IQ. The mean age was 20 years.	The ASDs showed less activation in the left dorsolateral prefrontal and inferior parietal areas and more activation in the visuospatial (bilateral superior parietal extending to inferior parietal and right occipital) area. Also, ASDs showed lower functional connectivity between higher-order working memory/executive areas and visuospatial regions (between the frontal and parietal–occipital).
Just et al [44]	Sentence comprehension	Participants were asked to read either a passive or active sentence and then respond to a probe to identify either the taker or recipient of an action.	17 high functioning ASDs and 17 TDs with IQs of 80 and above. The two groups were matched in age and IQ.	The ASD group had higher activation than the TD group in the left laterosuperior temporal area and less activation than the TDs in Broca's (left inferior frontal gyrus) area.
Gomot et al [46]	Auditory event related task	The response of participants to three different sound frequencies was recorded. The generated sounds were divided into three groups (standard, deviant, and novel) with a probability of occurrence of 0.82, 0.09, and 0.09, respectively.	12 male high functioning ASDs and 12 TDs matched in age, gender, and IQ. Participant ages varied between 10 and 15 years and IQ varied between 113 and 132 for both groups.	ASDs showed reduced activation in the left anterior cingulate cortex during both deviance and novelty detection. During novelty detection, ASDs also showed reduced activation in the bilateral temporo-parietal region and in the right inferior and middle frontal areas.

Reference	Task	Description	Participants	Results
Koshino et al [47]	Working memory task	Participants were asked to respond to one of three questions, either remember the first letter in a sequence (0-back) or remember the letter that appeared twice in a row (1-back) or remember a letter matched to one that had been presented two letters ago (2-back).	14 high functioning ASDs (13 males and 1 female) and 14 TDs (13 males and 1 female). All had IQs over 80 and were matched in age (mean age for ASDs was 25.7 years and for TDs 29.8 years),	TDs showed more activation in the left than the right parietal regions, while ASDs showed more right lateralized activation in the prefrontal and parietal regions. ASDs also showed more activation in the posterior regions including inferior temporal and occipital regions.
Grelotti et al [48]	Faces experiment	Participants were asked to select either a true or false label when a familiar or unfamiliar face or cartoon character was displayed with an associated label.	Two ASDs of ages 10 and 12 and 1 TD aged 17 years	ASDs showed more activation in the amygdala and fusiform gyrus for perceptual discriminations for cartoon faces but not for familiar or unfamiliar faces. This pattern of activation was not noticed in the TD subject.
Weng et al [49]	Facial expressions experiment	The participants were shown a sequence of sad and happy face images and they were asked to press the thumb button if it was a male face and the index button if it was a female face.	22 ASDs and 20 TDs with ages between 10 and 18 years old. The two groups were matched in terms of gender and IQ with no significance difference.	ASDs showed greater activation in response to faces (specifically for sad facial expressions) than TDs in the amygdala, vPFC, and striatum.
Allen and Courchesne [50]	Motor experiment	Participants were asked to complete a simple motor task that involved pressing the button repeatedly at a comfortable pace.	8 ASDs aged between 14 and 38 years and 8 TDs aged between 13 and 39 years. All participants had IQ above 70.	ASDs showed significantly greater cerebellar motor activation and significantly less cerebellar attention activation.

(Continued)

Table 16.1. (*Continued*)

Study	Experiment type	Experiment summary	Population	Findings
Zeeland *et al* [15]	Rewards experiment	Participants were asked to respond to classifying a set of shapes to either group 1 or 2 and two types of rewards were offered: monetary or social.	16 high functioning autism with normal IQ and 16 TDs with matched IQ and age. The mean age for both groups was 12.3 years.	Reduced neural response to both monetary and social rewards in ASD group. Also, deficit in frontostriatal response during social but not monetary reward in the ASD group was noticed.

16 matched ASDs and TDs, decreased FC was noted in ASDs compared to TDs, in particular between the precuneus and medial prefrontal cortex/anterior cingulate cortex DMN core areas. It was reported that FC magnitude is inversely correlated with ASD symptom severity, defined by the autism diagnosis observation schedule (ADOS) and the social responsiveness scale.

Reduced FC was also reported in [56], where the salience network (which includes insula and a medial temporal lobe network, and the amygdala) showed inter-network connectivity. This reduced connectivity is suggested to caused the difficul-ties in communication and integration of information across these networks, which could contribute to the ASD related social signal processing impairments.

Another study [35] showed alternations in functional connectivity between ASDs and TDs. An increased functional connectivity between primary sensory networks and subcortical networks was reported. In addition, decreased functional connec-tivity in the subcortico-cortical area was also reported.

The reduced connectivity in autism was also reported in [57], where a dataset of 14 ASDs and 15 TDs was used. For ASDs, anterior and posterior regions of the insular cortex showed reduced FC. This study suggests that this reduced functional connectivity is responsible for the emotional and interoceptive awareness related impairments in ASDs.

The study in [58] reported less functional connectivity in ASD than typically developing subjects. In this study a dataset of 46 ASDs and 46 TDs matched in terms of age and gender was used. The connectivity was measured in terms of clustering coefficients and characteristic path length. Another recent study [59] supported the under-connectivity hypothesis, where the between group comparison showed significantly reduced connectivity in the visuospatial and superior parietal areas in ASDs compared to typically developing individuals.

The study of [60] analyzed the connectivity in the dentate nucleus and cerebello-thalamo-cortical circuits that is known to be part of default mode networks, and decreased connectivity was reported in ASDs when compared to TDs. Although most studies support the under-connectivity hypothesis, mixed connectivity patterns were observed in the study [61]. In this study, under-connectivity was reported for males only, while hyper-connectivity was reported for females. The mixed con-nectivity patterns were also similarly studied by Hahamy *et al* [62]. Some brain areas in autistic subjects showed reduced connectivity and some other regions showed increased or ectopic connectivity. The mixed connectivity pattern observed in [62] was correlated with behavioral symptoms. Hence, this finding emphasizes the role of resting state fMRI in understanding functional connectivity patterns and identifying corresponding behaviors in autistic subjects.

The rest associated functional activity was compared between ASDs and TDs in [63]. The hypothesis in this study is that some areas in the brain, including the medial prefrontal cortex, rostral anterior cingulate, posterior cingulate, and precuneus, are known to have high metabolic activity during rest, which is suppressed during cognitively demanding tasks (the deactivation effect). This deactivation indicates a mental activity interruption during rest. Based on this hypothesis, the deactivation was measured in ASDs and compared to that in TDs, and the ASDs showed less deactivation, in addition, a strong correlation was found between social impairments and functional activity, particularly within the ventral medial prefrontal cortex. Table 16.2 summarizes the basic RfMRI findings discussed above.

16.5 Conclusions and future work

In this review, the two main types of fMRI are discussed. For task based fMRI, the most common analysis method is the GLM model and for RfMRI the most common analysis method is the ICA. Although these methods achieved very promising results in many applications, as shown in the case study applications to autism discussed above for example, there are still some analytical limitations that narrow the capabilities of these methods.

In the application to ASD, one of the ultimate goals to reach is to build a personalized ASD diagnosis system that utilizes the mentioned analysis techniques and previous hypotheses and provide a detailed personalized report for each subject. This report includes the overall diagnosis, symptom severity, and to what extend these symptoms will affect the subject's behavioral aspects.

Our future work in fMRI analysis will be focusing on the following aspects:

1. Build a model-free task based fMRI analysis framework and evaluate its performance compared to GLM.
2. Build a generative adversarial network (GAN) model for RfMRI analysis.
3. Evaluate the proposed algorithm on both synthetic data and real datasets.
4. Utilize the developed algorithms to build a personalized diagnosis system.
5. Correlate findings from diagnosis applications to behavioral reports.
6. Identify the highly affected neurocircuits in the brain and generate a detailed report for each subject identifying the corresponding affected functionalities.

This work could also be applied to various other applications in medical imaging, such as the kidney, the heart, the prostate, the lung, and the retina, as well as several non-medical applications [64–67].

One application is renal transplant functional assessment, in particular developing noninvasive CAD systems for renal transplant function assessment, utilizing different image modalities (e.g. ultrasound, computed tomography (CT), MRI, etc). Accurate assessment of renal transplant function is critically important for graft survival. Although transplantation can improve a patient's wellbeing, there is a potential post-transplantation risk of kidney dysfunction that, if not treated in a timely manner, can lead to the loss of the entire graft, and even patient death. In particular, dynamic and diffusion MRI-based systems have been clinically used to assess transplanted kidneys with the advantage of providing information on each kidney separately. For more details about renal transplant functional assessment, see [68–85, 85–95].

The heart is also an important application of this work. The clinical assessment of myocardial perfusion plays a major role in the diagnosis, management, and prognosis of ischemic heart disease patients. Thus, there have been ongoing efforts to develop automated systems for accurate analysis of myocardial perfusion using first-pass images [96–112].

Table 16.2. Summary of some RfMRI studies with the study reference, the population, and the findings.

Study	Population	Findings
Assaf *et al* [55]	16 high functioning ASDs (aged between 11 and 20 years; 15 males and 1 female) and 16 TDs (aged between 13 and 23 years; 14 males and 2 females). Groups were matched in IQ.	Reduced functional connectivity in DMN areas, in particular between the precuneus and medial prefrontal cortex/anterior cingulate cortex, for the ASD group.
von dem Hagen *et al* [56]	15 ASDs (aged between 19 and 36 years) and 24 TDs (aged between 19 and 40 years). IQ was over 85 for all subjects.	Reduced functional connectivity in DMN areas; in addition two more networks had reduced connectivity: the salience network, incorporating the insula, and medial temporal lobe network, incorporating the amygdala.
Cerliani *et al* [35]	166 ASDs (aged between 7 and 50 years) and 193 TDs (aged between 7 and 39). The mean IQ was 109.6 for ASDs and 111 for TDs.	ASDs showed increased functional connectivity between primary sensory networks and subcortical networks (thalamus and basal ganglia). In addition, ASDs showed decreased functional connectivity in subcortico-cortical area.
Ebisch *et al* [57]	15 ASDs (mean age = 15.95 years and mean IQ = 104.9) and 14 TDs (mean age = 15.79 and mean IQ =106.5)	Anterior and posterior regions of the insular cortex showed reduced functional connectivity in ASD.
Itahashi *et al* [58]	46 ASDs (7 females) and 46 TDs (7 females). The age range for both groups is between 19 and 50 years and all subjects have IQ above 83.	Reduced functional connectivity organization that tends to be random in ASDs, in particular in the bilateral superior temporal sulcus, right dorsolateral prefrontal cortex, and precuneus.
Rausch *et al* [59]	20 ASDs (1 female) and 22 TDs (3 females). The mean IQ in ASDs was 102.3 and for TDs was 103.72.	Reduced connectivity in visuospatial and superior parietal areas in ASDs compared to TDs.
Olivito *et al* [60]	10 high functioning ASDs with mean age of 23.8 years and 36 TDs with mean age of 26.5 years.	ASDs showed decreased connectivity between the cerebellar dentate nucleus (DN) and the CTC circuit targets compared to TDs.
Alaerts *et al* [61]	84 ASDs (42 males) and 150 TDs (75 males). The mean age was 13 years in ASDs and 13.7 years for TDs. The mean IQ was 101 for ASDs and 110.5 for TDs.	Reduced functional connectivity in ASD males compared to TD males and increased functional connectivity in ASD females compared to TD females.
Hahamy *et al* [62]	68 ASDs (mean age 26.6, age range 18–42, 62 males) and 73 TDs (mean age 25.82, age range 18–44, 59 males).	Mixed personalized connectivity patterns in ASDs that are correlated with behavioral attributes.

Another application for this work could be the detection of retinal abnormalities. The majority of ophthalmologists depend on visual interpretation for the identification of disease types. However, inaccurate diagnosis will affect the treatment procedure which may lead to fatal results. Hence, there is a crucial need for computer automated diagnosis systems that yield highly accurate results. Optical coherence tomography (OCT) has become a powerful modality for the noninvasive diagnosis of various retinal abnormalities such as glaucoma, diabetic macular edema, and macular degeneration. The problem with diabetic retinopathy (DR) is that the patient is not aware of the disease until the changes in the retina have progressed to a level that treatment tends to be less effective. Therefore, automated early detection could limit the severity of the disease and assist ophthalmologists in investigating and treating it more efficiently [113–115].

Abnormalities of the lung could also be another promising area of research and an application related to this work. Radiation-induced lung injury is the main side effect of radiation therapy for lung cancer patients. Although higher radiation doses increase the effectiveness of radiation therapy for tumor control, this can lead to lung injury as a greater quantity of normal lung tissues is included in the treated area. Almost 1/3 of patients who undergo radiation therapy develop lung injury following radiation treatment. The severity of radiation-induced lung injury ranges from ground-glass opacities and consolidation at the early phase to fibrosis and traction bronchiectasis in the late phase. Early detection of lung injury will thus help to improve the management of treatment [116–158].

This work can also be applied to other brain abnormalities, such as dyslexia in addition to autism. Dyslexia is one of the most complicated developmental brain disorders that affect children's learning abilities. Dyslexia leads to the failure to develop age-appropriate reading skills in spite of a normal intelligence level and adequate reading instructions. Neuropathological studies have revealed the abnormal anatomy of some structures, such as the corpus callosum, in dyslexic brains. There has been a lot of work in the literature that aims to develop CAD systems for diagnosing such a disorder, along with other brain disorders [159–181].

For the vascular system [182], this work could also be applied for the extraction of blood vessels, e.g. from phase contrast (PC) magnetic resonance angiography (MRA). Accurate cerebrovascular segmentation using noninvasive MRA is crucial for the early diagnosis and timely treatment of intracranial vascular diseases [164, 165, 183–188].

References

[1] Craddock R C, James G A, Holtzheimer P E, Hu X P and Mayberg H S 2012 A whole brain fMRI atlas generated via spatially constrained spectral clustering *Hum. Brain Mapp.* **33** 1914–28

[2] Feinberg D A, Moeller S, Smith S M, Auerbach E, Ramanna S, Glasser M F, Miller K L, Ugurbil K and Yacoub E 2010 Multiplexed echo planar imaging for sub-second whole brain fMRI and fast diffusion imaging *PLoS One* **5** e15710

[3] Kwong K K, Belliveau J W, Chesler D A, Goldberg I E, Weisskoff R M, Poncelet B P, Kennedy D N, Hoppel B E, Cohen M S and Turner R 1992 Dynamic magnetic resonance

imaging of human brain activity during primary sensory stimulation *Proc. Natl Acad. Sci.* **89** 5675–9

[4] Monti M M 2011 Statistical analysis of fMRI time-series: a critical review of the GLM approach *Front. Hum. Neurosci.* **5** 28

[5] Van Horn J D, Grethe J S, Kostelec P, Woodward J B, Aslam J A, Rus D, Rockmore D and Gazzaniga M S 2001 The Functional Magnetic Resonance Imaging Data Center (FMRIDC): the challenges and rewards of large-scale databasing of neuroimaging studies *Philos. Trans. R. Soc. Lond.* B **356** 1323–39

[6] Cole D M, Smith S M and Beckmann C F 2010 Advances and pitfalls in the analysis and interpretation of resting-state fMRI data *Front. Syst. Neurosci.* **4** 8

[7] MacIntosh B J, Mraz R, McIlroy W E and Graham S J 2007 Brain activity during a motor learning task: an fMRI and skin conductance study *Hum. Brain Mapp.* **28** 1359–67

[8] Stephan K, Magnotta V, White T, Arndt S, Flaum M, O'leary D and Andreasen N 2001 Effects of olanzapine on cerebellar functional connectivity in schizophrenia measured by fMRI during a simple motor task *Psychol. Med.* **31** 1065–78

[9] Yoo S-S and Jolesz F A 2002 Functional MRI for neurofeedback: feasibility study on a hand motor task *Neuroreport* **13** 1377–81

[10] Clark V P, Maisog J M and Haxby J V 1998 fMRI study of face perception and memory using random stimulus sequences *J. Neurophysiol.* **79** 3257–65

[11] Keizer A W, Nieuwenhuis S, Colzato L S, Teeuwisse W, Rombouts S A and Hommel B 2008 When moving faces activate the house area: an fMRI study of object-file retrieval *Behav. Brain Funct.* **4** 50

[12] Birbaumer N, Grodd W, Diedrich O, Klose U, Erb M, Lotze M, Schneider F, Weiss U and Flor H 1998 fMRI reveals amygdala activation to human faces in social phobics *Neuroreport* **9** 1223–6

[13] Richards J M, Plate R C and Ernst M 2013 A systematic review of fMRI reward paradigms used in studies of adolescents vs adults: the impact of task design and implications for understanding neurodevelopment *Neurosci. Biobehav. Rev.* **37** 976–91

[14] Knutson B, Fong G W, Adams C M, Varner J L and Hommer D 2001 Dissociation of reward anticipation and outcome with event-related fMRI *Neuroreport* **12** 3683–7

[15] Zeeland S-V, Ashley A, Dapretto M, Ghahremani D G, Poldrack R A and Bookheimer S Y 2011 Reward processing in autism *Autism Res.* **3** 53–67

[16] Beckmann C F, Jenkinson M and Smith S M 2003 General multilevel linear modeling for group analysis in fMRI *Neuroimage* **20** 1052–63

[17] Searle S R, Casella G and McCulloch C E 2009 *Variance Components* vol 391 (New York: Wiley)

[18] Zang Y, Jiang T, Lu Y, He Y and Tian L 2004 Regional homogeneity approach to fMRI data analysis *Neuroimage* **22** 394–400

[19] Plackett R L 1950 Some theorems in least squares *Biometrika* **37** 149–57

[20] Boynton G M, Engel S A, Glover G H and Heeger D J 1996 Linear systems analysis of functional magnetic resonance imaging in human V1 *J. Neurosci.* **16** 4207–21

[21] Worsley K J and Friston K J 1995 Analysis of fMRI time-series revisited—again *Neuroimage* **2** 173–81

[22] Marchini J L and Ripley B D 2000 A new statistical approach to detecting significant activation in functional MRI *NeuroImage* **12** 366–80

[23] Woolrich M W, Ripley B D, Brady M and Smith S M 2001 Temporal autocorrelation in univariate linear modeling of fMRI data *Neuroimage* **14** 1370–86

[24] Friston K, Josephs O, Zarahn E, Holmes A, Rouquette S and Poline J-B 2000 To smooth or not to smooth? Bias and efficiency in fMRI time-series analysis *NeuroImage* **12** 196–208

[25] Lund T E, Madsen K H, Sidaros K, Luo W-L and Nichols T E 2006 Non-white noise in fMRI: does modelling have an impact? *Neuroimage* **29** 54–66

[26] Jutten C and Herault J 1991 Blind separation of sources, part I: an adaptive algorithm based on neuromimetic architecture *Sign. Process* **24** 1–10

[27] Rombouts S A, Damoiseaux J S, Goekoop R, Barkhof F, Scheltens P, Smith S M and Beckmann C F 2009 Model-free group analysis shows altered bold fMRI networks in dementia *Hum. Brain Mapp.* **30** 256–66

[28] Jolliffe I T 1986 Principal component analysis and factor analysis *Principal Component Analysis* (Berlin: Springer) pp 115–28

[29] Wall M E, Rechtsteiner A and Rocha L M 2003 Singular value decomposition and principal component analysis *A Practical Approach to Microarray Data Analysis* (Berlin: Springer) pp 91–109

[30] Cha S-M and Chan L-W 2000 Applying independent component analysis to factor model in finance *Int. Conf. on Intelligent Data Engineering and Automated Learning* (Berlin: Springer) pp 538–44

[31] Hyvärinen A and Oja E 2000 Independent component analysis: algorithms and applications *Neural Netw* **13** 411–30

[32] Beckmann C F and Smith S M 2004 Probabilistic independent component analysis for functional magnetic resonance imaging *IEEE Trans. Med. Imaging* **23** 137–52

[33] Huber P J 1985 Projection pursuit *Ann. Stat.* 435–75

[34] Hyvärinen A 1998 New approximations of differential entropy for independent component analysis and projection pursuit *Advances in Neural Information Processing Systems* pp 273–9

[35] Cerliani L, Mennes M, Thomas R M, Di Martino A, Thioux M and Keysers C 2015 Increased functional connectivity between subcortical and cortical resting-state networks in autism spectrum disorder *JAMA Psychiatry* **72** 767–77

[36] Hinton G E 2012 A practical guide to training restricted Boltzmann machines *Neural Networks: Tricks of the Trade* (Berlin: Springer) pp 599–619

[37] Hinton G E 2002 Training products of experts by minimizing contrastive divergence *Neural Comput.* **14** 1771–800

[38] Hjelm R D, Calhoun V D, Salakhutdinov R, Allen E A, Adali T and Plis S M 2014 Restricted Boltzmann machines for neuroimaging: an application in identifying intrinsic networks *NeuroImage* **96** 245–60

[39] Mossel E and Vigoda E 2006 Limitations of Markov chain Monte Carlo algorithms for Bayesian inference of phylogeny *Ann. Appl. Probab.* 2215–34

[40] van Ravenzwaaij D, Cassey P and Brown S D 2016 A simple introduction to Markov chain Monte-Carlo sampling *Psychon. Bull. Rev.* 1–12

[41] Mengersen K L, Robert C P and Guihenneuc-Jouyaux C 1999 MCMC convergence diagnostics: a review *Bayesian Stat.* **6** 415–40

[42] Casanova M F, Buxhoeveden D P, Switala A E and Roy E 2002 Minicolumnar pathology in autism *Neurology* **58** 428–32

[43] Just M A, Cherkassky V L, Keller T A, Kana R K and Minshew N J 2006 Functional and anatomical cortical underconnectivity in autism: evidence from an fMRI study of an executive function task and corpus callosum morphometry *Cereb. Cortex* **17** 951–61

[44] Just M A, Cherkassky V L, Keller T A and Minshew N J 2004 Cortical activation and synchronization during sentence comprehension in high-functioning autism: evidence of underconnectivity *Brain* **127** 1811–21

[45] Damarla S R, Keller T A, Kana R K, Cherkassky V L, Williams D L, Minshew N J and Just M A 2010 Cortical underconnectivity coupled with preserved visuospatial cognition in autism: evidence from an fMRI study of an embedded figures task *Autism Res.* **3** 273–9

[46] Gomot M, Bernard F A, Davis M H, Belmonte M K, Ashwin C, Bullmore E T and Baron-Cohen S 2006 Change detection in children with autism: an auditory event-related fMRI study *Neuroimage* **29** 475–84

[47] Koshino H, Carpenter P A, Minshew N J, Cherkassky V L, Keller T A and Just M A 2005 Functional connectivity in an fMRI working memory task in high-functioning autism *Neuroimage* **24** 810–21

[48] Grelotti D J, Klin A J, Gauthier I, Skudlarski P, Cohen D J, Gore J C, Volkmar F R and Schultz R T 2005 fMRI activation of the fusiform gyrus and amygdala to cartoon characters but not to faces in a boy with autism *Neuropsychologia* **43** 373–85

[49] Weng S-J, Carrasco M, Swartz J R, Wiggins J L, Kurapati N, Liberzon I, Risi S, Lord C and Monk C S 2011 Neural activation to emotional faces in adolescents with autism spectrum disorders *J. Child Psychol. Psychiatry* **52** 296–305

[50] Allen G and Courchesne E 2003 Differential effects of developmental cerebellar abnormality on cognitive and motor functions in the cerebellum: an fMRIi study of autism *Am. J. Psychiatry* **160** 262–73

[51] Dichter G S, Richey J A, Rittenberg A M, Sabatino A and Bodfish J W 2012 Reward circuitry function in autism during face anticipation and outcomes *J. Autism Dev. Disord.* **42** 147–60

[52] Cox A, Kohls G, Naples A J, Mukerji C E, Coffman M C, Rutherford H J, Mayes L C and McPartland J C 2015 Diminished social reward anticipation in the broad autism phenotype as revealed by event-related brain potentials *Soc. Cogn. Affect. Neurosci.* **10** 1357–64

[53] Delmonte S, Balsters J H, McGrath J, Fitzgerald J, Brennan S, Fagan A J and Gallagher L 2012 Social and monetary reward processing in autism spectrum disorders *Mol. Autism* **3** 7

[54] Minshew N J and Keller T A 2010 The nature of brain dysfunction in autism: functional brain imaging studies *Curr. Opin. Neurol.* **23** 124

[55] Assaf M, Jagannathan K, Calhoun V D, Miller L, Stevens M C, Sahl R, O'boyle J G, Schultz R T and Pearlson G D 2010 Abnormal functional connectivity of default mode sub-networks in autism spectrum disorder patients *Neuroimage* **53** 247–56

[56] von dem Hagen E A, Stoyanova R S, Baron-Cohen S and Calder A J 2012 Reduced functional connectivity within and between social resting state networks in autism spectrum conditions *Soc. Cogn. Affect. Neurosci.* **8** 694–701

[57] Ebisch S J, Gallese V, Willems R M, Mantini D, Groen W B, Romani G L, Buitelaar J K and Bekkering H 2011 Altered intrinsic functional connectivity of anterior and posterior insula regions in high-functioning participants with autism spectrum disorder *Hum. Brain Mapp.* **32** 1013–28

[58] Itahashi T, Yamada T, Watanabe H, Nakamura M, Jimbo D, Shioda S, Toriizuka K, Kato N and Hashimoto R 2014 Altered network topologies and hub organization in adults with autism: a resting-state fMRI study *PLoS One* **9** e94115

[59] Rausch A, Zhang W, Haak K V, Mennes M, Hermans E J, van Oort E, van Wingen G, Beckmann C F, Buitelaar J K and Groen W B 2016 Altered functional connectivity of the amygdaloid input nuclei in adolescents and young adults with autism spectrum disorder: a resting state fMRI study *Mol. Autism* **7** 13

[60] Olivito G, Clausi S, Laghi F, Tedesco A M, Baiocco R, Mastropasqua C, Molinari M, Cercignani M, Bozzali M and Leggio M 2017 Resting-state functional connectivity changes between dentate nucleus and cortical social brain regions in autism spectrum disorders *Cerebellum* **16** 283–92

[61] Alaerts K, Swinnen S P and Wenderoth N 2016 Sex differences in autism: a resting-state fMRI investigation of functional brain connectivity in males and females *Soc. Cogn. Affect. Neurosci.* **11** 1002–16

[62] Hahamy A, Behrmann M and Malach R 2015 The idiosyncratic brain: distortion of spontaneous connectivity patterns in autism spectrum disorder *Nat. Neurosci.* **18** 302

[63] Kennedy D P, Redcay E and Courchesne E 2006 Failing to deactivate: resting functional abnormalities in autism *Proc. Natl Acad. Sci.* **103** 8275–80

[64] Mahmoud A H 2014 Utilizing radiation for smart robotic applications using visible, thermal, and polarization images *PhD Dissertation* University of Louisville, KY

[65] Mahmoud A, El-Barkouky A, Graham J and Farag A 2014 Pedestrian detection using mixed partial derivative based histogram of oriented gradients *2014 IEEE Int. Conf. on Image Processing (ICIP)* (Piscataway, NJ: IEEE) pp 2334–7

[66] El-Barkouky A, Mahmoud A, Graham J and Farag A 2013 An interactive educational drawing system using a humanoid robot and light polarization *2013 IEEE Int. Conf. on Image Processing* (Piscataway, NJ: IEEE) pp 3407–11

[67] Mahmoud A H, El-Melegy M T and Farag A A 2012 Direct method for shape recovery from polarization and shading *2012 19th IEEE Int. Conf. on Image Processing* (Piscataway, NJ: IEEE) pp 1769–72

[68] Ali A M, Farag A A and El-Baz A 2007 Graph cuts framework for kidney segmentation with prior shape constraints *Proc. of Int. Conf. on Medical Image Computing and Computer-Assisted Intervention, (MICCAI'07) (Brisbane, Australia, 29 October–2 November 2007)* **vol 1** pp 384–92

[69] Chowdhury A S, Roy R, Bose S, Elnakib F K A and El-Baz A 2012 Non-rigid biomedical image registration using graph cuts with a novel data term *Proc. of IEEE Int. Symp. on Biomedical Imaging: From Nano to Macro, (ISBI'12) (Barcelona, Spain, 2–5 May 2012)* pp 446–9

[70] El-Baz A, Farag A A, Yuksel S E, El-Ghar M E, Eldiasty T A and Ghoneim M A 2007 Application of deformable models for the detection of acute renal rejection *Deformable Models* (New York: Springer) pp 293–333

[71] El-Baz A, Farag A, Fahmi R, Yuksel S, El-Ghar M A and Eldiasty T 2006 Image analysis of renal DCE MRI for the detection of acute renal rejection *Proc. of IAPR Int. Conf. on Pattern Recognition (ICPR'06) (Hong Kong, 20–24 August 2006)* pp 822–5

[72] El-Baz A, Farag A, Fahmi R, Yuksel S, Miller W, El-Ghar M A, El-Diasty T and Ghoneim M 2006 A new CAD system for the evaluation of kidney diseases using DCE-MRI *Proc. of Int. Conf. on Medical Image Computing and Computer-Assisted Intervention, (MICCAI'08) (Copenhagen, Denmark, 1–6 October 2006)* pp 446–53

[73] El-Baz A, Gimel'farb G and El-Ghar M A 2008 A novel image analysis approach for accurate identification of acute renal rejection *Proc. of IEEE Int. Conf. on Image Processing, (ICIP'08) (San Diego, CA, USA, 12–15 October 2008)* pp 1812–5

[74] El-Baz A, Gimel'farb G and El-Ghar M A 2008 Image analysis approach for identification of renal transplant rejection *Proc. of IAPR Int. Conf. on Pattern Recognition, (ICPR'08) (Tampa, Florida, USA, 8–11 December 2008)* pp 1–4

[75] El-Baz A, Gimel'farb G and El-Ghar M A 2007 New motion correction models for automatic identification of renal transplant rejection *Proc. of Int. Conf. on Medical Image Computing and Computer-Assisted Intervention, (MICCAI'07) (Brisbane, Australia, 29 October–2 November 2007)* pp 235–43

[76] Farag A, El-Baz A, Yuksel S, El-Ghar M A and Eldiasty T 2006 A framework for the detection of acute rejection with dynamic contrast enhanced magnetic resonance imaging *Proc. of IEEE Int. Symp. on Biomedical Imaging: From Nano to Macro, (ISBI'06) (Arlington, VA, USA, 6–9 April 2006)* pp 418–21

[77] Khalifa F, Beache G M, El-Ghar M A, El-Diasty T, Gimel'farb G, Kong M and El-Baz A 2013 Dynamic contrast-enhanced MRI-based early detection of acute renal transplant rejection *IEEE Trans. Med. Imaging* **32** 1910–27

[78] Khalifa F, El-Baz A, Gimel'farb G and El-Ghar M A 2010 Non-invasive image-based approach for early detection of acute renal rejection *Proc. Int. Conf. Medical Image Computing and Computer-Assisted Intervention, (MICCAI'10) (Beijing, China, 20–24 September 2010)* pp 10–8

[79] Khalifa F, El-Baz A, Gimel'farb G, Ouseph R and El-Ghar M A 2010 Shape-appearance guided level-set deformable model for image segmentation *Proc. of IAPR Int. Conf. on Pattern Recognition, (ICPR'10) (Istanbul, Turkey, 23–26 August 2010)* pp 4581–4

[80] Khalifa F, El-Ghar M A, Abdollahi B, Frieboes H, El-Diasty T and El-Baz A 2013 A comprehensive non-invasive framework for automated evaluation of acute renal transplant rejection using DCE-MRI *NMR Biomed.* **26** 1460–70

[81] Khalifa F, El-Ghar M A, Abdollahi B, Frieboes H B, El-Diasty T and El-Baz A 2014 Dynamic contrast-enhanced MRI-based early detection of acute renal transplant rejection *2014 Annu. Scientific Meeting and Educational Course Brochure of the Society of Abdominal Radiology, (SAR'14) (Boca Raton, FL, 23–28 March 2014)* CID: 1855912

[82] Khalifa F, Elnakib A, Beache G M, Gimel'farb G, El-Ghar M A, Sokhadze G, Manning S, McClure P and El-Baz A 2011 3D kidney segmentation from CT images using a level set approach guided by a novel stochastic speed function *Proc. of Int. Confl Medical Image Computing and Computer-Assisted Intervention, (MICCAI'11)* (Toronto, Canada)(18–22 September 2011) pp 587–94

[83] Khalifa F, Gimel'farb G, El-Ghar M A, Sokhadze G, Manning S, McClure P, Ouseph R and El-Baz A 2011 A new deformable model-based segmentation approach for accurate extraction of the kidney from abdominal CT images *Proc. of IEEE Int. Conf. on Image Processing, (ICIP'11) (Brussels, Belgium, 11–14 September 2011)* pp 3393–6

[84] Mostapha M, Khalifa F, Alansary A, Soliman A, Suri J and El-Baz A 2014 Computer-aided diagnosis systems for acute renal transplant rejection: challenges and methodologies *Abdomen and Thoracic Imaging* ed A El-Baz, L Saba and J Suri (Berlin: Springer) pp 1–35

[85] Shehata M, Khalifa F, Hollis E, Soliman A, Hosseini-Asl E, El-Ghar M A, El-Baz M, Dwyer A C, El-Baz A and Keynton R 2016 A new non-invasive approach for early

classification of renal rejection types using diffusion-weighted MRI *IEEE Int. Conf. on Image Processing (ICIP), 2016* (Piscataway, NJ: IEEE) pp 136–40

[86] Khalifa F, Soliman A, Takieldeen A, Shehata M, Mostapha M, Shaffie A, Ouseph R, Elmaghraby A and El-Baz A 2016 Kidney segmentation from CT images using a 3D NMF-guided active contour model *IEEE 13th Int. Symp. on Biomedical Imaging (ISBI), 2016* (Piscataway, NJ: IEEE) pp 432–5

[87] Shehata M, Khalifa F, Soliman A, Takieldeen A, El-Ghar M A, Shaffie A, Dwyer A C, Ouseph R, El-Baz A and Keynton R 2016 3D diffusion MRI-based CAD system for early diagnosis of acute renal rejection *2016 IEEE 13th Int. Symp. on Biomedical Imaging (ISBI)* (Piscataway, NJ: IEEE) pp 1177–80

[88] Shehata M, Khalifa F, Soliman A, Alrefai R, El-Ghar M A, Dwyer A C, Ouseph R and El-Baz A 2015 A level set-based framework for 3D kidney segmentation from diffusion MR images *IEEE Int. Conf. on Image Processing (ICIP), 2015* (Piscataway, NJ: IEEE) pp 4441–5

[89] Shehata M, Khalifa F, Soliman A, El-Ghar M A, Dwyer A C, Gimel'farb G, Keynton R and El-Baz A 2016 A promising non-invasive CAD system for kidney function assessment *Int. Conf. on Medical Image Computing and Computer-Assisted Intervention* (Berlin: Springer) pp 613–21

[90] Khalifa F, Soliman A, Elmaghraby A, Gimel'farb G and El-Baz A 2017 3D kidney segmentation from abdominal images using spatial-appearance models *Comput. Math. Methods Med.* **2017** 1–10

[91] Hollis E, Shehata M, Khalifa F, El-Ghar M A, El-Diasty T and El-Baz A 2016 Towards non-invasive diagnostic techniques for early detection of acute renal transplant rejection: a review *Egypt. J. Radiol. Nucl. Med.* **48** 257–69

[92] Shehata M, Khalifa F, Soliman A, El-Ghar M A, Dwyer A C and El-Baz A 2017 Assessment of renal transplant using image and clinical-based biomarkers *Proc. of 13th Annual Scientific Meeting of American Society for Diagnostics and Interventional Nephrology (ASDIN'17) (New Orleans, LA, USA, 10–12 February 2017)*

[93] Shehata M, Khalifa F, Soliman A, El-Ghar M A, Dwyer A C and El-Baz A 2017 Early assessment of acute renal rejection *Proc. of 12th Annual Scientific Meeting of American Society for Diagnostics and Interventional Nephrology (ASDIN'16) (Phoenix, AZ, USA, 19–21 February 2016)*

[94] Eltanboly A, Ghazal M, Hajjdiab H, Shalaby A, Switala A, Mahmoud A, Sahoo P, El-Azab M and El-Baz A 2019 Level sets-based image segmentation approach using statistical shape priors *Appl. Math. Comput.* **340** 164–79

[95] Shehata M, Mahmoud A, Soliman A, Khalifa F, Ghazal M, El-Ghar M A, El-Melegy M and El-Baz A 2018 3D kidney segmentation from abdominal diffusion MRI using an appearance-guided deformable boundary *PLoS One* **13** e0200082

[96] Khalifa F, Beache G, El-Baz A and Gimel'farb G 2010 Deformable model guided by stochastic speed with application in cine images segmentation *Proc. of IEEE Int. Conf. on Image Processing, (ICIP'10) (Hong Kong, 26–29 September 2010)* pp 1725–8

[97] Khalifa F, Beache G M, Elnakib A, Sliman H, Gimel'farb G, Welch K C and El-Baz A 2013 A new shape-based framework for the left ventricle wall segmentation from cardiac first-pass perfusion MRI *Proc. of IEEE Int. Symp. on Biomedical Imaging: From Nano to Macro, (ISBI'13) (San Francisco, CA, 7–11 April 2013)* pp 41–4

[98] Khalifa F, Beache G M, Elnakib A, Sliman H, Gimel'farb G, Welch K C and El-Baz A 2012 A new nonrigid registration framework for improved visualization of transmural perfusion gradients on cardiac first-pass perfusion MRI *Proc. of IEEE Int. Symp. on Biomedical Imaging: From Nano to Macro, (ISBI'12) (Barcelona, Spain, 2–5 May 2012)* pp 828–31

[99] Khalifa F, Beache G M, Firjani A, Welch K C, Gimel'farb G and El-Baz A 2012 A new nonrigid registration approach for motion correction of cardiac first-pass perfusion MRI *Proc. of IEEE Int. Conf. on Image Processing, (ICIP'12) (Lake Buena Vista, FL, 30 September–3 October 2012)* pp 1665–8

[100] Khalifa F, Beache G M, Gimel'farb G and El-Baz A 2012 A novel CAD system for analyzing cardiac first-pass MR images *Proc. of IAPR Int. Conf. on Pattern Recognition (ICPR'12) (Tsukuba Science City, Japan, 11–15 November 2012)* pp 77–80

[101] Khalifa F, Beache G M, Gimel'farb G and El-Baz A 2011 A novel approach for accurate estimation of left ventricle global indexes from short-axis cine MRI *Proc. of IEEE Int. Conf. on Image Processing, (ICIP'11) (Brussels, Belgium, 11–14 September 2011)* pp 2645–9

[102] Khalifa F, Beache G M, Gimel'farb G, Giridharan G A and El-Baz A 2011 A new image-based framework for analyzing cine images *Handbook of Multi Modality State-of-the-Art Medical Image Segmentation and Registration Methodologies* vol 2 ed A El-Baz, U R Acharya, M Mirmedhdi and J S Suri (New York: Springer) ch 3 69–98

[103] Khalifa F, Beache G M, Gimel'farb G, Giridharan G A and El-Baz A 2012 Accurate automatic analysis of cardiac cine images *IEEE Trans. Biomed. Eng.* **59** 445–55

[104] Khalifa F, Beache G M, Nitzken M, Gimel'farb G, Giridharan G A and El-Baz A 2011 Automatic analysis of left ventricle wall thickness using short-axis cine CMR images *Proc. of IEEE Int. Symp. on Biomedical Imaging: From Nano to Macro, (ISBI'11) (Chicago, IL, 30 March–2 April 2011)* pp 1306–9

[105] Nitzken M, Beache G, Elnakib A, Khalifa F, Gimel'farb G and El-Baz A 2012 Accurate modeling of tagged CMR 3D image appearance characteristics to improve cardiac cycle strain estimation *2012 19th IEEE Int. Conf. on Image Processing (ICIP)* (Piscataway, NJ: IEEE) pp 521–4

[106] Nitzken M, Beache G, Elnakib A, Khalifa F, Gimel'farb G and El-Baz A 2012 Improving full-cardiac cycle strain estimation from tagged CMR by accurate modeling of 3D image appearance characteristics *2012 9th IEEE Int. Symp. on Biomedical Imaging (ISBI)* (Piscataway, NJ: IEEE) pp 462–5

[107] Nitzken M J, El-Baz A S and Beache G M 2012 Markov–Gibbs random field model for improved full-cardiac cycle strain estimation from tagged CMR *J. Cardiovasc. Magn. Reson.* **14** 1–2

[108] Sliman H, Elnakib A, Beache G, Elmaghraby A and El-Baz A 2014 Assessment of myocardial function from cine cardiac MRI using a novel 4D tracking approach *J. Comput. Sci. Syst. Biol.* **7** 169–73

[109] Sliman H, Elnakib A, Beache G M, Soliman A, Khalifa F, Gimel'farb G, Elmaghraby A and El-Baz A 2014 A novel 4D PDE-based approach for accurate assessment of myocardium function using cine cardiac magnetic resonance images *Proc. of IEEE Int. Conf. on Image Processing (ICIP'14) (Paris, France, 27–30 October 2014)* pp 3537–41

[110] Sliman H, Khalifa F, Elnakib A, Beache G M, Elmaghraby A and El-Baz A 2013 A new segmentation-based tracking framework for extracting the left ventricle cavity from cine

cardiac MRI *Proc. of IEEE Int. Conf. on Image Processing, (ICIP'13) (Melbourne, Australia, 15–18 September 2013)* pp 685–9

[111] Sliman H, Khalifa F, Elnakib A, Soliman A, Beache G M, Elmaghraby A, Gimel'farb G and El-Baz A 2013 Myocardial borders segmentation from cine MR images using bi-directional coupled parametric deformable models *Med. Phys.* **40** 1–13

[112] Sliman H, Khalifa F, Elnakib A, Soliman A, Beache G M, Gimel'farb G, Emam A, Elmaghraby A and El-Baz A 2013 Accurate segmentation framework for the left ventricle wall from cardiac cine MRI *Proc. of Int. Symp. on Computational Models for Life Science, (CMLS'13) (Sydney, Australia, 27–29 November 2013)* **vol 1559** pp 287–96

[113] Eladawi N, Elmogy M, Ghazal M, Helmy O, Aboelfetouh A, Riad A, Schaal S and El-Baz A 2018 Classification of retinal diseases based on OCT images *Front. Biosci.* **23** 247–64

[114] ElTanboly A, Ismail M, Shalaby A, Switala A, El-Baz A, Schaal S, Gimel'farb G and El-Azab M 2017 A computer-aided diagnostic system for detecting diabetic retinopathy in optical coherence tomography images *Med. Phys.* **44** 914–23

[115] Sandhu H S, El-Baz A and Seddon J M 2018 Progress in automated deep learning for macular degeneration *JAMA Ophthalmol.* **136** 1366–7

[116] Abdollahi B, Civelek A C, Li X-F, Suri J and El-Baz A 2014 PET/CT nodule segmentation and diagnosis: a survey *Multi Detector CT Imaging* ed L Saba and J S Suri (London: Taylor and Francis) ch 30, pp 639–51

[117] Abdollahi B, El-Baz A and Amini A A 2011 A multi-scale non-linear vessel enhancement technique *2011 Annu. Int. Conf. of the IEEE Engineering in Medicine and Biology Society, EMBC* (Piscataway, NJ: IEEE) pp 3925–9

[118] Abdollahi B, Soliman A, Civelek A, Li X-F, Gimel'farb G and El-Baz A 2012 A novel Gaussian scale space-based joint MGRF framework for precise lung segmentation *Proc. of IEEE Int. Conf. on Image Processing, (ICIP'12)* (Piscataway, NJ: IEEE) pp 2029–32

[119] Abdollahi B, Soliman A, Civelek A, Li X-F, Gimel'farb G and El-Baz A A novel 3D joint MGRF framework for precise lung segmentation *Machine Learning in Medical Imaging* (Berlin: Springer) pp 86–93

[120] Ali A M, El-Baz A S and Farag A A 2007 A novel framework for accurate lung segmentation using graph cuts *Proc. of IEEE Int. Symp. on Biomedical Imaging: From Nano to Macro, (ISBI'07)* (Piscataway, NJ: IEEE) pp 908–11

[121] El-Baz A, Beache G M, Gimel'farb G, Suzuki K and Okada K 2013 Lung imaging data analysis *Int. J. Biomed. Imaging* **2013** 1–2

[122] El-Baz A, Beache G M, Gimel'farb G, Suzuki K, Okada K, Elnakib A, Soliman A and Abdollahi B 2013 Computer-aided diagnosis systems for lung cancer: challenges and methodologies *Int. J. Biomed. Imaging* **2013** 1–46

[123] El-Baz A, Elnakib A, Abou El-Ghar M, Gimel'farb G, Falk R and Farag A 2013 Automatic detection of 2D and 3D lung nodules in chest spiral CT scans *Int. J. Biomed. Imaging* **2013** 1–11

[124] El-Baz A, Farag A A, Falk R and La Rocca R 2003 A unified approach for detection, visualization, and identification of lung abnormalities in chest spiral CT scans *Int. Congress Series* vol 1256 (Amsterdam: Elsevier) pp 998–1004

[125] El-Baz A, Farag A A, Falk R and La Rocca R 2002 Detection, visualization and identification of lung abnormalities in chest spiral CT scan: Phase-I *Proc. of Int. Conf. on Biomedical Engineering Cairo, Egypt,* vol 12

[126] El-Baz A, Farag A, Gimel'farb G, Falk R, El-Ghar M A and Eldiasty T 2006 A framework for automatic segmentation of lung nodules from low dose chest CT scans *Proc. of Int. Conf. on Pattern Recognition, (ICPR'06)* **vol 3** (Piscataway, NJ: IEEE) pp 611–4

[127] El-Baz A, Farag A, Gimel'farb G, Falk R and El-Ghar M A 2011 A novel level set-based computer-aided detection system for automatic detection of lung nodules in low dose chest computed tomography scans *Lung Imaging and Computer Aided Diagnosis* (London: Taylor and Francis) ch 10, pp 221–38

[128] El-Baz A, Gimel'farb G, Abou El-Ghar M and Falk R 2012 Appearance-based diagnostic system for early assessment of malignant lung nodules *Proc. of IEEE Int. Conf. on Image Processing, (ICIP'12)* (Piscataway, NJ: IEEE) pp 533–6

[129] El-Baz A, Gimel'farb G and Falk R 2016 A novel 3D framework for automatic lung segmentation from low dose CT images *Lung Imaging and Computer Aided Diagnosis* ed A El-Baz and J S Suri (London: Taylor and Francis) ch 1 pp 1–16

[130] El-Baz A, Gimel'farb G, Falk R and El-Ghar M 2010 Appearance analysis for diagnosing malignant lung nodules *Proc. of IEEE Int. Symp. on Biomedical Imaging: From Nano to Macro (ISBI'10)* (Piscataway, NJ: IEEE) pp 193–6

[131] El-Baz A, Gimel'farb G, Falk R and El-Ghar M A 2011 A novel level set-based CAD system for automatic detection of lung nodules in low dose chest CT scans *Lung Imaging and Computer Aided Diagnosis* vol 1 ed A El-Baz and J S Suri (London: Taylor and Francis) pp 221–38

[132] El-Baz A, Gimel'farb G, Falk R and El-Ghar M A 2008 A new approach for automatic analysis of 3D low dose CT images for accurate monitoring the detected lung nodules *Proc. of Int. Conf. on Pattern Recognition, (ICPR'08)* (Piscataway, NJ: IEEE) pp 1–4

[133] El-Baz A, Gimel'farb G, Falk R and El-Ghar M A 2007 A novel approach for automatic follow-up of detected lung nodules *Proc. of IEEE Int. Conf. on Image Processing, (ICIP'07)* **vol 5** (Piscataway, NJ: IEEE) pp V-501

[134] El-Baz A, Gimel'farb G, Falk R and El-Ghar M A 2007 A new CAD system for early diagnosis of detected lung nodules *ICIP 2007. 2007 IEEE Int. Conf. on Image Processing* **vol 2** (Piscataway, NJ: IEEE) pp II-461

[135] El-Baz A, Gimel'farb G, Falk R, El-Ghar M A and Refaie H 2008 Promising results for early diagnosis of lung cancer *Proc. of IEEE Int. Symp. on Biomedical Imaging: From Nano to Macro, (ISBI'08)* (Piscataway, NJ: IEEE) pp 1151–4

[136] El-Baz A, Gimel'farb G L, Falk R, Abou El-Ghar M, Holland T and Shaffer T 2008 A new stochastic framework for accurate lung segmentation *Proc. of Medical Image Computing and Computer-Assisted Intervention, (MICCAI'08)* pp 322–30

[137] El-Baz A, Gimel'farb G L, Falk R, Heredis D and Abou El-Ghar M 2008 A novel approach for accurate estimation of the growth rate of the detected lung nodules *Proc. of Int. Workshop on Pulmonary Image Analysis* pp 33–42

[138] El-Baz A, Gimel'farb G L, Falk R, Holland T and Shaffer T 2008 A framework for unsupervised segmentation of lung tissues from low dose computed tomography images *Proc. of British Machine Vision, (BMVC'08)* pp 1–10

[139] El-Baz A, Gimel'farb G, Falk R and El-Ghar M A 2011 3D MGRF-based appearance modeling for robust segmentation of pulmonary nodules in 3D LDCT chest images *Lung Imaging and Computer Aided Diagnosis* (London: Taylor and Francis) ch 3, pp 51–63

[140] El-Baz A, Gimel'farb G, Falk R and El-Ghar M A 2009 Automatic analysis of 3D low dose CT images for early diagnosis of lung cancer *Pattern Recogn* **42** 1041–51

[141] El-Baz A, Gimel'farb G, Falk R, El-Ghar M A, Rainey S, Heredia D and Shaffer T 2009 Toward early diagnosis of lung cancer *Proc. of Medical Image Computing and Computer-Assisted Intervention, (MICCAI'09)* (Berlin: Springer) pp 682–9

[142] El-Baz A, Gimel'farb G, Falk R, El-Ghar M A and Suri J 2011 Appearance analysis for the early assessment of detected lung nodules *Lung Imaging and Computer Aided Diagnosis* (London: Taylor and Francis) ch 17, pp 395–404

[143] El-Baz A, Khalifa F, Elnakib A, Nitkzen M, Soliman A, McClure P, Gimel'farb G and El-Ghar M A 2012 A novel approach for global lung registration using 3D Markov Gibbs appearance model *Proc. of Int. Conf. Medical Image Computing and Computer-Assisted Intervention, (MICCAI'12) (Nice, France, 1–5 October 2012)* pp 114–21

[144] El-Baz A, Nitzken M, Elnakib A, Khalifa F, Gimel'farb G, Falk R and El-Ghar M A 2011 3D shape analysis for early diagnosis of malignant lung nodules *Proc. of Int. Conf. Medical Image Computing and Computer-Assisted Intervention, (MICCAI'11) (Toronto, Canada, 18–22 September 2011)* pp 175–82

[145] El-Baz A, Nitzken M, Gimel'farb G, Van Bogaert E, Falk R, El-Ghar M A and Suri J 2011 Three-dimensional shape analysis using spherical harmonics for early assessment of detected lung nodules *Lung Imaging and Computer Aided Diagnosis* (London: Taylor and Francis) ch 19, pp 421–38

[146] El-Baz A, Nitzken M, Khalifa F, Elnakib A, Gimel'farb G, Falk R and El-Ghar M A 2011 3D shape analysis for early diagnosis of malignant lung nodules *Proc. of Int. Conf. on Information Processing in Medical Imaging, (IPMI'11) (Monastery Irsee, Germany, 3–8 July 2011)* pp 772–83

[147] El-Baz A, Nitzken M, Vanbogaert E, Gimel'Farb G, Falk R and Abo El-Ghar M 2011 A novel shape-based diagnostic approach for early diagnosis of lung nodules *2011 IEEE Int. Symp. on Biomedical Imaging: From Nano to Macro* (Piscataway, NJ: IEEE) pp 137–40

[148] El-Baz A, Sethu P, Gimel'farb G, Khalifa F, Elnakib A, Falk R and El-Ghar M A 2011 Elastic phantoms generated by microfluidics technology: validation of an imaged-based approach for accurate measurement of the growth rate of lung nodules *Biotechnol. J.* **6** 195–203

[149] El-Baz A, Sethu P, Gimel'farb G, Khalifa F, Elnakib A, Falk R and El-Ghar M A 2010 A new validation approach for the growth rate measurement using elastic phantoms generated by state-of-the-art microfluidics technology *Proc. of IEEE Int. Conf. on Image Processing, (ICIP'10) (Hong Kong, 26–29 September 2010)* 4381–3

[150] El-Baz A, Sethu P, Gimel'farb G, Khalifa F, Elnakib A, Falk R and Suri M A E-G J 2011 Validation of a new imaged-based approach for the accurate estimating of the growth rate of detected lung nodules using real CT images and elastic phantoms generated by state-of-the-art microfluidics technology *Handbook of Lung Imaging and Computer Aided Diagnosis* vol 1 ed A El-Baz and J S Suri (New York: Taylor and Francis) ch 18 405–20

[151] El-Baz A, Soliman A, McClure P, Gimel'farb G, El-Ghar M A and Falk R 2012 Early assessment of malignant lung nodules based on the spatial analysis of detected lung nodules *Proc. of IEEE Int. Symp. on Biomedical Imaging: From Nano to Macro, (ISBI'12)* (Piscataway, NJ: IEEE) pp 1463–6

[152] El-Baz A, Yuksel S E, Elshazly S and Farag A A 2005 Non-rigid registration techniques for automatic follow-up of lung nodules *Proc. of Computer Assisted Radiology and Surgery, (CARS'05)* **vol 1281** (Amsterdam: Elsevier) pp 1115–20

[153] El-Baz A S and Suri J S 2011 *Lung Imaging and Computer Aided Diagnosis* (Boca Raton, FL: CRC Press)

[154] Soliman A, Khalifa F, Dunlap N, Wang B, El-Ghar M and El-Baz A 2016 An iso-surfaces based local deformation handling framework of lung tissues *2016 IEEE 13th Int. Symp. on Biomedical Imaging (ISBI)* (Piscataway, NJ: IEEE) pp 1253–9

[155] Soliman A, Khalifa F, Shaffie A, Dunlap N, Wang B, Elmaghraby A and El-Baz A 2016 Detection of lung injury using 4D-CT chest images *2016 IEEE 13th Int. Symp. on Biomedical Imaging (ISBI)* pp 1274–7

[156] Soliman A, Khalifa F, Shaffie A, Dunlap N, Wang B, Elmaghraby A, Gimel'farb G, Ghazal M and El-Baz A 2017 A comprehensive framework for early assessment of lung injury *2017 IEEE Int. Conf. on Image Processing (ICIP)* (Piscataway, NJ: IEEE) pp 3275–9

[157] Shaffie A, Soliman A, Ghazal M, Taher F, Dunlap N, Wang B, Elmaghraby A, Gimel'farb G and El-Baz A 2017 A new framework for incorporating appearance and shape features of lung nodules for precise diagnosis of lung cancer 2017 IEEE *Int. Conf. on Image Processing (ICIP)* (Piscataway, NJ: IEEE) pp 1372–6

[158] Soliman A, Khalifa F, Shaffie A, Liu N, Dunlap N, Wang B, Elmaghraby A, Gimel'farb G and El-Baz A 2016 Image-based CAD system for accurate identification of lung injury *2016 IEEE Int. Conf. on Image Processing (ICIP)* (Piscataway, NJ: IEEE) pp 121–5

[159] Dombroski B, Nitzken M, Elnakib A, Khalifa F, El-Baz A and Casanova M F 2014 Cortical surface complexity in a population-based normative sample *Transl. Neurosci.* **5** pp 17–24

[160] El-Baz A, Casanova M, Gimel'farb G, Mott M and Switala A 2008 An MRI-based diagnostic framework for early diagnosis of dyslexia *Int. J. Comput. Assist. Radiol. Surg.* **3** 181–9

[161] El-Baz A, Casanova M, Gimel'farb G, Mott M, Switala A, Vanbogaert E and McCracken R 2008 A new CAD system for early diagnosis of dyslexic brains *Proc. Int. Conf. on Image Processing (ICIP'2008)* (Piscataway, NJ: IEEE) pp 1820–3

[162] El-Baz A, Casanova M F, Gimel'farb G, Mott M and Switwala A E 2007 A new image analysis approach for automatic classification of autistic brains *Proc. IEEE Int. Symp. on Biomedical Imaging: From Nano to Macro (ISBI'2007)* (Piscataway, NJ: IEEE) pp 352–5

[163] El-Baz A, Elnakib A, Khalifa F, El-Ghar M A, McClure P, Soliman A and Gimel'farb G 2012 Precise segmentation of 3-D magnetic resonance angiography *IEEE Trans. Biomed. Eng.* **59** 2019–29

[164] El-Baz A, Farag A, Gimel'farb G, El-Ghar M A and Eldiasty T 2006 Probabilistic modeling of blood vessels for segmenting MRA images *18th Int. Conf. on Pattern Recognition (ICPR'06)* **vol 3** (Piscataway, NJ: IEEE) pp 917–20

[165] El-Baz A, Farag A A, Gimel'farb G, El-Ghar M A and Eldiasty T 2006 A new adaptive probabilistic model of blood vessels for segmenting MRA images *Medical Image Computing and Computer-Assisted Intervention-MICCAI 2006* vol 4191 (Berlin: Springer) pp 799–806

[166] El-Baz A, Farag A A, Gimel'farb G and Hushek S G 2005 Automatic cerebrovascular segmentation by accurate probabilistic modeling of TOF-MRA images *Medical Image Computing and Computer-Assisted Intervention-MICCAI 2005* (Berlin: Springer) pp 34–42

[167] El-Baz A, Farag A, Elnakib A, Casanova M F, Gimel'farb G, Switala A E, Jordan D and Rainey S 2011 Accurate automated detection of autism related corpus callosum abnormalities *J. Med. Syst.* **35** 929–39

[168] El-Baz A, Farag A and Gimelfarb G 2005 Cerebrovascular segmentation by accurate probabilistic modeling of TOF-MRA images *Image Analysis* vol 3540 (Berlin: Springer) pp 1128–37

[169] El-Baz A, Gimel'farb G, Falk R, El-Ghar M A, Kumar V and Heredia D 2009 A novel 3D joint Markov–Gibbs model for extracting blood vessels from PC-MRA images *Medical Image Computing and Computer-Assisted Intervention-MICCAI 2009* vol 5762 (Berlin: Springer) pp 943–50

[170] Elnakib A, El-Baz A, Casanova M F, Gimel'farb G and Switala A E 2010 Image-based detection of corpus callosum variability for more accurate discrimination between dyslexic and normal brains *Proc IEEE Int. Symp. on Biomedical Imaging: From Nano to Macro (ISBI'2010)* (Piscataway, NJ: IEEE) pp 109–12

[171] Elnakib A, Casanova M F, Gimel'farb G, Switala A E and El-Baz A 2011 Autism diagnostics by centerline-based shape analysis of the corpus callosum *Proc IEEE Int. Symp. on Biomedical Imaging: From Nano to Macro (ISBI'2011)* (Piscataway, NJ: IEEE) pp 1843–6

[172] Elnakib A, Nitzken M, Casanova M, Park H, Gimel'farb G and El-Baz A 2012 Quantification of age-related brain cortex change using 3D shape analysis *2012 21st Int. Conf. on Pattern Recognition (ICPR)* (Piscataway, NJ: IEEE) pp 41–4

[173] Mostapha M, Soliman A, Khalifa F, Elnakib A, Alansary A, Nitzken M, Casanova M F and El-Baz A 2014 A statistical framework for the classification of infant DT images *2014 IEEE Int. Conf. on Image Processing (ICIP)* (Piscataway, NJ: IEEE) pp 2222–6

[174] Nitzken M, Casanova M, Gimel'farb G, Elnakib A, Khalifa F, Switala A and El-Baz A 2011 3D shape analysis of the brain cortex with application to dyslexia *2011 18th IEEE Int. Conf. on Image Processing (ICIP)* (Piscataway, NJ: IEEE) pp 2657–60

[175] El-Gamal F E-Z A, Elmogy M M, Ghazal M, Atwan A, Barnes G N, Casanova M F, Keynton R and El-Baz A S 2017 A novel CAD system for local and global early diagnosis of Alzheimer's disease based on PiB-PET scans *2017 IEEE Int. Conf. on Image Processing (ICIP)* (Piscataway, NJ: IEEE) pp 3270–4

[176] Ismail M, Soliman A, Ghazal M, Switala A E, Gimel'farb G, Barnes G N, Khalil A and El-Baz A 2017 A fast stochastic framework for automatic MR brain images segmentation *PloS one* **12** e0187391

[177] Ismail M M, Keynton R S, Mostapha M M, ElTanboly A H, Casanova M F, Gimel'farb G L and El-Baz A 2016 Studying autism spectrum disorder with structural and diffusion magnetic resonance imaging: a survey *Front. Hum. Neurosc.* **10** 211

[178] Alansary A *et al* 2016 Infant brain extraction in T1-weighted MR images using BET and refinement using LCDG and MGRF models *IEEE J. Biomed. Health Inform.* **20** 925–35

[179] Ismail M, Soliman A, ElTanboly A, Switala A, Mahmoud M, Khalifa F, Gimel'farb G, Casanova M F, Keynton R and El-Baz A 2016 Detection of white matter abnormalities in MR brain images for diagnosis of autism in children *2016 IEEE 13th International Symposium on Biomedical Imaging (ISBI)* pp 6–9

[180] Ismail M, Mostapha M, Soliman A, Nitzken M, Khalifa F, Elnakib A, Gimel'farb G, Casanova M and El-Baz A 2015 Segmentation of infant brain MR images based on

adaptive shape prior and higher-order MGRF *2015 IEEE Int. Conf. on Image Processing (ICIP)* pp 4327–31

[181] Asl E H, Ghazal M, Mahmoud A, Aslantas A, Shalaby A, Casanova M, Barnes G, Gimel'farb G, Keynton R and El-Baz A 2018 Alzheimer's disease diagnostics by a 3D deeply supervised adaptable convolutional network *Front. Biosci.* **23** 584–96

[182] Mahmoud A, El-Barkouky A, Farag H, Graham J and Farag A 2013 A non-invasive method for measuring blood flow rate in superficial veins from a single thermal image *Proc. of the IEEE Conf. on Computer Vision and Pattern Recognition Workshops* pp 354–9

[183] El-baz A, Shalaby A, Taher F, El-Baz M, Ghazal M, El-Ghar M A, Takieldeen A and Suri J 2017 Probabilistic modeling of blood vessels for segmenting magnetic resonance angiography images *Med. Res. Arch.* **5**

[184] Chowdhury A S, Rudra A K, Sen M, Elnakib A and El-Baz A 2010 Cerebral white matter segmentation from MRI using probabilistic graph cuts and geometric shape priors *ICIP* pp 3649–52

[185] Gebru Y, Giridharan G, Ghazal M, Mahmoud A, Shalaby A and El-Baz A 2018 Detection of cerebrovascular changes using magnetic resonance angiography *Cardiovascular Imaging and Image Analysis* (Boca Raton, FL: CRC Press) pp 1–22

[186] Mahmoud A, Shalaby A, Taher F, El-Baz M, Suri J S and El-Baz A 2018 Vascular tree segmentation from different image modalities *Cardiovascular Imaging and Image Analysis* (Boca Raton, FL: CRC Press) pp 43–70

[187] Taher F, Mahmoud A, Shalaby A and El-Baz A 2018 A review on the cerebrovascular segmentation methods *2018 IEEE Int. Symp. on Signal Processing and Information Technology (ISSPIT)* (Piscataway, NJ: IEEE) pp 359–64

[188] Kandil H, Soliman A, Fraiwan L, Shalaby A, Mahmoud A, ElTanboly A, Elmaghraby A, Giridharan G and El-Baz A 2018 A novel MRA framework based on integrated global and local analysis for accurate segmentation of the cerebral vascular system *2018 IEEE 15th Int. Symp. on Biomedical Imaging (ISBI 2018)* (Piscataway, NJ: IEEE) pp 1365–8

IOP Publishing

Neurological Disorders and Imaging Physics, Volume 3
Application to autism spectrum disorders and Alzheimer's
Ayman El-Baz and Jasjit S Suri

Chapter 17

Autism diagnosis using task-based functional MRI

Reem Haweel, Omar Dekhil, Ahmed Shalaby, Ali Mahmoud, Mohammed Ghazal, Hassan Hajjdiab, Said Ghniemy, Robert Keynton, Jasjit S Suri, Gregory N Barnes and Ayman El-Baz

Autism is a developmental disorder associated with difficulties in communication and social interaction and is defined over a wide spectrum. Currently, the gold standard in autism diagnosis is the autism diagnostic observation schedule (ADOS) interview. Alterations in functional activity are believed to be important in explaining the causative factors of autism. In this chapter, a novel framework for grading the severity level of autistic subjects using task-based functional MRI data is presented. A speech fMRI experiment is used to obtain local features related to the functional activity of the brain. According to ADOS reports, the adopted dataset is classified into three groups: mild, moderate, and severe. Our analysis is divided into two parts: (i) individual subject analysis and (ii) higher level group analysis. We use the individual analysis to extract the features used in classification, while the higher level analysis is used to infer the statistical differences between groups. The obtained classification accuracy is 72% and the group activation showed how informative our features are.

17.1 Introduction

Autism spectrum disorder (ASD) is a neuronal developmental disorder associated with a range of symptoms such as social, sensorimotor, and communicative deficits that differ in severity [1]. Autism is typically diagnosed at the age of three, however, some characteristics can sometimes be observed as early as 12 months old, in particular with the advent of medical imaging modalities and the recent state-of-the-art machine learning and deep learning algorithms [2]. Structural magnetic resonance imaging (sMRI), functional magnetic resonance imaging (fMRI), and

doi:10.1088/978-0-7503-1793-1ch17

diffusion tensor imaging (DTI) are widely adopted for analyzing the brain's structural and functional characteristics [3].

In general, task-based fMRI is used for assessing evoked blood oxygen level-dependence (BOLD) in all brain regions with the response to certain tasks belonging to a range of different task domains [4]. Basic fMRI tasks include motor tasks, visual processing tasks, auditory and language tasks, and basic social processing tasks [5].

Studies in the literature have statistically examined the hemodynamic effects associated with task-control using the general linear model (GLM) [6]. In recent publications, group analysis has been applied to address commonly activated brain areas of both ASD subjects and normal controls in response to auditory and language tasks. Studies in motor task experiments [7–9] showed that the left culmen and the right superior temporal gyrus have greater activation in neurotypicals while the inferior frontal gyri and the bilateral precentral gyri are more activated in subjects with autism. Muller *et al* [7] statistically examined the hemodynamic effects associated with task-control using the GLM. Less prefrontal activation in autistic adults and an enhanced activation in the premotor cortex and right pericentral were revealed in comparison to neurotypicals. The authors of [9] recorded the results of a simple hand movement task and analyzed the GLM modeled BOLD signal using the SPM package. The same areas, the inferior parietal, prefrontal, and superior temporal regions, were activated in both ASDs and control young adults. However, higher activation in the dorsal premotor cortex and less activation in the middle temporal gyrus and cuneus were recorded for ASD. fMRI results for visual processing tasks showed that the medial frontal gyrus and thalamus are more activated in autism, while in neurotypicals, the occipital region and the cingulate gyrus are more activated [7–12]. The study in [11] performed pursuit and saccadic eye movement and included the FIASCO package for motion artifact reduction and created F-statistic maps to record differences at each task. The maps were calculated by dividing voxel-wise chi-square values from within-group activation maps by the corresponding degrees of freedom that are appropriate for each group. Greater activation in the dorsolateral, caudate nucleus, prefrontal cortex, anterior and posterior cingulate cortex, and the medial thalamus were recorded for autistic individuals. Clery *et al* [12] used fMRI to evaluate the impact of unexpected visual changes. The BOLD signal was modeled using GLM and then analyzed using the SPM package. The frontal and occipital regions recorded similar changes for both groups, but more changes in the bilateral occipital cortex were exhibited.

Auditory and language tasks for adolescents and children in the literature have revealed greater activation in the right precentral gyrus region for ASD, while in neurotypicals the bilateral superior and temporal gyri were more activated [10–15]. In autistic adults, more activation was shown in the bilateral decline than in neurotypicals. In all age groups, neurotypicals experienced greater activation in the left cingulate gyrus. Gomot *et al* [14] used theSPM2 package to examine extreme repetitive behavior. They used a conical HRF for separate modeling of the events using its first-order temporal derivative. In autistic subjects, the left inferior parietal regions and the right prefrontal-premotor were more activated than in neuro-typicals. The authors of [15] made a study of three groups of typically developed toddlers, ASD toddlers of good outcome, and those of relatively lower language outcome by comparing language sensitive superior temporal cortices activity for each using the SPM package in analysis. The results showed the same activation in both groups of higher language in the language sensitive superior temporal cortices, whereas less activation was detected for those of lower language outcome.

Examples of basic fMRI social processing tasks are face processing, emotion processing, and motion in relation to social stimuli [13–15]. It is revealed in the literature that the bilateral superior temporal gyri exhibits more activation in ASD, whereas neurotypicals exhibit greater activation in the right inferior occipital gyrus and left fusiform gyrus. The study in [15] exposed subjects to some fearful faces and houses and compared their reactions. The BOLD signal was modeled using the FSL tool and GLM. It was shown that a slower response was exhibited by the autistic group.

Brain image modalities have been a powerful tool for autism diagnosis. Structural MRI and resting state fMRI are the commonly incorporated image modalities used for classification. Although task fMRI provides indicative information about brain functional impairment and biomarkers, only few attempts of classification using task fMRI are recorded in literature. In [16] the authors applied multivariate pattern analysis (MVPA) to two different fMRI experiments with social stimuli (faces and bodies). They used support vector machines (SVMs) and recursive feature elimination (RFE) for classification between 15 ASD and 14 control subjects and achieved an accuracy range between 69% and 92.3%. Another attempt in [17] demonstrated the effectiveness of different methods to train generalizable recurrent neural networks from small datasets to classify children with ASD ($N = 21$) versus typical control subjects ($N = 40$) from task fMRI scans. The classification accuracy ranged from 51.8% to 69.8%.

The focus of previous studies was to analyze or diagnose autism disorder in two groups, ASD or normal, which is not sufficient for addressing differences among autistic subjects across the wide autism spectrum. The goal of this chapter is to utilize brain processing and analysis tools as well as machine learning algorithms for developing more specialized objective computer aided diagnosis (CAD), hence providing a better diagnosis and treatment plan for each autistic subject individually, i.e. the concept of personalized medicine.

17.2 Materials and methods

Figure 17.1 shows the general framework of our proposed system for grading ASD subject severity level to mild, moderate, or severe using task-based fMRI images. The calibrated severity scores (CSS) for the toddler module, obtained from raw total domain scores of the autism diagnostic observation schedule (ADOS) [18], varied from 0 to 10. The CSS is divided into three classes: (i) mild (CSS 1–4); (ii) moderate (CSS 5–7); and (iii) severe (CSS 8–10). Three matched sets, one for each class, are included in this study, each with the size of 13 subjects. The task fMRI images were recorded in a speech experiment that includes four audio stimuli, complex forward speech, simple forward speech, backward speech, and silence, which alternate repeatedly over 6 min 20 s. Figure 17.2 illustrates the fMRI experiment design.

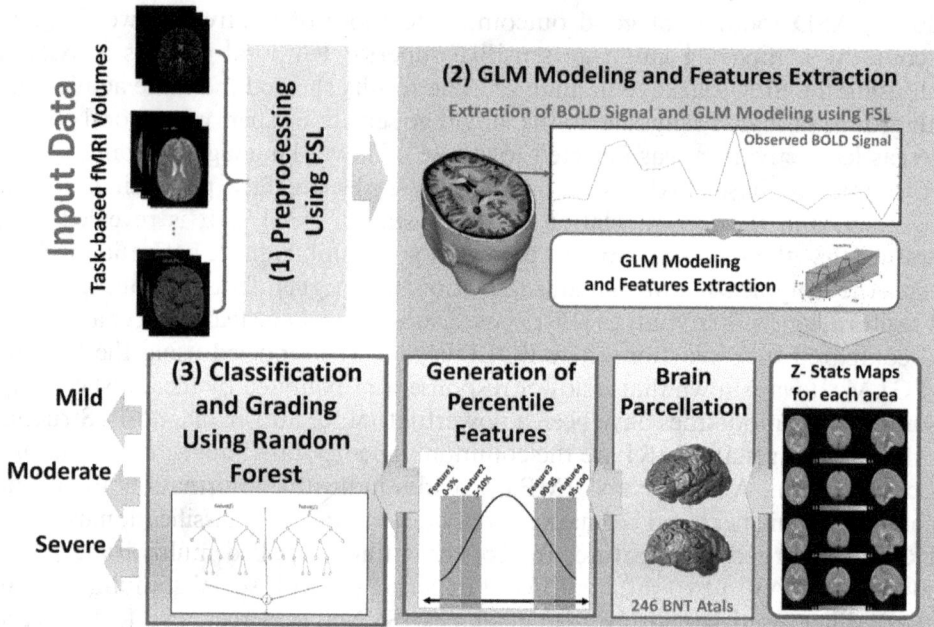

Figure 17.1. A block diagram of the proposed framework for grading autistic subject severity level to mild, moderate, or severe using task-based fMRI images.

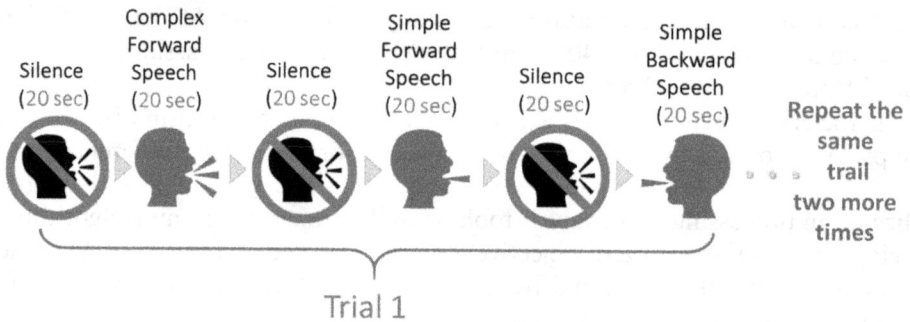

Figure 17.2. Illustration of the speech experiment that includes four audio stimuli: complex forward speech, simple forward speech, backward speech, and silence, which are alternatingly repeatedly over 6 min 20 s.

17.2.1 Data preprocessing

According to our framework for grading the ASD subject's severity level to mild, moderate, or severe in figure 17.1, first the preprocessing pipeline is performed using the FMRI Expert Analysis Tool (FEAT) [19] included in FMRIB's Software Library (FSL) [20]. The following steps are applied: (i) interleaved slice timing correction to correct for the effect of recording slices at different time points within the volume; (ii) motion correction to correct for the effect of subject motion in the scanner by applying rigid-body transformations with six degrees of freedom (DOF), the motion correction is done using MCFLIRT [21]; (iii) spatial smoothing using a

Gaussian window with full width at half maximum (FWHM) of 5 mm; (iv) high pass temporal filtering (100 s) to remove low frequency artifacts and scanner drifts; (v) brain extraction using BET; and (vi) two step registration. The first step is to register the functional volume to its high resolution anatomical scan, then register the anatomical scan to MNI-152 space with 12 DOF [22].

17.2.2 Multi level general linear model (GLM) in fMRI

Consider an experiment with N subjects, and for each subject K there is a vector of T time points, Y_k. The first level GLM is defined as

$$Y = X\beta + \epsilon, \tag{17.1}$$

where X_k is the design matrix, β_k are the parameter estimates, and ϵ_k is the error term (subject residuals). The residual is assumed to have zero mean $E(\epsilon_k) = 0$. Putting equation (17.1) in its matrix form it becomes

$$Y = \begin{bmatrix} X_1 & 0 & \cdots & 0 & 0 \\ 0 & X_2 & & \cdots & 0 \\ \vdots & \vdots & \ddots & \vdots & \vdots \\ 0 & 0 & & \cdots & X_N \end{bmatrix} * \begin{bmatrix} \beta_1 \\ \beta_2 \\ \vdots \\ \beta_N \end{bmatrix} + \begin{bmatrix} \epsilon_1 \\ \epsilon_2 \\ \vdots \\ \epsilon_N \end{bmatrix}. \tag{17.2}$$

By adding the second level of analysis on the groups level we have

$$\beta = X_G B_G + \eta, \tag{17.3}$$

where X_G is the group level design matrix, and it separates the two groups (controls and patients), β_G is the vector of the group level parameters, and η is the group level residuals. Also η is considered to have zero mean, $E(\eta) = 0$. Substituting with equation (17.3) in (17.1) we obtain

$$Y = XX_G \beta_G + \gamma, \tag{17.4}$$

where

$$\gamma = X\eta + \epsilon. \tag{17.5}$$

Also γ has zero mean, $E(\gamma) = 0$. The covariance of γ is $\mathrm{cov}(\gamma) = w = XV_G X^T + V$, where V is the covariance of ϵ and V_G is the covariance of η.

Using the general least squares approach [23], the first level parameters could be estimated as

$$\hat{\beta} = (X^T V^{-1} X) X^T V^{-1} Y \tag{17.6}$$

and

$$\mathrm{cov}(\hat{\beta}) = (X^T V^{-1} X)^{-1}. \tag{17.7}$$

Using the same approach but on the group level, the group parameters are given by

$$\hat{\beta}_G = (X_G^T V_G^{-1} X_G)^{-1} X^T V^{-1} Y \tag{17.8}$$

and

$$\text{cov}(\hat{\beta}_G) = (X_G^T V_G^{-1} X_G)^{-1}. \tag{17.9}$$

After estimating the parameters for each voxel, it is required to define some contrasts and check which voxels are significant with respect to these conditions. The most common techniques for testing the voxel significance are the paired t-test and paired z-test [24, 25]. The statistical analysis result could be expressed as a map of corrected P-values at each voxel. For more details about GLM parameter estimates, see [24].

As mentioned above, the speech task has four conditions: complex forward speech, simple forward speech, backward speech, and silence. In this study, we apply both the first level analysis and the higher level analysis to model these four regressors in GLM. The first level analysis is used to quantify the activation differences between the subjects for classification purposes, while the higher level analysis is applied to have more insightful analysis of the overall group differences.

17.2.3 Feature selection and classification

In this study, we utilize the output z-stats of each subject to extract features. The Brainnetome Atlas (BNT) [26] is applied to map the brain to 246 areas. A histogram with six bins is constructed to define the percent of z-stat intensity value within each interval, for each brain area. Since the z-stat map varied between 0 and 6, the used intervals are $(0 \leqslant |z| < 1, 1 \leqslant |z| < 2, 2 \leqslant |z| < 3, 3 \leqslant |z| < 4, 4 \leqslant |z| < 5, |z| \geqslant 5)$. This creates a feature vector of 246 areas by six features per subject. With the limited number of subjects (13 per group), there is a need to use a feature selection algorithm prior to classification to reduce the feature space dimensionality.

The algorithm used for feature selection is recursive feature elimination (RFE). This algorithm uses a random forest classifier to fit a model for the data and sort the feature importance, then starts removing the least features recursively [27]. To avoid overfitting over the small dataset, the RFE is run in a ten-fold cross validation experiment to extract the important features from the training folds and validate on the testing fold. The extracted features are then fed to a random forest classifier for classification.

17.3 Experimental results and conclusion

This study is conducted on the 'Biomarkers of Autism at 12 Months: From Brain Overgrowth to Genes' dataset obtained by the National Database for Autism Research (NDAR: http://ndar.nih.gov) that was collected starting in August 2007 and ending in June 2014 [22]. This dataset mainly contains autism diagnostic observation schedule (ADOS), genomics samples and subjects, sMRI (T1) and (T2), resting state fMRI, task-based fMRI, DTI and single-shell DTI data, with a varying number of included subjects for each scan. Six experiments were applied in this

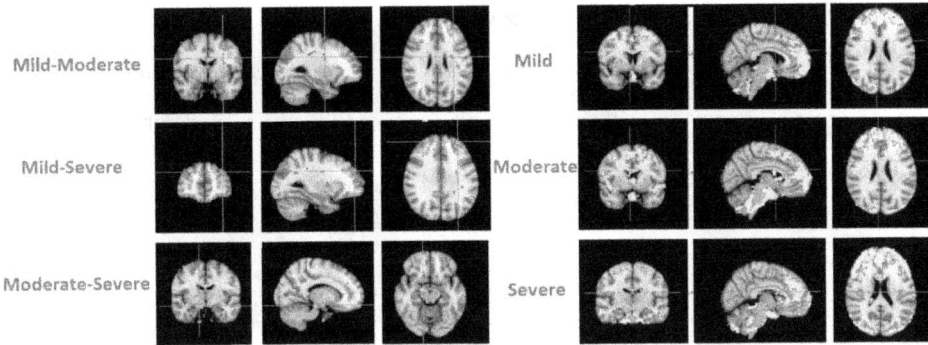

Figure 17.3. The activation pattern of mild, moderate, and severe (2nd column). Also, the significant differences in activation between three contrasts (1st column).

NDAR dataset: gene expression analysis, resting fMRI, speech fMRI, social orienting fMRI, word language fMRI, and emotion fMRI. This chapter analyzes the recorded fMRI scans for the speech experiment. The data selection criterion is to include the subjects that have the ADOS toddler module available. We have included 39 (13 mild, 13 moderate, and 13 severe) with ages ranging from 12 to 27 months (mean : 20.13 months and STD: 5.07 months). Each subject has (T1) structural MRI and speech fMRI scans that were acquired using a Signa HDxt, 1.5 T GE Healthcare scanner. Each fMRI scan consists of 154 volumes with TR = 2.5 s and TE = 30 ms. Each volume is constructed of 31 slices with an alternating in the plus direction slice acquisition pattern. Results are reported for both the statistical inference for each group and between all groups (mild, moderate, and severe) as well as the classification accuracies.

17.3.1 Higher level analysis

The higher level modeling is done using FLAME (FMRIB's Local Analysis of Mixed Effects). Group analysis reveals activation patterns between the three groups but with different significance levels. Figure 17.3 shows the activation pattern for each of the three groups. In addition it shows the significant differences in activation between them with three contrasts (mild > moderate, mild > severe, moderate > severe).

To show how the histogram features match the group activation, we extracted the mean percent of activated voxels out of the most significant histogram bins ($|z| > 3$) per group for each area. Table 17.1 shows the top five areas per group. The intersections between the top features and the group analysis is shown in figure 17.4. It is obvious that our top informative areas match the group analysis results. This is important to check the quality of feature representation. Moreover, such analysis gives insightful information about significant activation over brain areas rather than independent voxels as in GLM group analysis.

Table 17.1. The top five areas with the percent of significant voxels for each group.

| | Mild | Moderate | Severe | Percent of significant voxels | | |
				Mild	Moderate	Severe
First area	14	13	14	25.5	24.8	32.1
Second area	13	5	19	20.7	23.5	31.5
Third area	19	11	20	19.9	23.3	30.9
Forth area	20	19	13	18.5	23.1	28.6
Fifth area	32	20	24	14.5	22.1	24.7

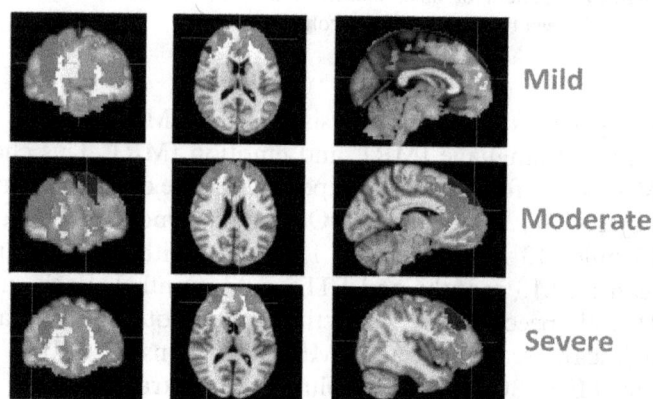

Figure 17.4. A cross section showing some of the top informative areas in terms of percent of significant voxels intersecting with the group level analysis results (yellow). This plan shows BNT atlas areas 5, 11, 13, 19, and 20.

Table 17.2. The classification confusion matrix of the three groups.

| | Predicted label | | |
	Mild	Moderate	Severe
True label	**9**	2	2
	1	**10**	2
	2	2	**9**

17.3.2 Classification results

In this experiment, we use the random forest classifier in both RFE and classification. We used a ten-fold cross validation technique and calculated the accuracy of classification. The achieved accuracy is 71.79%. In addition to calculating the accuracy, the confusion matrix between the three classes is also calculated. Table 17.2 shows the confusion matrix. The model hyperparameters for both

RFE and random forest classifier are both selected using a grid search. The optimal performance is obtained when using 11 trees with a maximum depth of 9 for RFE and 238 with maximum depth of 28 for random forest. Different classifiers other random forest were tested including: (i) linear SVM (accuracy = 54%) and (ii) SVM with RBF kernel (accuracy 59%) and neural network (accuracy = 64%). The random forest outperformed the other tested classifiers.

17.3.3 Conclusion and future work

In this study, we introduce a machine learning based approach for autism grading on the spectrum. To the best of our knowledge, this is the first effort to utilize task-based fMRI for this goal. With a limited number of subjects ($n = 39$), our algorithm achieved an accuracy of 72%. Our future work will focus mainly on integrating more data from different experiments to obtain more comprehensive understanding of how brain activation abnormalities in response to different tasks explain autism.

This work could also be applied to various other applications in medical imaging, such as the kidney, the heart, the prostate, the lung, and the retina, as well as several non-medical applications [28–31].

One application is renal transplant functional assessment, in particular in developing noninvasive CAD systems for renal transplant function assessment, utilizing different image modalities (e.g. ultrasound, computed tomography (CT), MRI, etc). Accurate assessment of renal transplant function is critically important for graft survival. Although transplantation can improve a patient's wellbeing, there is a potential post-transplantation risk of kidney dysfunction that, if not treated in a timely manner, can lead to the loss of the entire graft, and even patient death. In particular, dynamic and diffusion MRI-based systems have been clinically used to assess transplanted kidneys with the advantage of providing information on each kidney separately. For more details about renal transplant functional assessment, see [32–49, 49–59].

The heart is also an important application of this work. The clinical assessment of myocardial perfusion plays a major role in the diagnosis, management, and prognosis of ischemic heart disease patients. Thus, there have been ongoing efforts to develop automated systems for accurate analysis of myocardial perfusion using first-pass images [60–76].

Another application for this work could be the detection of retinal abnormalities. The majority of ophthalmologists depend on visual interpretation for the identification of diseases types. However, inaccurate diagnosis will affect the treatment procedure which may lead to fatal results. Hence, there is a crucial need for computer automated diagnosis systems that yield highly accurate results. Optical coherence tomography (OCT) has become a powerful modality for the noninvasive diagnosis of various retinal abnormalities such as glaucoma, diabetic macular edema, and macular degeneration. The problem with diabetic retinopathy (DR) is that the patient is not aware of the disease until the changes in the retina have progressed to a level that treatment tends to be less effective. Therefore, automated early detection could limit the severity of the disease and assist ophthalmologists in investigating and treating it more efficiently [77–79].

Abnormalities of the lung could also be another promising area of research and a related application for this work. Radiation-induced lung injury is the main side effect of radiation therapy for lung cancer patients. Although higher radiation doses increase the radiation therapy effectiveness for tumor control, this can lead to lung injury as a greater quantity of normal lung tissues is included in the treated area. Almost 1/3 of patients who undergo radiation therapy develop lung injury following radiation treatment. The severity of radiation-induced lung injury ranges from ground-glass opacities and consolidation at the early phase to fibrosis and traction bronchiectasis in the late phase. Early detection of lung injury will thus help to improve management of the treatment [80–122].

This work can also be applied to other brain abnormalities, such as dyslexia in addition to autism. Dyslexia is one of the most complicated developmental brain disorders that affect children's learning abilities. Dyslexia leads to the failure to develop age-appropriate reading skills in despite a normal intelligence level and adequate reading instruction. Neuropathological studies have revealed the abnormal anatomy of some structures, such as the corpus callosum, in dyslexic brains. There has been a lot of work published in the literature that aims to develop CAD systems for diagnosing such disorders, along with other brain disorders [123–144].

For the vascular system [145], this work could also be applied for the extraction of blood vessels, e.g. from phase contrast (PC) magnetic resonance angiography (MRA). Accurate cerebrovascular segmentation using noninvasive MRA is crucial for the early diagnosis and timely treatment of intracranial vascular diseases [128, 129, 146–151].

References

[1] Amaral D G, Schumann C M and Nordahl C W 2008 Neuroanatomy of autism *Trends Neurosci.* **31** 137–45

[2] Casanova M F *et al* 2017 *Autism Imaging and Devices* (Boca Raton, FL: CRC Press)

[3] Ismail M M, Keynton R S, Mostapha M M, El Tanboly A H, Casanova M F, Gimel'farb G L and El-Baz A 2016 Studying autism spectrum disorder with structural and diffusion magnetic resonance imaging: a survey *Front. Hum. Neurosci.* **10** 211

[4] Van Horn J D, Grethe J S, Kostelec P, Woodward J B, Aslam J A, Rus D, Rockmore D and Gazzaniga M S 2001 The Functional Magnetic Resonance Imaging Data Center (FMRIDC): the challenges and rewards of large–scale databasing of neuroimaging studies *Philos. Trans. R. Soc. Lond.* B **356** 39

[5] Casanova M F *et al* 2013 *Imaging the Brain in Autism* (Berlin: Springer)

[6] Friston K J, Holmes A P, Worsley K J, Poline J -P, Frith C D and Frackowiak R S 1994 Statistical parametric maps in functional imaging: a general linear approach *Hum. Brain Mapp.* **2** 189–210

[7] Müller R -A, Cauich C, Rubio M A, Mizuno A and Courchesne E 2004 Abnormal activity patterns in premotor cortex during sequence learning in autistic patients *Biol. Psychiatry* **56** 323–32

[8] Müller R -A, Kleinhans N, Kemmotsu N, Pierce K and Courchesne E 2003 Abnormal variability and distribution of functional maps in autism: an fMRI study of visuomotor learning *Am. J. Psychiatry* **160** pp 1847–62

[9] Perkins T J, Bittar R G, McGillivray J A, Cox I I and Stokes M A 2015 Increased premotor cortex activation in high functioning autism during action observation *J. Clin. Neurosci.* **22** 664–9

[10] Thakkar K N, Polli F E, Joseph R M, Tuch D S, Hadjikhani N, Barton J J and Manoach D S 2008 Response monitoring, repetitive behaviour and anterior cingulate abnormalities in autism spectrum disorders (ASD) *Brain* **131** 2464–78

[11] Takarae Y, Minshew N J, Luna B and Sweeney J A 2007 Atypical involvement of frontostriatal systems during sensorimotor control in autism *Psychiatry Res.: Neuroimaging* **156** 117–27

[12] Clery H, Andersson F, Bonnet-Brilhault F, Philippe A, Wicker B and Gomot M 2013 fMRI investigation of visual change detection in adults with autism *NeuroImage* **2** 303–12

[13] Gomot M, Bernard F A, Davis M H, Belmonte M K, Ashwin C, Bullmore E T and Baron-Cohen S 2006 Change detection in children with autism: an auditory event-related fMRI study *Neuroimage* **29** 475–84

[14] Gomot M, Belmonte M K, Bullmore E T, Bernard F A and Baron-Cohen S 2008 Brain hyper-reactivity to auditory novel targets in children with high-functioning autism *Brain* **131** 2479–88

[15] Lombardo M V, Pierce K, Eyler L T, Barnes C C, Ahrens-Barbeau C, Solso S, Campbell K and Courchesne E 2015 Different functional neural substrates for good and poor language outcome in autism *Neuron* **86** 567–77

[16] Chanel G, Pichon S, Conty L, Berthoz S, Chevallier C and Grèzes J 2016 Classification of autistic individuals and controls using cross-task characterization of fMRI activity *NeuroImage* **10** 78–88

[17] Dvornek N C, Yang D, Ventola P and Duncan J S 2018 Learning generalizable recurrent neural networks from small task-fMRI datasets *Int. Conf. on Medical Image Computing and Computer-Assisted Intervention* (Berlin: Springer) pp 329–37

[18] Gotham K, Pickles A and Lord C 2009 Standardizing ADOS scores for a measure of severity in autism spectrum disorders *J. Autism Dev. Disord.* **39** 693–705

[19] Woolrich M W, Ripley B D, Brady M and Smith S M 2001 Temporal autocorrelation in univariate linear modeling of fMRI data *Neuroimage* **14** 1370–86

[20] Jenkinson M, Beckmann C F, Behrens T E, Woolrich M W and Smith S M 2012 FSL *Neuroimage* **62** 782–90

[21] Jenkinson M, Bannister P, Brady M and Smith S 2002 Improved optimization for the robust and accurate linear registration and motion correction of brain images *Neuroimage* **17** 825–41

[22] Lancaster J L, Tordesillas-Gutirrez D, Martinez M, Salinas F, Evans A, Zilles K, Mazziotta J C and Fox P T 2007 Bias between MNI and Talairach coordinates analyzed using the ICBM-152 brain template *Hum. Brain Mapp.* **28** 1194–205

[23] Searle S R, Casella G and McCulloch C E 2009 *Variance Components* vol 391 (New York: Wiley)

[24] Beckmann C F, Jenkinson M and Smith S M 2003 General multilevel linear modeling for group analysis in fMRI *Neuroimage* **20** 1052–63

[25] Zang Y, Jiang T, Lu Y, He Y and Tian L 2004 Regional homogeneity approach to fMRI data analysis *Neuroimage* **22** 394–400

[26] Fan L *et al* 2016 The Human Brainnetome Atlas: a new brain atlas based on connectional architecture *Cereb. Cortex* **26** 3508–26

[27] Granitto P M, Furlanello C, Biasioli F and Gasperi F 2006 Recursive feature elimination with random forest for PTR-MA analysis of agroindustrial products *Chemometr. Intell. Lab. Syst.* **83** 83–90

[28] Mahmoud A H 2014 Utilizing radiation for smart robotic applications using visible, thermal, and polarization images *PhD Dissertation* University of Louisville

[29] Mahmoud A, El-Barkouky A, Graham J and Farag A 2014 Pedestrian detection using mixed partial derivative based histogram of oriented gradients *2014 IEEE Int. Conf. on Image Processing (ICIP)* (Piscataway, NJ: IEEE) pp 2334–7

[30] El-Barkouky A, Mahmoud A, Graham J and Farag A 2013 An interactive educational drawing system using a humanoid robot and light polarization *2013 IEEE Int. Conf. on Image Processing* (Piscataway, NJ: IEEE) pp 3407–11

[31] Mahmoud A H, El-Melegy M T and Farag A A 2012 Direct method for shape recovery from polarization and shading *2012 19th IEEE Int. Conf. on Image Processing* (Piscataway, NJ: IEEE) pp 1769–72

[32] Ali A M, Farag A A and El-Baz A 2007 Graph cuts framework for kidney segmentation with prior shape constraints *Proc. of Int. Conf. on Medical Image Computing and Computer-Assisted Intervention, (MICCAI'07) (Brisbane, Australia, 29 October–2 November 2007)* **vol 1** pp 384–92

[33] Chowdhury A S, Roy R, Bose S, Elnakib F K A and El-Baz A 2012 Non-rigid biomedical image registration using graph cuts with a novel data term *Proc. of IEEE Int. Symp. on Biomedical Imaging: From Nano to Macro, (ISBI'12) (Barcelona, Spain, 2–5 May 2012)* pp 446–9

[34] El-Baz A, Farag A A, Yuksel S E, El-Ghar M E, Eldiasty T A and Ghoneim M A 2007 Application of deformable models for the detection of acute renal rejection *Deformable Models* (New York: Springer) pp 293–333

[35] El-Baz A, Farag A, Fahmi R, Yuksel S, El-Ghar M A and Eldiasty T 2006 Image analysis of renal DCE MRI for the detection of acute renal rejection *Proc. of IAPR Int. Conf. on Pattern Recognition (ICPR'06) (Hong Kong, 20–24 August 2006)* pp 822–5

[36] El-Baz A, Farag A, Fahmi R, Yuksel S, Miller W, El-Ghar M A, El-Diasty T and Ghoneim M 2006 A new CAD system for the evaluation of kidney diseases using DCE-MRI *Proc. of Int. Conf. on Medical Image Computing and Computer-Assisted Intervention, (MICCAI'08) (Copenhagen, Denmark, 1–6 October 2006)* pp 446–53

[37] El-Baz A, Gimel'farb G and El-Ghar M A 2008 A novel image analysis approach for accurate identification of acute renal rejection *Proc. of IEEE Int. Conf. on Image Processing, (ICIP'08) (San Diego, CA, USA, 12–15 October 2008)* pp 1812–5

[38] El-Baz A, Gimel'farb G and El-Ghar M A 2008 Image analysis approach for identification of renal transplant rejection *Proc. of IAPR Int. Conf. on Pattern Recognition, (ICPR'08) (Tampa, Florida, USA, 8–11 December 2008)* pp 1–4

[39] El-Baz A, Gimel'farb G and El-Ghar M A 2007 New motion correction models for automatic identification of renal transplant rejection *Proc. of Int. Conf. on Medical Image Computing and Computer-Assisted Intervention, (MICCAI'07) (Brisbane, Australia, 29 October–2 November 2007)* pp 235–43

[40] Farag A, El-Baz A, Yuksel S, El-Ghar M A and Eldiasty T 2006 A framework for the detection of acute rejection with dynamic contrast enhanced magnetic resonance imaging *Proc. of IEEE Int. Symp. on Biomedical Imaging: From Nano to Macro, (ISBI'06) (Arlington, VA, USA, 6–9 April 2006)* pp 418–21

[41] Khalifa F, Beache G M, El-Ghar M A, El-Diasty T, Gimel'farb G, Kong M and El-Baz A 2013 Dynamic contrast-enhanced MRI-based early detection of acute renal transplant rejection *IEEE Trans. Med. Imaging* **32** 1910–27

[42] Khalifa F, El-Baz A, Gimel'farb G and El-Ghar M A 2010 Non-invasive image-based approach for early detection of acute renal rejection *Proc. of Int. Conf. Medical Image Computing and Computer-Assisted Intervention, (MICCAI'10) (Beijing, China, 20–24 September 2010)* pp 10–8

[43] Khalifa F, El-Baz A, Gimel'farb G, Ouseph R and El-Ghar M A 2010 Shape-appearance guided level-set deformable model for image segmentation *Proc. of IAPR Int. Conf. on Pattern Recognition, (ICPR'10) (Istanbul, Turkey, 23–26 August 2010)* pp 4581–4

[44] Khalifa F, El-Ghar M A, Abdollahi B, Frieboes H, El-Diasty T and El-Baz A 2013 A comprehensive non-invasive framework for automated evaluation of acute renal transplant rejection using DCE-MRI *NMR Biomed.* **26** 1460–70

[45] Khalifa F, El-Ghar M A, Abdollahi B, Frieboes H B, El-Diasty T and El-Baz A 2014 Dynamic contrast-enhanced MRI-based early detection of acute renal transplant rejection *2014 Annu. Scientific Meeting and Educational Course Brochure of the Society of Abdominal Radiology, (SAR'14) (Boca Raton, FL, 23–28 March 2014)* CID: 1855912

[46] Khalifa F, Elnakib A, Beache G M, Gimel'farb G, El-Ghar M A, Sokhadze G, Manning S, McClure P and El-Baz A 2011 3D kidney segmentation from CT images using a level set approach guided by a novel stochastic speed function *Proc. of Int. Conf. Medical Image Computing and Computer-Assisted Intervention, (MICCAI'11) (Toronto, Canada, 18–22 September 2011)* pp 587–94

[47] Khalifa F, Gimel'farb G, El-Ghar M A, Sokhadze G, Manning S, McClure P, Ouseph R and El-Baz A 2011 A new deformable model-based segmentation approach for accurate extraction of the kidney from abdominal CT images *Proc. of IEEE Int. Conf. on Image Processing, (ICIP'11) (Brussels, Belgium, 11–14 September 2011)* pp 3393–6

[48] Mostapha M, Khalifa F, Alansary A, Soliman A, Suri J and El-Baz A 2014 Computer-aided diagnosis systems for acute renal transplant rejection: challenges and methodologies *Abdomen and Thoracic Imaging* ed A El-Baz, L saba and J Suri (Berlin: Springer) pp 1–35

[49] Shehata M, Khalifa F, Hollis E, Soliman A, Hosseini-Asl E, El-Ghar M A, El-Baz M, Dwyer A C, El-Baz A and Keynton R 2016 A new non-invasive approach for early classification of renal rejection types using diffusion-weighted MRI *IEEE Int. Conf. on Image Processing (ICIP), 2016* (Piscataway, NJ: IEEE) pp 136–40

[50] Khalifa F, Soliman A, Takieldeen A, Shehata M, Mostapha M, Shaffie A, Ouseph R, Elmaghraby A and El-Baz A 2016 Kidney segmentation from CT images using a 3D NMF-guided active contour model *IEEE 13th Int. Symp. on Biomedical Imaging (ISBI), 2016* (Piscataway, NJ: IEEE) pp 432–5

[51] Shehata M, Khalifa F, Soliman A, Takieldeen A, El-Ghar M A, Shaffie A, Dwyer A C, Ouseph R, El-Baz A and Keynton R 2016 3D diffusion MRI-based CAD system for early diagnosis of acute renal rejection *2016 IEEE 13th Int. Symp. on Biomedical Imaging (ISBI)* (Piscataway, NJ: IEEE) pp 1177–80

[52] Shehata M, Khalifa F, Soliman A, Alrefai R, El-Ghar M A, Dwyer A C, Ouseph R and El-Baz A 2015 A level set-based framework for 3D kidney segmentation from diffusion MR images *IEEE Int. Conf. on Image Processing (ICIP), 2015* (Piscataway, NJ: IEEE) pp 4441–5

[53] Shehata M, Khalifa F, Soliman A, El-Ghar M A, Dwyer A C, Gimel'farb G, Keynton R and El-Baz A 2016 A promising non-invasive CAD system for kidney function assessment *Int. Conf. on Medical Image Computing and Computer-Assisted Intervention* (Berlin: Springer) pp 613–21

[54] Khalifa F, Soliman A, Elmaghraby A, Gimel'farb G and El-Baz A 2017 3D kidney segmentation from abdominal images using spatial-appearance models *Comput. Math. Methods Med.* **2017** 1–10

[55] Hollis E, Shehata M, Khalifa F, El-Ghar M A, El-Diasty T and El-Baz A 2016 Towards non-invasive diagnostic techniques for early detection of acute renal transplant rejection: a review *Egypt. J. Radiol. Nucl. Med.* **48** 257–69

[56] Shehata M, Khalifa F, Soliman A, El-Ghar M A, Dwyer A C and El-Baz A 2017 Assessment of renal transplant using image and clinical-based biomarkers *Proc. of 13th Annual Scientific Meeting of American Society for Diagnostics and Interventional Nephrology (ASDIN'17) (New Orleans, LA, USA, 10–12 February, 2017)*

[57] Shehata M, Khalifa F, Soliman A, El-Ghar M A, Dwyer A C and El-Baz A 2017 Early assessment of acute renal rejection *Proc. of 12th Annual Scientific Meeting of American Society for Diagnostics and Interventional Nephrology (ASDIN'16) (Phoenix, AZ, USA, 19–21 February 2016)*

[58] Eltanboly A, Ghazal M, Hajjdiab H, Shalaby A, Switala A, Mahmoud A, Sahoo P, El-Azab M and El-Baz A 2019 Level sets-based image segmentation approach using statistical shape priors *Appl. Math. Comput.* **340** 164–79

[59] Shehata M, Mahmoud A, Soliman A, Khalifa F, Ghazal M, El-Ghar M A, El-Melegy M and El-Baz A 2018 3D kidney segmentation from abdominal diffusion MRI using an appearance-guided deformable boundary *PLoS One* **13** e0200082

[60] Khalifa F, Beache G, El-Baz A and Gimel'farb G 2010 Deformable model guided by stochastic speed with application in cine images segmentation *Proc. of IEEE Int. Conf. on Image Processing, (ICIP'10) (Hong Kong, 26–29 September 2010)* pp 1725–8

[61] Khalifa F, Beache G M, Elnakib A, Sliman H, Gimel'farb G, Welch K C and El-Baz A 2013 A new shape-based framework for the left ventricle wall segmentation from cardiac first-pass perfusion MRI *Proc. of IEEE Int. Symp. on Biomedical Imaging: From Nano to Macro, (ISBI'13) (San Francisco, CA, 7–11 April 2013)* pp 41–4

[62] Khalifa F, Beache G M, Elnakib A, Sliman H, Gimel'farb G, Welch K C and El-Baz A 2012 A new nonrigid registration framework for improved visualization of transmural perfusion gradients on cardiac first-pass perfusion MRI *Proc. of IEEE Int. Symp. on Biomedical Imaging: From Nano to Macro, (ISBI'12) (Barcelona, Spain, 2–5 May 2012)* pp 828–31

[63] Khalifa F, Beache G M, Firjani A, Welch K C, Gimel'farb G and El-Baz A 2012 A new nonrigid registration approach for motion correction of cardiac first-pass perfusion MRI *Proc. of IEEE Int. Conf. on Image Processing, (ICIP'12) (Lake Buena Vista, FL, 30 September–3 October 2012)* pp 1665–8

[64] Khalifa F, Beache G M, Gimel'farb G and El-Baz A 2012 A novel CAD system for analyzing cardiac first-pass MR images *Proc. of IAPR Int. Conf. on Pattern Recognition (ICPR'12) (Tsukuba Science City, Japan, 11–15 November 2012)* pp 77–80

[65] Khalifa F, Beache G M, Gimel'farb G and El-Baz A 2011 A novel approach for accurate estimation of left ventricle global indexes from short-axis cine MRI *Proc. of IEEE Int. Conf. on Image Processing, (ICIP'11) (Brussels, Belgium, 11–14 September 2011)* pp 2645–9

[66] Khalifa F, Beache G M, Gimel'farb G, Giridharan G A and El-Baz A 2011 A new image-based framework for analyzing cine images *Handbook of Multi Modality State-of-the-Art Medical Image Segmentation and Registration Methodologies* vol 2 ed A El-Baz, U R Acharya, M Mirmedhdi and J S Suri (New York: Springer) ch 3, pp 69–98

[67] Khalifa F, Beache G M, Gimel'farb G, Giridharan G A and El-Baz A 2012 Accurate automatic analysis of cardiac cine images *IEEE Trans. Biomed. Eng.* **59** 445–55

[68] Khalifa F, Beache G M, Nitzken M, Gimel'farb G, Giridharan G A and El-Baz A 2011 Automatic analysis of left ventricle wall thickness using short-axis cine CMR images *Proc. of IEEE Int. Symp. on Biomedical Imaging: From Nano to Macro, (ISBI'11) (Chicago, Illinois, 30 March–2 April 2011)* pp 1306–9

[69] Nitzken M, Beache G, Elnakib A, Khalifa F, Gimel'farb G and El-Baz A 2012 Accurate modeling of tagged CMR 3D image appearance characteristics to improve cardiac cycle strain estimation *2012 19th IEEE Int. Conf. on Image Processing (ICIP)* (Piscataway, NJ: IEEE) pp 521–4

[70] Nitzken M, Beache G, Elnakib A, Khalifa F, Gimel'farb G and El-Baz A 2012 Improving full-cardiac cycle strain estimation from tagged CMR by accurate modeling of 3D image appearance characteristics *2012 9th IEEE Int. Symp. on Biomedical Imaging (ISBI)* (Piscataway, NJ: IEEE) pp 462–65

[71] Nitzken M J, El-Baz A S and Beache G M 2012 Markov–Gibbs random field model for improved full-cardiac cycle strain estimation from tagged CMR *J. Cardiovasc. Magn. Reson.* **14** 1–2

[72] Sliman H, Elnakib A, Beache G, Elmaghraby A and El-Baz A 2014 Assessment of myocardial function from cine cardiac MRI using a novel 4D tracking approach *J. Comput. Sci. Syst. Biol.* **7** 169–73

[73] Sliman H, Elnakib A, Beache G M, Soliman A, Khalifa F, Gimel'farb G, Elmaghraby A and El-Baz A 2014 A novel 4D PDE-based approach for accurate assessment of myocardium function using cine cardiac magnetic resonance images *Proc. of IEEE Int. Conf. on Image Processing (ICIP'14) (Paris, France, 27–30 October 2014)* pp 3537–41

[74] Sliman H, Khalifa F, Elnakib A, Beache G M, Elmaghraby A and El-Baz A 2013 A new segmentation-based tracking framework for extracting the left ventricle cavity from cine cardiac MRI *Proc. of IEEE Int. Conf. on Image Processing, (ICIP'13) (Melbourne, Australia, 15–18 September 2013)* pp 685–9

[75] Sliman H, Khalifa F, Elnakib A, Soliman A, Beache G M, Elmaghraby A, Gimel'farb G and El-Baz A 2013 Myocardial borders segmentation from cine MR images using bi-directional coupled parametric deformable models *Med. Phys.* **40** 1–13

[76] Sliman H, Khalifa F, Elnakib A, Soliman A, Beache G M, Gimel'farb G, Emam A, Elmaghraby A and El-Baz A 2013 Accurate segmentation framework for the left ventricle wall from cardiac cine MRI *Proc. of Int. Symp. on Computational Models for Life Science, (CMLS'13) (Sydney, Australia, 27–29 November 2013)* **vol 1559** pp 287–96

[77] Eladawi N, Elmogy M, Ghazal M, Helmy O, Aboelfetouh A, Riad A, Schaal S and El-Baz A 2018 Classification of retinal diseases based on OCT images *Front. Biosci.* **23** 247–64

[78] ElTanboly A, Ismail M, Shalaby A, Switala A, El-Baz A, Schaal S, Gimelfarb G and El-Azab M 2017 A computer-aided diagnostic system for detecting diabetic retinopathy in optical coherence tomography images *Med. Phys.* **44** 914–23

[79] Sandhu H S, El-Baz A and Seddon J M 2018 Progress in automated deep learning for macular degeneration *JAMA Ophthalmol.* **136** 1366–7

[80] Abdollahi B, Civelek A C, Li X-F, Suri J and El-Baz A 2014 PET/CT nodule segmentation and diagnosis: a survey *Multi Detector CT Imaging* ed L Saba and J S Suri (London: Taylor and Francis) ch 30, pp 639–51

[81] Abdollahi B, El-Baz A and Amini A A 2011 A multi-scale non-linear vessel enhancement technique *2011 Annu. Int. Conf. of the IEEE Engineering in Medicine and Biology Society, EMBC* pp 3925–9

[82] Abdollahi B, Soliman A, Civelek A, Li X-F, Gimel'farb G and El-Baz A 2012 A novel Gaussian scale space-based joint MGRF framework for precise lung segmentation *Proc. of IEEE Int. Conf. on Image Processing, (ICIP'12)* (Piscataway, NJ: IEEE) pp 2029–32

[83] Abdollahi B, Soliman A, Civelek A, Li X-F, Gimel'farb G and El-Baz A 2012 A novel 3D joint MGRF framework for precise lung segmentation *Machine Learning in Medical Imaging* (Berlin: Springer) pp 86–93

[84] Ali A M, El-Baz A S and Farag A A 2007 A novel framework for accurate lung segmentation using graph cuts *Proc. of IEEE Int. Symp. on Biomedical Imaging: From Nano to Macro, (ISBI'07)* (Piscataway, NJ: IEEE) pp 908–11

[85] El-Baz A, Beache G M, Gimel'farb G, Suzuki K and Okada K 2013 Lung imaging data analysis *Int. J. Biomed. Imaging* **2013** 1–2

[86] El-Baz A, Beache G M, Gimel'farb G, Suzuki K, Okada K, Elnakib A, Soliman A and Abdollahi B 2013 Computer-aided diagnosis systems for lung cancer: challenges and methodologies *Int. J. Biomed. Imaging* **2013** 1–46

[87] El-Baz A, Elnakib A, Abou El-Ghar M, Gimel'farb G, Falk R and Farag A 2013 Automatic detection of 2D and 3D lung nodules in chest spiral CT scans *Int. J. Biomed. Imaging* **2013** 1–11

[88] El-Baz A, Farag A A, Falk R and La Rocca R 2003 A unified approach for detection, visualization, and identification of lung abnormalities in chest spiral CT scans *International Congress Series* vol 1256 (Amsterdam: Elsevier) pp 998–1004

[89] El-Baz A, Farag A A, Falk R and La Rocca R 2002 Detection, visualization and identification of lung abnormalities in chest spiral CT scan: Phase-I *Proc. of Int. Conf. on Biomedical Engineering* (Cairo, Egypt) **vol 12**

[90] El-Baz A, Farag A, Gimel'farb G, Falk R, El-Ghar M A and Eldiasty T 2006 A framework for automatic segmentation of lung nodules from low dose chest CT scans *Proc. of Int. Conf. on Pattern Recognition, (ICPR'06)* **vol 3** (Piscataway, NJ: IEEE) pp 611–4

[91] El-Baz A, Farag A, Gimel'farb G, Falk R and El-Ghar M A 2011 A novel level set-based computer-aided detection system for automatic detection of lung nodules in low dose chest computed tomography scans *Lung Imaging and Computer Aided Diagnosis* (London: Taylor and Francis) vol 10 pp 221–38

[92] El-Baz A, Gimel'farb G, Abou El-Ghar M and Falk R 2012 Appearance-based diagnostic system for early assessment of malignant lung nodules *Proc. of IEEE Int. Conf. on Image Processing, (ICIP'12)* (Piscataway, NJ: IEEE) pp 533–6

[93] El-Baz A, Gimel'farb G and Falk R 2011 A novel 3D framework for automatic lung segmentation from low dose CT images *Lung Imaging and Computer Aided Diagnosis* ed A El-Baz and J S Suri (London: Taylor and Francis) ch 1, pp 1–16

[94] El-Baz A, Gimel'farb G, Falk R and El-Ghar M 2010 Appearance analysis for diagnosing malignant lung nodules *Proc. of IEEE Int. Symp. on Biomedical Imaging: From Nano to Macro (ISBI'10)* (Piscataway, NJ: IEEE) pp 193–6

[95] El-Baz A, Gimel'farb G, Falk R and El-Ghar M A 2011 A novel level set-based CAD system for automatic detection of lung nodules in low dose chest CT scans *Lung Imaging and Computer Aided Diagnosis* vol 1 ed A El-Baz and J S Suri (London: Taylor and Francis) ch 10, pp 221–38

[96] El-Baz A, Gimel'farb G, Falk R and El-Ghar M A 2008 A new approach for automatic analysis of 3D low dose CT images for accurate monitoring the detected lung nodules *Proc. of Int. Conf. on Pattern Recognition, (ICPR'08)* (Piscataway, NJ: IEEE) pp 1–4

[97] El-Baz A, Gimel'farb G, Falk R and El-Ghar M A 2007 A novel approach for automatic follow-up of detected lung nodules *Proc. of IEEE Int. Conf. on Image Processing, (ICIP'07)* **vol 5** (Piscataway, NJ: IEEE) pp V–501

[98] El-Baz A, Gimel'farb G, Falk R and El-Ghar M A 2007 A new CAD system for early diagnosis of detected lung nodules *IEEE Int. Conf. on Image Processing, 2007. ICIP 2007* **vol 2** (Piscataway, NJ: IEEE) pp II–461

[99] El-Baz A, Gimel'farb G, Falk R, El-Ghar M A and Refaie H 2008 Promising results for early diagnosis of lung cancer *Proc. of IEEE Int. Symp. on Biomedical Imaging: From Nano to Macro, (ISBI'08)* (Piscataway, NJ: IEEE) pp 1151–4

[100] El-Baz A, Gimel'farb G L, Falk R, Abou El-Ghar M, Holland T and Shaffer T 2008 A new stochastic framework for accurate lung segmentation *Proc. of Medical Image Computing and Computer-Assisted Intervention, (MICCAI'08)* pp 322–30

[101] El-Baz A, Gimel'farb G L, Falk R, Heredis D and Abou El-Ghar M 2008 A novel approach for accurate estimation of the growth rate of the detected lung nodules *Proc. of Int. Workshop on Pulmonary Image Analysis* pp 33–42

[102] El-Baz A, Gimel'farb G L, Falk R, Holland T and Shaffer T 2008 A framework for unsupervised segmentation of lung tissues from low dose computed tomography images *Proc. of British Machine Vision, (BMVC'08)* pp 1–10

[103] El-Baz A, Gimel'farb G, Falk R and El-Ghar M A 2011 3D MGRF-based appearance modeling for robust segmentation of pulmonary nodules in 3D LDCT chest images *Lung Imaging and Computer Aided Diagnosis* (London: Taylor and Francis) ch 3, pp 51–63

[104] El-Baz A, Gimel'farb G, Falk R and El-Ghar M A 2009 Automatic analysis of 3D low dose CT images for early diagnosis of lung cancer *Pattern Recogn.* **42** 1041–51

[105] El-Baz A, Gimel'farb G, Falk R, El-Ghar M A, Rainey S, Heredia D and Shaffer T 2009 Toward early diagnosis of lung cancer *Proc. of Medical Image Computing and Computer-Assisted Intervention, (MICCAI'09)* (Berlin: Springer) pp 682–9

[106] El-Baz A, Gimel'farb G, Falk R, El-Ghar M A and Suri J 2011 Appearance analysis for the early assessment of detected lung nodules *Lung Imaging and Computer Aided Diagnosis* (London: Taylor and Francis) ch 17, pp 395–404

[107] El-Baz A, Khalifa F, Elnakib A, Nitkzen M, Soliman A, McClure P, Gimel'farb G and El-Ghar M A 2012 A novel approach for global lung registration using 3D Markov Gibbs appearance model *Proc. of Int. Conf. Medical Image Computing and Computer-Assisted Intervention, (MICCAI'12) (Nice, France, 1–5 October 2012)* pp 114–21

[108] El-Baz A, Nitzken M, Elnakib A, Khalifa F, Gimel'farb G, Falk R and El-Ghar M A 2011 3D shape analysis for early diagnosis of malignant lung nodules *Proc. of Int. Conf. Medical Image Computing and Computer-Assisted Intervention, (MICCAI'11) (Toronto, Canada, 18–22 September 2011)* pp 175–82

[109] El-Baz A, Nitzken M, Gimel'farb G, Van Bogaert E, Falk R, El-Ghar M A and Suri J 2011 Three-dimensional shape analysis using spherical harmonics for early assessment of detected lung nodules *Lung Imaging and Computer Aided Diagnosis* (London: Taylor and Francis) ch 19, pp 421–38

[110] El-Baz A, Nitzken M, Khalifa F, Elnakib A, Gimel'farb G, Falk R and El-Ghar M A 2011 3D shape analysis for early diagnosis of malignant lung nodules *Proc. of Int. Conf. on Information Processing in Medical Imaging, (IPMI'11) (Monastery Irsee, Germany (Bavaria), 3–8 July 2011)* pp 772–83

[111] El-Baz A, Nitzken M, Vanbogaert E, Gimel'Farb G, Falk R and Abo El-Ghar M 2011 A novel shape-based diagnostic approach for early diagnosis of lung nodules *2011 IEEE Int. Symp. on Biomedical Imaging: From Nano to Macro* (Piscataway, NJ: IEEE) pp 137–40

[112] El-Baz A, Sethu P, Gimel'farb G, Khalifa F, Elnakib A, Falk R and El-Ghar M A 2011 Elastic phantoms generated by microfluidics technology: validation of an imaged-based approach for accurate measurement of the growth rate of lung nodules *Biotechnol. J.* **6** 195–203

[113] El-Baz A, Sethu P, Gimel'farb G, Khalifa F, Elnakib A, Falk R and El-Ghar M A 2010 A new validation approach for the growth rate measurement using elastic phantoms generated by state-of-the-art microfluidics technology *Proc. of IEEE Int. Conf. on Image Processing, (ICIP'10) (Hong Kong, 26–29 September 2010)* pp 4381–3

[114] El-Baz A, Sethu P, Gimel'farb G, Khalifa F, Elnakib A, Falk R and Suri M A E-G J 2011 Validation of a new imaged-based approach for the accurate estimating of the growth rate of detected lung nodules using real CT images and elastic phantoms generated by state-of-the-art microfluidics technology *Handbook of Lung Imaging and Computer Aided Diagnosis* vol 1 ed A El-Baz and J S Suri (New York: Taylor and Francis) ch 18, pp 405–20

[115] El-Baz A, Soliman A, McClure P, Gimel'farb G, El-Ghar M A and Falk R 2012 Early assessment of malignant lung nodules based on the spatial analysis of detected lung nodules *Proc. of IEEE Int. Symp. on Biomedical Imaging: From Nano to Macro, (ISBI'12)* (Piscataway, NJ: IEEE) pp 1463–6

[116] El-Baz A, Yuksel S E, Elshazly S and Farag A A 2005 Non-rigid registration techniques for automatic follow-up of lung nodules *Proc. of Computer Assisted Radiology and Surgery, (CARS'05)* **vol 1281** (Amsterdam: Elsevier) pp 1115–20

[117] El-Baz A S and Suri J S 2011 *Lung Imaging and Computer Aided Diagnosis* (Boca Raton, FL: CRC Press)

[118] Soliman A, Khalifa F, Dunlap N, Wang B, El-Ghar M and El-Baz A 2016 An iso-surfaces based local deformation handling framework of lung tissues *2016 IEEE 13th Int. Symp. on Biomedical Imaging (ISBI)* (Piscataway, NJ: IEEE) pp 1253–9

[119] Soliman A, Khalifa F, Shaffie A, Dunlap N, Wang B, Elmaghraby A and El-Baz A 2016 Detection of lung injury using 4D-CT chest images *2016 IEEE 13th Int. Symp. on Biomedical Imaging (ISBI)* (Piscataway, NJ: IEEE) pp 1274–7

[120] Soliman A, Khalifa F, Shaffie A, Dunlap N, Wang B, Elmaghraby A, Gimel'farb G, Ghazal M and El-Baz A 2017 A comprehensive framework for early assessment of lung injury *2017 IEEE Int. Conf. on Image Processing (ICIP)* (Piscataway, NJ: IEEE) pp 3275–9

[121] Shaffie A, Soliman A, Ghazal M, Taher F, Dunlap N, Wang B, Elmaghraby A, Gimel'farb G and El-Baz A 2017 A new framework for incorporating appearance and shape features of lung nodules for precise diagnosis of lung cancer *2017 IEEE Int. Conf. on Image Processing (ICIP)* (Piscataway, NJ: IEEE) pp 1372–6

[122] Soliman A, Khalifa F, Shaffie A, Liu N, Dunlap N, Wang B, Elmaghraby A, Gimel'farb G and El-Baz A 2016 Image-based CAD system for accurate identification of lung injury *2016 IEEE Int. Conf. on Image Processing (ICIP)* (Piscataway, NJ: IEEE) pp 121–5

[123] Dombroski B, Nitzken M, Elnakib A, Khalifa F, El-Baz A and Casanova M F 2014 Cortical surface complexity in a population-based normative sample *Transl. Neurosci.* **5** 17–24

[124] El-Baz A, Casanova M, Gimel'farb G, Mott M and Switala A 2008 An MRI-based diagnostic framework for early diagnosis of dyslexia *Int. J. Comput. Assist. Radiol. Surg.* **3** 181–9

[125] El-Baz A, Casanova M, Gimel'farb G, Mott M, Switala A, Vanbogaert E and McCracken R 2008 A new CAD system for early diagnosis of dyslexic brains *Proc. Int. Conf. on Image Processing (ICIP'2008)* (Piscataway, NJ: IEEE) pp 1820–3

[126] El-Baz A, Casanova M F, Gimel'farb G, Mott M and Switwala A E 2007 A new image analysis approach for automatic classification of autistic brains *Proc. IEEE Int. Symp. on Biomedical Imaging: From Nano to Macro (ISBI'2007)* (Piscataway, NJ: IEEE) pp 352–5

[127] El-Baz A, Elnakib A, Khalifa F, El-Ghar M A, McClure P, Soliman A and Gimel'farb G 2012 Precise segmentation of 3D magnetic resonance angiography *IEEE Trans. Biomed. Eng.* **59** 2019–29

[128] El-Baz A, Farag A, Gimel'farb G, El-Ghar M A and Eldiasty T 2006 Probabilistic modeling of blood vessels for segmenting MRA images *18th Int. Conf. on Pattern Recognition (ICPR'06)* **vol 3** (Piscataway, NJ: IEEE) pp 917–20

[129] El-Baz A, Farag A A, Gimel'farb G, El-Ghar M A and Eldiasty T 2006 A new adaptive probabilistic model of blood vessels for segmenting MRA images *Medical Image Computing and Computer-Assisted Intervention–MICCAI 2006* vol 4191 (Berlin: Springer) pp 799–806

[130] El-Baz A, Farag A A, Gimel'farb G and Hushek S G 2005 Automatic cerebrovascular segmentation by accurate probabilistic modeling of TOF-MRA images *Medical Image Computing and Computer-Assisted Intervention–MICCAI 2005* (Berlin: Springer) pp 34–42

[131] El-Baz A, Farag A, Elnakib A, Casanova M F, Gimel'farb G, Switala A E, Jordan D and Rainey S 2011 Accurate automated detection of autism related corpus callosum abnormalities *J. Med. Syst.* **35** 929–39

[132] El-Baz A, Farag A and Gimelfarb G 2005 Cerebrovascular segmentation by accurate probabilistic modeling of TOF-MRA images *Image Analysis* vol 3540 (Berlin: Springer) pp 1128–37

[133] El-Baz A, Gimel'farb G, Falk R, El-Ghar M A, Kumar V and Heredia D 2009 A novel 3D joint Markov–Gibbs model for extracting blood vessels from PC–MRA images *Medical Image Computing and Computer-Assisted Intervention–MICCAI 2009* vol 5762 (Berlin: Springer) pp 943–50

[134] Elnakib A, El-Baz A, Casanova M F, Gimel'farb G and Switala A E 2010 Image-based detection of corpus callosum variability for more accurate discrimination between dyslexic and normal brains *Proc. IEEE Int. Symp. on Biomedical Imaging: From Nano to Macro (ISBI'2010)* (Piscataway, NJ: IEEE) pp 109–12

[135] Elnakib A, Casanova M F, Gimel'farb G, Switala A E and El-Baz A 2011 Autism diagnostics by centerline-based shape analysis of the corpus callosum *Proc. IEEE Int. Symp. on Biomedical Imaging: From Nano to Macro (ISBI'2011)* (Piscataway, NJ: IEEE) pp 1843–6

[136] Elnakib A, Nitzken M, Casanova M, Park H, Gimel'farb G and El-Baz A 2012 Quantification of age-related brain cortex change using 3D shape analysis *2012 21st Int. Conf. on Pattern Recognition (ICPR)* (Piscataway, NJ: IEEE) pp 41–4

[137] Mostapha M, Soliman A, Khalifa F, Elnakib A, Alansary A, Nitzken M, Casanova M F and El-Baz A 2014 A statistical framework for the classification of infant DT images *2014 IEEE Int. Conf. on Image Processing (ICIP)* (Piscataway, NJ: IEEE) pp 2222–6

[138] Nitzken M, Casanova M, Gimel'farb G, Elnakib A, Khalifa F, Switala A and El-Baz A 2011 3D shape analysis of the brain cortex with application to dyslexia *2011 18th IEEE Int. Conf. on Image Processing (ICIP)* (Piscataway, NJ: IEEE) pp 2657–60

[139] El-Gamal F E-Z A, Elmogy M M, Ghazal M, Atwan A, Barnes G N, Casanova M F, Keynton R and El-Baz A S 2017 A novel CAD system for local and global early diagnosis of Alzheimer's disease based on PiB-PET scans *2017 IEEE Int. Conf. on Image Processing (ICIP)* (Piscataway, NJ: IEEE) pp 3270–4

[140] Ismail M, Soliman A, Ghazal M, Switala A E, Gimelfarb G, Barnes G N, Khalil A and El-Baz A 2017 A fast stochastic framework for automatic MR brain images segmentation *PloS one* **12** e0187391

[141] Alansary A *et al* 2016 Infant brain extraction in T1-weighted MR images using BET and refinement using LCDG and MGRF models *IEEE J. Biomed. Health Inform* **20** 925–35

[142] Ismail M, Soliman A, ElTanboly A, Switala A, Mahmoud M, Khalifa F, Gimel'farb G, Casanova M F, Keynton R and El-Baz A 2016 Detection of white matter abnormalities in MR brain images for diagnosis of autism in children *2016 IEEE 13th International Symposium on Biomedical Imaging (ISBI)* pp 6–9

[143] Ismail M, Mostapha M, Soliman A, Nitzken M, Khalifa F, Elnakib A, Gimel'farb G, Casanova M and El-Baz A 2015 Segmentation of infant brain MR images based on adaptive shape prior and higher-order MGRF *2015 IEEE Int. Conf. on Image Processing (ICIP)* pp 4327–31

[144] Asl E H, Ghazal M, Mahmoud A, Aslantas A, Shalaby A, Casanova M, Barnes G, Gimelfarb G, Keynton R and El-Baz A 2018 Alzheimers disease diagnostics by a 3D deeply supervised adaptable convolutional network *Front. Biosci.* **23** 584–96

[145] Mahmoud A, El-Barkouky A, Farag H, Graham J and Farag A 2013 A non-invasive method for measuring blood flow rate in superficial veins from a single thermal image *Proc. of the IEEE Conf. on Computer Vision and Pattern Recognition Workshops* pp 354–9

[146] El-baz A, Shalaby A, Taher F, El-Baz M, Ghazal M, El-Ghar M A, Takieldeen A and Suri J 2017 Probabilistic modeling of blood vessels for segmenting magnetic resonance angiography images *Med. Res. Arch.* **5**

[147] Chowdhury A S, Rudra A K, Sen M, Elnakib A and El-Baz A 2010 Cerebral white matter segmentation from MRI using probabilistic graph cuts and geometric shape priors *ICIP* 3649–52

[148] Gebru Y, Giridharan G, Ghazal M, Mahmoud A, Shalaby A and El-Baz A 2018 Detection of cerebrovascular changes using magnetic resonance angiography *Cardiovascular Imaging and Image Analysis* (Boca Raton, FL: CRC Press) pp 1–22

[149] Mahmoud A, Shalaby A, Taher F, El-Baz M, Suri J S and El-Baz A 2018 Vascular tree segmentation from different image modalities *Cardiovascular Imaging and Image Analysis* (Boca Raton, FL: CRC Press) pp 43–70

[150] Taher F, Mahmoud A, Shalaby A and El-Baz A 2018 A review on the cerebrovascular segmentation methods *2018 IEEE Int. Symp. on Signal Processing and Information Technology (ISSPIT)* (Piscataway, NJ: IEEE) pp 359–64

[151] Kandil H, Soliman A, Fraiwan L, Shalaby A, Mahmoud A, ElTanboly A, Elmaghraby A, Giridharan G and El-Baz A 2018 A novel MRA framework based on integrated global and local analysis for accurate segmentation of the cerebral vascular system *2018 IEEE 15th Int. Symp. on Biomedical Imaging (ISBI 2018)* (Piscataway, NJ: IEEE) pp 1365–8

www.ingramcontent.com/pod-product-compliance
Lightning Source LLC
Chambersburg PA
CBHW071941220326
41599CB00031BA/5839